Catchment Hydrology

Catchment Hydrology

Edited by Jason Hogan

SYRAWOOD
PUBLISHING HOUSE
New York

Published by Syrawood Publishing House,
750 Third Avenue, 9th Floor,
New York, NY 10017, USA
www.syrawoodpublishinghouse.com

Catchment Hydrology
Edited by Jason Hogan

International Standard Book Number: 978-1-64740-136-8 (Hardback)

Cataloging-in-Publication Data

Catchment hydrology / edited by Jason Hogan.
    p. cm.
Includes bibliographical references and index.
ISBN 978-1-64740-136-8
1. Hydrology. 2. Hydrography. 3. Earth sciences. 4. Water. I. Hogan, Jason.
GB661.2 .C38 2022
551.48--dc23

TABLE OF CONTENTS

PREFACE

Over the recent decade, advancements and applications have progressed exponentially. This has led to the increased interest in this field and projects are being conducted to enhance knowledge. The main objective of this book is to present some of the critical challenges and provide insights into possible solutions. This book will answer the varied questions that arise in the field and also provide an increased scope for furthering studies.

The scientific study of the movement, distribution and quality of water on earth and other planets is known as hydrology. The area of land where precipitation collects and drains off into any common outlet such as a river or bay is known as catchment. It includes all the surface water from rain runoff, groundwater and snowmelt. The catchment is the most significant factor in determining the amount or likelihood of flooding. Catchment hydrology primarily focuses on the study of drainage basins with respect to hydrology. It is based on the principle of continuity, which is used to perform a water balance on a catchment. This book provides comprehensive insights into the field of catchment hydrology. It traces the progress of this field and highlights some of its key concepts and applications. The topics covered in this book offer the readers new insights in this domain.

I hope that this book, with its visionary approach, will be a valuable addition and will promote interest among readers. Each of the authors has provided their extraordinary competence in their specific fields by providing different perspectives as they come from diverse nations and regions. I thank them for their contributions.

Editor

Water displacement by sewer infrastructure in the Grote Nete catchment, Belgium, and its hydrological regime effects

D. Vrebos[1], **T. Vansteenkiste**[2], **J. Staes**[1], **P. Willems**[2,3], and **P. Meire**[1]

[1]Department of Biology, Universiteit Antwerpen, Universiteitsplein 1, 2610 Wilrijk, Belgium
[2]Department of Civil Engineering, KU Leuven, Kasteelpark Arenberg 40, 3001 Leuven, Belgium
[3]Department of Hydrology and Hydraulic Engineering, Vrije Universiteit Brussel, Pleinlaan 2, 1050 Brussel, Belgium

Correspondence to: D. Vrebos (dirk.vrebos@uantwerpen.be)

Abstract. Urbanization and especially increases in impervious areas, in combination with the installation of wastewater treatment infrastructure, can impact the runoff from a catchment and river flows in a significant way. These effects were studied for the Grote Nete catchment in Belgium based on a combination of empirical and model-based approaches. Effective impervious area, combined with the extent of the wastewater collection regions, was considered as an indicator for urbanization pressure. It was found that wastewater collection regions ranging outside the boundaries of the natural catchment boundaries caused changes in upstream catchment area between −16 and +3 %, and upstream impervious areas between −99 and +64 %. These changes lead to important intercatchment water transfers. Simulations with a physically based and spatially distributed hydrological catchment model revealed not only significant impacts of effective impervious area on seasonal runoff volumes but also low and peak river flows. Our results show the importance, as well as the difficulty, of explicitly accounting for these artificial pressures and processes in the hydrological modeling of urbanized catchments.

1 Introduction

Urbanization significantly impacts flow regimes and water quality of river systems (Paul and Meyer, 2001; Jacobson, 2011). In particular, impervious areas exert several pressures on the hydrological cycle of catchments (Shuster et al., 2005). They affect infiltration, surface runoff and evapotranspiration, making the lateral processes potentially more important in urban settings than the vertical processes (Arnold and Gibbons, 1996; Becker and Braun, 1999; Brabec, 2009). These alterations in hydrological processes increase runoff peak flows and flood flashiness in rivers (Sheeder et al., 2002; Baker et al., 2004).

The effects of urbanization on peak flows have been studied in more detail compared to its effects on baseflow and low-flow events (Price, 2011). Baseflow represents stream flow fed from deep subsurface and delayed shallow subsurface storage (Ward and Robinson, 1989), while low flow addresses dry season minimum flows (Smakhtin, 2001; Price, 2011). As urbanization reduces infiltration and recharge, it is generally expected that river baseflow is affected as well (Simmons and Reynolds, 1982; Kauffman et al., 2009). Baseflow can, however, also be strongly influenced by various types of anthropogenic activities in the catchment, such as water abstractions, sewer leakage or groundwater intrusion (Seiler and Rivas, 1999; Smakhtin, 2001; Brandes et al., 2005; Wittenberg and Aksoy, 2010). Hence, the impact of urbanization on the different flow regimes is difficult to detect due to the strong temporal flow variations (weak signal-to-noise ratio). Although a weak tendency in baseflow decline and peak flow increase has been identified by some authors (Price, 2011), the combined effect of several anthropogenic and natural processes that influence baseflow and differences in assessment methodologies result in many remaining uncertainties in our understanding of baseflow behavior in peri-urban catchments (Hamel et al., 2013).

Sewers collect wastewater and, for combined sewer systems, also rainstorm water from pavements. They can also receive groundwater or leak wastewater to the groundwater system (Dirckx et al., 2009). The collected water is transported to a wastewater treatment plant (WWTP) and, after treatment, discharged into a receiving river. The WWTP thus aggregates water from the entire wastewater collection region (WWCR) and returns it to the environment at one single river location. Moreover, the WWCRs usually do not coincide with the catchment boundaries as they are mostly based on administrative borders. Consequently the associated sewer infrastructure might transfer water between different natural (sub)catchments and further affect the natural hydrological processes in the catchment (Simmons and Reynolds, 1982). Recent research on catchment delineation considered incorporation of these changes in hydrological flow paths using semi-automated procedures (Jankowfsky et al., 2013). Such delineations, however, remain largely data dependent and time consuming.

Total impervious area (TIA) is considered to be an important indicator of the urban disturbance and an important land use characteristic. Imperviousness of urban areas is, however, very heterogeneous. Infiltration of impervious areas may not always be zero (Ragab et al., 2003). Impervious areas that are directly connected to the receiving river have a much larger effect on that receiving river (Boyd et al., 1994; Walsh et al., 2009). Some studies therefore suggest that the subset of impervious surfaces that route storm water runoff directly to streams via storm water pipes, also called effective impervious area (EIA), may be a better predictor of stream flow alteration (Shuster et al., 2005; Roy and Shuster, 2009). Measurements of EIA are, however, much more difficult to obtain and therefore less commonly used in hydrological studies (Walsh et al., 2009).

Some studies have accounted for the difference between TIA and EIA in impact studies (Lee and Heaney, 2003; Shuster et al., 2005). The traditional calculation of TIA and EIA might, however, be erroneous since the difference in boundaries between the natural river catchment and the WWCRs is typically disregarded. When impervious areas are situated within a river catchment, the surface runoff from these areas might drain to a WWTP located outside the catchment. The impervious areas in that case do not contribute to the runoff of the considered catchment.

Next to empirical statistical analysis, hydrological models can offer additional methods for studying the impact of changes in pervious and impervious areas on catchment hydrology. Such models can indeed help in complementing existing data and obtaining a better insight in the hydrological behavior of a catchment and the hydrological impact of urbanization. To allow the impact of spatial (e.g., land-use-related) scenarios to be assessed, fully spatially distributed hydrological process models (FDPM) are required. Such models give a spatially detailed and potentially reliable description of the hydrological processes in the catchment (Abbott and Refsgaard, 1996; Refsgaard and Knudsen, 1996; Boyle et al., 2001; Ajami et al., 2004; Carpenter and Georgakakos, 2006), but require a high amount of spatially

explicit input data. After calibration of the large set of parameters in such models, a better match between simulated and observed hydrological variables may be obtained, but this does not necessarily mean that the model has a good accuracy. Model over parameterization and related parameter identifiability problems are well-known pitfalls (Beven, 1989; Jakeman and Letcher, 2003; Muleta and Nicklow, 2005). These problems limit the applicability of such models. The FDPMs perform well in catchments where the hydrological processes are still close to natural runoff conditions, but are typically less accurate in urban areas due to the several (unknown or difficult to model) human influences (Vansteenkiste et al., 2013).

Discarding these anthropogenic influences can lead to a significant model bias and related impact assessments. In an urbanized environment with extensive sewer infrastructure, this might not only affect the performance of the catchment runoff and river flow simulation; it can also have indirect effects on parameterization of other land uses and over- or underestimate individual runoff components (Vansteenkiste et al., 2013). It has previously been demonstrated that if one does not differentiate between TIA and EIA in the hydrological model, this may result in a large model bias (Alley and Veenhuis, 1983; Brabec, 2009). EIA is the most sensitive flow parameter in urban drainage models. Some authors have shown that calibration of this parameter may completely eliminate the bias in the results of these models (Willems and Berlamont, 1999; Kleidorfer et al., 2009).

This paper aims to quantify the importance of interbasin transfers and WWTP effluent flow contributions to downstream river flows for a selected river catchment in Belgium. The study makes use of measured river flows and effluent discharges from the different WWTPs installed in and outside the catchment. We evaluate the relative contribution of these WWTPs to the river flow, including peak and low flows. To understand the origin of the WWTP effluent discharges and the WWTP-induced water transfers between catchments, the sewer infrastructure and the EIA are assessed in a GIS environment, and compared with FDPM simulations for the study catchment. When implementing the FDPM, the above-mentioned modeling issues (e.g., impervious area calculation and interbasin transfers) are considered. Based on empirical data analysis, model-based results and the comparison between both; we demonstrate the magnitude and importance of

- water transfers across the catchment boundaries,

- water transfers across subcatchments within the catchment,

- the impact of EIA on river high and low flows and the performance of an FDPM.

We also discuss the implications the water transfers have on the FDPM-based impact analysis.

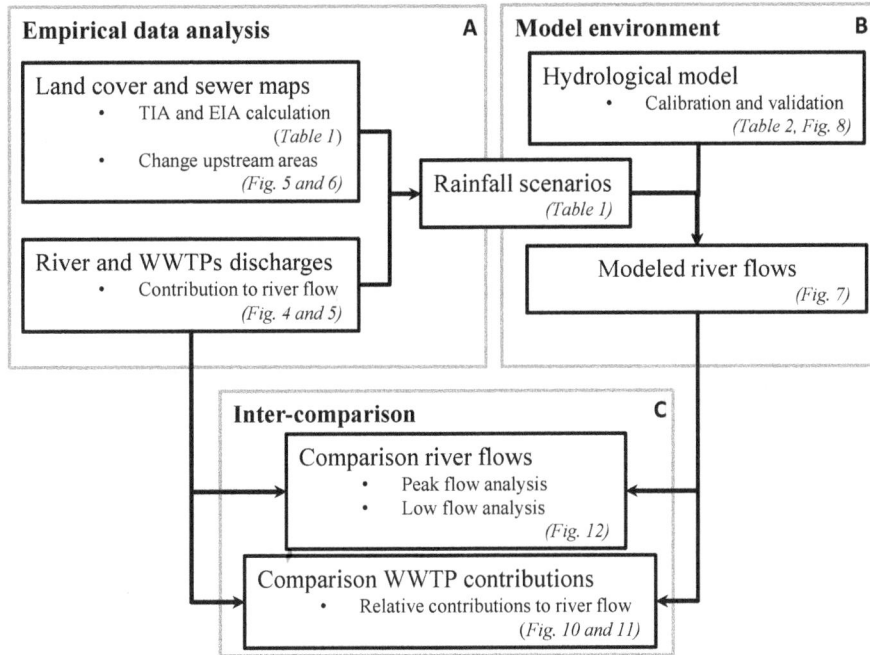

Fig. 1. General overview of the different steps of the research methodology and their interlinks.

2 Material and methods

2.1 Study area

The study catchment the Grote Nete river ($350\,\mathrm{km^2}$) is situated in the north of Belgium. It has a maritime, temperate climate with average precipitation of $800\,\mathrm{mm\,year^{-1}}$. The catchment is composed of a mosaic of semi-natural, agricultural and urbanized areas, with a total population of 218 815 (Statistics Belgium, 2011). Urbanized areas are mainly situated around the town centers, but important parts of the urbanization are spread along the main roads connecting the different towns. As a result the development of the sewer infrastructure is difficult, costly and time consuming.

Although the first WWTP in the catchment dates back from 1964, major investments in the sewer infrastructure only started 15 to 20 years ago (Dirckx et al., 2009). Nevertheless, large numbers of households are yet to be connected to the sewer infrastructure. The sewer system consists mostly of a combined system that collects both rain- and wastewater and is connected to sewer overflow devices (SOD) that are present at several locations in the catchment. Only a small, more recent part of the sewer system separates rain- and wastewater. Houses that are not yet connected to a sewer usually have a sceptic tank for basic treatment, after which the overflow drains to the nearest stream. The historical developments in the region and of the sewer system have led to a complex situation of connected and non-connected houses, roads and other impervious areas with or without rainwater separation.

2.2 Overview of the research approach

The approach and procedure to demonstrate the impacts of EIA connected to WWTPs and how they interlink in order to answer the research questions is visualized in Fig. 1. (A) As a first step, empirical quantification and analysis of the impervious areas as well as the river and WWTP discharges was carried out (further referred to as the empirical analysis). (B) The empirical data of both EIA and WWTP discharges were used to develop three reduced rainfall scenarios and simulated in an FDPM to model the river flows in the catchment. (C) The empirical and modeled river flows and impacts of the WWTP impacts were intercompared in order to obtain an improved understanding of both catchment and model behavior.

2.3 River flow and WWTP discharges

The Flemish Environment Agency (FEA) provided both hourly and daily mean river flow data ($\mathrm{m^3\,s^{-1}}$) for the river gauging station situated at the outlet of the catchment (Varendonk) for the period 2004–2008, as well as effluent discharge data for the different WWTPs that are related to the catchment (Fig. 2). To evaluate the overall impact of the WWTPs that discharge into the catchment (Mol and Geel), relative contributions of the WWTPs effluent discharges to the daily discharge of the Grote Nete river were calculated for the period 2004–2008. No discharge data were available for the WWTP of the military camp of Leopoldsburg. However, because of its small size (0.7 % of total EIA), its impact on the

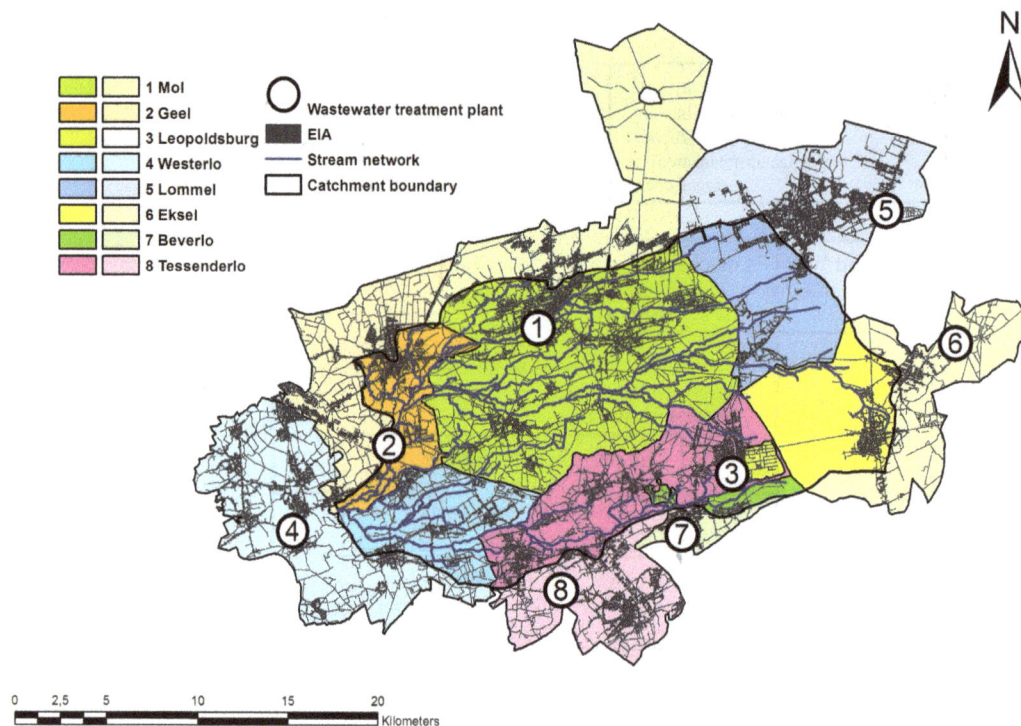

Fig. 2. Overview of the different WWTPs and their WWCRs that are situated within the Grote Nete catchment (1–3) as well as the WWTPs that receive wastewater from impervious areas that are situated within the Grote Nete catchment but discharge into another catchment (4–8). Parts of the WWCR that are situated within the natural catchment boundary are shown with a brighter shade, illustrating the discrepancy between the areas covered by the natural catchment and the sewer systems.

river system is considered to be negligible. The hourly river flow data were used for model calibration and validation.

2.4 Land use map and impervious area

The land use map used in this study was obtained from the National Geographical Institute (NGI) (NGI, 2007) and has a spatial accuracy of 1 m. This land use map (1 : 10 000 vector layers) is based on aerial photographs from 1998 (1 : 21 000), ground-truthed and adjusted in the following years until 2007, when the map was published. It contains 47 different land uses. For the hydrological simulation purposes, a reduced set of nine categories was considered: evergreen needleleaf forest, broad-leaved woodland, mixed forest, open scrublands, grasslands, permanent wetlands, croplands, impervious area and water bodies (Table 1). These nine classes follow the IGBP classification system based on their relationship to the modeled hydrological processes (Liu and De Smedt, 2004). Impervious areas include only completely sealed soils. Areas that can have reduced infiltration rates such as sandy roads (sand) or gardens are not considered to be part of the impervious areas. They are classified based on the most common characteristics of each land use type in the region (e.g., gardens are most often lawns).

The EIA is estimated on the basis of two thematic GIS layers representing the sewer system areas and are based on field

observations done by different administrations. Houses connected to the sewer system are shown in zoning maps (one for each municipality). These maps indicate the connection of all individual households to the sewer systems and which ones drain directly to a nearby stream. The zoning maps also indicate which houses will be connected in the future and which buildings will have to install individual wastewater treatment plants (FEA, 2008b). In order to conduct all analyses based on the same input data, the zoning maps were used to identify the buildings, present in the NGI land use map, connected to each sewer system.

Streets that are connected to the sewer system were identified based on the polylines describing each sewer system (FEA, 2008a). Streets in the NGI land use map that overlap with the sewer system were assumed to contribute to the EIA. Sewers that separate waste- and rainwater were left out from this part of the analysis. Combining both methods resulted in one map from which the EIA of each WWCR was derived.

2.5 Upstream area calculations

Subcatchments were delineated based on a 1 : 5000 digital elevation model (DEM) expressed as a 5 m raster (FEA, 2005, 2006). For each stream junction ($n = 131$), upstream areas were calculated using the method discussed in Jenson and Domingue (1988) (further referred to as the runoff method).

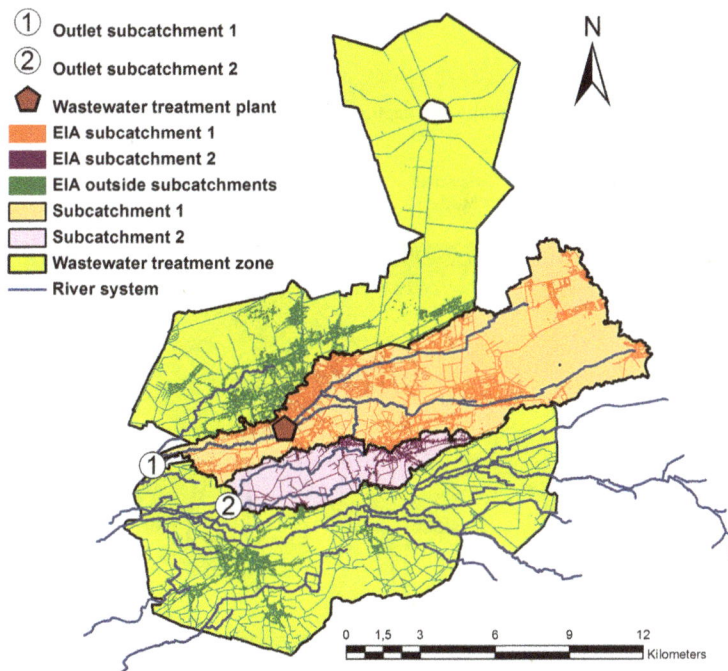

Fig. 3. Example on the calculation of the upstream areas. The WWTP discharges into subcatchment 1 (orange colored). Therefore the EIAs within the WWCR, but outside subcatchment 1 (green and purple colored), are included in the upstream area of subcatchment 1. As a result the area of this subcatchment increases by 404 ha of impervious area or by 5.1 % of the total area. Because the EIA is removed from subcatchment 2 (purple color), the area of subcatchment 2 decreases by 69 ha or 4.3 % of the total area.

By combining these upstream areas with the 1 m raster of the land use map, we calculated upstream impervious area for each stream junction.

Next the sewer infrastructure was considered (further referred to as the sewer method) (Fig. 3). In this method, the upstream areas were recalculated by removing the EIAs from their natural subcatchments and adding these EIAs to the subcatchment of the river reach into which the WWTP discharges. As zoning maps also indicate which houses yet have to be connected to the WWTPs, expected upstream areas for the near future were obtained as well. Differences in upstream impervious areas and total areas between the runoff method and the sewer method were considered as indicators of how strongly the catchment is affected by the sewer infrastructure. All GIS calculations were performed in ArcGIS 9.3 (ESRI Inc., 2009).

To make an evaluation of the impact of the sewer infrastructure on the catchment's overall water balance, the changes in upstream area and upstream impervious areas (TIA) between both methods were calculated for the 131 stream junctions in the catchment. The relative changes (%) were analyzed by means of histograms.

2.6 MIKE SHE model setup

MIKE SHE is a spatially distributed, physically based hydrological model (Abbott et al., 1986). It simulates the terrestrial water cycle including evapotranspiration (ET), overland flow, unsaturated soil water and groundwater storage and movements (Refsgaard and Storm, 1995; Feyen et al., 2000; DHI Water and Environment, 2008). The MIKE SHE model has been used worldwide for a wide range of applications (Refsgaard, 1997; Sun et al., 1998; Thompson et al., 2004; Sahoo et al., 2006; Zhang et al., 2008). For this study, a spatially distributed, physically based hydrological model was selected over other types of hydrological models as it can simulate the effect of spatially differentiated scenarios. It allowed us to evaluate the effect of changes in spatial patterns of surface runoff on the hydrological regime. The model was also used for other research purposes (Vansteenkiste et al., 2013) The representation of catchment characteristics and input data (digital terrain model, land use, soil) in MIKE SHE are provided through raster information. The MIKE SHE model for the Grote Nete catchment was built on a 250 m grid. It was developed with physics-based flow descriptions only for those processes that are relevant for the purposes of this study, i.e., overland and unsaturated flow. Given that the study focusses on spatial scenarios at the surface (changes in surface runoff), groundwater flow processes are considered to be secondary. Therefore the saturated zone was implemented through simplified lumped process descriptions, while surface processes were modeled in a spatially variable way (see Vansteenkiste et al., 2013, for details). The applied

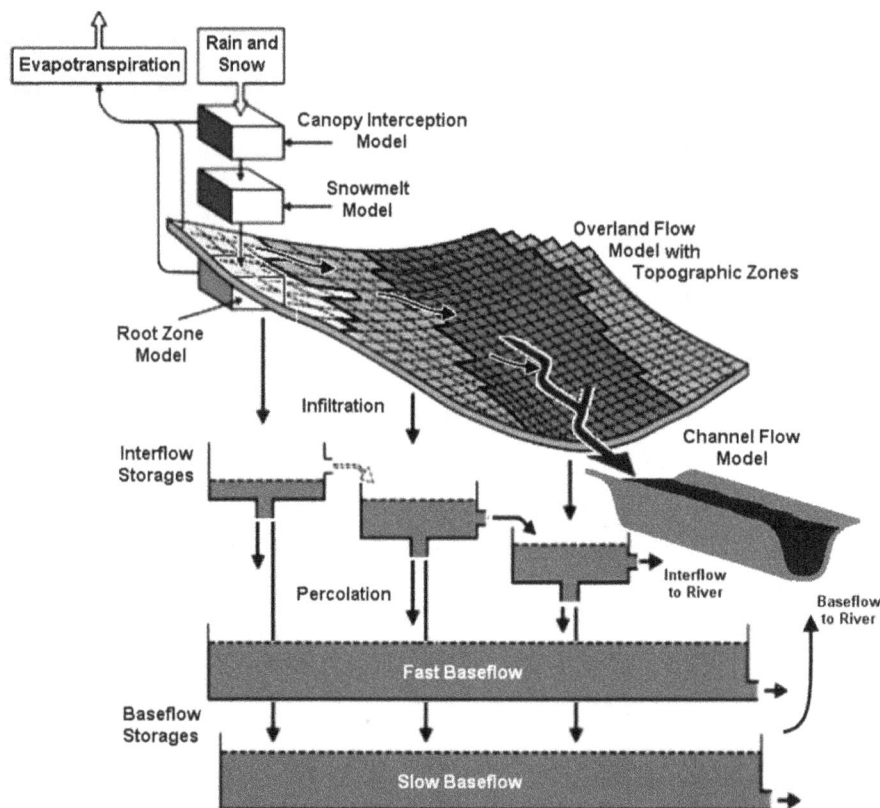

Fig. 4. Schematic representation of the applied MIKE SHE model configuration (Graham and Butts, 2006).

model configuration is schematized in Fig. 4 (Graham and Butts, 2006).

Hourly data from six rain gauges were used to describe the spatial variability of the rainfall over the catchment and used for meteorological input after applying Thiessen polygons. Only one potential evapotranspiration series was acquired from the national meteorological station located at Uccle, 30 km west of the study area, and applied. The growing cycle of the different crops was considered by means of a vegetation database that included leaf area index (LAI) and root depth (RD) and was based on Rubarenzya et al. (2007). Additional empirical parameters for determining the evapotranspiration of the crops were assessed from the literature (Kristensen and Jensen, 1975; DHI Water and Environment, 2008). The overland flow component was determined by the Strickler roughness coefficient, detention storage and initial water depths. The surface roughness was based on values from the literature (Chow, 1964) as a function of land use. Standard values were taken for the detention storage and initial water depths and are considered constant over the entire catchment (DHI Water and Environment, 2008). The MIKE SHE model was coupled to a full hydrodynamic river model, implemented in MIKE 11 (DHI Water and Environment, 2009) to route MIKE SHEs overland flow to the catchment outlet and account for the hydraulic effects of the river network and its infrastructure. The river network comprised

the main branches in the catchment, which were extracted from the Flemish hydrological atlas (FEA, 2005). The geometry of each river branch was specified in terms of cross sections obtained from field survey data. All infrastructures that were expected to have a significant impact on the river flow, such as bridges, culverts and weirs, were implemented in the model. For the unsaturated soil water component of the catchment model, soil moisture characteristics were defined by means of the model by Brooks and Corey (1966) for soil retention curves, and the equation by Averjanov (1950) for the soil hydraulic conductivities. The unsaturated zone parameters, needed to identify these relations, were based on the USDA – United States Department of Agriculture – soil information database. As mentioned above, the saturated zone was implemented through baseflow reservoirs applying simplified lumped storage and flow descriptions and parameters. More specifically, the entire groundwater system was divided into a series of shallow interflow reservoirs plus two deep baseflow reservoirs. These reservoirs allowed for differentiating between fast and slow components of baseflow discharge and storage. An overview of the most important model parameters in the considered model configuration is presented in Table 2.

Soil characteristics were derived from the USDA soil parameters classification system (Graham and Butts, 2006). Saturated zone flow was simulated using the linear reservoir

Table 1. Overview of the different land use classes used in the NGI map and the reclassification to 9 categories.

NGI description	IGBP vegetation
Coniferous trees	Evergreen needleleaf forest
Orchard	Evergreen needleleaf forest
Tree nursery	Evergreen needleleaf forest
Deciduous trees	Broad-leaved woodland
Poplar plantation	Broad-leaved woodland
Mixed deciduous and coniferous trees without dominant	Mixed forest
Mixed deciduous and coniferous trees with dominance of deciduous trees	Mixed forest
Mixed deciduous and coniferous trees with dominance of coniferous trees	Mixed forest
Sand	Open scrublands
Bare ground	Open scrublands
Coppice	Open scrublands
Heath	Open scrublands
Heath with deciduous trees	Open scrublands
Heath with coniferous trees	Open scrublands
Scrubs	Open scrublands
Brushwood	Open scrublands
Brushwood with scrubs	Open scrublands
Cemetery	Grasslands
Beds	Grasslands
Pasture	Grasslands
Gardens	Grasslands
Deep swamp	Permanent wetlands
Reedland	Permanent wetlands
Cropland	Croplands
Transformer station	Impervious area
Railway	Impervious area
Road	Impervious area
Crossroad	Impervious area
Industrial building (in use)	Impervious area
Industrial building (abandoned)	Impervious area
Warehouse	Impervious area
Silo	Impervious area
Greenhouse	Impervious area
Cooling tower	Impervious area
Non-university hospital	Impervious area
Town hall	Impervious area
Schoolhouse	Impervious area
Firehouse	Impervious area
Commercial building	Impervious area
Religious building	Impervious area
Sports hall	Impervious area
Covered grandstand	Impervious area
Non-covered grandstand	Impervious area
Indoor swimming pool	Impervious area
Building for drinking water supply	Impervious area
Normal building	Impervious area
Building for public use	Impervious area
Watercourse	Water bodies
Pond	Water bodies
Sluice	Water bodies

method. The entire groundwater system was divided into a series of shallow interflow reservoirs plus two deep base-flow reservoirs. These reservoirs allowed for differentiating between fast and slow components of baseflow discharge and storage. Water was routed through the linear reservoirs as interflow and baseflow and subsequently added to the MIKE 11 river network as lateral inflow in the lowest interflow reservoir (Graham and Butts, 2006).

2.7 Implementing the hydrological influence of the sewer infrastructure in MIKE SHE

To model the effect of the EIA on the hydrological regime of the Grote Nete catchment, the detailed land use map (1 : 5000) and EIA had to be resampled to the MIKE SHE model grid specifications (250 m). Despite the careful resampling for preserving the catchment land use in the model, an overestimation of TIA by 4.2 % remained. For each WWCR, the urban area and EIA were extracted and the percentage EIA per urban area WWCR calculated. These calculations were used in combination with the resampled urban area per WWCR in MIKE SHE to define the fraction of rainfall discharged by the sewer infrastructure to the river. Table 3 presents the percentage of EIA per WWCR and its urban area.

Incorporating the impact of the sewer infrastructure within the MIKE SHE model can be done in two different ways. This basically involves the removal of the surface runoff that is going to the sewer network from the total catchment runoff. This surface runoff to the sewage system can then be added to the river network at the WWTP discharge location as a point source, after accounting for the sewer WWTP routing time delay. To remove the sewer runoff from the catchment runoff, one of the first solutions is to take out, from the modeling domain, the grid cells that cover the impervious areas and that contribute to the sewer system. The problem encountered here in this study is that none of the 250 m grid cells are fully covered by that type of impervious surfaces. Only fractions of the grid cell areas contribute to the sewer system, making the removal of the grid cells impossible. Therefore we opted for the second solution: reducing the rainfall input proportional to the fraction of the sewer runoff contribution. This allowed us to take better in to account the fractions of impervious areas within each grid cell.

Three different rainfall scenarios were developed to assess the impact of the sewer system. For each scenario, the total measured rainfall in the catchment was reduced in relation to the assessed WWTP discharges. The rainfall reductions were spatially differentiated within the catchment by reducing the different rainfall series based on the overlap between the Thiessen polygons (different point rainfall input series) and the different WWCR regions (amount of EIA). The different scenarios of reduced rainfall were applied within the model to assess its impact in the model. Scenario 1 considered a reduction in rainfall within the WWCRs that discharge

Table 2. Overview of the MIKE SHE model parameters in the considered model configuration.

Component	Parameter	Description	Unit
Evaporation	C_{int}	Canopy interception	mm
	$c1, c2, c3$	Evapotranspiration empirical parameters	$mm\,day^{-1}$
	A_{root}	Root distribution index	1/m
	K_c	Crop coefficient	–
	LAI	Leaf area index	–
	RD	Root depth	mm
Overland flow	M	Strickler roughness coefficient	$(m^3\,s^{-1})-1$
	DS	Detention storage of the ground surface	mm
	H_{ini_OF}	Initial water depth on the ground surface	mm
Unsaturated flow	θ_{sat}	Saturated soil water content	$m^3\,m^{-3}$
	θ_{FC}	Soil water content at field capacity	$m^3\,m^{-3}$
	θ_{WP}	Soil water content at field wilting point	$m^3\,m^{-3}$
	θ_{res}	Residual soil water content	$m^3\,m^{-3}$
	K_{sat}	Saturated hydraulic conductivity	$m\,s^{-1}$
	n	Averjanov empirical constant	–
Channel flow	H_{ini_CF}	Initial water level	mm
	Q_{ini_CF}	Initial discharge	$m^3\,s^{-1}$
	n	Bed resistance	$(m^3\,s^{-1})-1$
Groundwater flow	S_{y_IF}	Specific yield for interflow reservoir	–
	H_{ini_IF}	Initial depth of the interflow reservoir	m
	$H_{treshold\_IF}$	Threshold depth of the interflow reservoir	m
	H_{bottom_IF}	Bottom depth of the interflow reservoir	m
	$t_{percolation}$	Percolation time	days
	RC_{IF}	Interflow time constant	days
	α_{SZ}	Fraction of percolation to the baseflow reservoir	–
	S_{y_BF}	Specific yield for the baseflow reservoir	–
	RC_{BF}	Baseflow time constant	days
	α_{UZ}	Unsaturated zone feedback fraction for baseflow	–
	H_{ini_BF}	Initial depth of the baseflow reservoir	m
	$H_{treshold\_BF}$	Threshold depth of the baseflow reservoir	m
	H_{bottom_BF}	Bottom depth of the baseflow reservoir	m

within the catchment to assess the impact of the sewer infrastructure on the river flows. Scenario 2 implemented a reduction in rainfall within the WWCRs that discharge outside the catchment to assess the impact of water transport outside the catchment. Scenario 3 took a reduction in rainfall across the entire catchment to evaluate the impact of all the sewers on the river system (Table 3). The original measured rainfall input series, applied to calibrate the model, is further referred to as the reference scenario.

The differences in runoff discharges between the initial model result and the simulations with reduced rainfall input gave us indications of the impact of the sewer infrastructure on the catchment runoff. The model results were compared for the different scenarios and assessed on an hourly, daily and monthly basis. The reductions in flow because of reduced rain were compared to the measured WWTP discharges as well as their relative contributions to the total river flows. Differences in relative contributions were calculated between the reference scenario and the rainfall scenarios 1 and 3. Changes in peak and low flows were evaluated in relation to the empirical return period (mean recurrence interval of these flows).

2.8 MIKE SHE model calibration

After completing the model setup, the MIKE SHE model was calibrated. Note that the MIKE SHE model code comprises numerous free parameters, whereas the guiding principle for complex models like MIKE SHE is to calibrate the model on as few free parameters as possible (Refsgaard and Storm, 1995). Therefore, the calibration parameters were reduced by a parameter sensitivity analysis similar to Xevi et al. (1997) and Thompson et al. (2004). The results of this sensitivity analysis for the Grote Nete model are not presented here, but can be found in Vansteenkiste et al. (2013). They show that the most sensitive parameters are the surface roughness and saturated zone parameters. These parameters are mainly related to the groundwater computations, but also have a strong influence on both low-flow and peak-flow magnitudes. In the end the model was calibrated using 14 parameters per

Table 3. Different variables used to implement the reduced rain scenarios: EIA in the catchment (ha), EIA per WWCR (%) and EIA per urban area unit (%) based on the NGI data.

WWTP	EIA (ha) per WWCR	EIA per WWCR (%)	EIA per urban area WWCR (%)	Scenario 1	Scenario 2	Scenario 3
Mol	782.07	5.17	72.77	*		*
Geel	284.31	7.2	67.42	*		*
Leopoldsburg	32.04	6.99	48.5	*		*
Tessenderlo	418.85	6.58	76.21		*	*
Westerlo	182.96	4.77	67.22		*	*
Beverlo	41.14	5.43	87.34		*	*
Lommel	145.43	2.75	47.57		*	*
Eksel	101.6	2.27	74.04		*	*

* indicates the WWRCs for which the rain was reduced in each scenario.

grid cell related to the distributed raster information, and 20 catchment-wide parameters related to the groundwater flow and evapotranspiration processes.

Calibration of the model was done against hourly stream flow measurements at the catchment outlet for the time period 2004–2006, while the years 2007 and 2008 were used for model validation. The most sensitive parameters were iteratively and manually adjusted between predefined limits until maximal correspondence between measured and predicted discharge runoff downstream of the catchment was achieved. The predefined parameter value limits represent the physically acceptable intervals and have been assessed on the basis of previous modeling studies of the Grote Nete catchment (Rubarenzya et al., 2007; Woldeamlak et al., 2007) and the literature (Chow et al., 1988; Anderson and Woessner, 1991; DHI Water and Environment, 2008).

The model correspondence was evaluated both qualitatively by visual inspection of the runoff results and quantitatively using goodness-of-fit statistics, including mean error (ME), root-mean-squared error (RMSE), correlation coefficient (R) and Nash–Sutcliffe efficiency (NSE) (Nash and Sutcliffe, 1970). Because the aim of this study was to investigate the impact on both high and low river flow conditions, independent peak and low flows, extracted from the time series using the method of Willems (2009), were also explicitly validated. This was done in scatterplots of simulated versus observed values as well as by means of empirical frequency distributions (peak and low flows versus return periods). The return periods of peak and low flows were calculated empirically as the total length of the available time series (in years) over the peak- and low-flow rank (1 for highest, 2 for second highest, etc.). Box–Cox transformation was applied to the simulated and observed peak and low flows to reach homoscedastic model residuals (Willems, 2009). This means that the model residuals can be represented by one distribution and equal weight is given to the peak- and low-flow values. The RMSE of the model residuals after transformation was optimized during model calibration.

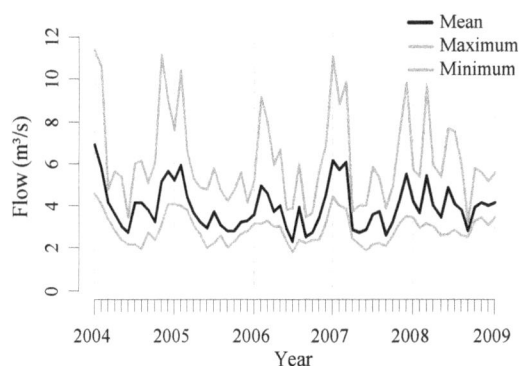

Fig. 5. Monthly mean, minimum and maximum of the daily measured flow ($m^3 s^{-1}$) for the period 2004–2008 at the outlet station (Varendonk) of the Grote Nete catchment.

3　Results

3.1　River flow contribution of WWTPs

Between January 2004 and December 2008, the Grote Nete had an average observed discharge of $3.95 \, m^3 \, s^{-1}$ at the catchment outlet. The upstream WWTPs discharged for the same period on average $0.31 \, m^3 \, s^{-1}$ of wastewater to the Grote Nete, or 7.9 % of the river flow. Discharges of both the river and WWTPs, however, varied substantially in time (Fig. 5). Rain events always lead to strong changes in river flow. For example, in 2007 there was a noticeable reduction in baseflow during spring and summer, followed by a strong increase during the winter period. In 2008 several rain periods led to a higher average flow during spring and summer. Monthly mean discharges of WWTPs and monthly mean river flows were found to be well correlated ($r^2 = 0.72$, $p < 0.001$). Correlation between daily mean WWTP and daily mean river discharges was, however, lower ($r^2 = 0.60$, $p < 0.001$).

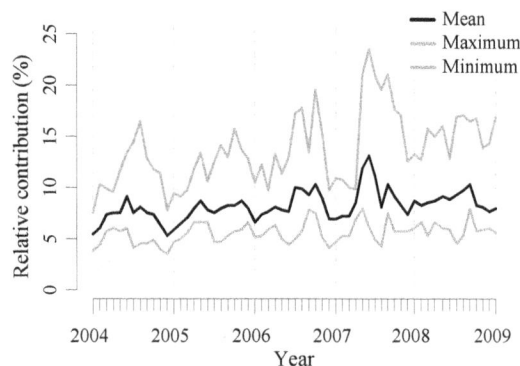

Fig. 6. Monthly mean, minimum and maximum relative contribution of the WWTPs (Mol and Geel) to the river flow at the outlet station (Varendonk) of the Grote Nete catchment.

In general, the WWTPs were found to contribute between 5.5 and 13.1 % of the monthly average river flow at the Grote Nete catchment outlet (Fig. 6). On a daily basis the contribution of the WWTPs to the river flow can decrease to 5.5 % during wet periods or increase up to 23.6 % during dry summer periods. The highest relative contributions were observed for rain events that occur during low river flow periods (e.g., convective thunderstorm periods after long, dry summer periods).

3.2 Water transfers between catchments and subcatchments

3.2.1 Current situation

From the analysis of the WWCRs we conclude that there are significant water transfers between the Grote Nete catchment and adjacent catchments. Of the total of 2836 ha TIA in the catchment, 1661 ha are currently connected to the WWTPs. This gives an initial ratio of 0.6 between TIA and EIA. Only 54.0 % of the EIA drains water that remains inside the catchment, the rest, mostly situated in the southern part of the catchment, drains water outside the catchment. At the same time waste-, ground- and rainwater from 461 ha, mostly from the north, is transported from outside to inside the catchment. If the difference in boundaries between catchment and WWCRs is taken into account, the EIA for the catchment is considered to be 1361 ha. This represents 4.0 % of the entire catchment area.

Upstream impervious areas and total upstream area change substantially when the WWCRs are incorporated into the calculations. A comparison between both calculation methods for 131 river junctions illustrates this impact. By taking the WWCRs into account, total upstream impervious areas decrease up to 99 %, as for most subcatchments impervious areas are connected to a WWTP located outside the subcatchment. For other river junctions, the upstream areas increase up to 64 % because a WWTP is situated upstream of the

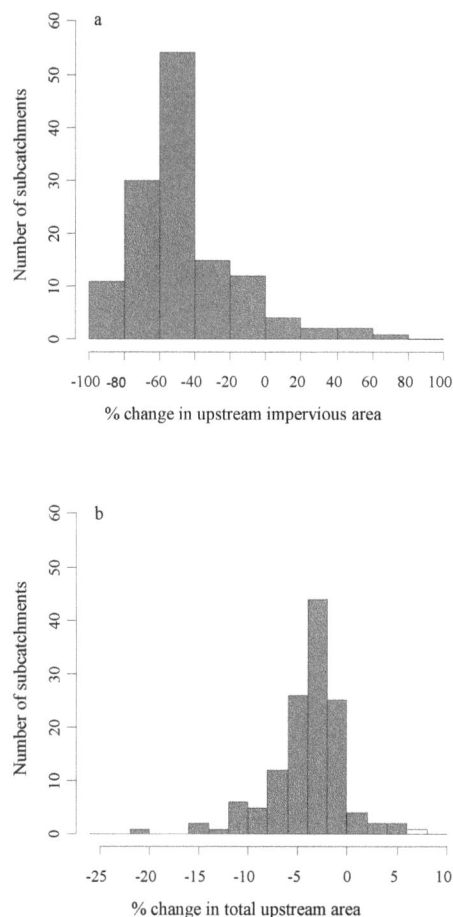

Fig. 7. (a) Histogram of the change in upstream impervious areas that demonstrates the impact of the sewer system on both upstream area calculations. For different stream junction points ($n = 131$) in the catchment, upstream impervious areas were calculated based on the natural catchment and after taking the sewer system into account. Differences between both types of upstream impervious areas were calculated as percentage change for each junction. **(b)** Histogram of the change in total upstream areas that demonstrates the impact of the sewer system on both upstream area calculations. For different stream junction points ($n = 131$) in the catchment, total upstream areas were calculated based on the natural catchment and after taking the sewer system into account. Differences between both types total upstream areas were calculated as percentage change for each junction.

junction (Fig. 7a). For the same reason, the change in total upstream area varies strongly (Fig. 7b).

3.2.2 Future developments

When the WWCR zoning plans are fully implemented in the future, another 245 ha of impervious areas will be connected to the WWTPs. Of those 245 ha, the surface runoff of 141 ha will be transported to other catchments, while the surface runoff of 148 ha will be imported from neighboring catchments.

Table 4. MIKE SHE calibrated parameters for the saturated zone.

	Interflow reservoir	Baseflow reservoir 1	Baseflow reservoir 2
Specific yield [−]	0.22	0.2	0.2
Initial depth [m]	15	30	30
Bottom depth [m]	15.065	40	40
Threshold depth [m]	15	30	30
Time constant [days]	4	12	12
Percolation time constant [days]	1.5	–	–
Fraction of percolation [−]	0.82	–	–
UZ feedback fraction [−]	–	0.32	0.12

When all subcatchments are evaluated, it is seen that most of the river junctions will experience an extra reduction in upstream area by 1 or 2 % (Fig. 8). Ten river junctions will, however, experience an increase of their upstream area by 1 or 2 % because of the upstream presence of a WWTP.

3.3 Model calibration and validation results

In comparison with Vansteenkiste et al. (2013) the main difference in the parameterization of the MIKE SHE model was related to the saturated zone component (Table 4) and the fine tuning of the overland flow roughness parameters (Table 5). Other parameters related to the overland flow, river flow and unsaturated zone were identical to Vansteenkiste et al. (2013).

Table 6 shows the model performance statistics ME, RMSE, R and NSE. These demonstrate the good general model performance. The statistics, however, demonstrate that the model performance is slightly better in the calibration period than in the validation period. Figure 9 shows the observed and simulated hourly runoff series for the calibration period.

Additional verification of the model performance for the high- and low-flow extremes is presented in Fig. 10. The observed independent high- and low-flow extremes are plotted against the simulated ones after Box–Cox transformation. These validation plots allow for evaluation of the model for its ability to predict extreme conditions. The model is able to simulate the extreme peak flows well, while the low-flow extremes are slightly overestimated by the model. The ME is very small for the peak flows ($0.05 \, \text{m}^3 \, \text{s}^{-1}$) and larger for the independent low-flow extremes ($0.14 \, \text{m}^3 \, \text{s}^{-1}$). Based on the good general model performance for total flows in both calibration and validation periods and for peak flows, the model was considered applicable for assessing the impact of the water transfers on these flow variables as a result of the sewer infrastructure.

Table 5. MIKE SHE calibrated parameters for the surface.

	Strickler roughness coefficient [$\text{m}^{1/3} \, \text{s}^{-1}$]
Deciduous needleleaf forest	2.5
Deciduous broadleaf forest	1.25
Mixed forest	1.82
Grasslands	6
Permanent wetlands	2
Croplands	13
Urban and built-up	90
Water bodies	90

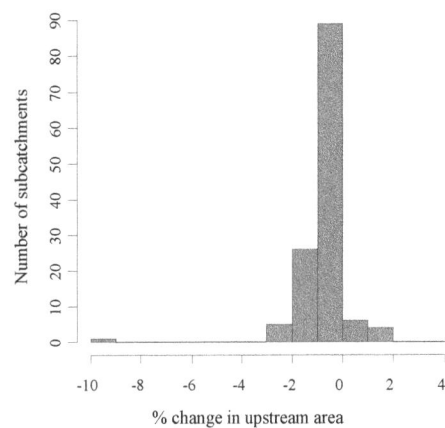

Fig. 8. Histogram of the change in upstream areas after full implementation of the zoning plans. For different stream junction points ($n = 131$) in the catchment, total upstream areas were calculated by comparing the current sewer system and the future sewer system after full implementation of the zoning plans. Differences in total upstream areas are given as percentage change for each stream junction.

3.4 Comparison with model impact results

3.4.1 River flow impact of WWTPs

As a first step, the different rainfall scenarios are compared with the reference scenario. Figure 11 shows the model-based differences in mean monthly river flows between the reference scenario and the adjusted rainfall scenarios. Based on this difference, the relative contributions of the EIA to the total river flow were obtained. These relative contributions vary between 2.2 and 7.2 % for scenario 1. For scenario 2 these contributions vary between 2.8 and 6.1 %.

3.4.2 Seasonal variation in river flow impact

A seasonal change in relative contribution of the WWTP infrastructure to the river flow was found. The largest contributions to the overall flow were found during summer and

Table 6. Statistical performance of total hourly river flows for the model calibration and validation periods at the outlet station of the Grote Nete catchment.

	Calibration	Validation
ME [$m^3\,s^{-1}$]	0.6	0.72
RMSE [$m^3\,s^{-1}$]	0.84	0.93
R [$-$]	0.88	0.84
NSE [$-$]	0.72	0.63

lowest during winter periods. The effect is, however, again less pronounced compared to the relative contribution based on the empirical analysis (Fig. 11).

A comparison is made between the model-based impact results and the empirical analysis of Sect. 3.1, where the river flows at the outlet station are adjusted for the connected areas. Results of that comparison show strong seasonal patterns in differences between the model-based and empirical analysis results (Fig. 12). Especially during the period of declining flows (flow recession periods) in spring and the beginning of summer, the model simulates much lower relative contributions of WWTP discharges compared to the empirical analysis. This difference is less pronounced or absent for the summer of 2008.

3.4.3 Impact on peak and low flows

The model-based impact results of the scenarios result in a decrease of both peak (Fig. 13a) and low flows (Fig. 13b) for given return periods. For events with an empirical return period higher than 1 year, both peak and low flows decrease proportional to the reduced rain scenarios. The effects thus are stronger for the low-flow event compared to the peak flow events (Table 7).

4 Discussion

4.1 Impact of WWTPs on the river baseflow

The overall impact of the WWTPs on the river flow depends on the timescale (daily, monthly or yearly). On average, the WWTPs are responsible for about 10 % of the catchment's discharge, but the relative WWTP contributions to the river flow can be significantly higher at short timescales and during dry periods (Fig. 6). The sewer infrastructure in the catchment is hence found to be an important point source of water in the rivers.

The high WWTPs effluent contribution to the total mean river discharge is due to the combined effect of different sewer-infrastructure-related processes: wastewater collection, rainwater runoff and groundwater intrusion. That parasitic groundwater, due to the groundwater intrusion, can be high in the region as has also been shown before by Dirckx

Table 7. Absolute and relative changes in peak and low flows at the outlet station of the Grote Nete catchment for empirical return periods higher than 1 year and the different rainfall scenarios compared with the reference scenario.

	Peak flows		Low flows	
	$m^3\,s^{-1}$	%	$m^3\,s^{-1}$	%
Scenario 1	0.33	3	0.3	5.5
Scenario 2	0.3	2.8	0.11	4.9
Scenario 3	0.62	6	0.23	10.6

et al. (2009). Due to this draining of the groundwater table, drought-related problems induced by the urbanization will further increase. Climate change scenarios for Flanders predict a strong decrease in river low flows during summer (Baguis et al., 2010; Vansteenkiste et al., 2013). The impact of the WWTPs on the overall flow is thus expected to increase in the future.

While the impact of the WWTPs on a river flow can be evaluated rather easily, the impact of connected impervious area on the flow regime of smaller reaches within the WWCR is more difficult to quantify. Often the roofs of buildings and pavements of a catchment are connected to sewers that transport storm and wastewater to a WWTP, which might be located outside the natural catchment boundary. If we would like to evaluate these changes, long-term river flow data need to be available that encompass also river flow data prior to the sewer development. Also a detailed inventory of the gradual expansion of the sewer infrastructure would be required.

4.2 TIA versus EIA

Impervious area is a landscape metric that is widely used as an indicator of water quality, quantity and river ecosystem health (Jacobson, 2011). In the empirical analysis, conducted in this study, the impact of the sewer infrastructure on the impervious areas within the catchment was evaluated. The proposed method allowed us to make a distinction between TIA and EIA and to evaluate both transfers between catchments and subcatchments. Whereas both upstream TIA and EIA were found to be useful indicators of the hydrological and ecological disturbance, the EIA is in general considered to be a better indicator for the anthropogenic impact on the hydrological regime (Roy and Shuster, 2009). The EIA of the catchment decreased significantly when the different WWCRs were incorporated into the calculation. Large parts of the connected impervious areas within the catchment do not contribute to the river flows inside the catchment but are drained to a neighboring catchment. As a result, the overall impact of urbanization within the catchment can be over- or underestimated. At the same time large amounts of wastewater are transported from outside the catchment. These changes to the natural system affect both

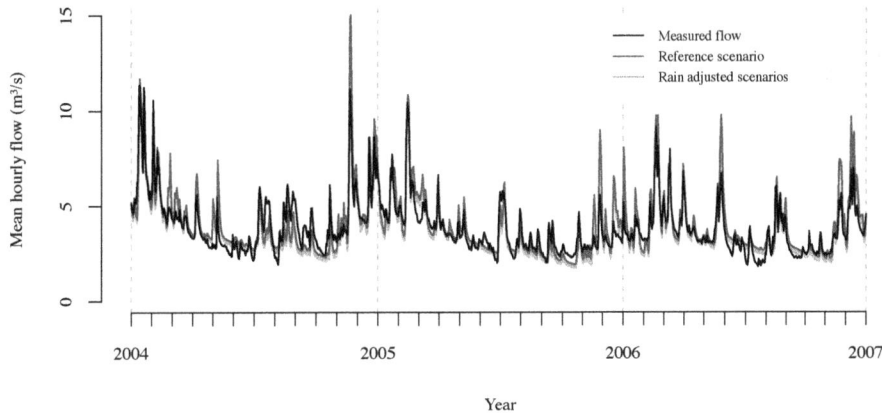

Fig. 9. Observed and simulated hourly river flow series for the model calibration period on a daily time step.

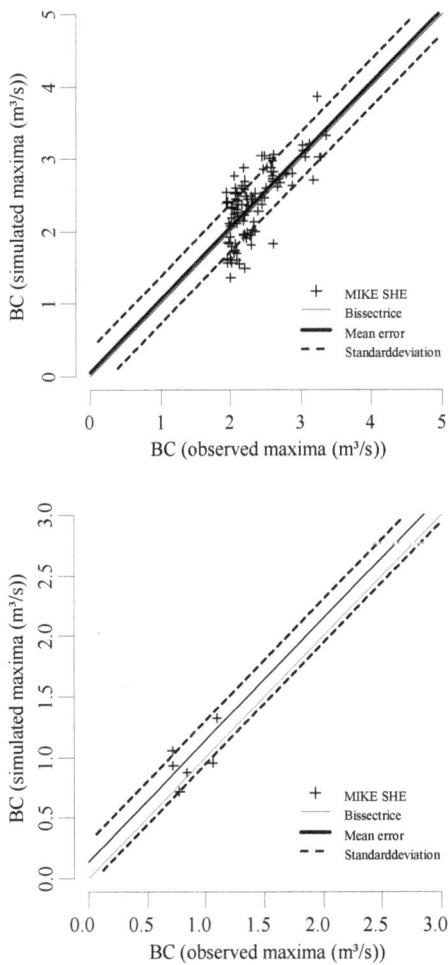

Fig. 10. (a) Scatter plot of simulated versus observed independent hourly peak flows at the outlet station of the Grote Nete catchment after Box–Cox transformation ($\lambda = 0.25$). (b) Scatter plot of simulated versus observed independent hourly low flows at the outlet station of the Grote Nete catchment after Box–Cox transformation ($\lambda = 0.25$).

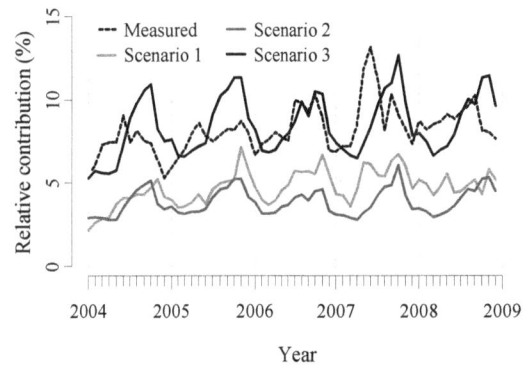

Fig. 11. Relative contribution (% monthly mean total flow) of the WWTPs to the total river flow. Measured refers to the empirical results obtained in Sect. 3.1.

the spatial and temporal distribution of runoff water in the catchments: from spatially distributed runoff to point inflows, from one catchment to another, increased surface runoff, reduced groundwater infiltration, shorter travel time and hence higher temporal and spatial aggregation, etc.

The changes in the total upstream areas, between -16 and $+3\%$, and upstream impervious areas, between -99 and $+64\%$, in our study were found to be large. For most of the subcatchments in our research area, both total upstream area and upstream impervious area decreased significantly. Although these subcatchments do not have actual upstream EIA, they are affected by the reduction in upstream impervious areas and the resulting decreased total upstream area. These reductions in impervious areas and the transfer of storm and wastewater to other reaches can lead to reductions in the flow regime and changes in related river characteristics. The actual absence of upstream impervious areas, due to wastewater allocation, might be more important than the presence of only a small portion of upstream impervious area.

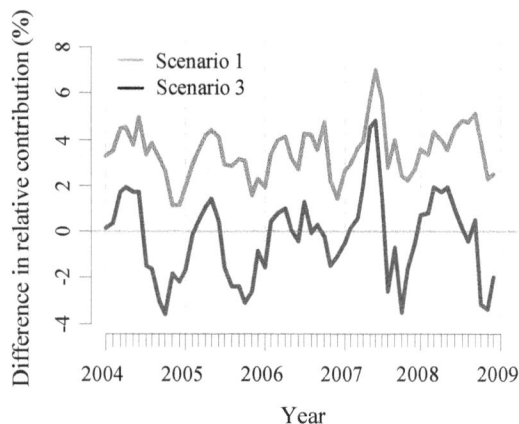

Fig. 12. Difference in relative contribution of the WWTPs to the total river flow at the outlet station of the Grote Nete catchment (% difference in monthly mean total flow) between the model-based and empirical analysis results (rainfall scenario 1 and 3). The difference in contribution increases during flow recession periods.

In contrast to many other studies we were able to use high-resolution data that are based on manual field observations instead of less accurate remote sensing data. The use of proxies for impervious areas, as used in other studies (Chabaeva et al., 2009), was not necessary. The same counts for the calculation of the EIA. Despite the high resolution of the data, some uncertainties remain in the impervious area classification (e.g., use of NGI classes and its influence on the TIA calculation) and EIA calculation (e.g., actual connection between road surfaces and the underlying sewer system). However, we expect that both metrics, TIA and EIA, are close to the actual situation in 2008 during low-flow periods. Nevertheless, due to increasing urbanization and sewer development, both indicators require a regular update.

SODs can have a profound impact on the hydrology, especially during extreme rain events. But their responses to these rain events can vary widely and are difficult to predict. Although relevant to the study area, the available data were not sufficient to incorporate SODs into the river flow analysis or into the EIA calculations or model development. It is expected that incorporating SODs in the analysis would result in a reduction of the EIA during extreme rain events. If such extreme rain events were to be analyzed explicitly, SODs should be integrated in the EIA calculations.

4.3 Model impact results

As opposed to the empirical analysis, the model-based results allow for explicit consideration of the catchment runoff dynamics, the highly non-linear hydrological responses to the changes in impervious areas and the interactions between different runoff components. However the use of hydrological models has, as is the case for all models, limitations. Traditional hydrological models impose restrictions on how to

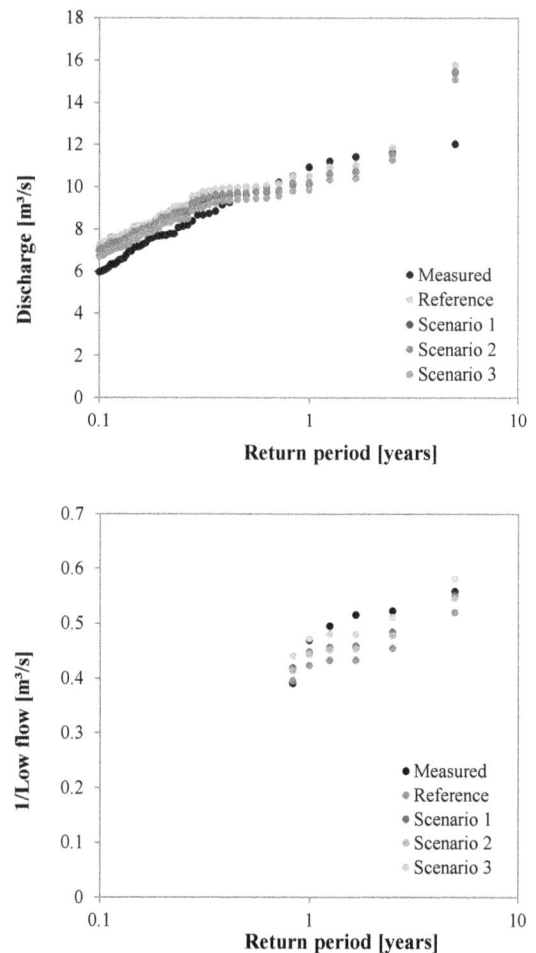

Fig. 13. (a) Return period of hourly peak flow extremes for the Grote Nete catchment between 2004 and 2008. **(b)** Return period of hourly low flow extremes for the Grote Nete catchment between 2004 and 2008.

deal with sewer infrastructure. In this study we evaluated the impact of the WWTPs by means of an existing, calibrated MIKE SHE hydrological model. Rainfall series were reduced proportional to the EIAs within the catchment to enable simulation of the impact of the surface runoff from impervious areas to the sewer system. This method is an alternative to the actual integration of a sewer model in the catchment model, which was not possible for reasons of data availability and model characteristics. Despite these simplifications of reality, the consistencies between the empirical and model-based results gave us confidence in the impact results. However, using this method it is not possible to evaluate the impact of the EIA situated outside the catchment. The latter impact evaluation would require sewer models to be integrated with the catchment hydrological model.

Previous studies have demonstrated the difficulties models can have to describe base flows in (peri-)urbanized areas (Elliott et al., 2010). Our model results revealed overestimations of low flow, while other recent studies have reported underestimations (Furusho et al., 2013). Hydrology in peri-urban catchments is typically a combination of fast and slow hydrological responses (Braud and Andrieu, 2013). The specific weighing of both responses is in general case specific and a result of the historical developments in the anthropogenic system. Although some characteristics in our catchment (e.g., fragmented land uses and related sewer system) are specific for the region, interbasin transfers of waste- and storm water are not rare and can play an important role at different scales in many peri-urban catchments. Impacts of these transfers in peri-urban catchments should therefore be analyzed and if necessary incorporated during model development. Bach and Ostrowski (2013) concluded, based on a semi-distributed model, that the representation of the flow processes in peri-urban models should take place at a high temporal and spatial resolution. Although FDPMs give a spatially detailed description of the hydrological processes, our results illustrate the challenges for sewer system integration in FDPMs, though they might be less if an open-source modeling system were used.

As expected, the different scenarios resulted in a decreased flow, compared to the reference scenario, proportional to the amount of EIA taken into account. The scenario with the lowest amount of EIA (scenario 2: transportation outside the catchment) resulted in the smallest change in flow. Besides an overall decrease in flow, both peak and low flows decreased. But low flows were proportionally more heavily affected by the rainfall reduction scenarios. These results confirm the higher impact of EIA during summer low flows found by the empirical analysis and again illustrate the high impact of EIA on the flow regime and its importance to be considered in the model. After consideration of the rainfall scenarios, the modeled rivers were higher than the measured river flow adjusted for the EIA, this despite the fact that the rainfall input was reduced for a similar amount within the different scenarios. Apparently, other processes like evapotranspiration within the model compensate for the reduced rainfall, leading to a lower reduction in flow.

Another problem is the coarse spatial resolution of the model. Due to this resolution, there is a general overestimation of both TIA and EIA implemented in the model compared to the high-resolution data. Nevertheless, similar values were obtained for the WWTP discharges and the reduction in flow in scenario 3 (full reduction based on all WWTPs). Because of the overestimation in the actual TIA and EIA, similar errors might be made to when no distinction is made between EIA and TIA (Alley and Veenhuis, 1983). At the same time the overestimation of the impervious surfaces may have biased the hydrological model parameters during the calibration (e.g., underestimation of the surface runoff coefficient). This means that if the model were to be used for impact analysis of urbanization and climate change scenarios, the impacts on peak flows and flood frequencies may be underestimated. This problem is further investigated

by Vansteenkiste et al. (2013) for the MIKE SHE model of the same catchment considered in this study.

An important aspect is the seasonal variation in the relative contribution of the EIA compared to the empirical data. Both scenario 1 (reduction for WWTPs inside the catchment) and 3 (full reduction based on all WWTPs) resulted in an underestimation of the EIA impact on the river flow during months with low flow and an overestimation during months with high flow. Hydrological models are often used to evaluate peak discharges and related flood risks. Climate change scenarios for Flanders, however, indicate an increase in frequency and duration of dry periods, making low-flow events more common (Boukhris et al., 2009). Therefore the importance of these low-flow events and their evaluation in hydrological models will become more important. A better incorporation of both impervious areas and WWTPs might be crucial for a better performance of the models in evaluating the effect of climate change on peak and low-flow events. Hydrological models are frequently used to predict changes in the hydrological regime. But if we want to use these to assess changes in climate, land use or other future developments within the catchment, consideration of the sewer transfers discussed in this paper becomes increasingly important. Our results show that the further development of the sewer infrastructure will have a profound impact on the upstream areas.

5 Conclusions

This paper presented a methodology to calculate EIA in a way that incorporates the effects of WWCRs that do not coincide with natural catchment boundaries. The methodology allows us to evaluate storm- and wastewater transfers between different catchments and indicate how strongly the catchment's hydrology is impacted by the sewer infrastructure. Comparisons between histograms or differences in histograms of catchment areas can display the vulnerability of the catchments to impervious area impacts and potential peak and low-flow events. The method also allows for study of how rivers that have no WWTP upstream are impacted by the upstream presence of EIA. These upstream impervious areas can have profound impacts on infiltration, surface runoff and the river flow regime.

We also simulated the impacts of the changes in impervious areas and WWTPs in FDPMs. By applying different rainfall scenarios, the impacts of wastewater transfers in the catchment were simulated and evaluated. At the same time, we were able analyze the impervious area parameterization within the model. Our results show that water displacements in and between catchments may severely impact the hydrological model results. Hence it may also be important to take these displacements into account in the hydrological model development. Although we used high-resolution data, the limited integration of all sewer processes (e.g., SODs) in the analysis prevented us from completely assessing the impact of the sewer system on the model performance.

The correct incorporation of impervious areas in models is of utmost importance as impervious areas have an impact on catchment delineation and different aspects of the flow regime. With increasing urbanization and sewer development, the impact of these processes on the hydrological regime are expected to further increase in the future. Important areas of further research remain, amongst others, as to (a) how to incorporate impervious areas from outside the catchment into the model, (b) how to remove the areas that are transported outside the catchment from the model domain, (c) how to better represent the seasonal variation in impervious area and WWTP impact in the model, and (d) how to integrate these processes based on less detailed data sets.

Acknowledgements. This research was supported by the University of Antwerp (UA-BOF fund), the Flanders Hydraulics Research (Waterbouwkundig Laboratorium, division of the Authorities of Flanders) and the Belgian Federal Science Policy Office (SUDEM-CLI cluster project, program "Science for a Sustainable Development").

Edited by: T. Kjeldsen

References

Abbott, M. B. and Refsgaard, J. C.: Distributed Hydrological Modelling, 1st Edn., Kluwer Academic Publishers, Dordrecht, 1996.

Abbott, M. B., Bathurst, J. C., Cunge, J. A., O'Connell, P. E., and Rasmussen, J.: An introduction to the European Hydrological System – Systeme Hydrologique Europeen, "SHE", 1: History and philosophy of a physically-based, distributed modelling system, J. Hydrol., 87, 45–59, 1986.

Ajami, N. K., Gupta, H., Wagener, T., and Sorooshian, S.: Calibration of a semi-distributed hydrologic model for streamflow estimation along a river system, J. Hydrol., 298, 112–135, 2004.

Alley, W. M. and Veenhuis, J. E.: Effective impervious area in urban runoff modeling, J. Hydraul. Eng.-ASCE, 109, 313–319, 1983.

Anderson, M. P. and Woessner, W. W.: Applied Groundwater Modeling: Simulation of Flow and Advective Transport, 1st Edn., Academic Press, Dordrecht, 1991.

Arnold, C. L. and Gibbons, C. J.: Impervious surface coverage – The emergence of a key environmental indicator, J. Am. Plan. Assoc., 62, 243–258, 1996.

Averjanov, S.: About permeability of subsurface soils in case of complete saturation, Engl. Collect., 7, 243–258, 1950.

Bach, M. and Ostrowski, M.: Analysis of intensively used catchments based on integrated modelling, J. Hydrol., 485, 148–161, 2013.

Baguis, P., Roulin, E., Willems, P., and Ntegeka, V.: Climate change scenarios for precipitation and potential evapotranspiration over central Belgium, Theor. Appl. Climatol., 99, 273–286, 2010.

Baker, D. B., Richards, R. P., Loftus, T. T., and Kramer, J. W.: A new flashiness index: Characteristics and applications to midwestern rivers and streams, J. Am. Water Resour. Assoc., 40, 503–522, 2004.

Becker, A. and Braun, P.: Disaggregation, aggregation and spatial scaling in hydrological, J. Hydrol., 217, 239–252, 1999.

Beven, K.: Changing ideas in hydrology – The case of physically-based models, J. Hydrol., 105, 157–172, 1989.

Boukhris, O. E. F., Willems, P., and Vanneuville, W.: Water and Urban Development Paradigms: Towards an Integration of Engineering, Design and Management Approaches, chap. The impact of climate change on the hydrology in highly urbanised Belgian areas, CRC Press, Boca Raton, 271–276, 2009.

Boyd, M. J., Bufill, M. C., and Knee, R. M.: Predicting pervious and impervious storm runoff from urban drainage basins, Hydrological Sciences Journal – Journal Des Sciences Hydrologiques, 39, 321–332, 1994.

Boyle, D. P., Gupta, H. V., Sorooshian, S., Koren, V., Zhang, Z. Y., and Smith, M.: Toward improved streamflow forecasts: Value of semidistributed modeling, Water Resour. Res., 37, 2749–2759, 2001.

Brabec, E. A.: Imperviousness and Land-Use Policy: Toward an Effective Approach to Watershed Planning, J. Hydrol. Eng., 14, 425–433, 2009.

Brandes, D., Cavallo, G. J., and Nilson, M. L.: Base flow trends in urbanizing watersheds of the Delaware River basin, J. Am. Water Resour. Assoc., 41, 1377–1391, 2005.

Braud, I. andFletcher, T. D. and Andrieu, H.: Hydrology of peri-urban catchments: Processes and modelling, J. Hydrol., 485, 1–4, 2013.

Brooks, R. H. and Corey, A. T.: Properties of porous media affecting fluid flow, J. Irrig. Drain. Eng., 92, 61–68, 1966.

Carpenter, T. M. and Georgakakos, K. P.: Intercomparison of lumped versus distributed hydrologic model ensemble simulations on operational forecast scales, J. Hydrol., 329, 174–185, 2006.

Chabaeva, A., Civco, D. L., and Hurd, J. D.: Assessment of Impervious Surface Estimation Techniques, J. Hydrol. Eng., 14, 377–387, 2009.

Chow, V. T.: Handbook of Applied Hydrology, McGraw-Hill Company, 1st Edn., New York, 1964.

Chow, V. T., Maidment, D. R., and Mays, L. W.: Applied Hydrology, 1st Edn., Tata McGraw-Hill Education, Singapore, 1988.

DHI Water and Environment: MIKE SHE user guide, DHI Software, Hørsholm, 2008.

DHI Water and Environment: MIKE11, a modeling system for rivers and channels, Reference Manual, DHI Software, Hørsholm, 2009.

Dirckx, G., Bixio, D., Thoeye, C., De Gueldre, G., and Van De Steene, B.: Dilution of sewage in Flanders mapped with mathematical and tracer methods, Urban Water J., 6, 81–92, 2009.

Elliott, A., Spigel, R., Jowett, I., Shankar, S., and Ibbitt, R.: Model application to assess effects of urbanisation and distributed flow controls on erosion potential and baseflow hydraulic habitat, Urban Water J., 7, 91–107, 2010.

ESRI Inc.: ArcGIS 9.3, ESRI Inc., Redlands, 2009.

FEA: Flemish Hygrological Atlas, Vlaamse Milieumaatschappij – afdeling Operationeel Waterbeheer, Brussel, 2005.

FEA: Digital Elevation Model Flanders, raster, 5 m, Agentschap voor Geografische Informatie Vlaanderen, Gent, 2006.

FEA: Sewer infrastructure of Flanders, Vlaamse Milieumaatschappij, Aalst, 2008a.

FEA: Zoning Map of Flanders, Vlaamse Milieumaatschappij, Aalst, 2008b.

Feyen, L., Vázquez, R., Christiaens, K., Sels, O., and Feyen, J.: Application of a distributed physically-based hydrological model to a medium size catchment, Hydrol. Earth Syst. Sci., 4, 47–63, doi:10.5194/hess-4-47-2000, 2000.

Furusho, C., Chancibault, K., and Andrieu, H.: Adapting the coupled hydrological model ISBA-TOPMODEL to the long-term hydrological cycles of suburban rivers: Evaluation and sensitivity analysis, J. Hydrol., 485, 139–147, 2013.

Graham, D. and Butts, M. B.: Watershed Models, chap. Flexible, integrated watershed modelling with MIKE SHE, 1st Edn., CRC Press, 245–272, 2006.

Hamel, P., Daly, E., and Fletcher, T. D.: Source-control stormwater management for mitigating the impacts of urbanisation on baseflow: A review, J. Hydrol., 485, 201–211, 2013.

Jacobson, C. R.: Identification and quantification of the hydrological impacts of imperviousness in urban catchments: A review, J. Environ. Manage., 92, 1438–1448, 2011.

Jakeman, A. J. and Letcher, R. A.: Integrated assessment and modelling: features, principles and examples for catchment management, Environ. Model. Softw., 18, 491–501, 2003.

Jankowfsky, S., Branger, F., Braud, I., Gironás, J., , and Rodriguez, F.: Comparison of catchment and network delineation approaches in complex suburban environments: application to the Chaudanne catchment, France, Hydrol. Process., 27, 3747–3761, doi:10.1002/hyp.9506, 2013.

Jenson, S. K. and Domingue, J. O.: Extracting topographic structure from digital elevation data for geographic information system analysis, Photogrammetric Engineering and Remote Sensing, 54, 1593–1600, 1988.

Kauffman, G. J., Belden, A. C., Vonck, K. J., and Homsey, A. R.: Link between Impervious Cover and Base Flow in the White Clay Creek Wild and Scenic Watershed in Delaware, J. Hydrol. Eng., 14, 324–334, 2009.

Kleidorfer, M., Deletic, A., Fletcher, T. D., and Rauch, W.: Impact of input data uncertainties on urban stormwater model parameters, Water Sci. Technol., 60, 1545–1554, 2009.

Kristensen, K. and Jensen, S.: A model for estimating actual evapotranspiration from potential evapotranspiration, Nord. Hydrol., 6, 170–188, 1975.

Lee, J. G. and Heaney, J. P.: Estimation of urban imperviousness and its impacts on storm water systems, J. Water Resour. Pl. Manage.-ASCE, 129, 419–426, 2003.

Liu, Y. B. and De Smedt, F.: WetSpa Extension, A GIS-ased Hydrologic Model fior Flood Prediction and Watershed Management, Documentation and User Manual, Vrije Universiteit Brussel – Department of Hydrology and Hydraulic Engineering, Brussel, 2004.

Muleta, M. K. and Nicklow, J. W.: Sensitivity and uncertainty analysis coupled with automatic calibration for a distributed watershed model, J. Hydrol., 306, 127–145, 2005.

Nash, J. E. and Sutcliffe, J. V.: River flow forecasting through conceptual models part I – A discussion of principles, J. Hydrol., 10, 282–290, 1970.

NGI: Top10Vector, Nationaal Geografisch Instituut, Brussel, 2007.

Paul, M. J. and Meyer, J. L.: Streams in the urban landscape, Annu. Rev. Ecol. System., 32, 333–365, 2001.

Price, K.: Effects of watershed topography, soils, land use, and climate on baseflow hydrology in humid regions: A review, Prog. Phys. Geogr., 35, 465–492, 2011.

Ragab, R., Rosier, P., Dixon, A., Bromley, J., and Cooper, J. D.: Experimental study of water fluxes in a residential area: 2. Road infiltration, runoff and evaporation, Hydrol. Process., 17, 2423–2437, 2003.

Refsgaard, A. and Storm, B.: Computer Models of Watershed Hydrology, chap. MIKE SHE, Water Resources Publications, Highlands Ranch, 809–846, 1995.

Refsgaard, J. C.: Parameterisation, calibration and validation of distributed hydrological models, J. Hydrol., 198, 69–97, 1997.

Refsgaard, J. C. and Knudsen, J.: Operational validation and intercomparison of different types of hydrological models, Water Resour. Res., 32, 2189–2202, 1996.

Roy, A. H. and Shuster, W. D.: Assessing impervious surface connectivity and applications for watershed management, J. Am. Water Resour. Assoc., 45, 198–209, 2009.

Rubarenzya, M. H., Graham, D., Feyen, J., Willems, P., and Berlamont, J.: A site-specific land and water management model in MIKE SHE, Nord. Hydrol., 38, 333–350, 2007.

Sahoo, G. B., Ray, C., and De Carlo, E. H.: Calibration and validation of a physically distributed hydrological model, MIKE SHE, to predict streamflow at high frequency in a flashy mountainous Hawaii stream, J. Hydrol., 327, 94–109, 2006.

Seiler, K. P. and Rivas, J. A.: Groundwater in the Urban Environment: Selected City Profiles, vol. 21 of IAH – International Contributions to Hydrogeology, chap. Recharge and discharge of the Caracas aquifer, Venezuela, Balkema Publishers, Rotterdam, 233–238, 1999.

Sheeder, S. A., Ross, J. D., and Carlson, T. N.: Dual urban and rural hydrograph signals in three small watersheds, J. Am. Water Resour. Assoc., 38, 1027–1040, 2002.

Shuster, W. D., Bonta, J., Thurston, H., Warnemuende, E., and Smith, D. R.: Impacts of impervious surface on watershed hydrology: A review, Urban Water J., 2, 263–275, 2005.

Simmons, D. L. and Reynolds, R. J.: Effects of urbanization on base-flow of selected south-shore streams, Long-Island, New-York, Water Resour. Bull., 18, 797–805, 1982.

Smakhtin, V. U.: Low flow hydrology: a review, J. Hydrol., 240, 147–186, 2001.

Statistics Belgium: Population, Nationaal Instituut voor de Statistiek, Brussel, 2011.

Sun, G., Riekerk, H., and Comerford, N. B.: Modeling the hydrologic impacts of forest harvesting on Florida flatwoods, J. Am. Water Resour. Assoc., 34, 843–854, 1998.

Thompson, J. R., Sorenson, H. R., Gavin, H., and Refsgaard, A.: Application of the coupled MIKE SHE/MIKE 11 modelling system to a lowland wet grassland in southeast England, J. Hydrol., 293, 151–179, 2004.

Vansteenkiste, T., Tavakoli, M., Ntegeka, V., Willems, P., De Smedt, F., and Batelaan, O.: Climate change impact on river flows and catchment hydrology: a comparison of two spatially distributed models, Hydrol. Process., 27, 3649–3662, 2013.

Walsh, C. J., Fletcher, T. D., and Ladson, A. R.: Retention Capacity: A Metric to Link Stream Ecology and Storm-Water Management, J. Hydrol. Eng., 14, 399–406, 2009.

Ward, R. and Robinson, M.: Principles of Hydrology, 3rd Edn., McGraw-Hill, Maidenhead, 1989.

Willems, P.: A time series tool to support the multi-criteria performance evaluation of rainfall-runoff models, Environ. Model.
Softw., 24, 311–321, 2009.

Willems, P. and Berlamont, J.: Probabilistic modelling of sewer system overflow emissions, Water Sci. Technol., 39, 47–54, 1999.

Wittenberg, H. and Aksoy, H.: Groundwater intrusion into leaky
sewer systems, Water Sci. Technol., 62, 92–98, 2010.

Woldeamlak, S. T., Batelaan, O., and De Smedt, F.: Effects of climate change on the groundwater system in the Grote-Nete catchment, Belgium, Hydrogeol. J., 15, 891–901, 2007.

Xevi, E., Christiaens, K., Espino, A., Sewnandan, W., Mallants, D.,
Sørensen, H., and Feyen, J.: Calibration, Validation and Sensitivity Analysis of the MIKE-SHE Model Using the Neuenkirchen
Catchment as Case Study, Water Resour. Manage., 11, 219–242,
1997.

Zhang, Z. Q., Wang, S. P., Sun, G., McNulty, S. G., Zhang, H. Y.,
Li, J. L., Zhang, M. L., Klaghofer, E., and Strauss, P.: Evaluation
of the MIKE SHE model for application in the Loess Plateau,
China, J. Am. Water Resour. Assoc., 44, 1108–1120, 2008.

A comparative assessment of rainfall-runoff modelling against regional flow duration curves for ungauged catchments

Daeha Kim[1], **Il Won Jung**[2], and **Jong Ahn Chun**[1]

[1] APEC Climate Center, Busan, 48058, South Korea
[2] Korea Infrastructure Safety & Technology Corporation, Jinju, Gyeongsangnam-do, 52852, South Korea

Correspondence to: Jong Ahn Chun (jachun@apcc21.org)

Abstract. Rainfall–runoff modelling has long been a special subject in hydrological sciences, but identifying behavioural parameters in ungauged catchments is still challenging. In this study, we comparatively evaluated the performance of the local calibration of a rainfall–runoff model against regional flow duration curves (FDCs), which is a seemingly alternative method of classical parameter regionalisation for ungauged catchments. We used a parsimonious rainfall–runoff model over 45 South Korean catchments under semi-humid climate. The calibration against regional FDCs was compared with the simple proximity-based parameter regionalisation. Results show that transferring behavioural parameters from gauged to ungauged catchments significantly outperformed the local calibration against regional FDCs due to the absence of flow timing information in the regional FDCs. The behavioural parameters gained from observed hydrographs were likely to contain intangible flow timing information affecting predictability in ungauged catchments. Additional constraining with the rising limb density appreciably improved the FDC calibrations, implying that flow signatures in temporal dimensions would supplement the FDCs. As an alternative approach in data-rich regions, we suggest calibrating a rainfall–runoff model against regionalised hydrographs to preserve flow timing information. We also suggest use of flow signatures that can supplement hydrographs for calibrating rainfall–runoff models in gauged and ungauged catchments.

1 Introduction

A standard method to predict daily streamflow is to employ a rainfall–runoff model that conceptualises catchment functional behaviours, and simulate synthetic hydrographs from atmospheric drivers (Wagener and Wheater, 2006; Blöschl et al., 2013). A prerequisite of this conceptual modelling approach is parameter identification to enable the rainfall–runoff model to imitate actual catchment behaviours. Conventionally, behavioural parameters are estimated via model calibration against observed hydrographs (referred to as the "hydrograph calibration" hereafter). The hydrograph calibration provides convenience to attain reproducibility of the predictand (i.e. streamflow time series), which is commonly used as a performance measure in rainfall–runoff modelling studies. Because the degree of belief in hydrological models is normally measured by how they can reproduce observations (Westerberg et al., 2011), use of the hydrograph calibration has a long tradition in runoff modelling (Hrachowitz et al., 2013).

The hydrograph calibration, however, can be challenged by epistemic errors in input and output data, sensitivity to calibration criteria, and inability under no or poor data availability (Westerberg et al., 2011; Zhang et al., 2008). Importantly, it is difficult to know whether the parameters optimised toward maximising hydrograph reproducibility are unique to represent actual catchment behaviours, since multiple parameter sets possibly show similar predictive performance (Beven, 2006, 1993). This low uniqueness of the optimal parameter set, namely the equifinality problem in conceptual hydrological modelling, can become a significant uncertainty source particularly when extrapolating the optimal parameters to ungauged catchments (Oudin et al., 2008).

To overcome or circumvent those disadvantages, distinctive flow signatures (i.e. metrics or auxiliary data representing catchment behaviours) in lieu of observed hydrographs can be used to identify model parameters (e.g. Yilmaz et al., 2008; Shafii and Tolson, 2015). The flow duration curve (FDC) has received particular attention in the signature-based model calibrations as a single criterion (e.g. Westerberg et al., 2014, 2011; Yu and Yang, 2000; Sugawara, 1979) or one of calibration constraints (e.g. Pfannerstill et al., 2014; Kavetski et al., 2011; Hingray et al., 2010; Blazkova and Beven, 2009; Yadav et al., 2007). The FDC, the relationship between flow magnitude and its frequency, provides a summary of temporal streamflow variations in a probabilistic domain (Vogel and Fennessey, 1994). Many FDC-related studies have found that climatological and geophysical characteristics within a catchment determine the shape of the FDC (e.g. Cheng et al., 2012; Ye et al., 2012; Yokoo and Sivaplan, 2011; Botter et al., 2007). With only few physical parameters, the shape of the period-of-record FDC could be analytically expressed (Botter et al., 2008). Based on this strong relationship between catchment physical properties and the FDC, one may hypothesise that model calibration against the FDC (referred to as the "FDC calibration" hereafter) can provide parameters that can sufficiently capture actual catchment behaviours. Sugawara (1979) is the first attempt at the FDC calibration, emphasising its advantage to reduce negative effects of epistemic errors in rainfall–runoff data. Westerberg et al. (2011) also showed that the FDC calibration may provide robust predictions to moderate disinformation such as the presence of event flows under inconsistency between inputs and outputs.

If it allows rainfall–runoff models to sufficiently capture functional behaviours of catchments, the FDC calibration would have an especial value in comparison to the parameter regionalisation for prediction in ungauged catchment. The parameter regionalisation, which transfers or extrapolates behavioural parameters from gauged to ungauged catchments (e.g. Kim and Kaluarachchi, 2008; Oudin et al., 2008; Parajka et al., 2007; Wagener and Wheater, 2006; Dunn and Lilly, 2011), conveniently provides a priori estimates of behavioural parameters and thus became a popular approach to parameter identification in ungauged catchments (see a comprehensive review in Parajka et al., 2013). However, it has a critical concern that regionalised parameters are highly dependent on model calibrations at gauged sites that may have substantial equifinality problems. Under no flow information in ungauged catchments, it is impossible to know whether regionalised parameters are behavioural. Thus, regionalised parameters might be insufficiently reliable and highly uncertain (Bárdossy, 2007; Oudin et al., 2008; Zhang et al., 2008).

On the other hand, the calibration against regional FDC (referred to as "RFDC_cal" hereafter) may reduce the primary concern in the classical parameter regionalisation scheme. The regional models predicting FDC at ungauged sites have showed strong performance – for instance, via regression analyses between quantile flows and catchment properties (e.g. Shu and Ouarda, 2012; Mohamoud, 2008; Smakhtin et al., 1997), geostatistical interpolation of quantile flows (e.g. Pugliese et al., 2014; Westerberg et al., 2014), and regionalisation of theoretical probability distributions (e.g. Atieh et al., 2017; Sadegh et al., 2016) among many variations. The parameters obtained from RFDC_cal are deemed behavioural, because a distinctive flow signature of the target ungauged catchment directly identifies them; however, predicted FDC should be reliable in this case. An FDC is a compact representation of runoff variability at all timescales from inter-annual to event scale, embedding various aspects of multiple flow signatures (Blöschl et al., 2013). Based on this strength, several studies have already showed promising predictive performance using RFDC_cal for ungauged catchments (e.g. Westerberg et al., 2014; Yu and Yang, 2000).

Nevertheless, practical questions arise when using RFDC_cal for ungauged catchments. First, the FDC is simplified information with flow magnitudes only; hence, the FDC calibration could worsen the equifinality problem relative to the hydrograph calibration. Due to no flow timing information in regional FDC, one may cast a concern that parameters obtained from RFDC_cal may provide poorer predictive performance than regionalised parameters gained from the hydrograph calibration. Indeed, there is additional uncertainty in predicted FDC possibly introduced by the regionalisation models (Westerberg et al., 2011; Yu et al., 2002). RFDC_cal may be undesirable when a simple parameter regionalisation can provide better performance, because regionalising observed FDC may require expensive efforts. Several comparative studies on parameter regionalisation (e.g. Parajka et al., 2013; Oudin et al., 2008) have suggested that the simple proximity-based parameter transfer can be competitive in many regions. Second, there may be additional flow signatures to improve predictive performance of the FDC calibration. Additional constraining can lead to better predictive performance of the RFDC (Westerberg et al., 2014); however, it is still an open question which flow signatures can supplement the FDC calibration.

As discussed, RFDC_cal seems promising for prediction in ungauged catchments. However, to our knowledge, RFDC_cal has never been evaluated in a comparative manner with classical parameter regionalisation except by Zhang et al. (2015), who assessed its performance in part. Therefore, this study aimed to evaluate predictive performance of RFDC_cal in comparison with a conventional parameter regionalisation. We focused on the absence of flow timing in the FDC and its impacts on rainfall–runoff modelling. In this work, a parsimonious four-parameter conceptual model was used to simulate daily hydrographs for 45 catchments in South Korea. To predict FDC in ungauged catchments, a geostatistical regional model was adopted here. The Monte Carlo sampling was used to identify model parameters and measure equifinality in the hydrograph and the FDC calibrations.

2 Description of the study area and data

For this study, we selected 45 catchments located across South Korea with no or negligible human-made influences on flow variations (Fig. 1). South Korea is characterised as a temperate and semi-humid climate with rainy summer seasons. North Pacific high pressure brings monsoon rainfall with high temperatures during summer seasons, while dry and cold weathers prevail in winter seasons due to Siberian high pressure. Typical ranges of annual precipitation are 1200–1500 and 1000–1800 mm in the northern and the southern areas respectively (Rhee and Cho, 2016). Annual mean temperatures in South Korea range between 10 and 15 °C (Korea Meteorological Administration, 2011). Approximately 60–70 % of precipitation falls in summer seasons between June and September (Bae et al., 2008). Streamflow usually peaks in the middle of summer seasons because of heavy rainfall or typhoons, and hence information of catchment behaviours is largely concentrated on summer-season hydrographs. Snow accumulation and ablation occurring at high elevations have minor influences on flow variations due to the relatively small amount of winter precipitation (Bae et al., 2008).

The study catchments were selected based on availability of streamflow data. High-quality daily streamflow data across South Korea have been produced since the establishment of the Hydrological Survey Centre in 2007 (Jung et al., 2010), though river stages have been monitored for an extensive length at a few gauging stations. Thus, we collected streamflow data at 29 river gauging stations from 2007 to 2015 together with inflow data of 16 multi-purpose dams for the same data period from the Water Resources Management Information System operated by the Ministry of Land, Infrastructure, and Transport of the South Korean government (available at http://www.wamis.go.kr/). The mean annual flow of the study catchments was $739\,\mathrm{mm\,yr^{-1}}$ with a standard deviation of $185\,\mathrm{mm\,yr^{-1}}$ during 2007–2015.

In addition, as atmospheric forcing inputs, we collected daily precipitation and maximum and minimum temperatures for 2005–2015 at 3 km grid resolution produced by spatial interpolations between 60 stations of the automated surface observing system (ASOS) maintained by the Korea Meteorological Administration (2011). The ASOS data were interpolated by the Parameter-elevation Regression on Independent Slope Model (PRISM; Daly et al., 2008), and overestimated pixels of the PRISM grid data were smoothed by the inverse distance method. Jung and Eum (2015) found that this combined method improved the spatial interpolation of precipitation and the temperatures in South Korea. The annual mean precipitation and temperature of the study catchments vary within ranges of 1145–$1997\,\mathrm{mm\,yr^{-1}}$ and 8.0–$13.8\,°\mathrm{C}$ during 2007–2015. Hydro-climatological features of the 45 catchments are summarised in Table 1.

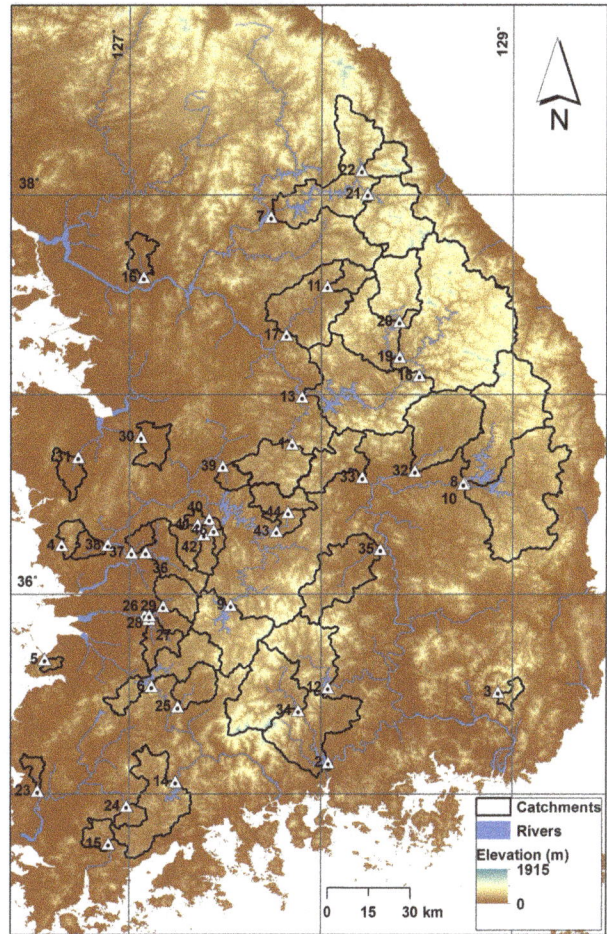

Figure 1. Locations of the study catchments in South Korea. The numbers are labelled at the outlet of each catchment.

3 Methodology

3.1 Hydrological model (GR4J)

A parsimonious rainfall–runoff model, GR4J (Perrin et al., 2003), was adopted to simulate daily hydrographs of the 45 catchments for 2007–2015. GR4J conceptualises functional catchment response to rainfall with four free parameters that regulate the water balance and water transfer functions. Figure 2 schematises the structure of GR4J. The four parameters (X1 to X4) conceptualise soil water storage, groundwater exchange, routing storage, and the base time of unit hydrograph respectively. Since its parsimonious and efficient structure allows robust calibration and reliable regionalisation of the parameters, GR4J has been frequently used for modelling daily hydrographs with various purposes under diverse climatic conditions (Zhang et al., 2015). The computation details and discussion are found in Perrin et al. (2003). The potential evapotranspiration (PE in Fig. 2) was estimated by the temperature-based model proposed by Oudin et al. (2005) for lumped rainfall–runoff modelling.

Table 1. Summary of hydrological features of the study catchments.

	Average	CV	Minimum	25 %	Median	75 %	Maximum
Area (km^2)	890	1.39	57	208	495	1013	6705
Elevation (m a.s.l.)	339	0.63	39	193	255	495	996
Mean annual prcp. (mm yr^{-1})	1359	0.14	1145	1247	1286	1388	1997
Mean annual temp. (°C)	11.9	0.13	7.9	11.3	12.3	13.0	13.8
Aridity index[a] (–)	0.66	0.11	0.44	0.61	0.68	0.71	0.76
P_{snow}[b]	35	0.66	6	23	28	50	141
Mean annual flow (mm yr^{-1})	739	0.25	232	624	740	838	1159
R_{PQ} (–)	0.55	0.27	0.18	0.45	0.54	0.63	0.91
I_{BF} (–)	0.49	0.16	0.27	0.44	0.49	0.56	0.62
D_{RL} (day^{-1})	0.63	0.10	0.50	0.60	0.63	0.66	0.77

[a] Ratio of potential ET to total precipitation. [b] Percentage of snowfall to total precipitation. Climatological features were calculated using spatial averages of the grid data, while the flow metrics were from the daily hydrographs for 2007–2015 as explained in Sect. 3.6.

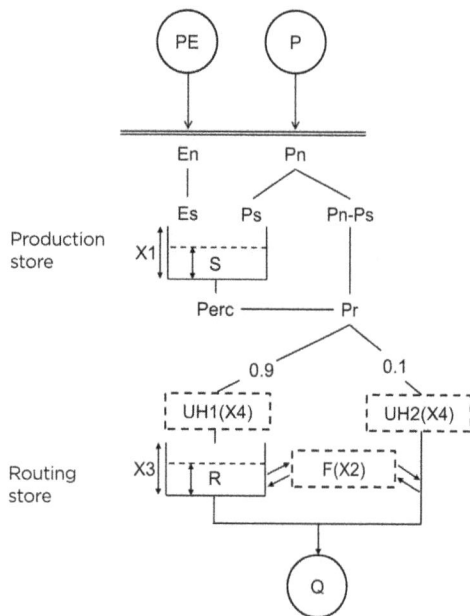

Figure 2. The schematised structure of GR4J (X1–X4: model parameters; PE: potential evapotranspiration; P: precipitation; Q: runoff; other letters indicate variables conceptualising internal catchment processes).

3.2 Preliminary data processing

Before rainfall–runoff modelling, we preliminarily processed the grid climatic data to convert precipitation data to liquid water forcing (i.e. rainfall and snowmelt depths) using a physics-based snowmelt model proposed by Walter et al. (2005). The preliminary snowmelt modelling was mainly for reducing systematic errors from no snow component in GR4J, which may affect model performance in catchments at relatively high elevations. We chose this preliminary processing to avoid adding more parameters (e.g. the temperature index) to the existing structure of GR4J. In the case of

GR4J, one additional parameter implies 25 % complexity increase in terms of the number of parameters. The snowmelt model uses the same inputs of GR4J to simulate point-scale snow accumulation and ablation processes (i.e. no additional inputs are required). The snowmelt model is a physics-based model but uses empirical methods to estimate its parameters for the energy balance simulation. As outputs, it produces the liquid water depths and the snow water equivalent. For lumped inputs to GR4J, we took spatially averaged pixel values of the liquid water depths and the maximum and minimum temperatures within the boundary of each catchment.

After the snowmelt modelling, consistency between the liquid water depths and the observed flows (i.e. input–output consistency) was checked using the current precipitation index (CPI; Smakhtin and Masse, 2000) defined as

$$I_t = I_{t-1} \cdot K + R_t, \tag{1}$$

where I_t is the CPI (mm) at day t, K is a decay coefficient (0.85 d^{-1}), and R_t is the liquid water depth (mm d^{-1}) at day t. CPI mimics temporal variations of typical streamflow data by converting intermittent precipitation data to a continuous time series with an assumption of the linear reservoir. The input–output consistency can be evaluated using correlation between CPI and observed streamflow as in Westerberg et al. (2014) and Kim and Kaluarachchi (2014). The Pearson correlation coefficients between CPI and streamflow data of the 45 catchments had an average of 0.67 with a range of 0.43–0.79, and no outliers were found in the box plot of the correlation coefficients. Hence, we assumed that consistency between climatic forcing and observed hydrographs was acceptable.

3.3 The hydrograph calibration in gauged catchments

To search behavioural parameter sets of GR4J against the streamflow observations (i.e. the hydrograph calibration), we used the objective function of Zhang et al. (2015) as the calibration criterion to consider the Nash–Sutcliffe efficiency

Table 2. Ranges of GR4J parameters used for parameter calibration (Demirel et al., 2013).

Parameter	Range
X1 (mm)	10 to 2000
X2 (mm)	−8 to +6
X3 (mm)	10 to 500
X4 (days)	0.5 to 4.0

(NSE) and the water balance error (WBE) together:

$$\text{OBJ} = (1 - \text{NSE}) + 5|\ln(1 + \text{WBE})|^{2.5}, \qquad (2)$$

$$\text{NSE} = 1 - \frac{\sum_{i=1}^{N}(Q_{\text{obs},i} - Q_{\text{sim},i})^2}{\sum_{i=1}^{N}(Q_{\text{obs},i} - \overline{Q_{\text{obs}}})^2}, \qquad (3)$$

$$\text{WBE} = \frac{\sum_{i=1}^{N}(Q_{\text{obs},i} - Q_{\text{sim},i})}{\sum_{i=1}^{N} Q_{\text{obs},i}}, \qquad (4)$$

where Q_{obs} and Q_{sim} are the observed and simulated flows respectively, $\overline{Q_{\text{obs}}}$ is the arithmetic mean of Q_{obs}, and N is the total number of flow observations. The best parameter set for each study catchment was obtained from minimisation of the OBJ using the Monte Carlo simulations described below.

To determine sufficient runs for the random simulations, we calibrated GR4J parameters using the shuffled complex evolution (SCE) algorithm (Duan et al., 1992) for one catchment with moderate input–output consistency with the parameter range given in Table 2 by Demirel et al. (2013). Then, the total number of random simulations was iteratively determined by adjusting the number of runs until the minimum OBJ of the random simulations became adequately close to the OBJ value from the SCE algorithm. We found that approximately 20 000 runs could provide the minimum OBJ value equivalent to that from the SCE algorithm. Subsequently, GR4J was calibrated by 20 000 runs of the Monte Carlo simulations for all 45 catchments, and the parameter sets with the minimum OBJ values were taken for runoff predictions. In addition, we sorted the 20 000 parameter sets in terms of corresponding OBJ values in ascending order, and the first 50 sets (0.25 % of the total samples) were taken to measure the degree of equifinality. We measured the equifinality simply by the prediction area between 2.5 and 97.5 % boundaries of runoff simulations given by the collected 50 parameter sets. This prediction area was later compared to that from the FDC calibration under the same Monte Carlo framework. Note that we estimated the prediction area to comparatively evaluate the degree of equifinality between the hydrograph and the FDC calibrations under the same sampling size and the same acceptance rate for all the catchments. For more sophisticated and reliable uncertainty estimation other methods are available, such as the generalised likelihood uncertainty estimation (GLUE; Beven and Bingley, 1992), the Bayesian total error analysis (BATEA;

Kavetski et al., 2006), and the differential evolution adaptive Metropolis (DREAM; Vrugt and Ter Braak, 2011).

For the hydrograph calibration, the 9-year streamflow data were divided into two parts for calibration (2011–2015) and for validity check (2007–2010), respectively. A 2-year warm-up period was used for initialising all runoff simulations in this study.

3.4 Model calibration against the regional FDC for ungauged catchments

Each catchment was treated ungauged for the comparative evaluation of RFDC_cal in the leave-one-out cross-validation (LOOCV) mode. For regionalising empirical FDC, the geostatistical method recently proposed by Pugliese et al. (2014) was used. Pugliese et al. (2014) employed the top-kriging method (Skøien et al., 2006) to spatially interpolate the total negative deviation (TND), which is defined as the area between the mean annual flow and below-average flows in a normalised FDC. The top-kriging weights that interpolate TND values were taken as weights to estimate flow quantiles of ungauged catchments from empirical FDC of surrounding gauged catchments. The FDC of an ungauged catchment in Pugliese et al. (2014) is estimated from normalised FDC of surrounding gauged catchments as

$$\hat{\Phi}(w_0, p) = \hat{\phi}(w_0, p) \cdot \overline{Q}(w_0), \qquad (5)$$

$$\hat{\phi}(w_0, p) = \sum_{i=1}^{n} \lambda_i \cdot \phi_i(w_i, p) \quad p \in (0, 1), \qquad (6)$$

where $\hat{\Phi}(w_0, p)$ is the estimated quantile flow $(\text{m}^3 \text{ s}^{-1})$ at an exceedance probability p (unitless) for an ungauged catchment w_0, $\hat{\phi}(w_0, p)$ is the estimated normalised quantile flow (unitless), $\overline{Q}(w_0)$ is the annual mean streamflow $(\text{m}^3 \text{ s}^{-1})$ of the ungauged catchment, and $\phi_i(w_i, p)$ and λ_i are normalised quantile flows (unitless) and corresponding top-kriging weights (unitless) of gauged catchment w_i, respectively. The unknown mean annual flow of an ungauged catchment, $\overline{Q}(w_0)$, can be estimated with a rescaled mean annual precipitation defined as

$$\text{MAP}^* = 3.171 \times 10^{-5} \cdot \text{MAP} \cdot A, \qquad (7)$$

where MAP^* is the rescaled mean annual precipitation $(\text{m}^3 \text{ s}^{-1})$, MAP is mean annual precipitation (mm yr^{-1}), and A is the area (km^2) of the ungauged catchment, and the constant 3.171×10^{-5} converts the units of MAP^* from $\text{mm yr}^{-1} \text{ km}^2$ to $\text{m}^3 \text{ s}^{-1}$.

A distinct advantage of the geostatistical method is its ability to estimate the entire flow quantiles in an FDC with a single set of top-kriging weights. Since a parametric regional FDC (e.g. Yu et al., 2002; Mohamoud, 2008) is obtained from independent models for each flow quantile in many cases – for instance, by multiple regressions between selected quantile flows and catchment properties – fundamental characteristics in an FDC continuum would be entirely or partly lost. The geostatistical method, on the other hand, treats all flow quantiles as a single object; thereby, features in an FDC continuum can be preserved. It showed promising performance to reproduce empirical FDC only using topological proximity between catchments. More details on the geostatistical method can be found in Pugliese et al. (2014).

For regionalising empirical FDC of the 45 catchments, we followed the same procedure of Pugliese et al. (2014). We obtained top-kriging weights (λ_i) by the geostatistical interpolation of TND values from observed FDC for the calibration period (2011–2015). Then, the top-kriging weights were used to interpolate empirical flow quantiles. The number of neighbours for the TND interpolation was iteratively determined as five, at which level additional neighbouring TND are unlikely to bring better agreement between the estimated and observed TND. In other words, normalised flow quantiles of five catchments surrounding the target ungauged catchment were interpolated with the top-kriging weights. Then, MAP* of the target ungauged catchment was multiplied. We predicted flow quantiles at 103 exceedance probabilities (p of 0.001, 0.005, 99 points between 0.01 and 0.99 at an interval of 0.01, 0.995, and 0.999) for rainfall–runoff modelling against regional FDC (i.e. RFDC_cal).

For runoff prediction in ungauged catchments, the GR4J parameters were identified by the same Monte Carlo sampling but towards minimisation of OBJ value between the regional and the modelled flow quantiles at the 103 exceedance probabilities. The best parameter set, which provided the minimum OBJ value, was taken as the best behavioural set of RFDC_cal for each catchment.

3.5 Proximity-based parameter regionalisation for ungauged catchments

We selected the proximity-based parameter transfer (referred to as "PROX_reg" hereafter) to comparatively evaluate predictive performance of RFDC_cal. The parameter regionalisation has three classical categories: (a) proximity-based parameter transfer (i.e. PROX_reg; e.g. Oudin et al., 2008); (b) similarity-based parameter transfer (e.g. McIntyre et al., 2005); and (c) regression between parameters and physical properties of gauged catchments (e.g. Kim and Kaluarachchi, 2008). A comprehensive review on the parameter regionalisation in Parajka et al. (2013) reported that PROX_reg has competitive performance under humid climate with low-complexity models relative to the other categories. Based on modelling conditions in this study (semihumid climate and four parameters), we chose PROX_reg to evaluate RFDC_cal.

To predict runoff at the 45 catchments in the LOOCV mode, we transferred the behavioural parameter sets obtained from the hydrograph calibration of the five donor catchments used for the FDC regionalisation. In other words, we used the same donor catchments for FDC regionalisation and PROX_reg. This allowed us to have consistency in transferring hydrological information from gauged to ungauged catchments between RFDC_cal and PROX_reg. Using the best behavioural parameter sets of the five donor catchments, we generated five runoff time series and took the arithmetic averages of them to represent runoff predictions by PROX_reg.

3.6 Performance evaluation

We used multiple performance metrics to evaluate predictive performance of all modelling approaches applied in this study. Predictive performance of each modelling approach was graphically evaluated using box plots of the performance metrics of the 45 catchments. In addition, we performed several paired t tests to check the statistical significance of performance differences between the modelling approaches. What follows is the description of the performance metrics.

To measure high- and low-flow reproducibility, we chose two traditional performance metrics: (1) the NSE between observed and predicted flows (Eq. 2b) and (2) the NSE of log-transformed flows (LNSE) respectively. LNSE is calculated as

$$\text{LNSE} = 1 - \frac{\sum_{i=1}^{N} \left(\ln \left(Q_{\text{obs},i} \right) - \ln \left(Q_{\text{sim},i} \right) \right)^2}{\sum_{i=1}^{N} \left(\ln \left(Q_{\text{obs},i} \right) - \overline{\ln \left(Q_{\text{obs}} \right)} \right)^2}. \tag{8}$$

Although NSE and LNSE are frequently used for performance evaluation, they may be sensitive to errors in flow observations (Westerberg et al., 2011). Hence, we additionally selected three typical flow metrics that embed dynamic flow variation in a compact manner: the runoff ratio (R_{QP}), the baseflow index (I_{BF}), and the rising limb density (D_{RL}). R_{QP}, I_{BF}, and D_{RL} are proxies of aridity and water-holding capacity, contribution of the baseflow to flow variations, and flashiness of catchment behaviours, respectively. They are defined as the ratio of runoff to precipitation, the ratio of baseflow to total runoff, and the inverse of average time to peak (d^{-1}) as

$$R_{\text{QP}} = \frac{\overline{Q}}{\overline{P}}, \tag{9}$$

$$I_{\text{BF}} = \sum_{t=1}^{T} \frac{Q_{\text{B},t}}{Q_t}, \tag{10}$$

$$D_{\text{RL}} = \frac{N_{\text{RL}}}{T_{\text{R}}}, \tag{11}$$

where $\overline{Q}$ and $\overline{P}$ are average flow and precipitation for a given period ($\text{mm}\,\text{d}^{-1}$), Q_t and $Q_{\text{B},t}$ ($\text{m}\,\text{d}^{-1}$) are the streamflow and the base flow at time t respectively, N_{RL} is the number of rising limb, and T_{R} is the total amount of time when the hydrograph is rising (days). $Q_{\text{B},t}$ can be calculated by sub-

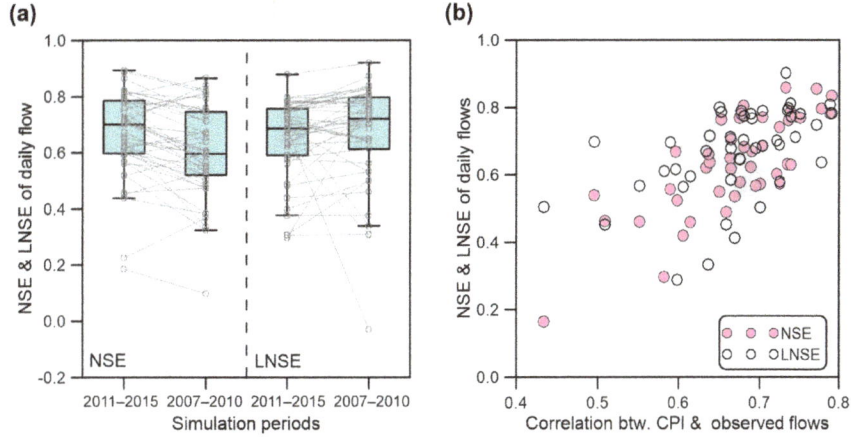

Figure 3. (a) Box plots of high flow (NSE) and low flow (LNSE) reproducibility of the behavioural parameters obtained from the hydrograph calibration at the 45 catchments. **(b)** The relationship between the input–output consistency and the model performance. The straight lines in the box plots connect the performance metrics for the calibration (2011–2015) and the validation periods (2007–2010) in each catchment.

tracting direct flow $Q_{D,t}$ from Q_t as

$$Q_{D,t} = c \cdot Q_{D,t-1} + 0.5 \cdot (1+c) \cdot (Q_t - Q_{t-1}),\tag{12}$$

$$Q_{B,t} = Q_t - Q_{D,t},\tag{13}$$

where c is the filter parameter, which was set to 0.925 (Brooks et al., 2011; Eckhardt, 2007).

Flow signature reproducibility of RFDC_cal and PROX_reg were evaluated by the relative absolute bias between modelled and observed signatures as

$$D_{FS} = \frac{|FS_{sim} - FS_{obs}|}{FS_{obs}},\tag{14}$$

where D_{FS} is the relative absolute bias, FS_{sim} is a flow signature of the modelled flows, and FS_{obs} is that of the observed flows.

4   Results

4.1   Hydrograph calibration and FDC regionalisation in gauged catchments

Figure 3a displays results of the parameter identification against the observed hydrographs (i.e. the hydrograph calibration). The 45 catchments had mean NSE and LNSE of 0.66 and 0.65 between the simulated and observed flows for the calibration period, respectively. The average NSE reduction from the calibration to the validation periods was 0.06 with a standard deviation of 0.10. The temporal transfer of the calibrated parameters did not decrease the mean LNSE value, while a wider LNSE range indicates that uncertainty of low-flow predictions may increase when temporally transferring the calibrated parameters.

The predictive performance was closely related to the input–output consistency (Fig. 3b), which was measured by

the Pearson correlation coefficient between the CPI and the observed flows. A low input–output consistency implies that the rainfall–runoff data may include significant epistemic errors such as minimal flow responses to heavy rainfall or excessive response to tiny rainfall. If the model calibration compensates disinformation from such errors, the parameters would be forced to have biases. Figure 3b shows that consistency in input–output data is a critical factor affecting parameter identification and thus performance. Perhaps screening catchments with low input–output consistency would provide better predictions in ungauged catchments. However, we did not consider it in the LOOCV for RFDC_cal and PROX_reg, since variation in input–output consistency would be a common situation. Rather, reducing the number of gauged catchments lowers spatial proximity, and thus can cause biases for ungauged catchments too. Overall, 27 catchments and 33 catchments showed NSE and LNSE values greater than 0.6. We assumed that the hydrograph calibration under the Monte Carlo framework, which was assisted by the SCE optimisation, was able to acceptably identify the behavioural parameters under given data quality.

Figure 4 illustrates the $1:1$ scatter plot between the observed and predicted flow quantiles of all the catchments, indicating high applicability of the top-kriging FDC regionalisation. The overall NSE and LNSE values between the observed and regionalised flow quantiles show good applicability of the geostatistical method. The NSE and LNSE values for individual catchments have averages of 0.83 and 0.91 with standard deviations of 0.25 and 0.11, respectively, implying that low-flow predictions were slightly better. The performance of the geostatistical method was relatively poor at locations where gauging density is low. Catchments 4, 10, 35, and 36, which recorded 0.6 or less NSE, are limitedly hatched with or adjacent to the other catchments; nonetheless, LNSE values of those catchments were still greater than

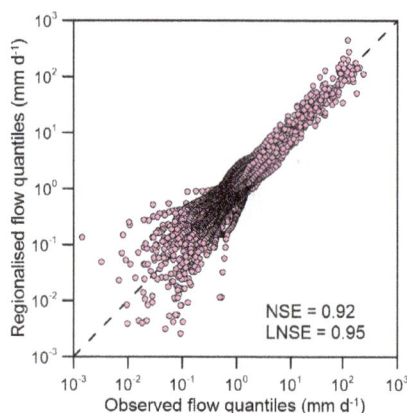

Figure 4. 1 : 1 scatter plot between the empirical flow quantiles and the flow quantiles predicted by the top-kriging FDC regionalisation method.

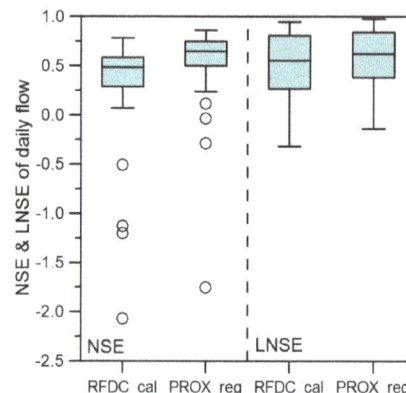

Figure 5. Box plots of NSE and LNSE values between the observed and the predicted hydrographs by RFDC_cal and PROX_reg for the 45 catchments under the cross-validation mode.

0.7. This result is consistent with a finding of Pugliese et al. (2016) that performance of the geostatistical method was sensitive to river gauging density. Transferring flow quantiles from remote catchments may not sufficiently capture functional similarity between donor and receiver catchments. In spite of the minor shortcomings, the geostatistical FDC regionalisation was deemed acceptable based on the high NSE and LNSE of flow quantiles. Topological proximity was generally a good predictor of flow quantiles for the study catchments.

4.2 Comparing hydrograph predictability between RFDC_cal and PROX_reg

Figure 5 compares the box plots of NSE and LNSE values between RFDC_cal and PROX_reg. PROX_reg generally outperforms RFDC_cal in predicting both high and low flows, suggesting that transferring parameters identified by observed hydrographs would be a better choice than a local calibration against predicted FDC. The differences between NSE values of PROX_reg and RFDC_cal have an average of 0.22 with a standard deviation of 0.34. Only eight catchments showed higher NSE with RFDC_cal. These higher NSE values of PROX_reg imply that PROX_reg is preferable when high-flow predictability is needed such as for flood analyses. In the case of LNSE, PROX_reg still had a higher median than RFDC_cal (0.53 and 0.62 for RFDC_cal and PROX_reg respectively). In 25 catchments, PROX_reg provided LNSE values greater than those of RFDC_cal.

The low performance of RFDC_cal was also found in the comparative assessment of Zhang et al. (2015), which evaluated RFDC_cal for 228 Australian catchments using the same GR4J model. Zhang et al. (2015) found that RFDC_cal was inferior to PROX_reg in the Australian catchments, because the FDC calibration poorly reproduced temporal flow variations relative to the hydrograph calibration. This study

confirms the difficulty of capturing dynamic catchment behaviours with FDC containing no flow timing information.

A major weakness of RFDC_cal is the absence of flow timing information in the parameter calibration process. Unlike RFDC_cal, PROX_reg did not discard the flow timing information. The regionalised parameters may be able to implicitly transfer the flow timing information from gauged to ungauged catchments (this hypothesis will be discussed in Sect. 4.4). Figure 6 illustrates how the absence of flow timing negatively influences predictive performance. For this comparison, the parameters were recalibrated against the observed FDC (not regional FDC) under the same Monte Carlo method to discard errors introduced by the FDC regionalisation (i.e. equivalent to calibration against perfectly regionalised FDC). The parameters identified by the observed hydrograph (Fig. 6a) brought a good predictability in both high and low flows, resulting in an excellent performance to reproduce the FDC. On the other hand, an excellent FDC reproducibility does not guarantee a good predictability in high flows (Fig. 6b). This indicates that reproducing FDC with rainfall–runoff models would be less able than the hydrograph calibration to capture functional catchment responses.

In addition, Fig. 6 shows that the prediction area of the 50 behavioural parameters from the Monte Carlo simulations (indicated by the grey areas and the blue arrows) became much larger when using the FDC calibration instead of the hydrograph calibration. We calculated the ratio of the prediction area of the FDC calibration to that of the hydrograph calibration, and refer to this as the equifinality ratio. It quantifies the degree of equifinality augmented by replacing the hydrograph calibration with the FDC calibration. Figure 7 displays the scatter plot between the equifinality ratio and the input–output consistency. The equifinality augmented by the loss of flow timing is likely to increase as the input–output consistency decreases. The average of the equifinality ratios was 1.96, implying that potential equifinality inherent in RFDC_cal could be substantial. This may suggest that the

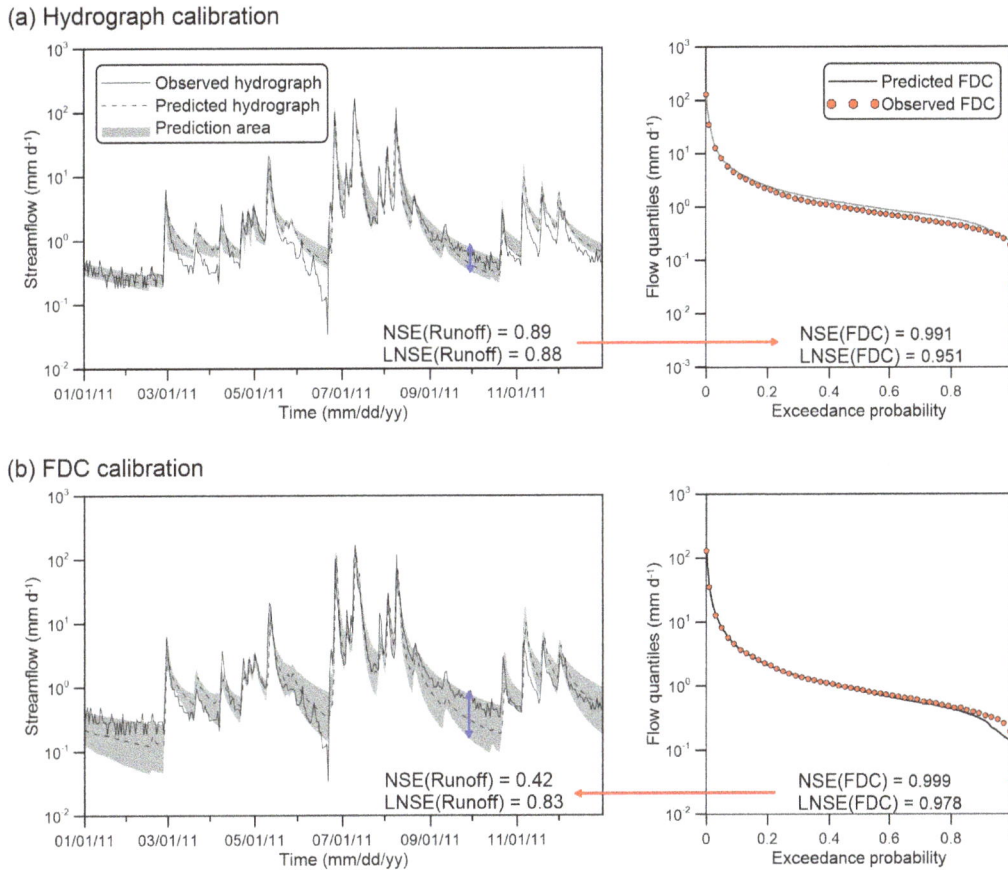

Figure 6. The observed and predicted hydrographs, the prediction areas, and the observed and predicted FDC given by **(a)** the hydrograph calibration and **(b)** the FDC calibration for Namgang Dam (Catchment 2 in Fig. 1).

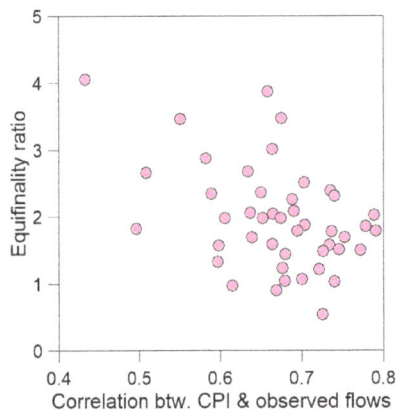

Figure 7. The input–output consistency vs. equifinality increased by replacing the hydrograph calibration with the FDC calibration. The equifinality ratio is defined as the ratio between the prediction areas of the 50 behavioural parameters gained from the FDC calibration and the hydrograph calibration.

equifinality problem embedded in RFDC_cal could be more significant than that in PROX_reg.

4.3 Comparing flow signature predictability between RFDC_cal and PROX_reg

Figure 8 summarises the performance of RFDC_cal and PROX_reg to regenerate the three flow signatures of R_{QP}, I_{BF}, and D_{RL}. RFDC_cal is competitive in reproducing the averaged-based signatures R_{QP} and I_{BF}, while it showed relatively weak ability to regenerate the event-based signature D_{RL}. R_{QP} and I_{BF} are flow metrics based on averages of long-term flow and precipitation in which no flow timing information is involved. In particular, RFDC_cal showed strong performance in reproducing I_{BF} relative to PROX_reg. This result can be explained by considering that baseflow has fewer temporal variations than direct runoff in the South Korean catchments under typical monsoonal climate. High seasonality of monsoonal precipitation causes high temporal variations in direct runoff during June to September, while relatively steady baseflow is dominant during dry seasons (October to May). In Namgang Dam (whose flow variation is displayed in Fig. 6), for example, the coef-

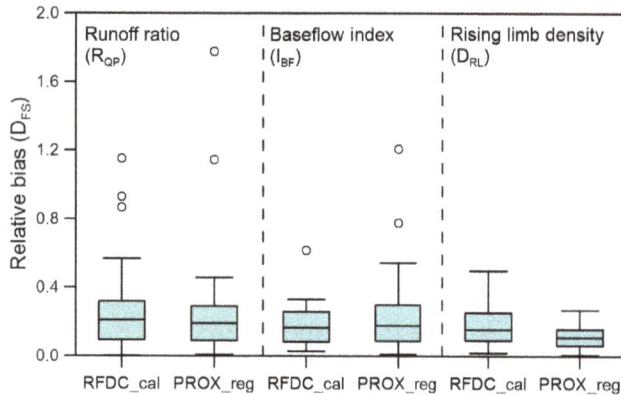

Figure 8. Flow signature reproducibility comparison between RFDC_cal and PROX_reg in terms of R_{QP} (**a**), I_{BF} (**b**), and D_{RL} (**c**).

ficient of variance (CV) of direct runoff was 5.86 for 2007–2015, which is approximately 3.5 times as high as the CV of the baseflow.

On the other hand, RFDC_cal was less able to reproduce D_{RL} than PROX_reg. This highlights the weakness of RFDC_cal in which only flow magnitudes were used for identifying model parameters. PROX_reg showed better performance in predicting D_{RL} than did RFDC_cal. Flow timing information gained from the observed hydrographs could be preserved, even after behavioural parameters were transferred to ungauged catchments. Overall, PROX_reg seems to be better than RFDC_cal to predict the three flow signatures together.

The box plots in Fig. 9 provide an indication that D_{RL} is likely to supplement the FDC calibration and thus improve RFDC_cal. From the collection of 50 behavioural parameter sets given by the FDC calibration, we chose the parameter set providing the lowest bias for each flow signature as the best behavioural sets, and simulated runoff again for all catchments. The high-flow predictability was fairly improved by additional constraining with D_{RL}, suggesting that flow metrics associated with flow timing make up for the weakness of the FDC calibration. Additional constraining with R_{QP} and I_{BF} did not bring appreciable improvement in the FDC calibration. However, PROX_reg was still better than the additional constraining with D_{RL}, indicating that a further study is needed for better constraining rainfall–runoff models using FDC together with additional flow metrics.

4.4 Paired t tests between the modelling approaches

For comparative evaluation in this study, we produced several runoff prediction sets using multiple rainfall modelling approaches. First, we calibrated GR4J against the observed hydrographs (referred to as Q_cal), and transferred the behavioural parameters to ungauged catchments in the LOOCV mode (PROX_reg). We constrained GR4J with the regional

FDC (RFDC_cal). To evaluate equifinality, we recalibrated the GR4J parameters against the observed FDC (referred to as FDC_cal). Additionally, we constrained the model with observed FDC plus the flow signatures, and significant performance improvement was found with D_{RL} (referred to as FDC + D_{RL}_cal). A paired t test using the performance metrics (NSE, LNSE, or D_{FS}) between these modelling approaches can answer various questions beyond the graphical evaluations with box plots. For paired t tests, we added one more case of transferring parameters gained from FDC_cal to ungauged catchments (referred to as FPROX_reg). FPROX_reg transfers behavioural parameters with no flow timing information from gauged to ungauged catchments. The mean NSE of FPROX_reg was 0.44 with a standard deviation of 0.49.

A primary hypothesis of this study was that RFDC_cal could outperform PROX_reg. This question can be addressed by looking at the NSE differences between RFDC_cal and PROX_reg. The mean NSE difference between them was −0.22 and the standard error was 0.051, providing an evaluation that the NSE differences were less than zero at a 95 % confidence level. The paired t test did not lend support to the hypothesis (i.e. PROX_reg outperformed RFDC_cal significantly). However, we can assume that D_{RL} can improve the predictive performance of FDC_cal. The mean NSE difference between FDC + DRL_cal and FDC_cal was 0.12 and the standard error was 0.025, confirming the significance at a 95 % confidence level.

Likewise, we tested several questions relevant to rainfall–runoff modelling in ungauged catchments using different combinations. In Table 3, we summarise the results of paired t tests for scientific questions that may arise from this study. One interesting question is, "Did the behavioural parameters from Q_cal contain flow timing information for ungauged catchments?" We addressed this question by comparing between PROX_reg and FPROX_reg with a hypothesis that predictability in ungauged catchments would decrease if the regionalised parameters were gained only from flow magnitudes. FPROX_reg uses FDC_cal for searching behavioural parameters at gauged catchments; thus, it cannot transfer flow timing information to ungauged catchments through the behavioural parameters. The mean NSE difference between PROX_reg and FPROX_reg was 0.10, and the standard error was 0.031. The NSE differences were greater than zero significantly. The behavioural parameters from Q_cal were likely to have flow timing information affecting predictability in ungauged catchments.

Table 3. Results of the paired t tests for potential questions on rainfall–runoff modelling in ungauged catchments.

Questions	Corresponding pair	[a]PM	[b]$\overline{\Delta\text{PM}}$	[c]SE	Answer
Q1. Did RFDC_cal outperform PROX_reg?	RFDC_cal–PROX_reg	NSE	−0.22	0.051	No*
Q2. Did D_{RL} improve FDC_cal?	FDC + DRL_cal–FDC_cal	NSE	0.12	0.025	Yes*
Q3. Did parameters from Q_cal contain flow timing information for ungauged catchments?	PROX_reg–FPROX_reg	NSE	0.10	0.031	Yes*
Q4. Did absence of flow timing affect model efficiency?	Q_cal–FDC_cal	NSE	0.23	0.026	Yes*
Q5. Did PROX_reg outperform RFDC_cal in predicting low flows?	PROX_reg–RFDC_cal	LNSE	0.09	0.031	Yes*
Q6. Did PROX_reg outperform RFDC_cal in reproducing I_{BF}?	PROX_reg–RFDC_cal	$D_{FS}(I_{BF})$	0.06	0.028	Unlikely
Q7. Did errors in regional FDC affect RFDC_cal significantly?	RFDC_cal–FDC_cal	NSE	−0.09	0.069	Unlikely

[a] Performance metric used for t test. [b] Mean PM difference between the corresponding pair. [c] Standard error of ΔPM. * ΔPM is significantly different from zero. The significance was evaluated at 95 % confidence levels.

5 Discussion and conclusions

5.1 RFDC_cal for rainfall–runoff modelling in ungauged catchments

The use of regional FDC as a single calibration criterion appears to be a good choice for searching behavioural parameters in ungauged sites. As discussed earlier, the FDC is a compact representation of runoff variability at all timescales, and thus able to embed multiple hydrological features in catchment dynamics (Blöschl et al., 2013). A pilot study of Yokoo and Sivapalan (2011) discovered that the upper part of an FDC is controlled by interaction between extreme rainfall and fast runoff, while the lower part is governed by baseflow recession behaviour during dry periods. The middle part connecting the upper and the lower parts is related to the mean within-year flow variations, which is controlled by interactions between water availability, energy, and water storage (Yaeger et al., 2012; Yokoo and Sivapalan, 2011). It is well documented that hydro-climatological processes within a catchment are reflected in the FDC (e.g. Cheng et al., 2012; Ye et al., 2012; Coopersmith et al., 2012; Yaeger et al., 2012; Botter et al., 2008), and therefore the model parameters identified solely by a regional FDC are expected to provide reliable predictions in ungauged catchments (e.g. Westerberg et al., 2014; Yu and Yang, 2000).

The comparative evaluation in this study provides another expected result, that the FDC calibration is able to reproduce the FDC itself, but it insufficiently captures functional responses of catchments due to the absence of flow timing information. A hydrograph is the most complete flow signature embedding numerous processes interacting within a catchment (Blöschl et al., 2013), being more informative than an FDC. Since any simplification of a hydrograph, including the FDC, loses some amount of flow information, it is no surprise that the FDC calibration worsens the equifinality. This study emphasises that the absence of flow timing in RFDC_cal may cause larger prediction errors than regionalised parameters gained from observed hydrographs. The paired t test between PROX_reg and FPROX_reg highlights that regionalised parameters gained from observed hydrographs were likely to contain intangible flow timing information even for ungauged catchments. The flow timing information implicitly transferred to ungauged catchments is a major difference between PROX_reg and RFDC_cal. The errors introduced by the FDC regionalisation were not significant due to the high performance of the geostatistical method in this study.

Because the hydrograph calibration can compensate for the errors in input–output data, one may convert the hydrograph into the FDC to avoid effects of disinformation on rainfall–runoff modelling. However, in this case, valuable flow timing information should be balanced in trade-off. For RFDC_cal in this study, we began with converting the observed hydrographs into the flow quantiles to regionalise them; thus, the flow timing information was initially lost. As shown, the performance of RFDC_cal was generally lower than that of PROX_reg. Therefore, when condensing observed hydrographs into flow signatures, preserving all available flow information in the hydrograph would be key for a successful rainfall–runoff modelling. This study shows that using only regionalised FDC could lead to less reliable rainfall–runoff modelling in ungauged catchments than regionalised parameters. An FDC is unlikely to preserve all flow information in a hydrograph necessary for rainfall–runoff modelling.

5.2 Suggestions for improving RFDC_cal

Westerberg et al. (2014) suggested the necessity of further constraining to reduce predictive uncertainty in RFDC_cal. This study found that RFDC_cal could provide comparable performance to regenerate the flow signatures within which only flow magnitudes are involved (i.e. R_{QP} and I_{BF}). To supplement regional FDC, flow signatures associated with flow timing seem to be essential. Figure 9 shows the potential of additional constraining with D_{RL}, and Q2 in Table 3 confirms it. Other flow signatures in temporal dimensions, such as the high- and the low-flow event durations in Westerberg and McMillan (2015), can be candidates to improve RFDC_cal. However, uncertainty in those flow signatures will be a challenge when it comes to building regional models for ungauged catchments (Westerberg et al., 2016).

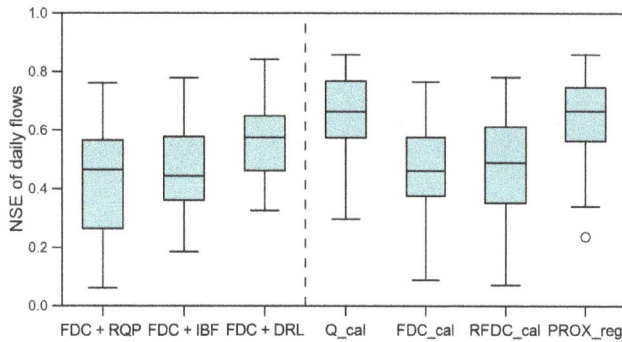

Figure 9. Predictive performance of the FDC calibrations additionally conditioned by R_{QP} (FDC + RQP), I_{BF} (FDC + IBF), and D_{RL} (FDC + DRL) in comparison to the other modelling approaches. Q_cal and FDC_cal refer to the hydrograph and the FDC calibration in gauged catchments respectively. Thirty-eight catchments with positive NSE for all the modelling approaches were used in the box plots.

An alternative method of RFDC_cal is to directly regionalise hydrographs to ungauged catchments (e.g. Viglione et al., 2013). In data-rich regions, topological proximity could better capture spatial variation of daily flows than rainfall–runoff modelling with regionalised parameters (Viglione et al., 2013). Although a dynamic model may be required for regionalising observed daily flows at an expensive computational cost, flow timing information would be contained in regionalised hydrographs. The parameter identification against the regional hydrographs may become a better approach than RFDC_cal and/or other signature-based calibrations.

5.3 Limitations and future research directions

There are caveats in our comparative evaluation. First, uncertainty in input–output data was not considered in our assessment. McMillan et al. (2012) reported typical ranges of relative errors in discharge data as 10–20 % for medium to high flow and 50–100 % for low flows. We assumed that quality of the discharge data was adequate. However, other methods objectively considering uncertainty could better estimate model performance and the equifinality (e.g. Westerberg et al., 2011, 2014). Second, we used a conceptual runoff model with a fixed structure for all the catchments. Uncertainty from the model structure would vary across the study catchments; nevertheless, the structural uncertainty was not measured here. Our comparative assessment was based on the basic premise that modelling conditions should be fixed for all study catchments. Third, we compared RFDC_cal and PROX_reg in a region with sufficient data lengths and quality at gauged catchments. The lessons from this study may not be expandable to ungauged catchments under poor data availability. Finally, though the proximity-based parameter regionalisation was good for the South Korean catchments, comparison between RFDC_cal and other regionalisation

methods, such as the regional calibration and the similarity-based parameter transfer, may provide beneficial information for rainfall–runoff modelling in ungauged catchments. Comparative assessment between RFDC_cal and other parameter regionalisation using more sample catchments under diverse climates will provide more meaningful lessons.

We can no longer hypothesise that the parameters gained against regionalised FDC would perform sufficiently, because an FDC contains less information than a hydrograph (i.e. the absence of flow timing). For improving RFDC_cal, we suggest supplementing RFDC_cal with flow signatures in temporal dimensions. Then, the question of how to make flow signatures more informative than (or equally informative to) hydrographs should be addressed. This may be impossible only using flow signatures originating from hydrographs (e.g. mean annual flow, baseflow index, recession rates, FDC). Combinations of those signatures are unlikely to be more informative than their origins (i.e. hydrographs), though it depends on how much disinformation is present in the observed flows. Future research topics could include finding new signatures that supplement hydrographs, and how to combine them with existing flow signatures for rainfall–runoff modelling in ungauged catchments.

5.4 Conclusions

While rainfall–runoff modelling against regional FDC appeared a good approach for prediction in ungauged catchments, this study highlights its weakness in the absence of flow timing information, which may cause poorer predictive performance than the simple proximity-based parameter regionalisation. The following conclusions are worth emphasising.

For ungauged catchments in South Korea, where spatial proximity well captured functional similarity between gauged catchments, the model calibration against regional FDC is unlikely to outperform the conventional proximity-based parameter transfer for daily runoff prediction. The absence of flow timing information in regional FDC seems to cause a substantial equifinality problem in the parameter identification process and thus lower predictability.

The model parameters gained from observed hydrographs contain flow timing information even for ungauged catchments. This intangible flow timing information should be discarded if one calibrates a rainfall–runoff model against regional FDC. This information loss may reduce predictability in ungauged catchments significantly.

To improve the calibration against regional FDC, flow metrics in temporal dimensions, such as the rising limb density, need to be included as additional constraints. As an alternative approach, if river gauging density is high, regionalised hydrographs preserving flow timing information can be used for local calibrations at ungauged catchments.

For better prediction in ungauged catchments, it is necessary to find new flow signatures that can supplement the observed hydrographs. How to combine them with existing information will be a future research topic for rainfall–runoff modelling in ungauged catchments.

Competing interests. The authors declare that they have no conflict of interest.

Acknowledgements. This study was supported by the APEC Climate Center. We send special thanks to Yoe-min Jeong and Hyung-Il Eum for the PRISM data sets. We greatly appreciate constructive comments and suggestions from the reviewers that significantly improved the paper.

Edited by: Fabrizio Fenicia

References

Atieh, M., Taylor, G., Sttar, A. M. A., and Gharadaghi, B.: Prediction of flow duration curves for ungauged basins, J. Hydrol., 545, 383–394, 2017.

Bae, D.-H., Jung, I.-W., and Chang, H: Long-term trend of precipitation and runoff in Korean river basins, Hydrol. Process., 22, 2644–2656, 2008.

Bárdossy, A.: Calibration of hydrological model parameters for ungauged catchments, Hydrol. Earth Syst. Sci., 11, 703–710, https://doi.org/10.5194/hess-11-703-2007, 2007.

Beven, K. J.: Prophecy, reality and uncertainty in distributed hydrological modelling, Adv. Water Resour., 16, 41–51, 1993.

Beven, K. J.: A manifesto for the equifanality thesis, J. Hydrol., 320, 18–36, 2006.

Beven, K. J. and Bingley, A.: The future of distributed models. Model calibration and uncertainty prediction, Hydrol. Process., 6, 279–298, 1992.

Blazkova, S. and Beven, K.: A limits of acceptability approach to model evaluation and uncertainty estimation in flood frequency estimation by continuous simulation: Skalka catchment, Czech Republic, Water Resour. Res., 45, W00b16, https://doi.org/10.1029/2007wr006726, 2009.

Blöschl, G., Sivapalan, M., Wagener, T., Viglione, A., and Savenije, H.: Runoff Prediction in Ungauged Basins, Simthesis across Processes, Places, and Scales, Cambridge University Press, New York, USA, 2013.

Botter, G., Porporato, A., Rodriguez-Iturbe, I., and Rinaldo, A.: Basin-scale soil moisture dynamics and the probabilistic characterization of carrier hydrologic flows: Slow, leaching-prone components of the hydrologic response, Water Resour. Res., 43, W02417, https://doi.org/10.1029/2006WR005043, 2007.

Botter, G., Zanardo, S., Porporato, A., Rodriguez-Iturbe, I., and Rinaldo, A.: Ecohydrological model of flow duration curves and annual minima, Water Resour. Res., 44, W08418, https://doi.org/10.1029/2008WR006814, 2008.

Brooks, P. D., Troch, P. A., Durcik, M., Gallo, E., and Schlegel, M.: Quantifying regional-scale ecosystem response to changes in precipitation: Not all rain is created equal, Water Resour. Res., 47, W00J08, https://doi.org/10.1029/2010WR009762, 2011.

Cheng, L., Yaeger, M., Viglione, A., Coopersmith, E., Ye, S., and Sivapalan, M.: Exploring the physical controls of re-

gional patterns of flow duration curves – Part 1: Insights from statistical analyses, Hydrol. Earth Syst. Sci., 16, 4435–4446, https://doi.org/10.5194/hess-16-4435-2012, 2012.

Coopersmith, E., Yaeger, M. A., Ye, S., Cheng, L., and Sivapalan, M.: Exploring the physical controls of regional patterns of flow duration curves – Part 3: A catchment classification system based on regime curve indicators, Hydrol. Earth Syst. Sci., 16, 4467–4482, https://doi.org/10.5194/hess-16-4467-2012, 2012.

Daly, C., Halbleib, M., Smith, J. I., Gibson, W. P., Doggett, M. K., Taylor, G. H., Curtis, J., and Pasteris, P. P.: Physiographically sensitive mapping of climatological temperature and precipitation across the conterminous United States, Int. J. Climatol., 28, 2031–2064, https://doi.org/10.1002/joc.1688, 2008.

Demirel, M. C., Booiji, M. J., and Hoekstra, A. Y.: Effect of different uncertainty sources on the skill of 10 day ensemble low flow forecasts for two hydrological models, Water Resour. Res., 49, 4035–4053, https://doi.org/10.1002/wrcr.20294, 2013.

Duan, Q., Sorooshian, S., and Gupta, V. K.: Effective and efficient global optimisation for conceptual rainfall–runoff models, Water Resour. Res., 28, 1015–1031, 1992.

Dunn, S. M. and Lilly, A.: Investigating the relationship between a soils classification and the spatial parameters of a conceptual catchment scale hydrological model, J. Hydrol., 252, 157–173, https://doi.org/10.1016/S0022-1694(01)00462-0, 2001.

Eckhardt, K.: A comparison of baseflow indices, which were calculated with seven different baseflow separation methods, J. Hydrol., 352, 168–173, https://doi.org/10.1016/j.jhydrol.2008.01.005, 2007.

Hingray, B., Schaefli, B., Mezghani, A., and Hamdi, Y.: Signature-based model calibration for hydrological prediction in mesoscale Alpine catchments, Hydrolog. Sci. J., 55, 1002–1016, https://doi.org/10.1080/02626667.2010.505572, 2010.

Hrachowitz, M., Savenije, H. H. G., Blöschl, G., McDonnell, J. J., Sivapalan, M., Pomeroy, J. W., Arheimer, B., Blume, T., Clark, M. P., Ehret, U., Fenicia, F., Freer, J. E., Gelfan, A., Gupta, H. V., Hughes, D. A., Hut, R. W., Montanari, A., Pande, S., Tetzlaff, D., Troch, P. A., Uhlenbrook, S., Wagener, T., Winsemius, H. C., Woods, R. A., Zehe, E., and Cudennec, C.: A decade of Predictions in Ungauged Basins (PUB) – a review, Hydrolog. Sci. J., 58, 1198–1255, https://doi.org/10.1080/02626667.2013.803183, 2013.

Jung, S.-W., Lee, J.-W., Kim, C.-Y., Hwang, S.-H., Oh, C.-Y., Lee, Y.-G., Kim, D., Park, Y.-H., Lee, S.-C., Yoo, S.-S., Kim, J.-C., Lee, S.-H., Kim, S.-E., Lee, C.-D., Hwangbo, J., Lee, G.-Y., Kwon, D.-S., Park, S.-H., Lee, K.-S., and Shim, E. J.: Standardisation of methods and criteria for hydrological survey in South Korea, Report Number: 11-1611492-00058-01, Ministry of Land, Transport, and Maritime Affairs, Gyeonggido, South Korea, Written in Korean, 2010.

Jung, Y. and Eum, H.-I.: Application of a statistical interpolation method to correct extreme values in high-resolution gridded climate variables, J. Clim. Chang. Res., 6, 331–334, 2015.

Kavetski, D., Fnicia, F., and Clark, M.: Impact of temporal data resolution on parameter inference and model identification in conceptual hydrological modeling: Insights from an experimental catchment, Water Resour. Res., 47, W05501, https://doi.org/10.1029/2010WR009525, 2011.

Kavetski, D., Kuczera, G., and Franks, S. W.: Bayesian analysis of input uncertainty in hydrological modeling: 1. Theory, Water Resour. Res., 42, W03407, https://doi.org/10.1029/2005WR004368, 2006.

Kim, D. and Kaluarachchi, J.: Predicting streamflows in snowmelt-driven watersheds using the flow duration curve method, Hydrol. Earth Syst. Sci., 18, 1679–1693, https://doi.org/10.5194/hess-18-1679-2014, 2014.

Kim, U. and Kaluarachchi, J. J.: Application of parameter estimation and regionalization methodologies to ungauged basins of the Upper Blue Nile River Basin, Ethiopia, J. Hydrol., 362, 39–56, https://doi.org/10.1016/j.jhydrol.2008.08.016, 2008.

Korea Meteorological Administration: Climatological normals of Korea (1981–2010), Publ. 11-1360000-000077-14, 678 pp., available at: http://www.kma.go.kr/down/Climatological_2010.pdf (last access: 16 January 2017), 2011.

McIntyre, N., Lee, H., Wheater, H., Young, A., and Wagener, T.: Ensemble predictions of runoff in ungauged catchments, Water Resour. Res., 41., W12434, https://doi.org/10.1029/2005WR004289, 2005.

McMillan, H., Krueger, T., and Freer, J.: Benchmarking observational uncertainties for hydrology: rainfall, river discharge, and water quality, Hydrol. Process., 26, 4078–4111, 2012.

Mohamoud, Y. M.: Prediction of daily flow duration curves and streamflow for ungauged catchments using regional flow duration curves, Hydrolog. Sci. J., 53, 706–724, 2008.

Oudin, L., Hervieu, F., Michel, C., Perrin, C., Andréassian, V., Anctil, F., and Loumagne, C.: Which potential evapotranspiration input for a lumped rainfall-runoff model? Part 2 – Towards a simple and efficient potential evapotranspiration model for rainfall-runoff modelling, J. Hydrol., 303, 290–306, 2005.

Oudin, L., Andréassian, V., Perrin, C., Michel, C., and Le Moine, N.: Spatial proximity, physical similarity, regression and ungaged catchments: a comparison between of regionalization approaches based on 913 French catchments, Water Resour. Res., 44, W03413, https://doi.org/10.1029/2007WR006240, 2008.

Parajka, J., Blöschl, G., and Merz, R.: Regional calibration of catchment models: potential for ungauged catchments, Water Resour. Res., 43, W06406, https://doi.org/10.1029/2006WR005271, 2007.

Parajka, J., Viglione, A., Rogger, M., Salinas, J. L., Sivapalan, M., and Blöschl, G.: Comparative assessment of predictions in ungauged basins – Part 1: Runoff-hydrograph studies, Hydrol. Earth Syst. Sci., 17, 1783–1795, https://doi.org/10.5194/hess-17-1783-2013, 2013.

Perrin, C., Michel, C., and Andréassian, V.: Improvement of a parsimonious model for streamflow simulation, J. Hydrol., 279, 275–289, 2003.

Pfannerstill, M., Guse, B., and Fohrer N.: Smart low flow signature metrics for an improved overall performance evaluation of hydrological models, J. Hydrol., 510, 447–458, 2014.

Pugliese, A., Castellarin, A., and Brath, A.: Geostatistical prediction of flow-duration curves in an index-flow framework, Hydrol. Earth Syst. Sci., 18, 3801–3816, https://doi.org/10.5194/hess-18-3801-2014, 2014.

Pugliese, A., Farmer, W. H., Castellarin, A., Archfield, S. A., and Vogel, R. M.: Regional flow duration curves: Geostatistical techniques versus multivariate regression, Adv. Water Resour., 96, 11–22, 2016.

Rhee, J. and Cho, J.: Future changes in drought characteristics: regional analysis for South Korea under CMIP5 projections, J. Hydrometeorol., 17, 437–450, 2016.

Sadegh, M., Vrugt, J. A., Gupta, H. V., and Xu, C.: The soil water characteristics as new class of closed-form parametric expressions for the flow duration curve, J. Hydrol., 535, 438–456, 2016.

Shafii, M. and Tolson, B. A.: Optimizing hydrological consistency by incorporating hydrological signatures into model calibration objectives, Water Resour. Res., 51, 3796–3814, https://doi.org/10.1002/2014WR016520, 2015.

Shu, C. and Ouarda, T. B. M. J.: Improved methods for daily streamflow estimates at ungauged sites, Water Resour. Res., 48, W02523, https://doi.org/10.1029/2011WR011501, 2012.

Skøien, J. O., Merz, R., and Blöschl, G.: Top-kriging – geostatistics on stream networks, Hydrol. Earth Syst. Sci., 10, 277–287, https://doi.org/10.5194/hess-10-277-2006, 2006.

Smakhtin, V. P. and Masse, B.: Continuous daily hydrograph simulation using duration curves of a precipitation index, Hydrol. Process., 14, 1083–1100, 2000.

Smakhtin, V. Y., Hughes, D. A., and Creuse-Naudine, E.: Regionalization of daily flow characteristics in part of the Eastern Cape, South Africa, Hydrolog. Sci. J., 42, 919–936, 1997.

Sugawara, M.: Automatic calibration of the tank model, Hydrological Sciences Bulletin, 24, 375–388, https://doi.org/10.1080/02626667909491876, 1979.

Viglione, A., Parajka, J., Rogger, M., Salinas, J. L., Laaha, G., Sivapalan, M., and Blöschl, G.: Comparative assessment of predictions in ungauged basins – Part 3: Runoff signatures in Austria, Hydrol. Earth Syst. Sci., 17, 2263–2279, https://doi.org/10.5194/hess-17-2263-2013, 2013.

Vogel, R. M. and Fennessey, N. M.: Flow duration curves. I: New interpretation and confidence intervals, J. Water Res. Plan. Man., 120, 485–504, 1994.

Vrugt, J. A. and Ter Braak, C. J. F.: DREAM$_{(D)}$: an adaptive Markov Chain Monte Carlo simulation algorithm to solve discrete, noncontinuous, and combinatorial posterior parameter estimation problems, Hydrol. Earth Syst. Sci., 15, 3701–3713, https://doi.org/10.5194/hess-15-3701-2011, 2011.

Wagener, T. and Wheater, H. S.: Parameter estimation and regionalization for continuous rainfall-runoff models including uncertainty, J. Hydrol., 320, 132–154, 2006.

Walter, M. T., Brooks, E. S., McCool, D. K., King, L. G., Molnau, M., and Boll, J.: Process-based snowmelt modeling: does it require more input data than temperature-index modeling?, J. Hydrol., 300, 65–75, https://doi.org/10.1016/j.jhydrol.2004.05.002, 2005.

Westerberg, I. K. and McMillan, H. K.: Uncertainty in hydrological signatures, Hydrol. Earth Syst. Sci., 19, 3951–3968, https://doi.org/10.5194/hess-19-3951-2015, 2015.

Westerberg, I. K., Guerrero, J.-L., Younger, P. M., Beven, K. J., Seibert, J., Halldin, S., Freer, J. E., and Xu, C.-Y.: Calibration of hydrological models using flow-duration curves, Hydrol. Earth Syst. Sci., 15, 2205–2227, https://doi.org/10.5194/hess-15-2205-2011, 2011.

Westerberg, I. K., Gong, L., Beven, K. J., Seibert, J., Semedo, A., Xu, C.-Y., and Halldin, S.: Regional water balance modelling using flow-duration curves with observational uncertainties, Hydrol. Earth Syst. Sci., 18, 2993–3013, https://doi.org/10.5194/hess-18-2993-2014, 2014.

Westerberg, I. K., Wagener, T., Coxon, G., McMillan, H. K., Castellarin, A., Montanari, A., and Freer, J.: Uncertainty in hydrological signatures for gauged and ungauged catchments, Water Resour. Res., 52, 1847–1865, https://doi.org/10.1002/2015WR017635, 2016.

Yadav, M., Wagener, T., and Gupta, H.: Regionalization of constraints on expected watershed response behavior for improved predictions in ungauged basins, Adv. Water Resour., 30, 1756–1774, 2007.

Yaeger, M., Coopersmith, E., Ye, S., Cheng, L., Viglione, A., and Sivapalan, M.: Exploring the physical controls of regional patterns of flow duration curves – Part 4: A synthesis of empirical analysis, process modeling and catchment classification, Hydrol. Earth Syst. Sci., 16, 4483–4498, https://doi.org/10.5194/hess-16-4483-2012, 2012.

Ye, S., Yaeger, M., Coopersmith, E., Cheng, L., and Sivapalan, M.: Exploring the physical controls of regional patterns of flow duration curves – Part 2: Role of seasonality, the regime curve, and associated process controls, Hydrol. Earth Syst. Sci., 16, 4447–4465, https://doi.org/10.5194/hess-16-4447-2012, 2012.

Yilmaz, K. K., Gupta, H. V., and Wagener, T.: A process-based diagnostic approach to model evaluation: Application to the NWS distributed hydrologic model, Water Resour. Res., 44, W09417, https://doi.org/10.1029/2007WR006716, 2008.

Yokoo, Y. and Sivapalan, M.: Towards reconstruction of the flow duration curve: development of a conceptual framework with a physical basis, Hydrol. Earth Syst. Sci., 15, 2805–2819, https://doi.org/10.5194/hess-15-2805-2011, 2011.

Yu, P.-S. and Yang, T.-C.: Using synthetic flow duration curves for rainfall–runoff model calibration at ungauged sites, Hydrol. Process., 14, 117–133, https://doi.org/10.1002/(SICI)1099-1085(200001)14:1<117::AID-HYP914>3.0.CO;2-Q, 2000.

Yu, P. S., Yang, T. C., and Wang, Y. C.: Uncertainty analysis of regional flow duration curves, J. Water Res. Pl.-ASCE, 128, 424–430, 2002.

Zhang, Y., Vaze, J., Chiew, F. H. S., and Li, M.: Comparing flow duration curve and rainfall-runoff modelling for predicting daily runoff in ungauged catchments, J. Hydrol., 525, 72–86, 2015.

Zhang, Z., Wagener, T., Reed, P., and Bhushan, R.: Reducing uncertainty in predictions in ungauged basins by combining hydrologic indices regionalization and multiobjective optimization, Water Resour. Res., 44, W00B04, https://doi.org/10.1029/2008WR006833, 2008.

3

Climate change and non-stationary flood risk for the upper Truckee River basin

L. E. Condon[1,2]**, S. Gangopadhyay**[1]**, and T. Pruitt**[1]

[1]Bureau of Reclamation, Technical Service Center, Denver, Colorado, USA
[2]Hydrologic Science and Engineering Program and Department of Geology and Geological Engineering, Colorado School of Mines, Golden, Colorado, USA

Correspondence to: L. E. Condon (lcondon@mymail.mines.edu)

Abstract. Future flood frequency for the upper Truckee River basin (UTRB) is assessed using non-stationary extreme value models and design-life risk methodology. Historical floods are simulated at two UTRB gauge locations, Farad and Reno, using the Variable Infiltration Capacity (VIC) model and non-stationary Generalized Extreme Value (GEV) models. The non-stationary GEV models are fit to the cool season (November–April) monthly maximum flows using historical monthly precipitation totals and average temperature. Future cool season flood distributions are subsequently calculated using downscaled projections of precipitation and temperature from the Coupled Model Intercomparison Project Phase 5 (CMIP-5) archive. The resulting exceedance probabilities are combined to calculate the probability of a flood of a given magnitude occurring over a specific time period (referred to as flood risk) using recent developments in design-life risk methodologies. This paper provides the first end-to-end analysis using non-stationary GEV methods coupled with contemporary downscaled climate projections to demonstrate the evolution of a flood risk profile over typical design life periods of existing infrastructure that are vulnerable to flooding (e.g., dams, levees, bridges and sewers). Results show that flood risk increases significantly over the analysis period (from 1950 through 2099). This highlights the potential to underestimate flood risk using traditional methodologies that do not account for time-varying risk. Although model parameters for the non-stationary method are sensitive to small changes in input parameters, analysis shows that the changes in risk over time are robust. Overall, flood risk at both locations (Farad and Reno) is projected to increase 10–20 % between the historical period 1950 to 1999 and the future period 2000 to 2050 and 30–50 % between the same historical period and a future period of 2050 to 2099.

1 Introduction

"Stationarity is dead" (Milly et al., 2008), yet the standard practice for flood frequency analysis is predicated on this very assumption. This discrepancy has not gone unnoticed within the scientific community and there is a growing body of research investigating (1) trends in observed floods (e.g., Franks, 2002; Vogel et al., 2011), (2) ways to incorporate non-stationarity into frequency distributions (e.g., Katz et al., 2002; Raff et al., 2009) and (3) methodologies to interpret risk and approach design within a non-stationary framework (e.g., Mailhot and Duchesne, 2010; Rootzén and Katz, 2013; Salas and Obeysekara, 2014). Both the frequency and intensity of extreme events are particularly susceptible to change because small shifts in the center of a distribution can potentially have much larger impacts on the tails (Meehl et al., 2000). Regardless of climate change, naturally occurring long-term climate oscillations, such as the El Niño–Southern Oscillation (ENSO), have been linked to low frequency variability in flood frequency (e.g., Cayan et al., 1999; Jain and Lall, 2001). Anthropogenic climate change has the potential to amplify natural climatic variability throughout the interconnected climate and hydrologic systems.

Already trends in many hydrologic variables have been observed across the western United States (as well as around the world). For example, clear increases in temperature have been measured across the west (e.g., Cayan et al., 2001; Dettinger and Cayan, 1995). Precipitation trends are more variable. Regonda et al. (2005) found increased total winter precipitation (rain and snow) from 1950 to 1999 in many sites across the western United States, although springtime snow water equivalent (SWE) was shown to decline over the same period. Similarly, Mote et al. (2005) analyzed snowpack trends in western North America and reported widespread declines in springtime SWE over the period 1925–2000, especially since the middle of the 20th century. They attribute this decline predominantly to climatic factors such as ENSO, Pacific Decadal Oscillation (PDO) and positive trends in regional temperature. Easterling et al. (2000) summarized previous studies on precipitation trends. They note that trends vary from region to region but, in general, increases in precipitation have occurred disproportionately in the extremes. Several subsequent studies have observed increasing trends in extreme precipitation events, although the changes are relatively small (Gutowski et al., 2008; Kunkel, 2003; Madsen and Figdor, 2007).

Research has also demonstrated increasing trends in flood frequency in some regions. Walter and Vogel (2010) and Vogel et al. (2011) observed increasing flood magnitudes across the United States using stream gauge records, and Franks (2002) showed statistically significant increases in flood frequency since the 1940s. Still, non-stationary flood behavior has been historically difficult to quantify and there has been some debate about the significance of flood frequency trends. For example, Hirsch (2011) noted both increasing and decreasing trends in annual flood magnitudes in different regions of the US. Also, Douglas et al. (2000) found that, if one takes into account spatial correlation, many previous findings of flood trends are not statistically significant. Difficulty in diagnosing flood trends is not unique to the western US; a literature review of historical flood studies across Europe also found spatial variability in flood trends (Hall et al., 2014).

Even when significant trends are found, the complexity of flooding mechanisms, which depend on many variables that can vary regionally and seasonally, makes it difficult to attribute trends to specific causes. Illustrating the importance of seasonality, Small et al. (2006) showed that if a high-precipitation event occurs in the fall, as opposed to the spring, it will contribute to baseflow rather than inducing flooding. Also, urbanization can drastically increase the impervious area of a basin, thus amplifying floods by decreasing infiltration and speeding runoff. The largest flood magnitude increases observed by both Walter and Vogel (2010) and Vogel et al. (2011) were in basins with urban development. The influence of development trends on flood behavior can be difficult to separate from other variables. For example, Villarini et al. (2009) could not conclusively tie reduced stationarity (i.e., changes in mean and/or variance) in peak discharge records to climate change because of variability in the other factors that influence runoff.

Merz et al. (2012) note that attributing changes in flood hazard is complicated by the complex array of drivers that can include land cover change and infrastructure development as well as natural climate variability and change. Here we set aside the impacts of development and management practices and focus on the role of climate change. However, even with this simplification, future extremes can still be influenced by a number of interrelated variables such as changes in temperature, precipitation efficiency and vertical wind velocity (Mullet et al., 2011; O'Gorman and Schneider, 2009). Analyzing global circulation model (GCM) outputs, Pierce et al. (2012) found total changes in precipitation to be small relative to the existing variability, but noted larger seasonal changes in storm intensity and frequency. Despite uncertainty, many studies agree that warming will increase the potential for intense rainfall (Allan, 2011; Gutowski et al., 2008; Pall et al., 2011; Sun et al., 2007). Furthermore, Min et al. (2011) found that some GCM simulations may underestimate extreme precipitation events in the Northern Hemisphere, indicating that projections of extreme precipitation based on GCM outputs may be conservative.

Studies have also predicted increases in flood frequency and magnitude with a warmer climate, especially in snowmelt-dominated basins (e.g., Das et al., 2011). As with historical flooding trends, translating forecasted climate variables to flood frequency is a complex process and several methodologies have been used. Downscaled GCM climate forcings can be used to drive hydrologic models and simulate future floods directly (e.g., Das et al., 2011; Vogel et al., 2011; Raff et al., 2009). With this approach, traditional stationary flood frequency distributions can be fit to the simulated floods to calculate return periods of interest (e.g., Raff et al., 2009; Vogel et al., 2011). This allows for return periods and flood magnitudes that change over time, as with the flood magnification and recurrence reduction factors calculated by Walter and Vogel (2010) and Vogel et al. (2011). While these approaches do capture temporal changes between analysis periods, they still assume that flood mechanisms are stationary within each period of analysis.

This limitation can be overcome using non-stationary generalized extreme value (GEV) distributions where the model parameters, like mean (i.e., location) and spread (i.e., scale), are allowed to vary as a function of time (e.g., Gilroy and McCuen, 2012) or with relevant covariates (e.g., Griffis and Stedinger, 2007; Katz et al., 2002; Towler et al., 2010). This approach has been gaining popularity for flood frequency estimation. Using this technique, it is not necessary to simulate future floods directly by forcing a hydrologic model with projected hydroclimate fields (e.g., precipitation and temperature). The parameters of the GEV model, like mean and spread, change with time (i.e., non-stationary) based on a linear combination of covariates like precipitation and temperature. Historical relationships between extreme events and hydroclimate fields are used to identify the weighting of covariates. These weights are then used to estimate parameters for future time periods using precipitation and temperature outputs from hydroclimate projections. For example, Gilroy and McCuen (2012) used non-stationary GEV models of flood frequency that incorporated a linear trend in the location parameter. Similarly, Griffis and Stedinger (2007) and Towler et al. (2010) used climate variables as covariates for the distribution parameters.

While non-stationary flood forecasting methods provide flexibility in analyzing flood variability, they are also incongruent with many of the traditional metrics used in water resources planning. Historically, most infrastructures that are vulnerable to flooding (e.g., dams, levees, sewers and bridges) have been designed to withstand flooding of a specified return period (e.g., the 100-year flood). However, these calculations rely on a flood frequency distribution which is assumed to remain stationary with time, and hence the return period design metric is also assumed to be stationary. When non-stationary methods are used, the underlying flood frequency distributions, and associated return periods, vary with time. Thus, under a non-stationary climate, the notion of a static return period flood event (e.g., 100-year flood, 200-year flood) is no longer a valid concept.

To address this issue, Rootzén and Katz (2013) introduced the concept of design life level to calculate the risk of a given flood magnitude occurring over a specified time period. Salas and Obeysekera (2014) further demonstrated the relevance of this technique to the hydrologic community using flood frequency examples. However, this methodology has yet to receive widespread attention within the hydrologic community. Here, we present a non-stationary flood frequency assessment for the upper Truckee River basin (UTRB) using contemporary downscaled climate projections and the non-stationary design life level technique introduced by Rootzén and Katz (2013) to quantify flood risk. Note that, following the convention of Rootzén and Katz (2013), we use the term flood risk to refer to the probability of an extreme event occurring and not as a quantification of expected losses. While the methodology used for this analysis is previously established, this paper provides the first end-to-end demonstration of non-stationary GEV analysis coupled with contemporary downscaled climate projections (specifically, downscaled climate projections from the Coupled Model Intercomparison Project Phase 5, or CMIP-5) to quantify how the flood risk profiles may evolve in the upper Truckee River basin over the 21st century. The flood analysis presented here is part of a larger study on climate change impacts in the Truckee River basin (Reclamation, 2010). This project is supported by local water managers and conducted by the Bureau of Reclamation through the WaterSMART Basin Studies Program authorized under US Public Law 111-11, Subtitle F (SECURE Water Act). The intent of this work is (1) to investigate potential flood risk changes over time in the UTRB and (2) to demonstrate the applicability of non-stationary techniques in a regional flood analysis to make these tools more accessible to the hydrologic community.

The paper is organized as follows. Section 2 provides background on the study area along with the data sets and models used. The methodologies of using non-stationary spatial GEV analysis in conjunction with climate projections and time-evolving risk assessment are described in Sect. 3. Results and discussions of findings are given in Sect. 4. Summary and conclusions from the analysis are presented in Sect. 5.

2 Background

This section provides background on the study area (Sect. 2.1), streamflow data and simulations (Sect. 2.2) and climate data and models (Sect. 2.3).

2.1 Upper Truckee River basin

The Truckee River originates in the northern Sierra Nevada in California (above Lake Tahoe) and flows northeast to Nevada, where it ends in the Pyramid Lake (Fig. 1). The total basin area is roughly 7900 km^2; however, the area upstream of Reno (2763 km^2), henceforth referred to as the upper Truckee River, provides the majority of the basin's precipitation through snowpack. The focus of this analysis is on the Farad and Reno gauge locations shown in Fig. 1, henceforth referred to as Farad and Reno. The Farad gauge is located roughly 1.5 km downstream of the Farad hydropower plant and provides a cumulative measure of all of the upper basin tributaries (Stokes, 2002). Most of the available water supply is generated upstream of the Farad gauge (USACE, 2013a). The Reno gauge is located downstream of Farad in the heart of Reno and is a good reference point for analyzing urban flooding. The intervening area between the Farad and Reno gauges is small, roughly 350 km^2 [km], and there are only two small tributaries that enter the main stem between Farad and Reno.

Flooding in the upper Truckee generally takes one of three forms. Some of the most severe floods have resulted from heavy rain events covering most of the basin and lasting 1 to 6 days. These storms generally occur from November to April and may be linked to atmospheric rivers (Ralph and Dettinger, 2012). Snowmelt floods are also common from April to July. Although snowmelt floods transmit large volumes of water for longer durations, they generally do not cause damage because they are typically well predicted and can be regulated with upstream reservoirs. Finally, in late summer (July–August), local cloudbursts can generate high-intensity precipitation over small areas. These storms can cause local damage to tributaries but generally do not have a large impact on the main stem of the Truckee.

In the 20th century, nine major floods have been recorded on the Truckee River, all of which occurred from November to April (USACE, 2013b). The flood of record occurred in January of 1997 and was caused by warm rain falling on

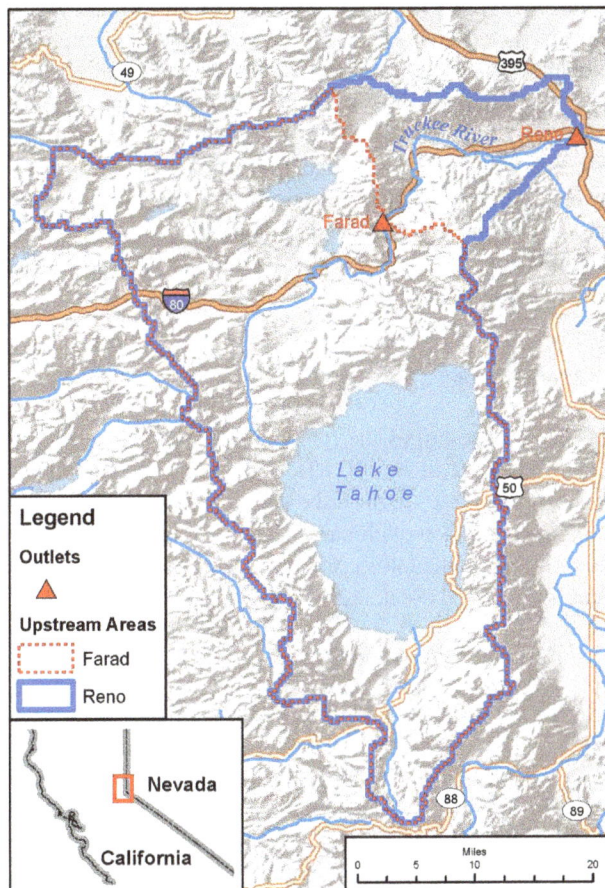

Figure 1. Map of model domain including the Farad and Reno gauges and their drainage areas.

a large snowpack (∼ 180 % of normal) and melting nearly all of the snowpack below ∼ 2100 m (USACE, 2013b). The floods of 1950, 1955 and 1963 were some of the most damaging due to the development of Reno along the river during this time period (USACE, 2013b). Subsequent flood damages have been, at least partially, mitigated by the implementation of flood infrastructure starting in the 1960s.

2.2 Streamflow data and simulations

Streamflow has been measured at both the Farad and Reno USGS gauges. However, gauge flows are not readily applicable to flood frequency analysis due to upstream developments of water supply and flood control structures. For example, upstream of Reno there are four dams with flood control capabilities (i.e., Martis Creek Dam, Prosser Creek Dam, Stampede Dam and Boca Dam); in addition to Tahoe, Donner and Independence lakes provide incidental flood regulation. Unregulated flow estimates were developed by the US Army Corps of Engineers (USACE) but are only available for historical flood events (USACE, 2013b). Therefore, we simulate unregulated flows from 1950 to 1999 using the three-

layer variable infiltration capacity (VIC) model and validate results using the available unregulated flow estimates.

A brief summary of the VIC model is provided here, and for additional technical specifications the reader is referred to Liang et al. (1994, 1996) and Nijssen et al. (1997). VIC is a gridded hydrologic model designed to simulate macroscale (spatial resolution is greater than 1 km) water balances using parameterized sub-grid infiltration and vegetation processes. In the VIC model, surface water infiltrates to the subsurface based on soil properties, and soil moisture is distributed vertically through three model layers extending up to about 2 m below the land surface. At the surface, potential evapotranspiration (PET) is simulated using the Penman–Monteith PET model (Maidment, 1993). Surface flows are determined in a two-step process. First, the water balance for each grid cell is calculated independently to determine surface runoff and baseflow, and subsequently runoff from each cell is routed to river channels and outlets using a predefined routing network. Here we drive VIC with daily weather forcings including precipitation, maximum and minimum temperature, and wind speed. Additional climate variables such as short- and long-wave radiation, relative humidity and vapor pressure are calculated within the model using established empirical relationships. The VIC model is well documented and has already been used in a number of hydrologic and climate change studies (e.g., Christensen and Lettenmaier, 2007; Christensen et al., 2004; Gangopadhyay et al., 2011; Maurer et al., 2007; Payne et al., 2004; Reclamation, 2011; Van Rheenen et al., 2004). Recently VIC has also been applied for real-time flood estimation (Wu et al., 2014).

The VIC model used for this analysis was part of the Bureau of Reclamation (Reclamation) West Wide Climate Risk Assessment (WWCRA) effort and is described in Reclamation (2011). The WWCRA VIC model encompasses the western US. Simulated and observed streamflows were compared at 152 locations primarily from the USGS Hydroclimatic Data Network (Slack et al., 1993) and 43 additional locations of importance to Reclamation's water management activities. Among the evaluated locations are several in the Truckee basin including the Truckee River at Farad. For details on model calibration and development we refer the reader to Reclamation (2011) and Gangopadhyay et al. (2011). While we do not discuss model calibration further here, in the subsequent sections we provide additional model verification for flood simulation in the UTRB.

2.3 Climate data and models

As noted in the previous section, the VIC model requires daily climate inputs to drive water balance simulations. We use the national 1/8° (roughly 12 km) gridded data set from Maurer et al. (2002) for historical (i.e., 1950–1999) climate observations. Additionally, monthly total precipitation and average temperature were aggregated for the upstream area of each gauge for every month of the flood season (i.e., November through April). These values are used as covariates for fitting non-stationary GEV models discussed in Sect. 3.

Future gridded precipitation and temperature values from 2000 to 2099 were generated from GCM outputs. We analyzed 234 projections generated by 37 different climate models from the CMIP-5 archive (Taylor et al., 2012). In the absence of objective guidance in contemporary climate literature to limit the number of projections, we chose to include all of the available CMIP-5 projections of future climate in this study. However, it should be noted that other studies have demonstrated that a subset of projections could provide comparable results for specific study objective (e.g., water supply) (Pierce et al, 2009; Harding et al., 2012). Projections span four representative concentration pathways (RCPs) for greenhouse gas emissions. Each GCM projection includes monthly gridded precipitation and temperature from 1950 to 2099 at a coarse grid resolution ranging between ~65 and 250 km.

Reclamation in collaboration with other federal and non-federal partners has developed a monthly archive of downscaled CMIP-5 projections at the finer 1/8° resolution using the two-step bias correction and spatial disaggregation (BCSD) algorithm described by Wood et al. (2004). For this analysis we extended the existing hydrology archive to cover the UTRB domain for all 234 BCSD CMIP-5 climate projections following the steps detailed below. A subset of the CMIP-5 hydrology projections is publically accessible through the archive of the downscaled CMIP3 and CMIP5 climate and hydrology projections at http://gdo-dcp.ucllnl.org/downscaled_cmipprojections/. Additional documentation on the archive and the methodology is provided in Reclamation (2014).

The downscaled climate variables include monthly total precipitation, monthly maximum and minimum temperatures and monthly average temperature. Before applying the BCSD algorithm, all 234 climate projections were first gridded from their respective native GCM scale to a common grid of 1° latitude by 1° longitude. Similarly, the observed 1/8° gridded data set (Maurer et al., 2002) was aggregated to the coarser 1° latitude by 1° longitude grid. Next, for a given climate variable, GCM and location (1° latitude by 1° longitude grid cell), the bias correction (BC) step uses quantile mapping between monthly cumulative distribution functions (CDFs) of historical simulated and historical observed values to identify biases over a common climatological period – in this case, 1950–1999. The projected future climate variables from the same GCM at the same location are then bias-corrected using the identified bias. The result of bias correction is an adjusted GCM data set (20th century and 21st century, linked together) that is statistically consistent with the observed data during the bias correction overlap period (i.e., 1950–1999 in this application). Note that the BC step happens at the coarse 1° latitude by 1° longitude grid. Next, multiplicative adjustment factors (ratio of bias-corrected GCM

to observed) for precipitation and offset adjustment factors (bias-corrected GCM minus observed) for temperature are calculated for each of the 1° latitude by 1° longitude grid cells (Reclamation, 2013). These adjustments are then spatially disaggregated (SD) to a 1/8° latitude by 1/8° longitude grid. Finally, the adjustments are applied (multiplicative for precipitation, additive for temperature) to the finer resolution, 1/8° gridded observed precipitation and temperature fields (Maurer et al., 2002) to derive the 1/8° gridded BCSD climate projections.

3 Methodology

This section describes the methodology used for flood frequency analysis in the UTRB. Discussion is divided into two sections. First, we describe the process of extreme value modeling using non-stationary GEV distributions (Sect. 3.1). Second, the methodology for design-life level risk assessment is described (Sect. 3.2)

3.1 Extreme value modeling

Extreme values analysis (EVA) deals with the examination of the tail (i.e., extreme) values of a distribution (as opposed to standard approaches which are generally more concerned with the average system behavior). EVA methods are standard practice for flood frequency analysis because they are designed to capture the behavior of low-frequency high-impact events. Furthermore, Katz (2010) points out that in climate change studies traditional approaches are not sufficient and extreme value statistics are needed. For this analysis, we use the GEV, which is commonly applied to flood frequency analysis to model block maxima from streamflow time series (e.g., Katz et al., 2002; Towler et al., 2010). The cumulative distribution function (CDF) for the GEV, F, is as follows:

$$F(z; \theta) = \exp \left\{ -\left[1 + \xi \left(\frac{z - \mu}{\sigma} \right) \right]^{-1/\xi} \right\}, \qquad (1)$$

where z is the streamflow maxima value of interest and θ is the parameter set (μ, σ, ξ) used to specify the distribution, such that the center is given by the location (μ), the spread by the scale (σ) and the behavior of the upper tail by the shape (ξ). Based on the shape parameter, the GEV can take one of three forms: Gumbel, or light tailed, when ξ is zero; Fréchet, or heavy tailed, if ξ is positive; and Weibull, or bounded, when ξ is negative. Following the methodology of Towler et al. (2010), GEV parameters (μ, σ, ξ) are fitted using the maximum likelihood estimation (MLE) technique.

In traditional stationary flood frequency analysis, it is assumed that observations are independent and identically distributed (IID), and therefore model parameters (μ, σ, ξ) derived from the observed flood record are assumed to remain constant across the period of record and into the future. Here, we introduce non-stationarity into the distribution by allowing location and scale parameters to change with relevant covariates, such that:

$$\mu(t) = \beta_{0,\mu} + \beta_{1,\mu}x_1 + \ldots + \beta_{n,\mu}x_n, \quad (2)$$

$$\sigma(t) = \beta_{0,\sigma} + \beta_{1,\sigma}x_1 + \ldots + \beta_{n,\sigma}x_n, \quad (3)$$

where the β variables represent the coefficients and the x variables are the covariates. In keeping with previous studies, the shape parameter, which is the most difficult to estimate, is assumed constant (e.g., Obeysekara and Salas, 2014; Salas and Obeysekera, 2013; Towler et al., 2010).

Some previous studies (e.g., Salas and Obeysekera, 2014; Stedinger and Griffis, 2011) have developed non-stationary location and scale parameters that are explicitly dependent on time. This approach requires first the derivation of temporal flooding trends and second the projection of this trend into the future. Here we derive location and scale parameters based on time-varying meteorological variables (i.e., temperature and precipitation). With the approach used here, temporal trends in flooding are introduced as a function of temporal variability in precipitation and temperature, but no explicit trend is specified a priori.

To determine the optimal set of covariates for a non-stationary model, additional statistical methods must be employed. The Akaike information criterion (AIC; Akaike, 1974), given in Eq. (4), weighs the goodness of the fit of a model with the level of complexity.

$$\text{AIC} = 2(\text{NLLH}) + 2K \quad (4)$$

Here, NLLH is the negative log-likelihood estimated for a model fitted with K parameters. In this formulation, higher-ranked models have lower AIC scores. For this analysis, the best model is selected using pairwise comparisons of NLLH scores following the methods of Salas and Obeysekera (2014) and others. Models are compared using the deviance statistic (D), which is equal to twice the difference in NLLH scores. The D is then tested for significance based on a chi-squared distribution with the degrees of freedom set equal to the difference in the number of parameters (K) between models. Finally, p values less than 0.05 indicate a statistically significant improvement in model performance at the 5 % significance level.

Following the methodology described above, GEV distributions are fitted to time series of maximum monthly historical (1950–1999) 1-day simulated streamflows (detailed in Sect. 2) for the cool season (November to April). Although there are some unregulated historical flow estimates, the available data set only covers six storms. Therefore to be consistent, we fit our model only to the simulated flows. The data set includes maximum daily streamflows for each month in the cool season defined by the block of months November through April, as opposed to the more traditional single value per year. This technique was also used by Towler et al. (2010), who noted that expanding the data set helps avoid the problems associated with using maximum likelihood estimate on small data sets. However, as noted by Towler et al. (2010), when multiple values are used per year the calculated probabilities must be adjusted appropriately to derive annual values. Floods during the cool season generally last between 1 and 4 days. Here we focus on the 1-day flood peak, as opposed to multi-day flood volumes, because this is a representative metric for flood damage. Additionally, using the 1-day flood maximum focuses the analysis on flood magnitude rather than duration.

Two covariates were considered, monthly total precipitation (P) and mean temperature (T), averaged over the upstream area for each gauge. As discussed in Sect. 2, precipitation is a relevant covariate because many of the floods in this season are rain-on-snow events or extreme rainfall events. Similarly, temperature drives snowmelt and is an important contributor to UTRB flood events (e.g., January 1997 event). Both stationary and non-stationary GEV models were evaluated using the extRemes package (Gilleland and Katz, 2011) in the R statistical computing environment (R Core Team, 2012).

3.2 Time-varying risk assessment

Traditional flood planning relies on the concept of return periods, which are usually calculated as the inverse of annual exceedance probability for a given flood magnitude, assuming a stationary distribution: for example, the log-Pearson Type III (LP3) distribution described by the Interagency Advisory Committee on Water Data bulletin 17B (IACWD, 1982). However, when non-stationary models are used, the distribution parameters, and hence the exceedance probabilities, vary with time. Table 1 compares various flood probability calculations between stationary and non-stationary approaches (Salas and Obeysekera, 2014). As shown here, when the flood distribution is stationary, the return period for a given flood magnitude is constant and relies only on the exceedance probability (Eq. 4a in Table 1). However, if distribution parameters are non-stationary, then the return period will vary based on the period of interest (Eq. 4b in Table 1). This concept is easily extended to flood risk (here defined as the probability of a flood of a given magnitude occurring, not expected losses). In traditional analyses, the risk of a flood occurring in a given period depends only on the length of the period (Eq. 5a), while in a non-stationary analysis, risk depends on both the length of time considered and the time period itself (Eq. 5b). This is the concept of design life level proposed by Rootzén and Katz (2013). Here we adopt the design-life level risk framework given by Eq. (5b) in Table 1 and calculate the risk of flood for a range of future periods and design life lengths.

Table 1. Flood calculations using stationary and non-stationary distributions (adapted from Salas and Obeysekera, 2014).

Eq. #	Description	(a) Stationary	(b) Non-stationary
1	Exceedance probability (probability of flood[1] occurring in year x)	p	p_x
2	Probability of the first flood occurring in year x[2]	$f(x) = (1 - p)^{x-1} p$	$f(x) = p_x \prod_{t=1}^{x-1} 1 - p_t$
3	Probability of a flood occurring before year x[3]	$F(x) = \sum_{i=1}^{x} f(i)$ $F(x) = 1 - (1 - p)^x$	$F(x) = 1 - \prod_{t=1}^{x} (1 - p_t)$
4	Return period (expected waiting time between flood occurrences[4,5])	$E(X) = \sum_{x=1}^{\infty} x \cdot P(X = x)$ $E(X) = 1/p$	$E(X) = 1 + \sum_{x=1}^{x_{max}} \prod_{t=1}^{x} (1 - P_t)$
5	Probability of a flood occurring before the design life n[6]	$R = P(X \le n) = F(n)$ $R = 1 - (1 - p)^n$	$R = 1 - \prod_{t=1}^{n} (1 - p_t)$

[1] Flood is defined as a flow exceeding a predefined threshold; [2] $f(x)$ is the probability density function of X; [3] $F(x)$ is the cumulative distribution function of X; [4] X is a random variable denoting the waiting time for the first flood occurrence; [5] x_{max} is the time when p_x equals 1; [6] n is the length of the time period over which flood risk is calculated.

4 Results and discussion

Results are grouped into three sections. First we present the development of the non-stationary GEV models (Sect. 4.1). Next the models are verified by comparing simulated results to observations (Sect. 4.2). Finally we present future projections of flood frequency analysis (Sect. 4.3).

4.1 Extreme value model development

A suite of models were fit to the logarithms of block cool season (November–April) maxima flows simulated by the calibrated VIC model with different non-stationary parameter combinations. The model structures tested include stationary, non-stationary location, non-stationary scale and non-stationary location and scale. For all model structures, model fit was tested using one or both covariates (i.e., precipitation and temperature). Models were also tested using the block maxima flows directly; however, performance was improved considerably with the logarithmic transformation. Validation of the VIC-simulated flows as well as the GEV models is presented in the following section.

Table 2 summarizes NLLH and AIC scores for each model configuration. The D for pairwise comparisons of NLLH scores and the p values calculated for each D based on a chi-squared distribution are also provided. The bottom row of Table 2 provides the number of parameters in each model

and the model number that was used for the pairwise comparisons. As shown here, the models with non-stationary location and scale relying on both precipitation and temperature as covariates have the best (i.e., lowest) NLLH scores for both stations and are a statistically significant improvement over the other models listed in Table 2. Figure 2 plots stationary and non-stationary location and scale models with histograms of observed flow for both gauges. Qualitatively, the stationary model fits well with the center of the distribution but overestimates the tails. The non-stationary models overestimate the median values but are a closer fit to the extreme values.

The coefficients for Eqs. (2) and (3) for the selected models are provided in Table 3. Using the coefficients determined above, the location and scale parameters are calculated for every climate projection (i.e., 234) and flood season month (i.e., November to April 1950 to 2099) based on the downscaled precipitation and temperature values detailed in Sect. 2 (note that the shape parameter remains fixed). Thus, for every future month there is a separate GEV distribution curve for each of the 234 climate projections.

To address uncertainty of model parameters (namely the model coefficients, β in Eqs. 2 and 3), models of the same form (i.e., non-stationary location and scale with precipitation and temperature as covariates) were also fit to the historical simulation period (1950–1999) using downscaled precipitation and temperature from all 234 climate projections.

Table 2. Negative log likelihood (NLLH) and Akaike information criterion (AIC) scores for each model, as well as the deviance statistics (D) of pairwise comparisons of different model configurations (P is precipitation only, T is temperature only and $P \& T$ are both) and the p values of each D score based on a chi-squared distribution. The number of parameters in each model and the models used for comparison are listed at the bottom of the table. The selected model for each station is in bold.

Station	Metric	Stationary	Non-stationary location			Non-stationary scale			Non-stationary location and scale		
			$P \& T$	P	T	$P \& T$	P	T	$P \& T$	P	T
		1	2	3	4	5	6	7	8	9	10
Farad	NLLH	508.9	422.9	467.1	499.7	487.3	500.9	506.5	**416.4**	462.2	496.9
	AIC	1023.7	855.9	942.3	1007.4	984.6	1009.8	1021.1	**846.8**	934.4	1003.8
	D		171.8	83.4	18.3	43.1	15.9	4.7	**13.0**	9.9	5.7
	p value of D		< 0.05	< 0.05	< 0.05	< 0.05	< 0.05	< 0.05	< 0.05	< 0.05	< 0.05
Reno	NLLH	505.4	418.4	462.5	496.0	484.4	497.6	503.1	**408.8**	457.4	493.2
	AIC	1016.8	846.8	932.9	1000.0	978.8	1003.2	1016.1	**831.7**	924.8	996.5
	D		174.0	85.9	18.8	42.0	15.6	4.7	**19.1**	10.1	5.5
	p value of D		< 0.05	< 0.05	< 0.05	< 0.05	< 0.05	< 0.05	< 0.05	< 0.05	< 0.05
# of model parameters		3	5	4	4	5	4	4	7	5	5
Model # compared to for p value			1	1	1	1	1	1	2	3	4

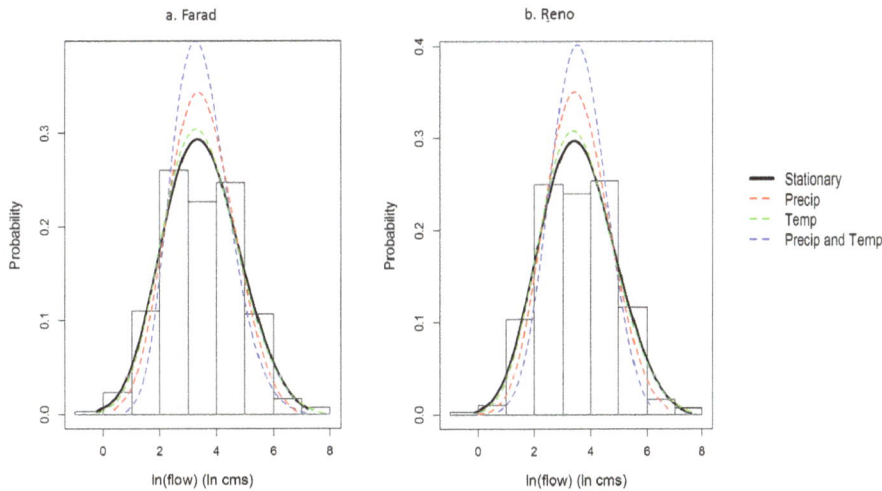

Figure 2. Probability density function of fitted stationary (solid black) and non-stationary (dashed) GEV models compared to historical VIC-simulated flow histogram.

Because each climate projection seeks to reproduce similar behavior over the historical 1950–1999 period, the variability between projections in this time frame is a measure of uncertainty in model coefficients given the representation of the same physical system. This differs from the variability between climate projections in future periods (i.e., after 1999), which is a measure of uncertainty in future forcing conditions. Table 3 shows the interquartile range of model coefficients calculated from the 234 historical GCM simulations.

Using these parameters, the return period of the design flood at Reno (37 600 cfs, 1065 cms) was calculated for every set of model parameters using observed historical precipitation and temperature. The observed model estimates a return

period of 45 years while the interquartile range (IQR) using the simulated model parameters (i.e., the model parameters estimated from each of the 234 historical GCMs) with observed precipitation and temperature varies from 28 to 247 years. Note that the return period of 45 years estimated from observed meteorology is within the IQR of 28 to 247 years. Although the IQR is large, it should be kept in mind that some of the uncertainty in this range is a result of the performance of individual GCMs in simulating historical climate and in the BCSD downscaling methodology. The monthly BCSD algorithm used for downscaling GCM climate only constrains the monthly precipitation and temperature statistics (total precipitation and mean monthly tempera-

Table 3. Summary of derived model covariates for Eqs. (2) and (3) based on historical observations (historical observed) and using historical simulated data from the 234 CMIP-5 projections (historical simulated interquartile range, IQR).

	Farad		Reno	
	Historical observed	Historical simulated IQR	Historical observed	Historical simulated IQR
$\beta_{0\mu}$	2.155	1.738 4.794	2.582	2.135 4.827
$\beta_{1\mu}$	0.175	0.053 0.148	0.180	0.066 0.152
$\beta_{2\mu}$	0.115	0.046 0.138	0.105	0.046 0.124
$\beta_{0\sigma}$	0.211	0.517 1.673	0.530	0.569 1.748
$\beta_{1\sigma}$	−0.013	−0.020 0.006	−0.018	−0.023 0.008
$\beta_{2\sigma}$	0.027	−0.012 0.022	0.017	−0.015 0.019
Shape (ξ)	−0.178	−0.389 −0.094	−0.275	−0.389 −0.070

ture) over the historical 1950–1999 period. Furthermore, uncertainty is introduced when monthly total precipitation and mean temperature are disaggregated to daily values. Thus, the estimated IQR implicitly captures climate simulation and downscaling uncertainties, in addition to explicitly representing model parameter uncertainty. The need to consider uncertainties at each and every step of the process starting with, for example, downscaling methods (statistical, dynamical or some combination of statistical and dynamical methods) is a topic of ongoing research.

4.2 Hydrologic and GEV model validation

Since we used modeled VIC flows for flood analysis, there are two considerations for model validation. First, we compare VIC-simulated 1-day flood events to the observed unregulated flow estimates (i.e., validating that our calibrated VIC model is accurately simulating flood flows). Second, we compare the GEV-modeled floods to the VIC-simulated 1-day flood events and the observed unregulated flow estimates (i.e., validating that the GEV models we fit to the simulated data match both the observed unregulated flows and the VIC-simulated flows).

Although unregulated flows are not available for the entire period of record, 1-day maximum unregulated flow estimates are available at Reno for six historical floods (USACE, 2013b). Figure 3 plots the observed flow (blue triangle) with the 1-day VIC flow that was simulated using historical observed forcings from Maurer et al. (2002) (red triangle) and a box plot of the non-stationary GEV distribution for the same month generated using the same monthly historical precipitation and temperature (i.e., Maurer et al., 2002). Comparing first the 1-day maximum VIC-simulated flow with the observed flow the maximum percent difference between the natural logarithm of simulated and observed flows is 12 %. There does appear to be a slight positive bias in the VIC simulations (i.e., VIC-simulated flows are greater than observed flood flows). Still, the simulated flood values (red circles) generally fall within the interquartile range of the GEV

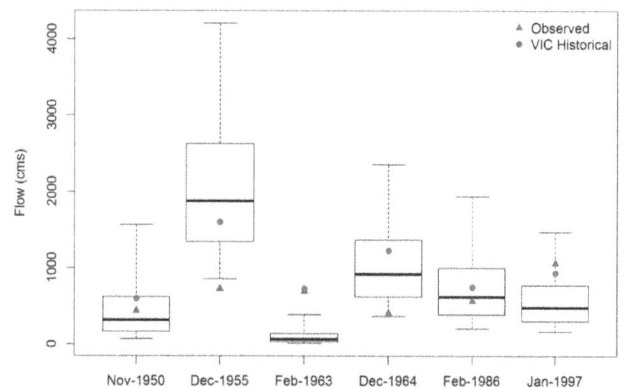

Figure 3. "Observed" unregulated flow estimated from gauge records (blue triangle) compared to VIC-simulated flow (red circles) and the simulated GEV distribution. Boxes span the 25th to 75th percentile of the GEV distribution for a given month and the whiskers extend to the 5th and 95th percentiles.

distribution except in the case of the 2 February 1963 flood and the 2 January 1997 flood.

In these instances the VIC simulation matches very closely (percent difference in the natural logarithm of flows are 0.5 and 1.2 %, respectively) with the observed flow; however, the GEV model underestimates the events. This discrepancy is caused by the flood timing. In both cases the flood occurs at the very beginning of the month. In the GEV framework, the precipitation and temperature are used as covariates for the flow of the same month. However, for these storms, flooding is linked to precipitation and temperature in the month of flooding and the preceding month. Therefore, the GEV model simulates the flood in the preceding month and/or underestimates the flood magnitude if the precipitation is split between 2 months. While this is a limitation for matching individual historical events (primarily timing), it should not be a major concern for future projections. This is because, for the purposes of risk calculations, it does not matter in what month the GEV model simulates the flood event as long as it realistically captures flood magnitude behavior.

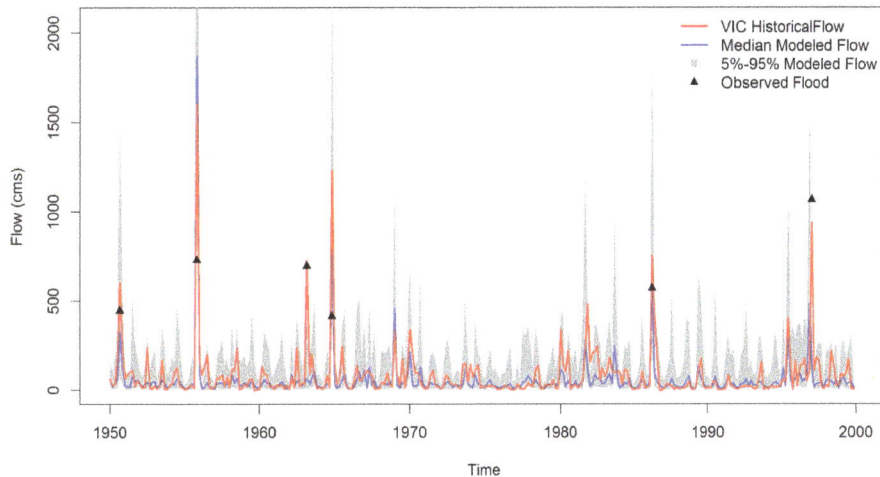

Figure 4. VIC-simulated 1-day flood maximums for November through April 1950 to 1999 (red lines) compared with the historical GEV distributions (blue line is median and grey shading is the 5th to 95th percentile range) and the six observed flow rates.

Comparing the GEV model distribution to the other observed floods (blue triangles), the distribution encompasses the observed flood magnitude (within the 5th and 95th percentile) for all except two of the floods (1955 and 1963). For 1963, the VIC-simulated and observed floods are in close agreement (the difference between the natural logarithm of simulated and observed flows is the smallest of any event at 0.5 %), and the discrepancy with the GEV model is consistent with the flood timing described above. The 1955 flood resulted from 38 cm of melted snow combined with 33 cm of rainfall over a 3-day period (O'Hara et al., 2007). In the historical forcings used to drive the VIC model, December 1955 has 75 cm of precipitation, which is the highest December precipitation value in the historical period. In this instance, the VIC-simulated flow falls within the interquartile range of the GEV model, but the high monthly precipitation results in an overestimation of the flood magnitude. Again, this is a limitation of using monthly forcings because the total December precipitation is used as a covariate and not a storm-specific value, though in many cases the storm-specific values constitute the bulk of the monthly precipitation totals.

Figure 4 is a time series plot of VIC historical simulated flow along with the median and 5th to 95th percentile flow of the GEV model. As would be expected from the model fit demonstrated in Figs. 2 and 3, Fig. 4 shows that the VIC-simulated flows are generally close to the median GEV-modeled flow and nearly always fall within the 5th to 95th percentile range. Although there are differences in the simulation of individual events discussed above, the median simulated flood magnitudes are only greater than the maximum observed flood in two instances of the 300 historical months simulated.

In general, Figs. 3 and 4 show that the VIC-simulated flows match closely with the observed floods (based on percent difference in the natural logarithm of flows) and that

the interquartile range of the GEV distributions encompasses the observed and simulated flows in most instances. Figure 3 does illustrate some of the complications in matching individual events. However, based on analysis of the driving forces behind each individual event we are able to explain and document the sources of these discrepancies. Based on this analysis we conclude that the VIC model behavior has a reasonable match with the natural system.

4.3 Future flood risk

Future flood risk is calculated using Eq. (5b) from Table 1. For the first part of this analysis we define "flood" as 1-day flow exceeding 1065 cms (37 600 cfs). This is the maximum historical unregulated flow at Reno from the 2 January 1997 event and is considered to be the design flood for flood protection infrastructure design. For each simulation month (1950–2099 November–April) exceedance probabilities are calculated for every climate projection (234 in total) using the selected non-stationary GEV models from Table 3 (fit to the historical observations) and the projected monthly precipitation and temperature. As detailed in Sect. 3.2, when exceedance probabilities are time dependent, the flood risk (refer to Eq. 5b, Table 1) is a function of both the length of the design life and the period of operation. Figure 5 plots the risk of flood versus project life for three time periods: 1950 to 1999, 2000 to 2049 and 2050 to 2099. In other words, this is the risk of a flood exceeding 1065 cms in the next n years from 1950, 2000 or 2050. The median and interquartile ranges show the distribution of the 234 climate projections simulated. Here we use the interquartile range, as opposed to the 5th and 95th percentile, to focus on the central tendencies of each time period. Note that the ranges presented here express the variability between climate projections. Uncertainty of future VIC model simulations is not investigated

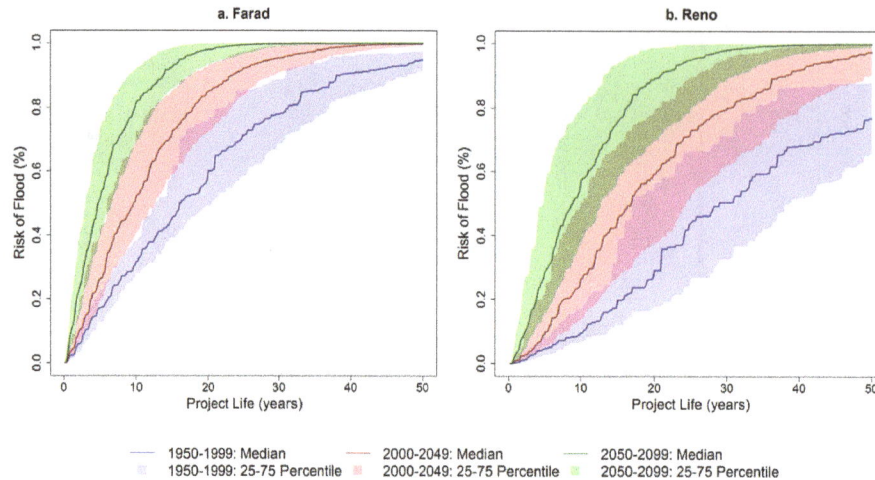

Figure 5. Probability of 1-day flood exceeding historical maximum of 1065 cms (risk) at Farad and Reno. Solid lines represent the median risk of the 234 climate projections and shading covers the interquartile range (i.e., 25th to 75th percentile).

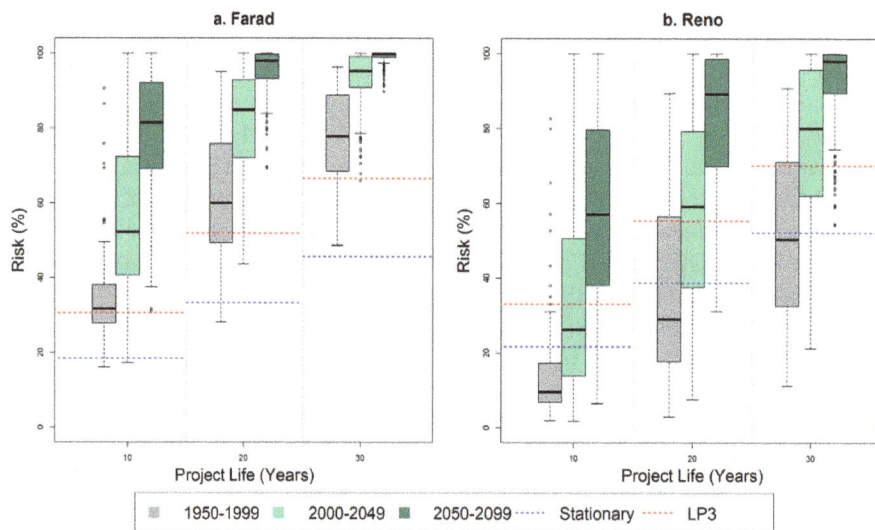

Figure 6. Box plots of the probability of a 1-day flood exceeding 1065 cms (risk) for three project life lengths (10, 20 and 30 years). Results are grouped by time period (1950–1999, 2000–2049 and 2050–2099). Blue dashed lines show the flood risk calculated from the stationary GEV model fit to the historical data.

here. For a detailed analysis of uncertainty in VIC simulations the reader is referred to Elsner et al. (2014).

For both Farad and Reno there is a clear positive shift in flood risk between the three time periods. In all cases the median risk for each subsequent time period falls outside the interquartile range of the preceding time period although the prediction spread for Reno is greater than that for Farad. It is important to note that the flood risk is actually higher at Farad than Reno in both the historical and future periods despite the fact that the observed flow distributions at the two stations are very similar (refer to Fig. 2). This shift between Farad and Reno is caused by the differences in the shape parameters (refer to Table 3). Farad has a relatively heavier tailed distribution (i.e., the GEV shape parameter for Farad is greater

than the shape parameter for Reno) and therefore flood risks are increased. The sensitivity of the model parameters (and the associated flood risk) to small differences in the flow and covariate distributions is further demonstrated by Fig. 6.

Figure 6 presents the project life risk from Fig. 5 for three project life periods (10, 20 and 30 years). Box plots show the non-stationary model results for the 234 climate projections with the different time periods compared side by side. Also, the risk calculated using a stationary GEV model and a stationary LP3 model (i.e., the distribution prescribed by Bulletin 17B and fitted using L-moments; IACWD, 1982) fit to the historical flow data are plotted for reference (blue and red dashed lines, respectively). Comparing these three approaches (non-stationary GEV, stationary GEV and stationary LP3) provides information on the sensitivity of results to the modeling approach and non-stationary parameters. For instance, both stationary models are fit to the same historical simulated flows (one using MLE and the other using L-moments) so differences between the stationary lines reflect the impact of model choice and fitting approach on estimated risk. Conversely the stationary GEV model (blue line) and the historical non-stationary models (grey box plot) have the same model form and cover the same time period; the only difference is the addition of covariates to estimate model parameters. Thus differences between these two show the effect of model parameter changes from the non-stationary approach. Finally, variability between the box plots for a given design period demonstrates the evolution of risk over time (i.e., the impact of climate change on risk). The latter (i.e., changing risk over time) is the purpose of this analysis; however, before assessing change over time, we must first discuss the impact of model choice and parameters on risk estimates.

For both of the stationary methods, the risk increases with project life following Eq. (5a) from Table 1. The distinction between these lines and the non-stationary approaches is that, with the stationary approach, a single exceedance probability is calculated for the given flood magnitude, and this probability is assumed to remain constant throughout the design life. Also, for both stationary approaches the model is fit directly to the historical 1-day maximum flow distribution and no covariates are required (note that stationary models are not fit to the future time periods because this would require future simulated flows). Comparing the GEV (blue line) and the LP3 (red line) stationary models, there is a 10–20 % increase in risk between the two models. This difference is purely a function of model form and highlights the sensitivity of the risk calculations to model choice.

Contrasting the difference between the stationary (blue line) and the non-stationary GEV for the historical time period (grey box plot) illustrates the effect of adding non-stationary parameters to a given model form. Recall that in both cases the GEV model is fit to the historical simulated flows. However, for the stationary approach, model fitting results in a single set of parameters (location, scale and shape), whereas with the non-stationary approach we derive the shape parameter and a set of coefficients for linear models to determine the location and scale parameters based on precipitation and temperature values. Thus, for the non-stationary approach, different location and scale parameters are calculated for every historical cool season month and GCM model (234).

Overall, there is close agreement between the stationary (S) and average non-stationary (NS) location parameters (6.55 S vs. 6.64 NS at Farad and 6.63 S vs. 6.78 NS at Reno). However, for both gauges the scale parameter is lower with the non-stationary approach (1.30 S vs. 0.94 NS at Farad and 1.28 S vs. 0.96 NS at Reno). At Reno the shape parameter is similar (−0.24 S vs. −0.27 NS), but at Farad the difference is somewhat larger (−0.24 S vs. −0.18 NS). Differ-

ences in model parameters are reflected in the distance between the stationary GEV model (blue line) and the median historical non-stationary GEV box plots (center of the grey box plots) in Fig. 6. For Reno, the stationary line is closer to the historical box plots. However, at Farad the non-stationary box plots are consistently higher than the stationary line. The larger differences between the stationary and non-stationary models for Farad result from changes in the shape parameter between the stationary and non-stationary model fits. This change demonstrates the sensitivity of model results to changes in model parameters.

As with Fig. 5, Fig. 6 shows significant increases in risk moving into the future and subsequently larger differences between the stationary and non-stationary approach. By the second future period the differences between the stationary and non-stationary models can be as much as 50 % or more. For both gauges, difference in risk between the non-stationary and stationary approaches grows over time, indicating greater potential to underestimate the future risk if non-stationary parameters are not considered.

Results were also grouped by RCPs to analyze connections between greenhouse gas emission rates and changes in flood risk. As shown in Fig. 7, we observed no clear trend in flood risk based on the different RCPs. This indicates that, for this flood statistic in this basin, the variability between GCM model form and initial conditions likely overwhelms the influence of greenhouse gas emissions when comparing between scenarios. Although we caution that this is not a general finding, for this application we show that the variability between projections within any RCP scenario is larger than the difference between RCP scenarios. Harding et al. (2012) also noted similar behavior in their study of the Colorado River basin.

Given the sensitivity of projected risk to model parameters, an obvious question is whether increases in risk over time are similarly sensitive. For the 1065 cms flood plotted in Fig. 6, the increased risk with added project life (i.e., 20 years vs. 10 years) is greater with the non-stationary models than the stationary one at both stations. This is intuitive, given the increased flood risk with time demonstrated in Fig. 5 for the non-stationary models. Although Farad has higher overall risk, the relative increase in risk between time periods is similar between the two stations. For example, the median 10-year flood risk when comparing the first (1950–1999) and second (2000–2049) time periods increases by 21 % for Farad and 29 % for Reno.

Next, analysis is expanded to a range of flood magnitudes. Figure 8 plots the flood risk over a 10-year project life starting in 1950, 2000 and 2050 for flood values ranging from 283 to 1416 cms (10 000 to 50 000 cfs). As would be expected, the 10-year flood risk decreases with increasing flood rate. The shapes of the curves are slightly different between Farad and Reno; flood risk decreases more sharply with increased flow at Reno than Farad. Again, this behavior is a function of the shape parameter of the respective GEV distri-

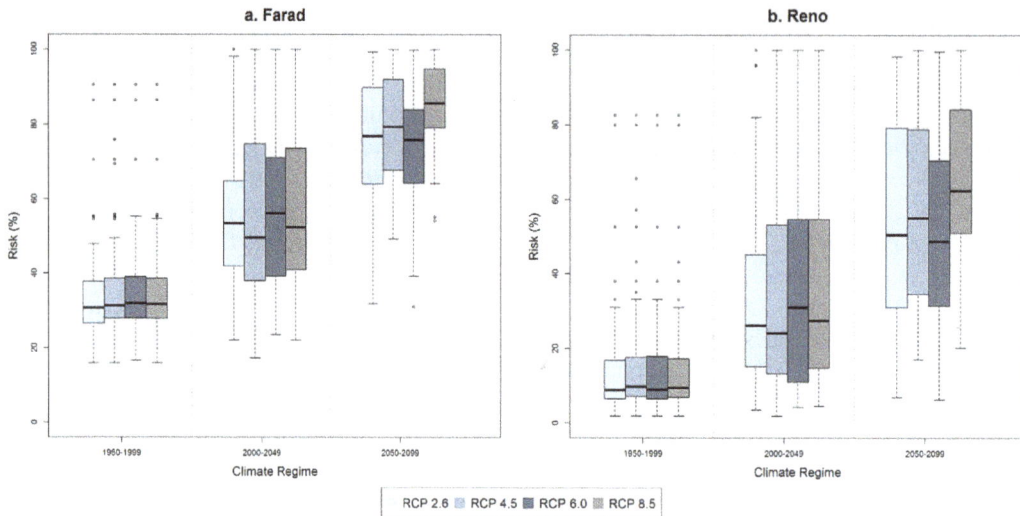

Figure 7. Box plots of the probability of a 1-day flood exceeding 1065 cms (risk) in 10 years for three 50-year periods. Results are grouped by the representative concentration pathways (RCPs) used to drive the GCM projection. RCP 8.5 has the largest increase in greenhouse gas concentrations and RCP 2.6 the smallest.

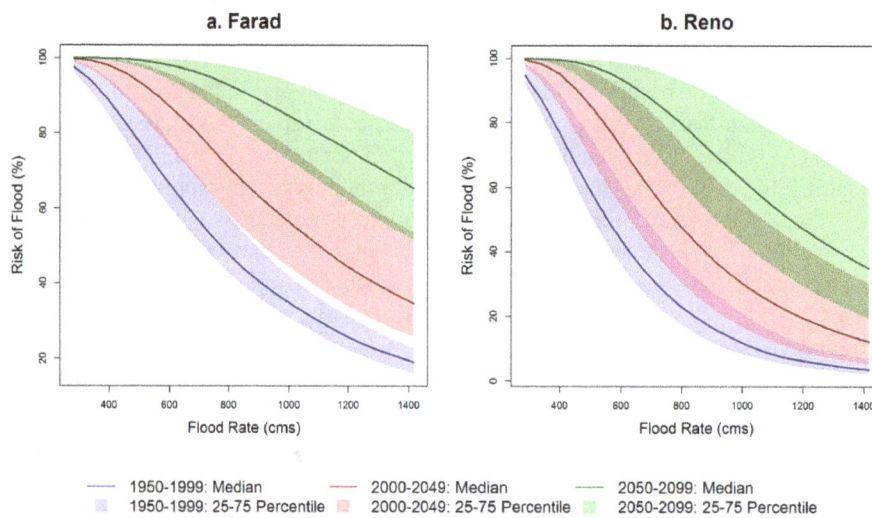

Figure 8. Probability of flood in a 10-year project life (risk) vs. median 1-day flood rate at (a) Farad and (b) Reno for three time periods 1950–1999 (blue), 2000–2049 (red) and 2050–2099 (green). Solid lines represent the median of the 234 climate projections and shading covers the interquartile range (i.e., 25th to 75th percentile).

butions. Despite these differences, both gauges display clear shifts between time periods similar to Fig. 5. Again, the median risk for each subsequent period consistently falls outside the interquartile range of the preceding period.

Changes in the median flood risk (i.e., differences between the solid lines in Fig. 8) between each future period and the historical period are plotted in Fig. 9 for both gauges. As would be expected based on the qualitative differences in Fig. 8, the shape of the Farad and Reno difference curves are slightly different. However, the salient point for this anal-

ysis is that the increased risk between periods and the two stations is generally within 10 %. Overall, the increased risk between the first future period (2000–2050) and the historical period (1950–1999) is between 10 and 20 % for flows from 600 to 1200 cms. Similarly, the increased risk from the historical period to the second future period (2050–2099) is between 30 and 50 %. Differences for the highest and lowest flows are difficult to assess because the median is skewed for high and low flow values by the fact that the risk values are bound between 0 and 100 %.

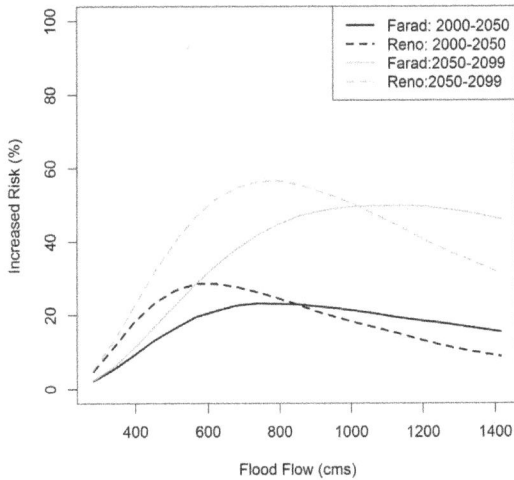

Figure 9. Increased probability of flood occurrence for a 10-year project life (risk) from the historical period (1950–1999) to each of the two future periods, 2000–2050 (black) and 2050–2099 (grey). Farad is plotted with a solid line and Reno is a dashed line.

5 Summary and conclusions

The analysis presented is unique in its incorporation of non-stationary GEV analysis using CMIP-5 projections and the design-life level risk assessment concepts. We present our findings as a relevant case study and an example application of recent developments in non-stationary flood assessment. Lacking sufficient unregulated flow data we simulate historical floods using the VIC model. Subsequently we use the simulated floods to fit non-stationary GEV models, with downscaled monthly precipitation and temperature as covariates. Although there are some discrepancies between individual simulated and observed flood events, we demonstrate that the VIC model adequately captures the range of flood magnitudes. Furthermore, we show that the GEV-modeled historical floods are in good agreement with both the VIC-simulated floods and the published flood events (USACE, 2013b).

Discrepancies between historical and simulated events often result from the monthly time step used for covariates. This can affect the ability to model floods that are generated by precipitation that occurs in 2 months. Also, because the climate variables are monthly aggregates, and not event based, large floods can be generated in months with high precipitation even if that precipitation does not occur in one concentrated event. Despite these differences, comparison with historical flood events demonstrates that the GEV model does reasonably well at simulating historical flood magnitudes, even if some individual historical events are not matched exactly.

Using the derived non-stationary GEV models, we generate flood distributions for 234 CMIP5 climate projections from 1950 to 2099. For the historical 1-day design flood magnitude of 1065 cms, results show significant increases in the frequency of high-flow events in the future. From a water management standpoint, this finding translates directly to increased flood risk. For example, we calculate a 21 % (29 %) increase risk of a 1065 cms flood over a 10-year design life for Farad (Reno) from the historical time period to the first future period and similar increases from the first future period to the second. Increased risk between time periods is also relatively consistent for longer design life periods and similar shifts in flood risk are noted across a range of flood magnitudes. For both stations, the increased risk from the historical to the first future period is between 10 and 20 % and from the historical to the second future period is between 30 and 50 % for floods ranging from 600 to 1200 cms.

The significant increases in flood risk through time indicate the importance of non-stationary flood frequency analysis for future infrastructure planning, and the potential to underestimate risk when stationarity is assumed. For both stations the difference between the stationary and non-stationary approach increases over time. By the second future period (2000–2049), differences in risk calculations between the stationary and non-stationary models can be 50 % or larger. This finding is in keeping with a number of recent studies (e.g., Griffis and Stedinger, 2007; Katz et al., 2002; Towler et al., 2010) that have highlighted potential applications for non-stationary analysis of flood frequency.

An important consideration for this approach is the sensitivity of results to model parameters. In all cases, the flood risk is higher at Farad than Reno due to the relatively heavier tailed distribution that was fit. Estimated model parameters differed by station despite the fact that the flow, precipitation and temperature distributions for both locations are very similar. While these changes affected the overall risk projections, the relative increase in risk over time remained consistent between stations. This indicates that the more robust metric from this analysis is the relative increase in flood risk and not the absolute values. This finding is further supported by the fact that absolute flood risk estimates could be impacted by model bias. By focusing on differences in risk, we specifically highlight the impact of non-stationarity on risk assessment as opposed to parameter sensitivity. Similarly, it is important to note that this analysis is based on natural flow estimates and does not include infrastructure development or operation. Therefore, results indicate the potential increase in the underlying natural flood risk in the UTRB over the 21st century.

Acknowledgements. The authors would like to acknowledge the Reclamation WaterSMART Basin Study program for funding the Truckee River basin study which supported this publication. In addition, we would like to thank the reviewers for their constructive comments that helped to improve the manuscript.

Edited by: R. Merz

References

Akaike, H.: New look at statistical-model identification, IEEE Trans. Autom. Control, 19, 716–723, 1974.

Allan, R. P.:Climate change: Human influence on rainfall, Nature, 470, 344–345, 2011.

Cayan, D. R., Redmond, K. T., and Riddle, L. G.: ENSO and Hydrologic Extremes in the Western United States, J. Climate, 12, 2881–2893, 1999.

Cayan, D. R., Kammerdiener, S. A., Dettinger, M. D., Caprio, J. M., and Peterson, D. H.: Changes in the Onset of Spring in the Western U.S., B. Am. Meteorol. Soc., 82, 399–415, 2001.

Christensen, N. S. and Lettenmaier, D. P.: A multimodel ensemble approach to assessment of climate change impacts on the hydrology and water resources of the Colorado River Basin, Hydrol. Earth Syst. Sci., 11, 1417–1434, doi:10.5194/hess-11-1417-2007, 2007.

Christensen, N. S., Wood, A. W., Lettenmaier, D. P., and Palmer, R. N.: Effects of climate change on the hydrology and water resources of the Colorado river basin, Climate Change, 62, 337–363, 2004.

Das, T., Pierce, D. W., Cayan, D. R., Vano, J. A., and Lettenmaier, D. P.: The importance of warm season warming to western U.S. streamflow changes, Geophys. Res. Lett., 38, L23403, doi:10.1029/2011GL049660, 2011.

Dettinger, M. D. and Cayan, D. R.: Large-scale Atmospheric Forcing of Recent Trends toward Early Snowmelt Runoff in California, J. Climate, 8, 606–623, 1995.

Douglas, E. M., Vogel, R. M., and Kroll, C. N.: Trends in floods and low flows in the United States: impact of spatial correlation, J. Hydrol., 240, 90–105, 2000.

Easterling, D. R., Meehl, G. A., Parmesan, C., Changnon, S. A., Karl, T. R., and Mearns, L. O.: Climate Extremes: Observations, Modeling and Impacts, Science, 289, 2068–2074, 2000.

Elsner, M. M., Gangopadhyay, S. G., Pruitt, T., Brekke, L. D., Mizukami, N., and Clark, M. P.: How does the choice of distributed meteorological data affect hydrologic model calibration and streamflow simulations?, J. Hydrometeorol., 15, 1384–1403, 2014.

Franks, S. W.: Identification of a change in climate state using regional flood data, Hydrol. Earth Syst. Sci., 6, 11–16, doi:10.5194/hess-6-11-2002, 2002.

Gangopadhyay, S., Pruitt, T., Brekke, L. D., and Raff, D. A.: Hydrologic Projections for the Western United States, Eos, 92, 441–452, 2011.

Gilleland, E. and Katz, R. W.: New software to analyze how extremes change over time, Eos, 92, 13–14, 2011.

Gilroy, K. L. and McCuen, R. H.: A nonstationary flood frequency analysis method to adjust for future climate change and urbanization, J. Hydrol., 414–415, 40–48, 2012.

Griffis, V. and Stedinger, J. R.: Incorporating climate change and variability into Bulletin 17B LP3 model, paper presented at ASCE World Env. & Water Resour. Congress, Tampa, Florida, USA, 2007.

Gutowski, W. J., Hegerl, G. C., Holland, G. J., Knutson, T. R., Mearns, L. O., Stouffer, R. J., Webster, P. J., Wehner, M. F., and Zwiers, F. W.: Causes of Observed Changes in Extremes and Projections of Future Changes in Weather and Climate Extremes in a Changing Climate, Regions of Focus: North America, Hawaii, Caribbean, and U.S. Pacific Islands Rep., Washington, D.C., 2008.

Hall, J., Arheimer, B., Borga, M., Brázdil, R., Claps, P., Kiss, A., Kjeldsen, T. R., Kriaučiūnienė, J., Kundzewicz, Z. W., Lang, M., Llasat, M. C., Macdonald, N., McIntyre, N., Mediero, L., Merz, B., Merz, R., Molnar, P., Montanari, A., Neuhold, C., Parajka, J., Perdigão, R. A. P., Plavcová, L., Rogger, M., Salinas, J. L., Sauquet, E., Schär, C., Szolgay, J., Viglione, A., and Blöschl, G.: Understanding flood regime changes in Europe: a state-of-the-art assessment, Hydrol. Earth Syst. Sci., 18, 2735–2772, doi:10.5194/hess-18-2735-2014, 2014.

Harding, B. L., Wood, A. W., and Prairie, J. R.: The implications of climate change scenario selection for future streamflow projection in the Upper Colorado River Basin, Hydrol. Earth Syst. Sci., 16, 3989–4007, doi:10.5194/hess-16-3989-2012 2012.

Hirsch, R. M. (2011), A perspective on nonstationarity and water management, *Journal of the American Water Resources Association,* 47(), 436-446.

IACWD – Interagency Advisory Committee on Water Data: Guidelines for determining flood flow frequency: Bulletin 17B of the Hydrology Subcommittee, Office of Water Data Coordination, US Geological Survey, Reston, VA, p. 183, 1982.

Jain, S. and Lall, U.: Floods in a changing climate: Does the past represent the future?, Water Resour. Res., 37, 3193–3205, 2001.

Katz, R. W.: Statistics of extremes in climate change, Climate Change, 100, 71–76, 2010.

Katz, R. W., Parlange, M. B., and Naveau, P.: Statistics of extremes in hydrology, Adv. Water Resour., 25, 1287–1304, 2002.

Kunkel, K. E.: North American Trends in Extreme Precipitation, Nat. Hazards, 29, 291–305, 2003.

Liang, X., Lettenmaier, D. P., Wood, E. F., and Burges, S. J.: A simple hydrologically based model of land surface water and energy fluxes for general circulation models, J. Geophys. Res., 99, 14415–14428, 1994.

Liang, X., Wood, E. F., and Lettenmaier, D. P.: Surface soil moisture parameterization of the VIC-2L model: Evaluation and modification, Global Planet. Change, 13, 195–206, 1996.

Madsen, T. and Figdor, E.: When it Rains it Pours – Global Warming and the Rising Frequency of Extreme Precipitaiton in the U.S., Environmental America Research and Policy Center, http://www.environmentamerica.org/reports/ (last access: January 2015), 2007.

Maidment, D. R.: Handbook of Hydrology, McGraw-Hill, 1993.

Mailhot, A. and Duchesne, S.: Design criteria of urban drainage infrastructures under climate change, J. Water Resour. Pl. Manage., 136, 201–208, 2010.

Maurer, E. P., Brekke, L. D., Pruitt, T., and Duffy, P. B.: Fine-resolution climate projections enhance regional climate change impact studies, Eos Tran. AGU, 88, 504, 2007.

Maurer, E. P., Wood, A. W., Adam, J. C., Lettenmaier, D. P., and Nijssen, B.: A Long-Term Hydrologically-Based Dataset of Land Surface Fluxes and States for the Conterminous United States, J. Climate, 15, 3237–3252, 2002.

Meehl, G. A., Karl, T., Easterling, D. R., Changnon, S., Pielke Jr., R., Changnon, D., Evans, J., Groisman, P. Y., Knutson, T. R., Kunkel, K. E., Mearns, L. O., Parmesan, C., Pulwarty, R., Root, T., Sylves, R. T., Whetton, P., and Zwiers, F.: An Introduction to Trends in Extreme Weather and Climate Events: Observations, Socioeconomic Impact, Terrestrial Ecological Impacts, and Model Projections, B. Am. Meteorol. Soc., 81, 413–416, 2000.

Merz, B., Vorogushyn, S., Uhlemann, S., Delgado, J., and Hundecha, Y.: HESS Opinions "More efforts and scientific rigour are needed to attribute trends in flood time series", Hydrol. Earth Syst. Sci., 16, 1379-1387, doi:10.5194/hess-16-1379-2012, 2012.

Milly, P. C. D., Betancourt, J., Falkenmark, M., Hirsch, R. M., Kundzewicz, Z. W., Lettenmaier, D. P., and Stouffer, R. J.: Stationarity Is Dead: Whither Water Management, Science, 319, 573–574, 2008.

Min, S.-K., Zhang, X., Zwiers, F. W., and Hegerl, G. C.: Human contribution to more-intense precipitation extremes, Nature, 470, 378–381, 2011.

Mote, P. W., Hamlet, A. F., Clark, M. P., and Lettenmaier, D. P.: Declining Mountain Snowpack In Western North America, B. Am. Meteorol. Soc., 86, 39–49, doi:10.1175/BAMS-86-1-39, 2005.

Mullet, C. J., O'Gorman, P. A., and Back, L. E.: Intensification of precipitaiton extremes with warming in a cloud resolving model, J. Climate, 24, 2784–2800, 2011.

Nijssen, B., Lettenmaier, D. P., Liang, X., Wetzel, S. W., and Wood, E. F.: Streamflow simulation for continental-scale river basins, Water Resour. Res., 33, 711–724, 1997.

Obeysekera, J. and Salas, J. D.: Quantifying the uncertainty of design floods under nonstationary conditions, J. Hydrol. Eng;, 19, 1438–1446, 2014.

O'Gorman, P. A. and Schneider, T.: The physical basis for increases in precipitaiton extremes in simulations of 21st century climate change, P. Natl. Acad. Sci., 106, 14773–14777, 2009.

O'Hara, B. F., Barbato, G. E., James, J. W., Angeloff, H. A., and Cylke, T.: Weather and climate of the Redno-Carson City-Lake Tahoe Region, Nevada Special Publication 34, Bureau of Mines and Geology, Reno, Nevada, USA, 2007.

Pall, P., Aina, T., Stone, D. A., Stott, P. A., Nozawa, T., Hilberts, A. G. J., Lohmann, D., and Allen, M. R.: Anthropogenic greenhouse gas contribution to flood risk in England and Wales in autumn 2000, Nature, 470, 382–385, 2011.

Payne, J. T., Wood, A. W., Hamlet, A. F., Palmer, R. N., and Lettenmaier, D. P.: Mitigating the effects of climate change on the water resources of the Columbia River basin, Climate Change, 62, 233–256, 2004.

Pierce, D. W., Barnett, T. P., Santer, B. D., and Gleckler, P. J.: Selecting global climate models for regional climate change studies, P. Natl. Acad. Sci., 106, 8441–8446, 2009.

Pierce, D. W., Das, T., Cayan, D. R., Maurer, E. P., Miller, N. L., Bao, Y., Kanamitsu, M., Yoshimura, K., Snyder, M. A., Sloan, L. C., Franco, G., and Tyree, M.: Probabilistic estimates of future changes in California temperature and precipitation using statistical and dynamical downscaling, Clim. Dynam., 40, 839–856, 2012.

Raff, D. A., Pruitt, T., and Brekke, L. D.: A framework for assessing flood frequency based on climate projection information, Hydrol. Earth Syst. Sci., 13, 2119–2136, doi:10.5194/hess-13-2119-2009, 2009.

Ralph, F. M. and Dettinger, M. D.: Historical and National Perspectives on Extreme West Coast Precipitation Associated with Atmospheric Rivers during December 2010, B. Am. Meteorol. Soc., 93, 783–790, 2012.

R Core Team: R: A language and environment for statistical computing. R Foundation for Statistical Computing, Vienna, Austria, http://www.R-project.org/ (last access: 26 November 2014), 2012.

Reclamation: Truckee River Basin Study, Fact Sheet, available at: http://www.usbr.gov/WaterSMART/bsp/docs/fy2010/Truckee_BasinFactsheetFinal.pdf (last access: 22 September 2014), 2010.

Reclamation: West-wide climate risk assessments: Bias-corrected and spatially downscaled surface water projections, Tech. Memo. 86-68210-2011-01, US Department of the Interior, Bureau of Reclamation, Technical Service Center, Denver, Colorado, 138 pp., 2011.

Reclamation: Downscaled CMIP3 and CMIP5 Climate Projections: Release of Downscaled CMIP5 Climate Projections, Comparison with Preceding Information, and Summary of User Needs, Downscaled CMIP3 and CMIP5 Climate and Hydrology Projections, US Department of the Interior, Bureau of Reclamation, Technical Service Center, Denver, Colorado, 47 pp., 2013.

Reclamation: Downscaled CMIP3 and CMIP5 Climate and Hydrology Projections: Release of Hydrology Projections, Comparison with preceding Information, and Summary of User Needs, prepared by the US Department of the Interior, Bureau of Reclamation, Technical Service Center, Denver, Colorado, 110 pp., 2014.

Regonda, S. K., Rajagopalan, B., Clark, M., and Pitlick, J.: Seasonal Cycle Shifts in Hydroclimatology Over the Western U.S., J. Climate, 18, 372–384, 2005.

Rootzén, H. and Katz, R. W.: Design Life Level: Quantifying risk in a changing climate, Water Resour. Res., 49, 5964–5972, 2013.

Salas, J. and Obeysekera, J.: Revisiting the Concepts of Return Period and Risk for Nonstationary Hydrologic Extreme Events, J. Hyrol. Eng., 19, 554–568, 2014.

Slack, J. R., Lumb, A. M., and Landwehr, J. M.: Hydroclimatic data network (HCDN): A U.S. Geological Survey streamflow data set for the United Sates for the study of climate variation 1874–1988, USGS Water Resour. Invest. Rep. 93-4076, USGS, Reston, Virginia, USA, 1993.

Small, D., Islam, S., and Vogel, R. M.: Trends in precipitation and streamflow in the eastern U.S.: Paradox or perception?, Geophys. Res. Lett., 33, L03403, doi:10.1029/2005GL024995, 2006.

Stedinger, J. R. and Griffis, V. W.: Getting from here to where? Flood frequency analysis and climate, J. Am. Water Resour. Ass., 47, 506–513, 2011.

Stokes, J.: Draft Farad Diversion Dam Replacement Project Environmental Impact Report Rep., State Water Resources Control Board, Sacramento, CA, 2002.

Sun, Y., Solomon, S., Dai, A., and Portmann, R. W.: How Often Will it Rain?, J. Climate, 20, 4801–4818, 2007.

Taylor, K. E., Stouffer, R. J., and Meehl, G. A.: A Summary of the CMIP5 Experiment Design, B. Am. Meteorol. Soc., 93, 485–498, 2012.

Towler, E., Rajagopalan, B., Gilleland, E., Summers, R. S., Yates, D., and Katz, R. W.: Modeling hydrologic and water quality extremes in a changing climate: A statistical approach based on extreme value theory, Water Resour. Res., 46, W11504, doi:10.1029/2009WR008876, 2010.

USACE: Final environmental impact statement for the Truckee Meadows Flood Control Project: General Reevaluation Report Volume 1, US Army Corps of Engineers, Sacramento, 2013a.

USACE: Truckee Meadows Flood Control Project, Nevada: Draft General Reevaluation Report Rep., US Army Corps of Engineers, Sacramento, 2013b.

Van Rheenen, N. T., Wood, A. W., Palmer, R. N., and Lettenmaier, D. P.: Potential implications of PCM climate change scenarios for Sacramento-San Joaquin River Bain hydrology and water resources, Climate Change, 62, 257–281, 2004.

Villarini, G., Serinaldi, F., Smith, J. A., and Krajewski, W. F.: On the stationarity of annual flood peaks in the continental United States during the 20th century, Water Resour. Res., 45, W08417, doi:10.1029/2008WR007645, 2009.

Vogel, R. M., Yaindl, C., and Walter, M.: Nonstationarity: Flood magnification and recurrence reduction factors in the United States, J. Am. Water Resour. Assoc., 47, 464–474, 2011.

Walter, M. and Vogel, R. M.: Increasing trends in peak flows in the northeastern United States and their impacts on design, paper presented at 2nd Joint Federal Interagency Conference, Las Vegas, NV, 2010.

Wood, A. W., Leung, L. R., Sridhar, V., and Lettenmaier, D. P.: Hydrologic implications of dynamical and statistical approaches to downscaling climate model outputs, Climatic Change, 62, 189–216, 2004.

Wu, H., Adler, R. F., Tian, Y., Juffman, G. J., Li, H., and Wang, J. J.: Real-time global flood estimation using satellite-based precipitation and a coupled land surface routing model, Water Resour. Res., 50, 2693–2717, 2014.

Hydrological modeling in glacierized catchments of central Asia – status and challenges

Yaning Chen, Weihong Li, Gonghuan Fang, and Zhi Li

State Key Laboratory of Desert and Oasis Ecology, Xinjiang Institute of Ecology and Geography, Chinese Academy of Sciences, Urumqi 830011, China

Correspondence to: Yaning Chen (chenyn@ms.xjb.ac.cn)

Abstract. Meltwater from glacierized catchments is one of the most important water supplies in central Asia. Therefore, the effects of climate change on glaciers and snow cover will have increasingly significant consequences for runoff. Hydrological modeling has become an indispensable research approach to water resources management in large glacierized river basins, but there is a lack of focus in the modeling of glacial discharge. This paper reviews the status of hydrological modeling in glacierized catchments of central Asia, discussing the limitations of the available models and extrapolating these to future challenges and directions. After reviewing recent efforts, we conclude that the main sources of uncertainty in assessing the regional hydrological impacts of climate change are the unreliable and incomplete data sets and the lack of understanding of the hydrological regimes of glacierized catchments of central Asia. Runoff trends indicate a complex response to changes in climate. For future variation of water resources, it is essential to quantify the responses of hydrologic processes to both climate change and shrinking glaciers in glacierized catchments, and scientific focus should be on reducing uncertainties linked to these processes.

1 Introduction

Climate change is widely anticipated to exacerbate water stress in central Asia in the near future (Siegfried et al., 2012), as the vast majority of the arid lowlands in the region are highly dependent on glacier meltwater supplied by the Tienshan Mountains, which are known as the "water tower" of central Asia (Hagg et al., 2007; Sorg et al., 2012; Lutz et al., 2014). In fact, in the alpine river basins of the northern Tienshans, glacier meltwater contributes 10 % of annual runoff and 20 % of runoff during the drought years (Aizen et al., 1997); therefore, climate-driven changes in glacier/snow-fed runoff regimes have significant effects on water supplies (Immerzeel et al., 2010; Kaser et al., 2010).

According to a study conducted by the Eurasian Development Bank, changes in temperature and precipitation in central Asia have led to rapid regression in glaciers (Ibatullin et al., 2009). The overall decrease in total glacier area and mass from 1961 to 2012 to be $18 \pm 6\,\%$ and $27 \pm 15\,\%$, respectively. These values correspond to a total area loss of $2960 \pm 1030\,\mathrm{km}^2$, and an average glacier mass change rate of $-5.4 \pm 2.8\,\mathrm{Gt\,yr}^{-1}$ (Farinotti et al., 2015). If the warming projections developed by the Intergovernmental Panel on Climate Change (IPCC) prove to be true, the glacierized river systems in central Asia will undergo unfavorable hydrological changes, e.g., altered seasonality, increased flood risk, higher and intense spring discharge and water deficiency, in hot and dry summer periods, especially given the sharp rise in water demand (Hagg et al., 2006; Siegfried et al., 2012). The development of hydrological models on accounting for changes in current and future runoff is therefore crucial for water resources allocation in river basins, and includes understanding climatic variability as well as the impact of human activities on climate (Bierkens, 2015).

Hydrological modeling is an indispensable approach to water resources research and management in large river basins. Such models help researchers understand past and current changes and provide a way to explore the implications of management decisions and imposed changes. The purpose of hydrological modeling on basin scale is primarily to support decision-making for water resources management, which can be summarized as resource assessment, vulnerability assessment, impact assessment, flood risk assessment, prediction and early warning (World Meteorological Organization, 2009). It is important to choose the most suitable hydrological model for a particular watershed based on the area's climate, hydrology and underlying surface conditions.

The Tienshan Mountains span several countries and subregions, creating a decentralized political entity of complex multi-national and multi-ethnic forms. There are three large transboundary international rivers that originate in the high mountains of central Asia. In an international river, hydrological changes are related to the interests of the abutting riparian countries (Starodubtsev and Truskavetskiy, 2011; Xie et al., 2011; Guo et al., 2015). However, as conflicts between political states may arise for any number of reasons (political, cultural, etc.), transboundary issues may result in fragmented research and thus limit the development of hydrological modeling.

Amid this potential hindrance to robust research efforts, the effect of climate change on glaciers, permafrost and snow cover is having increasing impacts on runoff in glacierized central Asian catchments. However, solid water is seldom explicitly considered within hydrological models due to the lack of complete glacier data. Our knowledge of snow/glacier changes and their responses to climate forcing is still mostly incomplete. Analysis of current and future water resources variations in central Asia may promote adaptation strategies to alleviate the negative impacts of expected increased variability in runoff changes resulting from climate change.

In this paper, we review hydrological modeling efforts in five major river basins originating from the Tienshan Mountains in central Asia, namely, the Tarim River basin, the watersheds in the northern slope of the Tienshan Mountains (which includes several small river basins), the Issyk Lake basin, the Ili River basin, and the Amu Darya and Syr Darya basins (Fig. 1). Their topographical characteristics, climate and vegetation together with the glacierized area are listed in Table 1. We examine the types, purpose and use of existing models and assess the constraints and gaps in knowledge. The current lack of understanding of high-altitude hydrological regimes is causing uncertainty in assessing the regional hydrological impacts of climate change (Miller et al., 2012). Snow and glacial melt as supplies of solid water is a key element in streamflow regimes (Lutz et al., 2014); therefore, it is necessary to include glacier mass balance estimates in the model calibration procedure (Schaefli et al., 2005; Stahl et al., 2008; Konz and Seibert, 2010; Mayr et al., 2013).

2 Modeling hydrological responses to climate change

Changes in the amount and seasonal distribution of river runoff may have severe implications for water resources management in central Asia. "Glacier runoff" is defined as the total runoff generated from the melting of glaciers (snow and glacier), but can also include liquid precipitation on glacierized areas (Unger-Shayesteh et al., 2013). A large number of hydrological models applied in glacierized catchments of central Asia are basin-scale models, which contain empirical hydrological models as well as physical hydrological models (Table 2). These glacio-hydrological models are useful tools for anticipating and evaluating the impacts of climate changes in the headwater catchments of the main Asian rivers (Miller et al., 2012).

2.1 Current and future runoff changes

River runoff responds in a complex way to variations in climate and the cryosphere. At the same time, runoff changes also depend on dominant runoff components. Table 2 shows that annual runoff anomalies have increased to some extent (except in the western Tienshan Mountains) and inconsistencies between changes in precipitation and runoff have occurred in heavily glacierized catchments. In rivers fed by snow and glaciers, runoff has increased (e.g., in northern Tienshan Mountains) and rising temperatures dominate the runoff changes by, for instance, increasing the snowmelt/glacier melt and decreasing snowfall fraction (ratio of solid precipitation to liquid precipitation) (Chen, 2014). Khan and Holko (2009) compared runoff changes with variations in snow cover area and snow depth. They suggested that the mismatch between decreasing trends in snow indicators and the increasing river runoff could be the result of enhanced glacier melting. Heavily glacierized river basins showed mainly positive runoff trends in the past few decades (simulated under different scenarios in the head rivers of the Tarim River basin), while those with less or no glacierization exhibited wide variations in runoff (Duethmann et al., 2015; Kaldybayev et al., 2016).

With further warming and the resulted acceleration of glacier retreat, glacier inflection points will or have already appeared. The amount of surface water will probably decline or keep high volatility due to glacial retreat and reduced storage capacity of glaciers (Chen et al., 2015). For instance, near-future runoffs are projected to increase to some extent, with increments of 13–35 % during 2011–2050 compared to 1960–2006 for the Yarkand River, −1–18 % in the 21st century compared to 1986–2020 under RCP4.5 for the Kaidu River, and 23 % in 2020 for the Hotan River (Table 2). For the long-term, however, total runoff is projected to be smaller than today. The hydrological responses to climate change around the world were discussed in Sect. 2.3.

2.2 Contribution of glacier melt/snowmelt water in river runoff

Kemmerikh (1972) estimated the contribution of groundwater, snowmelt and glacier melt to the total runoff of the alpine rivers in central Asia. Based on the hydrograph separation

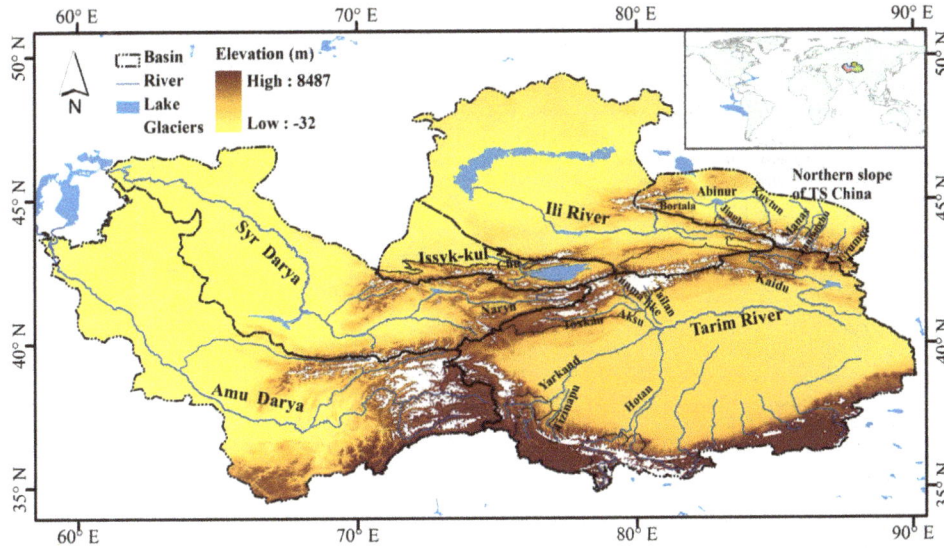

Figure 1. Map of central Asian headwaters with main river basins or hydrological regions, namely, the Tarim River basin, the watersheds in the northern slope of the Tienshan Mountains, the Issyk Lake basin, the Ili River basin, and the Amu Darya and the Syr Darya basins. Lake outlines are from Natural Earth (http://www.naturalearthdata.com/). The river system is derived based on elevations of the SRTM (Shuttle Radar Topography Mission) 90 m data. Glacier information was obtained from RGI (Randolph Glacier Inventory).

Table 1. Summary of climatic and underlying conditions of the basins. The topography is based on SRTM data, glacier data are from RGI (Randolph Glacier Inventory) and climate is based on the world map of the Köppen–Geiger climate classification. Vegetation is from the land use data from Xinjiang institute of Ecology and Geography.

Catchment	Tarim River basin	Catchments in northern Tienshan Mountains, China	Issyk Lake basin	Ili River basin	Amu Darya Basin	Syr Darya Basin
Location	Surrounded by the Tienshan Mountains and the Kunlun Mountains	Northern Tienshan	Western Tienshan	Western Tienshan Valley	Western Tienshan and Pamir	Western Tienshan
Topography						
Basin area (km^2)	868 811	126 463	102 396	429 183	674 848	442 476
Percentage of elevation > 3000 m (%)	28.00	13.80	14.50	4.60	20.50	9.50
Glaciation area (km^2)	15 789	1795	994	2170	9080	1850
Climate						
Dominant climate	arid cold	arid cold	arid cold; continental	arid cold; continental	arid cold; snow	arid cold
Vegetation						
Forest percent (%)	0.7	10.4	6.4	4.1	10.9	2.5
Pasture percent (%)	16.7	14.5	31.2	28.6	19.4	17.3
Percent of water, snow, ice (%)	5.4	3.9	7.8	5.3	5.3	2.8

methodology, the glacier melt contribution ranged between 5 and 40 % in the plains and around 70 % in upstream basins. The ratio of glacier melt contribution to runoff varies between 3.5 and 67.5 % with a mean of 24.0 % for the 24 catchments in the Tienshan Mountains based on hydrological modeling (Y. Zhang et al., 2016).

Distributed hydrological models provide a more useful tool for the investigation of changes in different runoff components. For example, the variable infiltration capacity (VIC)

Table 2. Summary of hydrological modeling in glacierized central Asian catchments.

Catchments	Models	Major conclusions	Innovations and limitations	References
Tarim River				
Tarim River basin	Modified two-parameter semi-distributed water balance model	Improved the original two-parameter monthly water balance model by incorporating the topographic indexes and could get comparable results to the TOPMODEL model and Xinanjiang model.	Less input data are required; Lack of glacier and snowmelt processes.	Peng and Xu (2010); Chen et al. (2006)
	TOPMODEL model	In the Aksu River, runoff was more closely related to precipitation, whereas in the Hotan River, it was more closely related to temperature.		
	Xinanjiang model	Runoffs of the Aksu, Yarkand and Hotan rivers exhibited increasing tendencies in 2010 and 2020 under different scenarios generated from the reference years, e.g., 23 % increase for the Hotan River.		
	PPR (Projection pursuit regression) model	If temperature rises 0.5–2.0 °C, runoff will increase with temperature for the Aksu, Yarkand and Hotan rivers.	Lack of physical basis.	Wu et al. (2003)
	VIC (variable infiltration capacity)	For the Tarim River, runoff will decrease slightly in 2020–2025 based on VIC forced by HadCM3 under A2 and B2 when not considering glacier melt.	Lack of glacier module.	Liu et al. (2010)
Tailan River	Modified degree-day model including potential clear sky direct solar radiation coupled with a linear reservoir model	Glacier runoff increases linearly with temperature over these ranges whether or not the debris layer is taken into consideration. The glacier runoff is less sensitive to temperature change in the debris-covered area than the debris-free area.	Considered the effect of solar radiation and quantified the debris effect.	Y. Zhang et al. (2007)
Aksu River including Kumalike and Toxkan rivers	Xinanjiang model	Precipitation has a weak relationship with runoff in the Kumalike River.	Joined the snowmelt module.	P. Wang et al. (2012)
			The model could not well capture the snowmelt-/precipitation-induced peak streamflow.	
	VIC-3L model	Glacier melt, snowmelt and rainfall accounted for 43.8, 27.7 and 28.5 % of the discharge for the Kumalike River and 23.0, 26.1 and 50.9 % for the Toxkan River.	The model performance was obviously improved through coupling a degree-day glacier melt scheme, but accurately estimating areal precipitation in alpine regions still remains.	Zhao et al. (2013, 2015)
		For the Kumalike River and the Toxkan River, the runoff has increased 13.6 and 44.9 % during 1970–2007, and 94.5 and 100 % of the increases were attributed by precipitation increase.		
		For the Kumalike River, glacier area will reduce by > 30 % resulting in decreased meltwater in summer and annual discharge (about 2.8–19.4 % in the 2050s).		
	SWIM model	The model is capable to reproduce the monthly discharge at the downstream gauge well, using the local irrigation information and the observed upstream inflow discharges.	Investigated the glacier lake outburst floods using a modeling tool. Inclusion of an irrigation module and a river transmission losses module of the SWIM model.	Huang et al. (2015); Wortmann et al. (2014)
		About 18 % of the incoming headwater resources consumed up to the gauge in Xidaqiao, and about 30 % additional water is consumed between Xidaqiao and Alar.	Model uncertainties are the largest in the snowmelt and glacier melt periods.	
		Different irrigation scenarios were developed and showed that the improvement of irrigation efficiency was the most effective measure for reducing irrigation water consumption and increasing river discharge downstream.		
	WASA model	Glacier melt contributes to 35–48 and 9–24 % for the Kumalike River and the Toxkan River.	The model considered changes in glacier geometry (e.g., glacier area and surface elevation).	Duethmann et al. (2015)
		For the Kumalike River, glacier geometry changes lead to a reduction of 14–23 % of streamflow increase compared to constant glacier geometry.	It used a multi-objective calibration based on glacier mass balance and discharge.	
	The temperature and precipitation revised AR(p) model;	The AR(p) model is capable of predicting the streamflow in the Aksu River basin while the NAM model is not ideal.	AR(p) needs less hydrological and meteorological data. Both model fails to model sudden floods such as ice dam collapse floods.	Ouyang et al. (2007)
	NAM (NedborAfstromnings Model) rainfall runoff model			

Table 2. Continued.

Catchments	Models	Major conclusions	Innovations and limitations	References
Tarim River				
Kaidu River	MIKE-SHE model	Compared remote sensing data and station-based data in simulating the hydrological processes. Remote sensing data are comparable to conventional data. Remote sensing data could partly overcome the lack of necessary hydrological model input data in developing or remote regions.	Missing glacier melt; Lack of observation to verify the meteorological condition in the mountainous regions.	T. Liu et al. (2012); Liu et al. (2013)
	HBV (Hydrologiska Byrans Vattenbalansavdelning) model	When the base runoff is $100\,\mathrm{m^3\,s^{-1}}$, the critical rainfall for primary and secondary warning floods are 50 and 30 mm, respectively, for the Kaidu River.	It underestimated the peak streamflow while overestimated the base flow.	Fan et al. (2014)
	SRM (snowmelt runoff model) including potential clear sky direct solar radiation and the effective active temperature.	Spring streamflow is projected to increase in the future based on HadCM3.	Limited observations resulted in low modeling precision. The APHRODITE precipitation performed well in hydrological modeling in the Kaidu River.	Y. C. Zhang et al. (2007); Ma et al. (2013); Li et al. (2014).
	SWAT	Precipitation and temperature lapse rates account for 64.0 % of model uncertainty.	Quantified uncertainty resulted from the meteorological inputs.	Fang et al. (2015a, b)
		Runoff increases (-1)–18 and 4–20 % in the 21st century under RCP4.5 and RCP8.5 compared to 1986–2005 based on a cascade of regional climate model RegCM, bias correction and SWAT model.		
	Modified system dynamics model	Simulations of low flow and normal flow are much better than the high flow, and spring peak flow is better than the summer pecks in the Kaidu River.	Applied the effective cumulative temperature to calculate snowmelt process and soil temperature for each layer to describe water movement in soil.	F. Y. Zhang et al. (2016)
Yarkand River	MIKE-SHE model	Simulated snowpack using station data differs significantly from that using remote sensing data.	Lack of glacier module	Liu et al. (2016b)
	Integrating Wavelet Analysis (WA) and back-propagation artificial neural network (BPANN)	Runoff presented an increasing trend similar with temperature and precipitation at the timescale of 32 years. But at the 2-, 4-, 8-, and 16-year timescale, runoff presented nonlinear variation.	Interpreted the nonlinear characteristics of the hydro-climatic process using statistic method.	Xu et al. (2014)
	Degree-day model	Decreasing rate of glacier mass was $4.39\,\mathrm{mm\,a^{-1}}$ resulting in a runoff increasing trend of $0.23 \times 10^8\,\mathrm{m^3\,a^{-1}}$ during 1961–2006. Sensitivity of mass balance to temperature is $0.16\,\mathrm{mm\,a^{-1}\,^{\circ}C^{-1}}$.	The glacier dynamics is considered and the area–volume scaling factor is calibrated using remote sensing data.	Xie et al. (2006); Zhang et al. (2012a, b)
		Glacier runoff will increase 13–35 % during 2011–2050 compared to 1960–2006 with obvious increase in summer.		
Tizinapu	SRM including snow albedo	It could well simulate the runoff of the Tizinapu River.	Lack of glacier module.	Li and Williams (2008)
		Runoff is dominated by precipitation and temperature lapse rates, and snow albedo.		
Hotan River	Integrating Wavelet Analysis (WA) and back-propagation artificial neural network (BPANN)	Runoff correlates well with the 0 °C level height in summer for the northern slope of Kunlun Mountains.	Interpreted the nonlinear characteristics of the hydro-climatic process.	Xu et al. (2011)
Catchments in northern slope of Tienshan Mountains, China				
Manas River	(1) SWAT model	Glacier area decreased by 11 % during 1961–1999 and glacier melt contributes 25 % of discharge.	Both the glacier melt module and two-reservoir method were included in the hydrological simulations.	Yu et al. (2011); Luo et al. (2012); Luo et al. (2013); Gan and Luo (2013)
	(2) SRM model	Better simulation of snowmelt runoff than rainfall–runoff by the SRM.	Snow cover calculation algorithm is added to validate model performance.	Yu et al. (2013)
	(3) EasyDHM model	EasyDHM model could reproduce the streamflow.	The validation is based on streamflow alone.	Xing et al. (2014)
Urumqi River	(1) Isotope hydrograph separation (IHS)	Glacier meltwater contributes to 9 % of runoff.	The IHS method has overwhelming potential in analyzing hydrological components for ungauged watersheds.	Kong and Pang (2012)
	(2) Water balance model	The cumulative mass balance of the glacier was -13.69 m during 1959–2008; proportion of glacier runoff increased from 62.8 to 72.1 %.	Foused on runoff generation on the glacierized and ablation area.	Sun et al. (2013)

Table 2. Continued.

Catchments	Models	Major conclusions	Innovations and limitations	References
Catchments in northern slope of Tienshan Mountains, China				
	(3) HBV model	For a glacierized catchment (glacierization ratio is 18 %), the discharge will increase by 66 ± 35 % or decrease by 40 ± 13 % if the glacier size keeps unchanged or glacier disappears in 2041–2060.	Considering future runoff under different glacier change scenarios.	Sun et al. (2015)
	(4) Exponential regression	Glacier runoff is critically affected by the ground temperature.	This study shed light on glacier runoff estimation based on ground temperature for data-scarce regions.	Chen et al. (2012)
	(5) SRM model	The degree day factor is not constant for different elevation bands.	Calculated the curve of snow cover shrinkage based on MODIS data.	Huai et al. (2013)
	(6) THModel (Thermodynamic Watershed Hydrological Model)	THModel can indeed simulate runoff processes in the glacier and snow-dominated catchment reasonably well.	An energy balance model is proposed to close the balance equation of soil freezing and thawing.	Mou et al. (2008)
Ebinur Lake catchment including Jinghe River, Kuytun River and Bortala River	(1) SWAT model and the sequential cluster method	For the Jinghe River, 85.7 % of the runoff reduction is caused by human activity and 14.3 % by climate change.	Identified the effects of human activities and climate change on runoff.	Dong et al. (2014); Yao et al. (2014)
	(2) Runoff CAR (Controlled AutoRegressive) model	The Jinghe River and Kuytun River exhibited a slightly increasing trend, but an adverse trend in the Bortala River. In a warm humid scenario, runoff in the Jinghe River and Bortala River will increase while it will decrease in the Kuytun River.	The CAR is based on past and present values without physical basis.	
Juntanghu Basin	DHSVM (Distributed Hydrology Soil Vegetation Model)	The coupled WRF (Weather Research and Forecasting) modelling system and DHSVM model could predict 24 h snowmelt runoff with relative error within 15 %.	MODIS snow cover and the calculated snow depth data are used in the snowmelt runoff modeling.	Zhao et al. (2009)
Issyk Lake Basin				
Small rivers around the Issyk Lake	Degree-day approach	Runoff contribution is varying in a broad range depending on the degree of glacierization in the particular sub-catchment. All rivers showed a relative increase in annual river runoff ranging between 3.2 and 36 %.	The glacier melt runoff fraction at the catchment outlet can be considerably overestimated.	Dikich and Hagg (2003)
Chu River	SWAT-RSG (RSG: rain, snow and glacier) model	General decrease was expected in glacier runoff (−26.6 to −1.0 %), snowmelt (−21.4 to +1.1 %) and streamflow (−27.7 to −6.6 %); Peak streamflow will be put forward for 1 month.	Use the glacier dynamics and assessed the model performance based on both streamflow and glacier area.	Ma et al. (2015)
Ili River basin				
Gongnaisi River	SRM model	Runoff is sensitive to snow cover area and temperature.	SRM is capable to model the snowmelt runoff.	Ma and Cheng (2003)
		If temperature increases 4 °C, the runoff will decrease by 9.7 % with snow coverage and runoff shifting forward.		
Tekes River	SWAT model	Glaciers have retreated about 22 % since 1970s, which was considerably higher than the Tienshan average (4.7 %) and China average (11.5 %), resulting in a decrease of proportion of precipitation recharged runoff from 9.8 % in 1966–1975 to 7.8 % in 2000–2008.	Using two land use data and two Chinese glacier inventories, the model could well reproduce streamflow.	Xu et al. (2015)
Ili River	DTVGM (multi-spatial data-based Distributed Time-Variant Gain Model) model	Daily runoff correlated closely with snowmelt, suggesting a snowmelt module is indispensable.	This method has less dependence on conventional observation.	Cai et al. (2014)
	Water balance model	Water decrease in 1911–1986 in the middle and lower reaches of the Lake Balkhash is due to decreased rainfall and reservoirs storage.	–	Kezer and Matsuyama (2006); Guo et al. (2011)
Amu Darya and Syr Darya basins				
Amu Darya and Syr Darya	STREAM	The runoff of the Syr Darya declined considerably over the last 9000 years, but show much smaller responses to future warming.	Simulated long-term discharge for the Holocene and future period.	Aerts et al. (2006)
		For the Amu Darya and Syr Darya basins, the glacier-covered areas have decrease 15 and 22 % in 2001–2010 compared to the baseline (1960–1990).	The model includes the calculation of rainwater, snowmelt water and glacier runoff (based on the glacier altitude and equilibrium lime altitude).	Savoskul et al. (2003); Savoskul et al. (2004); Savoskul and Smakhtin (2013)
		For the Amu River Basin, 20–25 % of the glaciers will retain under a temperature increment being 4–5 °C and precipitation increase rate being 3 %/°C. For the Syr Darya, runoff under the A2 and B2 scenarios will increase 3–8 % in 2010–2039, with sharpened spring peak and a slight lowered runoff from late June to August.		

Table 2. Continued.

Catchments	Models	Major conclusions	Innovations and limitations	References
Amu Darya and Syr Darya basins				
	AralMountain model	For the Amu Darya, glacier melt and snowmelt contribute to 38 and 26.9 % of runoff, while for the Syr Darya the proportions are 10.7 and 35.2 %.	Fully simulated the hydrological processes.	Immerzeel et al. (2012a)
		Glacier will retreat by 46.4–59.5 % by 2050 depending on selected GCM (General Circulation Model). For the Syr Darya, average water supply to the downstream will decrease by 15 % for 2021–2030 and 25 % for 2041–2050. For the Amu Darya the expected decreases are 13 % (2021–2030) and 31 % (2041–2050).		
Test sites "Abramov" in SyrDarya and "Oigaing"in Amu Darya	HBV-ETH (HBV model expanded at the Swiss Federal Institute of Technology) and OEZ (a water balance equation model)	Overall good model performances were achieved with the maximum discrepancy of simulated and observed monthly runoff within 20 mm.	It considered geographical, topographical and hydrometeorological features of test sites, and reduced modeling uncertainties.	Hagg et al. (2007)
		General enhanced snowmelt during spring and a higher flood risk in summer are predicted under a doubling atmospheric CO_2 concentration with greatest runoff increases occurring in August for the highly glaciated catchments and in June for the nival catchment.	This procedure requires a lot meteorological and land surface data and knowledge of the hydrological processes.	
Panj River	HBV-ETH	For the upper Panj catchment, the current glacier extent will decrease by 36 and 45 %, respectively, assuming temperature increment being 2.2 and 3.1 °C.	Application of glacier parameterization scheme.	Hagg et al. (2013)
Naryn River	SWAT-RSG model	Glacier area has decreased 7.3 % during 1973–2002.	Incorporated glacier dynamics and validated the model using two glacier inventories.	Gan et al. (2015)
		Glaciers will recede with only 8 % of the small glaciers retain by 2100 under RCP8.5 and net glacier melt runoff will reach peak in about 2040 and decrease later.		
Syr Darya	NAM model with a separate land-ice model	Glacier volume will lose 31 % ± 4 % under SRES A2 until 2050s, and the runoff peak will shift forward by 30–60 days from the current spring/early summer towards a late winter/early spring runoff regime.	The NAM model was improved to be robust using only five freely calibrated parameters.	Siegfried et al. (2012)

model was used to calculate the components of runoff in the source river for the Tarim River. The results showed that, in terms of runoff, glacier meltwater, snowmelt water and rainfall accounted for 43.8, 27.7 and 28.5 % of the Kumalike River, and 23.0, 26.1 and 50.9 % of the Toxkan River, respectively (Zhao et al., 2013); this result is comparable to the conclusion that glacier melt accounts for 31–36 % based on isotope tracer (Sun et al., 2016). However, accurately quantifying the contributions of glacier melt, snowmelt and rainfall to runoff in central Asian streams is challenging (Unger-Shayesteh et al., 2013).

2.3 Glacio-hydrological responses to climate change: a comparison

To analyze the hydrological responses to climate change of the glacierized Tienshan Mountains, the responses of several major glacierized mountainous regions are discussed. For the Himalaya–Hindu Kush region, investigations suggested that a regression of the maximum spring streamflow period in the annual cycle of about 30 days, and annual runoff decreased by about 18 % for the snow-fed basin, whereas it increased by about 33 % for the glacier-fed basin using the Satluj Basin as a typical region (Singh and Bengtsson, 2005). For the Tibetan Plateau, the glacier retreat could lead to an expansion of lakes; e.g., glacier mass loss between 1999 and 2010 con-

tributed to about 11.4–28.7 % of the lake level rise in the three glacier-fed lakes, namely, Siling Co, Nam Co and Pung Co (Lei et al., 2013). Analysis from groundwater storage indicated that the groundwater for the major basins in the Tibetan Plateau increased during 2003–2009 with a trend rate of $+1.86 \pm 1.69 \, \mathrm{Gt \, yr^{-1}}$ for the Yangtze River Source Region and $+1.14 \pm 1.39 \, \mathrm{Gt \, yr^{-1}}$ for the Yellow River Source Region (Xiang et al., 2016).

For the South American Andes, melting at the glacier summit has occurred. With the continually increase in temperature, although glacier melt was dominated by maybe other processes in some regions, the probability seems high that the current glacier melting will continue. With the loss of glacier water, the current dry-season water resources will be heavily depleted once the glaciers have disappeared (Barnett et al., 2005).

For the Alps, many investigations have been implemented, ranging from glacier-scale modeling to large basin-scale or region-scale modeling (Finger et al., 2015; Abbaspour et al., 2015). Glacier meltwater provided about $5.28 \pm 0.48 \, \mathrm{km^3 \, a^{-1}}$ of freshwater during 1980–2009. About 75 % of this volume occurred during July–September, providing water for large low-lying rivers including the Po, the Rhine and the Rhône (Farinotti et al., 2016). Under the context of climate change, decreases of glacier meltwater in both annual and summer runoff contributions are anticipated. For example, annual runoff contributions from presently glacierized surfaces are expected to decrease by 13 % by 2070–2099 compared to 1980–2009, despite of nearly unchanged contributions from precipitation under RCP 4.5 (Farinotti et al., 2016).

The hydrological processes in the glacierized regions have something in common; i.e., the annual runoff is likely to reduce in a warming climate with high spatial–temporal variation at the middle or end of the 21st century. Seasonally, increased snowmelt runoff and water shortage of summer runoff with the disappearing glaciers are expected. However, there are also differences in the responses of hydrological processes to climate change. For example, the contrasting climate change impact on river flows from glacierized catchments in the Himalayas and Andes (Ragettli et al., 2016). In the Langtang catchment in Nepal, increased runoff is expected with limited shifts between seasons, whereas for the Juncal catchment in Chile, the runoff has already been decreasing. These qualitative or quantitative differences are mainly caused by glaciation ratio, regional weather pattern and glacier property (Hagg and Braun, 2005).

However, for many glacierized catchment in the Tienshan Mountains, currently or for the next several decades, the runoff appears to be normal or even an increasing trend, giving an illusion of better prospects. It is particularly worth mentioning that, once the glacier storage (fossil water) melts away, the water system is likely to go from plenty to want, exacerbating water stress given the increasing water demand.

3   Limitations of the available hydrological models

3.1   Meteorological inputs in hydrological modeling and prediction

In mountainous regions of central Asia, meteorological input uncertainty could account for over 60 % of model uncertainty (Fang et al., 2015a). The greatest challenge in hydrological modeling has been lack of robust and reliable complete meteorological data, especially since the collapse of the Soviet Union in the late 1980s. In this section, the value and limitations of different data sets used in hydrological modeling (e.g., station data, remote sensing data) and future predictions (e.g., outputs of GCMs (General Circulation Models) and RCMs (Regional Climate Models) are discussed.

3.1.1   Observational data

Traditionally, hydrological models are forced by station-scale meteorological data in or near the studied watershed (e.g., Fang et al., 2015a; Peng and Xu, 2010). However, station-scale data can only describe the climate at a specific point in space, and most of them located at the foot of mountains. This limitation needs to be taken into consideration when interpolating station data into basin-scale over rugged terrain. Li et al. (2014) applied the interpolated gridded pre-

cipitation data set (APPRODITE) to force the SRM model. Applying in situ observational meteorological data is also associated with other challenges, as detailed below.

Lack of stations

One of the greatest challenges inherent in station-scale meteorological data is the low density of meteorological stations. As the mountainous regions of central Asia are characterized by complex terrain, it is inaccurate to represent the climatic conditions of basins using data from limited stations. Some researchers (Liu et al., 2016b; Fang et al., 2015a) have addressed this challenge by attempting to interpolate temperature/precipitation into a basin scale using elevation bands, based on the assumption that climate variables increase or decrease with elevation. Temperature lapse rates could also be validated using the Integrated Global Radiosonde Archive (IGRA) data set (Li and Williams, 2008). However, this modification could not take account of the source of water vapor and mountain aspect for basins with complex landform. Due to the fact that uniform precipitation gradients cannot be derived and temperature lapse rates are not constant throughout the year (Immerzeel et al., 2014), it is a challenge to use elevation bands to interpolate station-scale climate into basin-scale climate.

Lack of homogeneity test

Many hydrological modeling studies do not factor in errors in observations, even though homogeneous climate records are required in hydrological design. In central Asia, changes in regulation protocols or relocation of stations also lead to observational errors. Checking the input data should be the first step in hydrological modeling due to the rule of "garbage in yields garbage out".

3.1.2   Remote sensing data and reanalysis data

Remote sensing and reanalysis data are increasingly being used in hydrological modeling. T. Liu et al. (2012) and Liu et al. (2016b) evaluated remote sensing precipitation data of the Tropical Rainfall Measuring Mission (TRMM) and temperature data of Moderate Resolution Imaging Spectroradiometer (MODIS). The results indicated that snow storage and snowpack that were modeled using the remote sensing climate are different from those modeled using station-scale observational data. The model forced by the remote sensing data showed better performance in spring snowmelt (T. Liu et al., 2012). Huang et al. (2010a) analyzed the input uncertainty of remote sensing precipitation data interpreted from FY-2. In addition to meteorological data, surface information interpreted from satellite images, e.g., soil moisture, land use and snow cover, can also be used in hydrologic modeling (Cai et al., 2014).

As demonstrated in numerous research studies, data assimilation holds considerable potential for improving hydrological predictions (Y. Liu et al., 2012). Cai et al. (2014) used Global Land Data Assimilation System (GLDAS) 3 h air temperature data to force the MS-DTVGM model, while Duethmann et al. (2015) used the Watch Forcing Data based on ERA-40 (WFD-E40) to force the hydrological model.

Remote sensing and reanalysis data are supposed for use in large-scale hydrological modeling due to their low spatial resolution. Another limitation in using remote sensing and reanalysis data is that these data are biased to some extent. For example, the TRMM data are mostly valuable only for tropical regions, and reanalysis data, including ERA-40, NCEP/NCAR and GPCC (Global Precipitation Climatology Centre), fail to reveal any significant correlation with station data (Sorg et al., 2012).

Given the advantages and disadvantages of observation data, remote sensing data and reanalysis data, a better approach would be to combine observations and other data sets in hydrological modeling.

3.1.3 GCM or RCM outputs

GCMs or RCMs provide climate variables for evaluating future hydrological processes. However, the greatest challenges in applying these data sets are their low spatial resolutions (e.g., the spatial resolution of GCMs in CMIP5 ranges from 0.75 to 3.25°) and considerable biases. In addition, different GCMs or RCMs generally give different climate projections. Therefore, when forcing a hydrological model using the outputs of climate models, the evaluation results depend heavily on the selection of GCMs and consequently result in higher uncertainty in GCMs than that in other sources (emission scenarios, hydrological models, downscaling, etc.) (Bosshard et al., 2013).

Many downscaling methods have been developed to overcome these drawbacks. Although some statistical downscaling methods, such as SDSM (Wilby et al., 2002), are widely used in climate change impact studies, their use in the mountainous regions of central Asia is limited due to the lack of fine observational data to downscale GCM outputs. To overcome the data scarcity for this region, G. H. Fang et al. (2015) evaluated different bias correction methods in downscaling the outputs of one RCM model and used the bias-corrected climate to force a hydrological model in the data-scarce Kaidu River basin. Liu et al. (2011) used perturbation factors to downscale the GCM outputs and force the hydrological model.

3.2 Glacier melt modeling

Glacier melt accounts for a large part of the discharge for the alpine basins in central Asia as discussed above. However, most hydrological modeling does not include glacier melt and accumulation processes. For example, Liu et al. (2010) failed to account for the glacier processes in the VIC model in the Tarim River; Peng and Xu (2010) missed the

glacier module in Xinanjiang and TOPMODEL; and Fang et al. (2015a) failed to account for glacier processes, though the glacier melt could contribute up to 10 % of discharge of the Kaidu River basin. Similarly, in their research on the Yarkand River basin, Liu et al. (2016a) neglected the influence of glacier melt in the SWAT and MIKE-SHE models, even though the glacier covered an area of $5574\,km^2$. The most widely used hydrological models, such as the distributed SWAT, the MIKE-SHE model and the conceptual SRM model, as a rule do not calculate glacier melt processes, despite the fact that excluding the glacier processes could induce large errors in glacierized catchments. Glacier processes are complex in that glacier melt will at first increase due to the rise in ablation and lowering of glacier elevation, and then, after reaching its peak, will decrease due to the shrinking in glacier area (Xie et al., 2006). Moreover, simulation errors can be re-categorized as precipitation or glacier meltwater and consequently result in a greater uncertainty in the water balance in high mountain areas (Mayr et al., 2013).

During the last few decades, a large variety of melt models have been developed (Hock, 2005). Previous studies have investigated glacier dynamics for the mountainous regions. Among these studies, Hock (2005) reviewed glacier melt-related processes at the surface–atmosphere interface ranging from a simple temperature-index model to a sophisticated energy-balance model. Glacier models that are physically based (e.g., mass-energy fluxes and glacier flow dynamics) depend heavily on detailed knowledge of local topography and hydrometeorological data, which are generally limited in high mountain regions (Michlmayr et al., 2008). Hence, they mostly applied to well-documented glaciers and have few applications in basin-scale hydrological models.

The temperature-index method (or its variants), which only requires temperature for meteorological input, is widely used to calculate glacier melt (Konz and Seibert, 2010). As is illustrated by Oerlemans and Reichert (2000), glaciers can be reconstructed from long-term meteorological records, e.g., summer temperature is the dominant factor for glaciers in a dry climate (e.g., Abramov glacier). In recent years, hydrologists have been trying to add other meteorological variables into the calculations of glacier melt; e.g., Y. Zhang et al. (2007) included potential clear sky direct solar radiation in the degree-day model, and Yu et al. (2013) stated that accumulated temperature is more effective than daily average temperature for calculating the snowmelt runoff model. Using degree-day calculation is much simpler than using energy balance approaches and could actually produce comparable or better model performance when applied in mountainous basins (Ohmura, 2001).

More recently, the melt module has been incorporated into different kinds of hydrological models. Zhao et al. (2015) integrated a degree-day glacier melt algorithm into a macroscale hydrologic model (VIC) and indicated that annual and summer runoff would decrease by 9.3 and 10.4 %, respectively, for reductions in glacier areas of 13.2 % in the

Kumalike River basin. Hagg et al. (2013) analyzed anticipated glacier and runoff changes in the Rukhk catchment of the upper Amu Darya Basin, using the HBV-ETH model by including glacier melt and snowmelt processes. Their results showed that with temperature increases of 2.2 and 3.1 °C, the current glacier extent of 431 km^2 will reduce by 36 and 45 %, respectively. Luo et al. (2013), taking the Manas River basin as a case study, investigated glacier melt processes by including the algorithm of glacier melt, sublimation/evaporation, accumulation, mass balance and retreat in a SWAT model. The results showed that glacier melt contributed 25 % to streamflow, although the glacier area makes up only 14 % of the catchment drainage area.

3.2.1 Paucity of glacier variation data

The existing glacier data set, which includes the World Glacier Inventory (WGI), the Randolph Glacier Inventory (RGI) and global land-ice measurements from space (GLIMS), has been developed rapidly. These data, however, generally focus on glaciers in the present time or those existing in the former Soviet Union. For example, the source data of WGI were derived during 1940s–1960s, and the GLIMS for the Amu Darya Basin is from 1960 to 2004 (Donald et al., 2015). These data can depict the characteristics of the glacier status, but fail to reproduce glacier variation. Only a few glaciers (Abramov, Tuyuksu, Urumqi no. 1 Glacier, etc.) have long-term variation measurements (Savoskul and Smakhtin, 2013). CAWa (Central Asian Water; www.cawa-project.net/) are intended to contribute to a reliable regional data basis of central Asia from the monitoring stations, sampling and remote sensing. The missing glacier variation information leads to a misrepresentation of glacier dynamics.

3.2.2 Lack of glacier mass balance data

Glacier measurements reproduced by remote sensing data usually give glacier area instead of glacier water equivalent; therefore, errors will occur when converting glacier area to glacier mass. Glaciologists normally use a specified relation (e.g., empirical) between glacier volume and glacier area to estimate glacier mass balance (Stahl et al., 2008; Luo et al., 2013). Aizen et al. (2007) applied the radio-echo sounding approach to obtain glacier ice volume. Recently, ICESat (Ice, Cloud and land Elevation Satellite; http://icesat.gsfc.nasa.gov/) could provide multi-year elevation data needed to determine ice-sheet mass balance.

This paper focuses primarily on glacier melt modules. It does not discuss snowmelt processes, as hydrological models generally include them either in a degree-day approach or energy balance basis. Furthermore, this paper does not analyze water routing processes or evapotranspiration because there are several ways to simulate soil water storage change and model evapotranspiration (Bierkens, 2015).

3.3 Model calibration and validation

For model calibration, two important issues are discussed here: the length of the calibration period and objective functions.

Generally, hydrological modeling requires several years' calibration. For example, Yang et al. (2012) indicated that a 5-year warm-up is sufficient before hydrological model calibration and a 4-year calibration could obtain satisfactory model performance. More venturesomely, a 6-month calibration could lead to good model performance for an arid watershed (Sun et al., 2016). Konz and Seibert (2010) stated that one year's calibration of using glacier mass balances could effectively improve the hydrological model. Selecting the appropriate calibration period is significant, as model performance could depend on calibration data. Refsgaard (1997) used a split-sample procedure to obtain better model calibration and validation effectively and efficiently.

Most studies on calibration procedures in hydrology have examined goodness-of-fit measures based on simulated and observed runoff. However, as the hydrological sciences develop further, multi-objective calibration is emerging as the preferred approach. It not only includes multi-site streamflow (which has proved to be advantageous compared to single-site calibration (S. Wang et al., 2012), and multi-metrics of streamflow (Yang et al., 2014), but also involves multiple examined hydrological components (e.g., soil moisture). Most of the studies reviewed here use the discharge to calibrate and validate the hydrological model, yet Gupta et al. (1998) argued that a strong "equifinality effect" may exist due to the compensation effect, where an underestimation of precipitation may be compensated by an overestimation of glacier melt, and vice versa. Stahl et al. (2008) suggested that observations on mass balances should be used for model calibration, as large uncertainties exist in the data-scarce alpine regions. Therefore, multi-criteria calibration and validation is necessary, especially for glacier/snow recharged regions.

Many recent studies have attempted to include mass balance data into model calibration (Stahl et al., 2008; Huss et al., 2008; Konz and Seibert, 2010; Parajuli et al., 2009). Duethmann et al. (2015) used a multi-objective optimization algorithm that included objective functions of glacier mass balance and discharge to calibrate the hydrological model WASA (Model of Water Availability in Semi-Arid Environments). Another approach for improving model efficiency is to calibrate the glacier melt processes and the precipitation dominated processes separately (Immerzeel et al., 2012b). Further, in addition to the mass balance data used to calibrate the hydrological model, the glacier-area/glacier-volume scaling factor can also be calibrated with the observed glacier area change monitored by remote sensing data (Zhang et al., 2012b).

4 Future challenges and directions

Modeling hydrological processes and understanding hydrological changes in mountainous river basins will provide important insights into future water availability for downstream regions of the basins. In modeling the glacierized catchments of central Asia, the greatest challenge still remains the lack of reliable and complete data, including meteorological data, glacier data and surface conditions. This challenge is very difficult to overcome due to the inaccessibility of the terrain and the oftentimes conflicting politics of the countries that share the region. Even so, future efforts could be focused on constructing additional stations and doing more observations (e.g., the AKSU-TARIM project; http://www.aksu-tarim.de/).

For alpine basins with scarce data, knowledge about water generation processes and the future impact of climate change on water availability is also poor. Moreover, the contribution of glacier melt varies significantly among basins and even along river channels, adding even more complexity to hydrological responses to climate change.

Uncertainty should always be analyzed and calculated in hydrological modeling, especially when evaluating climate change impact studies that contain a cascade of climate models, downscaling, bias correction and hydrological modeling, whose uncertainties are currently insufficiently quantified (Johnston and Smakhtin, 2014). The evaluation contains uncertainty in each part of the cascade, such as climate modeling uncertainty or hydrological modeling uncertainty (i.e., input uncertainty, structure or module uncertainty and parameter uncertainty), all of which could lead to a considerably wide bandwidth compared to the changes of the water resources. In contrast, by taking into account all of these uncertainties, reliable evaluation of model confidence could be acquired by decision-makers and peers.

4.1 Publication of model setups and input data

As was suggested by Johnston and Smakhtin (2014), publication of model setups and input data is necessary for other researchers to replicate the modeling or build coherent nested models. From these setups and data, researchers can build their own models from existing work rather than starting from scratch. Another advantage of researchers sharing their work is to help each other evaluate existing models from other viewpoints.

4.2 Integration of different data sources

After appropriate preprocessing, several types of data, including remote sensing and reanalysis, could be used in hydrological modeling, as Liu et al. (2013) indicated that remote sensing data could reproduce comparable results to the traditional station data. In recent years, isotope data are increasingly used to define water components (Sun et al., 2016)

and it would be a fortune for hydrologists to validate their models, or even calibrate the models (Fekete et al., 2006). The overall idea here is to build and integrate more comprehensive data sets in order to improve hydrological modeling. An example of this approach can be found in Naegeli et al. (2013), who attempted to construct a worldwide data set of glacier thickness observations compiled entirely from a literature review.

4.3 Multi-objective calibration and validation

A hydrological model should not only "mimic" observed discharge but also well reproduce snow accumulation and melt dynamics or the glacier mass change (e.g., Konz and Seibert, 2010). As discussed previously, hydrological models that are calibrated based on discharge alone may be of high uncertainty and even "equifinality" for different parameters or inputs. This could happen especially when one or several modules are missing. For example, one might overestimate the mountainous precipitation or underestimate the evapotranspiration if the glacier melt module is missing. Therefore, it is suggested to account for each hydrological component as much as possible. We strongly suggest the use of multi-objective functions and multi-metrics to calibrate and evaluate hydrological models. Compared to single objective calibration, which was dependent on the initial starting location, multi-objective calibration provides more insight into parameter sensitivity and helps to understand the conflicting characteristics of these objective functions (Yang et al., 2014). Therefore, the use of different kinds of data and objective functions could improve a hydrological model and provide more realistic results.

For the data-scarce Tienshan Mountains, however, we do not recommend an overcomplex or physicalized modeling of each component as lack of validation data, which may result in equifinality discussed previously under stable climate. The more empirical models (enhanced temperature-index approaches) could reproduce comparable results to the sophisticated, fully physically based models (Hock, 2005). It is worth mentioning that the physically based glacier models are more advanced when quantify future dynamics of glaciers and glacier/snow redistribution when the climatic and hydrologic systems are not stable (Hock, 2005). The physical models should be further developed and used in glacier modeling as long as there is enough input and validation data.

Having a reliable hydrological model is important for understanding and modeling water changes, which are key issues of water resources management. The developments and associated challenges described in this paper are extrapolations of current trends and are likely to be the focus of research in the coming decades.

5   Data availability

As a review article, this paper does not include any research data. All the cited references could be found through the Internet.

Author contributions. Yaning Chen and Weihong Li wrote the main manuscript text; Gonghuan Fang and Zhi Li prepared Fig. 1 and gave some assistance to paper searching and reviewing. All authors reviewed the manuscript.

Acknowledgements. The research is supported by the National Natural Science Foundation of China (41630859; 41471030) and the CAS "Light of West China" Program (2015-XBQN-B-17).

Edited by: Q. Chen

References

Abbaspour, K. C., Rouholahnejad, E., Vaghefi, S., Srinivasan, R., Yang, H., and Kløve, B.: A continental-scale hydrology and water quality model for Europe: Calibration and uncertainty of a high-resolution large-scale SWAT model, J. Hydrol., 524, 733–752, doi:10.1016/j.jhydrol.2015.03.027, 2015.

Aerts, J., Renssen, H., Ward, P., De Moel, H., Odada, E., Bouwer, L., and Goosse, H.: Sensitivity of global river discharges under Holocene and future climate conditions, Geophys. Res. Lett., 33, L19401, doi:10.1029/2006GL027493, 2006.

Aizen, V., Aizen, E., and Kuzmichonok, V.: Glaciers and hydrological changes in the Tien Shan: simulation and prediction, Environ. Res. Lett., 2, 045019, doi:10.1088/1748-9326/2/4/045019, 2007.

Aizen, V. B., Aizen, E. M., Melack, J. M., and Dozier, J.: Climatic and hydrologic changes in the Tien Shan, central Asia, J. Climate, 10, 1393–1404, 1997.

Barnett, T. P., Adam, J. C., and Lettenmaier, D. P.: Potential impacts of a warming climate on water availability in snow-dominated regions, Nature, 438, 303–309, 2005.

Bierkens, M. F. P.: Global hydrology 2015: State, trends, and directions, Water Resour. Res., 51, 4923–4947, doi:10.1002/2015WR017173, 2015.

Bosshard, T., Carambia, M., Goergen, K., Kotlarski, S., Krahe, P., Zappa, M., and Schär, C.: Quantifying uncertainty sources in an ensemble of hydrological climate-impact projections, Water Resour. Res., 49, 1523–1536, doi:10.1029/2011WR011533, 2013.

Cai, M., Yang, S., Zeng, H., Zhao, C., and Wang, S.: A distributed hydrological model driven by multi-source spatial data and its application in the Ili River Basin of Central Asia, Water Resour. Manag., 28, 2851–2866, 2014.

Chen, R., Qing, W., Liu, S., Han, H., He, X., Wang, J., and Liu, G.: The relationship between runoff and ground temperature in glacierized catchments in China, Environ. Earth Sci., 65, 681–687, 2012.

Chen, Y.: Water resources research in Northwest China, Springer Science & Business Media, doi:10.1007/978-94-017-8017-9, 2014.

Chen, Y., Takeuchi, K., Xu, C. C., Chen, Y. P., and Xu, Z. X.: Regional climate change and its effects on river runoff in the Tarim Basin, China, Hydrol. Process, 20, 2207–2216, doi:10.1002/hyp.6200, 2006.

Chen, Y., Li, Z., Fan, Y., Wang, H., and Deng, H.: Progress and prospects of climate change impacts on hydrology in the arid region of northwest China, Environ. Res., 139, 11–19, 2015.

Dikich, A. and Hagg, W.: ABHANDLUNGEN-Climate driven changes of glacier runoff in the Issyk-Kul Basin, Kyrgyzstan, Zeitschrift fur Gletscherkunde und Glazialgeologie, 39, 75–86, 2003 (in Russian).

Donald, A., Ulrich, K., and Caleb, P.: The Role of Glaciers in the Hydrologic Regime of the Amu Darya and Syr Darya Basins, Washington, D.C., 2015.

Dong, W., Cui, B., Liu, Z., and Zhang, K.: Relative effects of human activities and climate change on the river runoff in an arid basin in northwest China, Hydrol. Process., 28, 4854–4864, 2014.

Duethmann, D., Bolch, T., Farinotti, D., Kriegel, D., Vorogushyn, S., Merz, B., Pieczonka, T., Jiang, T., Su, B. D., and Guntner, A.: Attribution of streamflow trends in snow and glacier melt-dominated catchments of the Tarim River, Central Asia, Water Resour. Res., 51, 4727–4750, doi:10.1002/2014wr016716, 2015.

Fan, J., Jiang, Y., Chen, Y., Chen, P., Bai, S., and Yu, X.: The Critical Rainfall Calculation in Kaidu River Based on HBV Hydrological Model, Desert and Oasis Meteorology, 8, 31–35, 2014.

Farinotti, D., Longuevergne, L., Moholdt, G., Duethmann, D., Mölg, T., Bolch, T., Vorogushyn, S., and Güntner, A.: Substantial glacier mass loss in the Tien Shan over the past 50 years, Nat. Geosci., 8, 716–722, doi:10.1038/ngeo2513, 2015.

Farinotti, D., Pistocchi, A., and Huss, M.: From dwindling ice to headwater lakes: could dams replace glaciers in the European Alps?, Environ. Res. Lett., 11, 054022, doi:10.1038/NGEO2513, 2016.

Fang, G., Yang, J., Chen, Y., Xu, C., and De Maeyer, P.: Contribution of meteorological input in calibrating a distributed hydrologic model in a watershed in the Tianshan Mountains, China, Environ. Earth Sci., 74, 2413–2424, doi:10.1007/s12665-015-4244-7, 2015a.

Fang, G., Yang, J., Chen, Y., Zhang, S., Deng, H., Liu, H., and De Maeyer, P.: Climate Change Impact on the Hydrology of a Typical Watershed in the Tianshan Mountains, Advances in Meteorology, 2015, 1–10, doi:10.1155/2015/960471, 2015b.

Fang, G. H., Yang, J., Chen, Y. N., and Zammit, C.: Comparing bias correction methods in downscaling meteorological variables for a hydrologic impact study in an arid area in China, Hydrol. Earth Syst. Sci., 19, 2547–2559, doi:10.5194/hess-19-2547-2015, 2015.

Fekete, B. M., Gibson, J. J., Aggarwal, P., and Vörösmarty, C. J.: Application of isotope tracers in continental scale hydrological modeling, J. Hydrol., 330, 444–456, doi:10.1016/j.jhydrol.2006.04.029, 2006.

Finger, D., Vis, M., Huss, M., and Seibert, J.: The value of multiple data set calibration versus model complexity for improving the performance of hydrological models in mountain catchments, Water Resour. Res., 51, 1939–1958, 2015.

Gan, R. and Luo, Y.: Using the nonlinear aquifer storage-discharge relationship to simulate the base flow of glacier- and snowmelt-dominated basins in northwest China, Hydrol. Earth Syst. Sci., 17, 3577–3586, doi:10.5194/hess-17-3577-2013, 2013.

Gan, R., Luo, Y., Zuo, Q. T., and Sun, L.: Effects of projected climate change on the glacier and runoff generation in the Naryn River Basin, Central Asia, J. Hydrol., 523, 240–251, doi:10.1016/j.jhydrol.2015.01.057, 2015.

Guo, L., Xia, Z., and Wang, Z.: Comparisons of hydrological variations and environmental effects between Aral Sea and Lake Balkhash, Adv. Water Sci., 22, 764–770, 2011.

Guo, L., Xia, Z., Zhou, H., Huang, F., and Yan, B.: Hydrological Changes of the Ili River in Kazakhstan and the Possible Causes, J. Hydraul. Eng.-ASCE, 20, 05015006, doi:10.1061/(asce)he.1943-5584.0001214, 2015.

Gupta, H. V., Sorooshian, S., and Yapo, P. O.: Toward improved calibration of hydrologic models: Multiple and noncommensurable measures of information, Water Resour. Res., 34, 751–763, 1998.

Hagg, W. and Braun, L.: The influence of glacier retreat on water yield from high mountain areas: comparison of Alps and Central Asia, Climate and Hydrology in Mountain Areas, 18, 263–275, 2005.

Hagg, W., Braun, L., Weber, M., and Becht, M.: Runoff modelling in glacierized Central Asian catchments for present-day and future climate, Nordic Hydrology, 37, 93–105, 2006.

Hagg, W., Braun, L. N., Kuhn, M., and Nesgaard, T. I.: Modelling of hydrological response to climate change in glacierized Central Asian catchments, J. Hydrol., 332, 40–53, doi:10.1016/j.jhydrol.2006.06.021, 2007.

Hagg, W., Hoelzle, M., Wagner, S., Mayr, E., and Klose, Z.: Glacier and runoff changes in the Rukhk catchment, upper Amu-Darya basin until 2050, Global Planet. Change, 110, 62–73, doi:10.1016/j.gloplacha.2013.05.005, 2013.

Hock, R.: Glacier melt: a review of processes and their modelling, Prog. Phys. Geog., 29, 362–391, 2005.

Huai, B., Li, Z., Sun, M., and Xiao, Y.: Snowmelt runoff model applied in the headwaters region of Urumqi River, Arid Land Geography, 36, 41–48, 2013 (in Chinese with English abstract).

Huang, S., Krysanova, V., Zhai, J., and Su, B.: Impact of Intensive Irrigation Activities on River Discharge Under Agricultural Scenarios in the Semi-Arid Aksu River Basin, Northwest China, Water Resour. Manag., 29, 945–959, doi:10.1007/s11269-014-0853-2, 2015.

Huang, Y., Chen, X., Bao, A., and Ma, Y.: Distributed Hydrological Modeling in Kaidu Basin: MIKE-SHE Model Calibration and Uncertainty Estimation, J. Glaciol. Geocryol., 32, 567–572, 2010a (in Chinese with English abstract).

Huss, M., Farinotti, D., Bauder, A., and Funk, M.: Modelling runoff from highly glacierized alpine drainage basins in a changing climate, Hydrol. Process., 22, 3888–3902, doi:10.1002/hyp.7055, 2008.

Ibatullin, S., Yasinsky, V., and Mironenkov, A.: Impacts of climate change on water resources in Central Asia, Sector report no. 6, Eurasian development bank, Almaty, 44 pp., 2009.

Immerzeel, W. W., Van Beek, L. P., and Bierkens, M. F.: Climate change will affect the Asian water towers, Science, 328, 1382–1385, 2010.

Immerzeel, W. W., Lutz, A., and Droogers, P.: Climate change impacts on the upstream water resources of the Amu and Syr Darya River basins, Wageningen, the Netherlands, 1–103, 2012a.

Immerzeel, W. W., Van Beek, L., Konz, M., Shrestha, A., and Bierkens, M.: Hydrological response to climate change in a glacierized catchment in the Himalayas, Climatic Change, 110, 721–736, 2012b.

Immerzeel, W. W., Petersen, L., Ragettli, S., and Pellicciotti, F.: The importance of observed gradients of air temperature and precipitation for modeling runoff from a glacierized watershed in the Nepalese Himalayas, Water Resour. Res., 50, 2212–2226, doi:10.1002/2013WR014506, 2014.

Johnston, R. and Smakhtin, V.: Hydrological Modeling of Large river Basins: How Much is Enough?, Water Resour. Manag., 28, 2695–2730, doi:10.1007/s11269-014-0637-8, 2014.

Kaldybayev, A., Chen, Y., Issanova, G., Wang, H., and Mahmudova, L.: Runoff response to the glacier shrinkage in the Karatal river basin, Kazakhstan, Arabian Journal of Geosciences, 9, 1–8, doi:10.1007/s12517-015-2106-y, 2016.

Kaser, G., Großhauser, M., and Marzeion, B.: Contribution potential of glaciers to water availability in different climate regimes, P. Natl. Acad. Sci. USA, 107, 20223–20227, 2010.

Kemmerikh, A. O.: The role of glaciers for river runoff in Central Asia [Rol' lednikov v stoke rek Sredney Azii], Data of Glaciological Studies, 20, 82–94, 1972 (in Russian).

Kezer, K. and Matsuyama, H.: Decrease of river runoff in the Lake Balkhash basin in Central Asia, Hydrol. Process., 20, 1407–1423, doi:10.1002/hyp.6097, 2006.

Khan, V. and Holko, L.: Snow cover characteristics in the Aral Sea Basin from different data sources and their relation with river runoff, J. Marine Syst., 76, 254–262, 2009.

Kong, Y. and Pang, Z.: Evaluating the sensitivity of glacier rivers to climate change based on hydrograph separation of discharge, J. Hydrol., 434, 121–129, 2012.

Konz, M. and Seibert, J.: On the value of glacier mass balances for hydrological model calibration, J. Hydrol., 385, 238–246, doi:10.1016/j.jhydrol.2010.02.025, 2010.

Lei, Y., Yao, T., Bird, B. W., Yang, K., Zhai, J., and Sheng, Y.: Coherent lake growth on the central Tibetan Plateau since the 1970s: Characterization and attribution, J. Hydrol., 483, 61–67, doi:10.1016/j.jhydrol.2013.01.003, 2013.

Li, L. and Simonovic, S.: System dynamics model for predicting floods from snowmelt in North American prairie watersheds, Hydrol. Process., 16, 2645–2666, 2002.

Li, L., Shang, M., Zhang, M., Ahmad, S., and Huang, Y.: Snowmelt runoff simulation driven by APHRODITE precipitation dataset, Adv. Water Sci., 25, 53–59, 2014 (in Chinese with English abstract).

Li, X. and Williams, M. W.: Snowmelt runoff modelling in an arid mountain watershed, Tarim Basin, China, Hydrol. Process., 22, 3931–3940, 2008.

Li, Z., Wang, W., Zhang M., Wang F., and Li, H.: Observed changes in streamflow at the headwaters of the Urumqi River, eastern Tianshan, central Asia, Hydrol. Process., 24, 217–224, doi:10.1002/hyp.7431, 2010.

Liu, J., Liu, T., Bao, A., De Maeyer, P., Feng, X., Miller, S. N., and Chen, X.: Assessment of Different Modelling Studies on the Spatial Hydrological Processes in an Arid Alpine Catchment, Water Resour. Manag., 30, 1757–1770, 2016a.

Liu, J., Liu, T., Bao, A., De Maeyer, P., Kurban, A., and Chen, X.: Response of Hydrological Processes to Input Data in High Alpine Catchment: An Assessment of the Yarkant River basin in China, Water, 8, 181, doi:10.3390/w8050181, 2016b.

Liu, T., Willems, P., Pan, X. L., Bao, An. M., Chen, X., Veroustraete, F., and Dong, Q. H.: Climate change impact on water resource extremes in a headwater region of the Tarim basin in China, Hydrol. Earth Syst. Sci., 15, 3511–3527, doi:10.5194/hess-15-3511-2011, 2011.

Liu, T., Willems, P., Feng, X. W., Li, Q., Huang, Y., Bao, A. M., Chen, X., Veroustraete, F., and Dong, Q. H.: On the usefulness of remote sensing input data for spatially distributed hydrological modelling: case of the Tarim River basin in China, Hydrol. Process., 26, 335–344, doi:10.1002/hyp.8129, 2012.

Liu, Y., Weerts, A. H., Clark, M., Hendricks Franssen, H.-J., Kumar, S., Moradkhani, H., Seo, D.-J., Schwanenberg, D., Smith, P., van Dijk, A. I. J. M., van Velzen, N., He, M., Lee, H., Noh, S. J., Rakovec, O., and Restrepo, P.: Advancing data assimilation in operational hydrologic forecasting: progresses, challenges, and emerging opportunities, Hydrol. Earth Syst. Sci., 16, 3863–3887, doi:10.5194/hess-16-3863-2012, 2012.

Liu, T., Fang, H., Willems, P., Bao, A. M., Chen, X., Veroustraete, F., and Dong, Q. H.: On the relationship between historical land-use change and water availability: the case of the lower Tarim River region in northwestern China, Hydrol. Process., 27, 251–261, doi:10.1002/hyp.9223, 2013.

Liu, Z., Xu, Z., Huang, J., Charles, S. P., and Fu, G.: Impacts of climate change on hydrological processes in the headwater catchment of the Tarim River basin, China, Hydrol. Process., 24, 196–208, doi:10.1002/hyp.7493, 2010.

Luo, Y., Arnold, J., Allen, P., and Chen, X.: Baseflow simulation using SWAT model in an inland river basin in Tianshan Mountains, Northwest China, Hydrol. Earth Syst. Sci., 16, 1259–1267, doi:10.5194/hess-16-1259-2012, 2012.

Luo, Y., Arnold, J., Liu, S., Wang, X., and Chen, X.: Inclusion of glacier processes for distributed hydrological modeling at basin scale with application to a watershed in Tianshan Mountains, northwest China, J. Hydrol., 477, 72–85, doi:10.1016/j.jhydrol.2012.11.005, 2013.

Lutz, A., Immerzeel, W., Shrestha, A., and Bierkens, M.: Consistent increase in High Asia's runoff due to increasing glacier melt and precipitation, Nature Climate Change, 4, 587–592, 2014.

Ma, H. and Cheng, G.: A test of Snowmelt Runoff Model (SRM) for the Gongnaisi River basin in the western Tianshan Mountains, China, Chinese Sci. Bull., 48, 2253–2259, 2003.

Ma, C., Sun, L., Liu, S., Shao, M. A., and Luo, Y.: Impact of climate change on the streamflow in the glacierized Chu River Basin, Central Asia, Journa of Arid Land, 7, 501–513, 2015.

Ma, Y., Huang, Y., Chen, X., Li, Y., and Bao, A.: Modelling Snowmelt Runoff under Climate Change Scenarios in an Ungauged Mountainous Watershed, Northwest China, Math. Probl. Eng., 2013, 808565, doi:10.1155/2013/808565, 2013.

Mayr, E., Hagg, W., Mayer, C., and Braun, L.: Calibrating a spatially distributed conceptual hydrological model using runoff, annual mass balance and winter mass balance, J. Hydrol., 478, 40–49, 2013.

Michlmayr, G., Lehning, M., Koboltschnig, G., Holzmann, H., Zappa, M., Mott, R., and Schoener, W.: Application of the Alpine 3D model for glacier mass balance and glacier runoff studies at Goldbergkees, Austria, Hydrol. Process., 22, 3941–3949, doi:10.1002/hyp.7102, 2008.

Miller, J. D., Immerzeel, W. W., and Rees, G.: Climate change impacts on glacier hydrology and river discharge in the Hindu Kush-Himalayas: a synthesis of the scientific basis, Mt. Res. Dev., 32, 461–467, 2012.

Mou, L., Tian, F., Hu, H., and Sivapalan, M.: Extension of the Representative Elementary Watershed approach for cold regions: constitutive relationships and an application, Hydrol. Earth Syst. Sci., 12, 565–585, doi:10.5194/hess-12-565-2008, 2008.

Naegeli, K., Gärtner-Roer, I., Hagg, W., Huss, M., Machguth, H., and Zemp, M.: Worldwide dataset of glacier thickness observations compiled by literature review, EGU General Assembly Conference Abstracts, 2013, 3077, 2013.

Oerlemans, J. and Reichert, B.: Relating glacier mass balance to meteorological data by using a seasonal sensitivity characteristic, J. Glaciol., 46, 1–6, 2000.

Ohmura, A.: Physical basis for the temperature-based melt-index method, J. Appl. Meteorol., 40, 753–761, 2001.

Ouyang, R., Cheng, W., Wang, W., Jiang, Y., Zhang, Y., and Wang, Y.: Research on runoff forecast approaches to the Aksu River basin, Sci. China Ser. D, 50, 16–25, 2007.

Parajuli, P. B., Nelson, N. O., Frees, L. D., and Mankin, K. R.: Comparison of AnnAGNPS and SWAT model simulation results in USDA-CEAP agricultural watersheds in south-central Kansas, Hydrol. Process., 23, 748–763, doi:10.1002/hyp.7174, 2009.

Peng, D. Z. and Xu, Z. X.: Simulating the Impact of climate change on streamflow in the Tarim River basin by using a modified semi-distributed monthly water balance model, Hydrol. Process., 24, 209–216, doi:10.1002/hyp.7485, 2010.

Ragettli, S., Immerzeel, W. W., and Pellicciotti, F.: Contrasting climate change impact on river flows from high-altitude catchments in the Himalayan and Andes Mountains, P. Natl. Acad. Sci. USA, 113, 9222–9227, 2016.

Refsgaard, J. C.: Parameterisation, calibration and validation of distributed hydrological models, J. Hydrol., 198, 69–97, doi:10.1016/s0022-1694(96)03329-x, 1997.

Savoskul, O. and Smakhtin, V.: Glacier systems and seasonal snow cover in six major Asian river basins: water storage properties under changing climate, International Water Management Institute, 2013.

Savoskul, O. S., Chevnina, E. V., Perziger, F. I., Vasilina, L. Y., Baburin, V. L., Danshin, A. I., Matyakubov, B., and Murakaev, R. R.: Water, climate, food, and environment in the Syr Darya Basin, Contribution to the project ADAPT, Adaptation strategies to changing environments, 2003.

Savoskul, O. S., Shevnina, E. V., Perziger, F., Barburin, V., and Danshin, A.: How Much Water will be Available for Irrigation in the Future? The Syr Darya Basin (Central Asia), in: Climate change in contrasting river basins: adaptation strategies for water, food and environment, edited by: Aerts, J. C. and Droogers, P., 93–113, 2004.

Schaefli, B., Hingray, B., Niggli, M., and Musy, A.: A conceptual glacio-hydrological model for high mountainous catchments, Hydrol. Earth Syst. Sci., 9, 95–109, doi:10.5194/hess-9-95-2005, 2005.

Siegfried, T., Bernauer, T., Guiennet, R., Sellars, S., Robertson, A. W., Mankin, J., Bauer-Gottwein, P., and Yakovlev, A.: Will climate change exacerbate water stress in Central Asia?, Climatic Change, 112, 881–899, doi:10.1007/s10584-011-0253-z, 2012.

Singh, P. and Bengtsson, L.: Impact of warmer climate on melt and evaporation for the rainfed, snowfed and glacierfed basins in the Himalayan region, J. Hydrol., 300, 140–154, doi:10.1016/j.jhydrol.2004.06.005, 2005.

Sorg, A., Bolch, T., Stoffel, M., Solomina, O., and Beniston, M.: Climate change impacts on glaciers and runoff in Tien Shan (Central Asia), Nature Climate Change, 2, 725–731, 2012.

Stahl, K., Moore, R., Shea, J., Hutchinson, D., and Cannon, A.: Coupled modelling of glacier and streamflow response to future climate scenarios, Water Resour. Res., 44, W02422, doi:10.1029/2007WR005956, 2008.

Starodubtsev, V. and Truskavetskiy, S.: Desertification processes in the Ili River delta under anthropogenic pressure, Water Resour., 38, 253–256, 2011.

Sun, C. J., Chen, Y. N., Li, X. G., and Li, W. H.: Analysis on the streamflow components of the typical inland river, Northwest China, Hydrol. Sci. J., 61, 970–981, doi:10.1080/02626667.2014.1000914, 2016.

Sun, M., Li, Z., Yao, X., and Jin, S.: Rapid shrinkage and hydrological response of a typical continental glacier in the arid region of northwest China–taking Urumqi Glacier No. 1 as an example, Ecohydrol, 6, 909–916, 2013.

Sun, M., Li, Z., Yao, X., Zhang, M., and Jin, S.: Modeling the hydrological response to climate change in a glacierized high mountain region, northwest China, J. Glaciol., 61, 127–136, 2015.

Sun, W., Wang, Y., Cui, X., Yu, J., Zuo, D., and Xu, Z.: Physically-based distributed hydrological model calibration based on a short period of streamflow data: case studies in two Chinese basins, Hydrol. Earth Syst. Sci. Discuss., doi:10.5194/hess-2016-192, in review, 2016.

Unger-Shayesteh, K., Vorogushyn, S., Farinotti, D., Gafurov, A., Duethmann, D., Mandychev, A., and Merz, B.: What do we know about past changes in the water cycle of Central Asian headwaters? A review, Global Planet. Change, 110, 4–25, 2013.

Wang, P., Jiang, H., and Mu, Z.: Simulation of runoff process in headstream of Aksu River, Journal of Water Resources and Water Engineering, 23, 51–57, 2012a (in Chinese with English abstract).

Wang, S., Zhang, Z., Sun, G., Strauss, P., Guo, J., Tang, Y., and Yao, A.: Multi-site calibration, validation, and sensitivity analysis of the MIKE SHE Model for a large watershed in northern China, Hydrol. Earth Syst. Sci., 16, 4621–4632, doi:10.5194/hess-16-4621-2012, 2012b.

Wilby, R. L., Dawson, C. W., and Barrow, E. M.: SDSM – a decision support tool for the assessment of regional climate change impacts, Environ. Modell. Softw., 17, 145–157, 2002.

World Meteorological Organization: Guide to Hydrological Practices, Volume II, Management of Water Resources and Application of Hydrological Practices, WMO-No. 168, 302 pp., 2009.

Wortmann, M., Krysanova, V., Kundzewicz, Z. W., Su, B., and Li, X.: Assessing the influence of the Merzbacher Lake outburst floods on discharge using the hydrological model SWIM in the Aksu headwaters, Kyrgyzstan/NW China, Hydrol. Process., 28, 6337–6350, doi:10.1002/hyp.10118, 2014.

Wu, S., Han, P., Li, Y., Xue, Y., and Zhu, Z.: Predicted Variation Tendency of the Water Resources in the Headwaters of the Tarim River, J. Glaciol. Geocryol., 26, 708–711, 2003 (in Chinese with English abstract).

Xiang, L., Wang, H., Steffen, H., Wu, P., Jia, L., Jiang, L., and Shen, Q.: Groundwater storage changes in the Tibetan Plateau and adjacent areas revealed from GRACE satellite gravity data, Earth Planet. Sc. Lett., 449, 228–239, 2016.

Xie, L., Long, A., Deng, M., Li, X., and Wang, J.: Study on Ecological Water Consumption in Delta of the Lower Reaches of Ili River, J. Glaciol. Geocryol., 33, 1330–1340, 2011.

Xie, Z.-C., Wang, X., Feng, Q.-H., Kang, E. S., Liu, C.-H., and Li, Q.-Y.: Modeling the response of glacier systems to climate warming in China, Ann. Glaciol., 43, 313–316, 2006.

Xing, K., Lei, X., Lei, X., and Jin, S.: Application of distributed hydrological model EsayDHM in runoff simulation of Manasi river basin, J. Water Res. Water Eng., 20–23, 2014 (in Chinese with English abstract).

Xu, B., Lu, Z., Liu, S., Li, J., Xie, J., Long, A., Yin, Z., and Zou, S.: Glacier changes and their impacts on the discharge in the past half-century in Tekes watershed, Central Asia, Phys. Chem. Earth, 89, 96–103, 2015.

Xu, J., Chen, Y., Li, W., Yang, Y., and Hong, Y.: An integrated statistical approach to identify the nonlinear trend of runoff in the Hotan River and its relation with climatic factors, Stoch. Env. Res. Risk A., 25, 223–233, 2011.

Xu, J., Chen, Y., Li, W., Nie, Q., Song, C., and Wei, C.: Integrating wavelet analysis and BPANN to simulate the annual runoff with regional climate change: a case study of Yarkand River, Northwest China, Water Resour. Manag., 28, 2523–2537, 2014.

Yang, J., Liu, Y., Yang, W., and Chen, Y.: Multi-objective sensitivity analysis of a fully distributed hydrologic model WetSpa, Water Resour. Manag., 26, 109–128, 2012.

Yang, J., Castelli, F., and Chen, Y.: Multiobjective sensitivity analysis and optimization of distributed hydrologic model MOBIDIC, Hydrol. Earth Syst. Sci., 18, 4101–4112, doi:10.5194/hess-18-4101-2014, 2014.

Yao, J. Q., Liu, Z. H., Yang, Q., Meng, X. Y., and Li, C. Z.: Responses of Runoff to Climate Change and Human Activities in the Ebinur Lake Catchment, Western China, Water Resour., 41, 738–747, doi:10.1134/s0097807814060220, 2014.

Yu, M., Chen, X., Li, L., Bao, A., and Paix, M. J.: Streamflow simulation by SWAT using different precipitation sources in large arid basins with scarce raingauges, Water Resour. Manag., 25, 2669–2681, 2011.

Yu, M., Chen, X., Li, L., Bao, A., and de la Paix, M. J.: Incorporating accumulated temperature and algorithm of snow cover calculation into the snowmelt runoff model, Hydrol. Process., 27, 3589–3595, doi:10.1002/hyp.9372, 2013.

Zhang, F. Y., Ahmad, S., Zhang, H. Q., Zhao, X., Feng, X. W., and Li, L. H.: Simulating low and high streamflow driven by snowmelt in an insufficiently gauged alpine basin, Stoch. Env. Res. Risk A., 30, 59–75, doi:10.1007/s00477-015-1028-2, 2016.

Zhang, Y., Luo, Y., Sun, L., Liu, S., Chen, X., and Wang, X.: Using glacier area ratio to quantify effects of melt water on runoff, J. Hydrol., 538, 269–277, 2016.

Zhang, S., Gao, X., Ye, B., Zhang, X., and Hagemann, S.: A modified monthly degree – model for evaluating glacier runoff changes in China. Part II: application, Hydrol. Process., 26, 1697–1706, 2012a.

Zhang, S., Ye, B., Liu, S., Zhang, X., and Hagemann, S.: A modified monthly degree – model for evaluating glacier runoff changes in China. Part I: model development, Hydrol. Process., 26, 1686–1696, 2012b.

Zhang, Y., Liu, S., and Ding, Y.: Glacier meltwater and runoff modelling, Keqicar Baqi glacier, southwestern Tien Shan, China, J. Glaciol., 53, 91–98, 2007.

Zhang, Y. C., Li, B. L., Bao, A. M., Zhou, C., Chen, X., and Zhang, X. R.: Study on snowmelt runoff simulation in the Kaidu River basin, Sci. China Ser. D, 50, 26–35, 2007.

Zhao, Q., Liu, Z., Ye, B., Qin, Y., Wei, Z., and Fang, S.: A snowmelt runoff forecasting model coupling WRF and DHSVM, Hydrol. Earth Syst. Sci., 13, 1897–1906, doi:10.5194/hess-13-1897-2009, 2009.

Zhao, Q., Ye, B., Ding, Y., Zhang, S., Yi, S., Wang, J., Shangguan, D., Zhao, C., and Han, H.: Coupling a glacier melt model to the Variable Infiltration Capacity (VIC) model for hydrological modeling in north-western China, Environ. Earth Sci., 68, 87–101, 2013.

Zhao, Q. D., Zhang, S. Q., Ding, Y. J., Wang, J., Han, H. D., Xu, J. L., Zhao, C. C., Guo, W. Q., and Shangguan, D. H.: Modeling Hydrologic Response to Climate Change and Shrinking Glaciers in the Highly Glacierized Kunma Like River Catchment, Central Tian Shan, J. Hydrometeorol., 16, 2383–2402, doi:10.1175/jhm-d-14-0231.1, 2015.

Fractal analysis of urban catchments and their representation in semi-distributed models: imperviousness and sewer system

Auguste Gires[1], Ioulia Tchiguirinskaia[1], Daniel Schertzer[1], Susana Ochoa-Rodriguez[2], Patrick Willems[3], Abdellah Ichiba[1,4], Li-Pen Wang[3], Rui Pina[2], Johan Van Assel[5], Guendalina Bruni[6], Damian Murla Tuyls[3], and Marie-Claire ten Veldhuis[6]

[1]HMCo, École des Ponts, UPE, Champs-sur-Marne, France
[2]Urban Water Research Group, Department of Civil and Environmental Engineering, Imperial College London, Skempton Building, London SW7 2AZ, UK
[3]Hydraulics Laboratory, KU Leuven, 3001, Heverlee (Leuven), Belgium
[4]Conseil Départemental du Val-de-Marne, Direction des Services de l'Environnement et de l'Assainissement (DSEA), Bonneuil-sur-Marne, 94381, France
[5]Aquafin NV, Dijkstraat 8, 2630 Aartselaar, Belgium
[6]Water Management Department, Delft University of Technology, P.O. Box 5048, 2600 GA Delft, the Netherlands

Correspondence to: Auguste Gires (auguste.gires@enpc.fr)

Abstract. Fractal analysis relies on scale invariance and the concept of fractal dimension enables one to characterize and quantify the space filled by a geometrical set exhibiting complex and tortuous patterns. Fractal tools have been widely used in hydrology but seldom in the specific context of urban hydrology. In this paper, fractal tools are used to analyse surface and sewer data from 10 urban or peri-urban catchments located in five European countries. The aim was to characterize urban catchment properties accounting for the complexity and inhomogeneity typical of urban water systems. Sewer system density and imperviousness (roads or buildings), represented in rasterized maps of $2\,\text{m} \times 2\,\text{m}$ pixels, were analysed to quantify their fractal dimension, characteristic of scaling invariance. The results showed that both sewer density and imperviousness exhibit scale-invariant features and can be characterized with the help of fractal dimensions ranging from 1.6 to 2, depending on the catchment. In a given area consistent results were found for the two geometrical features, yielding a robust and innovative way of quantifying the level of urbanization. The representation of imperviousness in operational semi-distributed hydrological models for these catchments was also investigated by computing fractal dimensions of the geometrical sets made up of the subcatchments with coefficients of imperviousness greater than a range of thresholds. It enables one to quantify how well spatial structures of imperviousness were represented in the urban hydrological models.

1 Introduction

The aim of this paper is to consistently characterize urban catchment properties accounting for the complexity and inhomogeneity typical of urban water systems. It is focused on two main properties of urban catchments, namely the geometry of the sewer system and the distribution of impervious surfaces. Such characterization is important to obtain insights in the urban catchment response behaviour at the various spatial scales that control the relation between rainfall and sewer flows; to develop convenient methods that allow for evaluation of the urban catchment characteristics implemented in urban drainage models (the ones that are of importance for obtaining reliable spatially variable urban catchment responses; e.g. spatial imperviousness structure); to develop methods that support the urban hydrological modeller in the decision making process with regard to spatial details required to obtain reliable model (impact) results. Achieving this has proved to be difficult using traditional tools, mostly based upon Euclidean geometry, due to the variability and in-homogeneity in catchment characteristics (inter alios Berne et al., 2004). An alternative to traditional tools could be the use of fractal geometry (Mandelbrot, 1983), which relies on the concept of scale invariance; i.e. similar structures are visible at all scales. The concept of fractal dimension enables one to characterize, in a scale-invariant way, the space filled by a geometrical set in its embedding space. Fractal analysis and more generally scaling analysis have been often and successfully used in geophysics, including hydrology, but seldom in the specific context of urban hydrology.

For example, fractal analyses have been used to characterize river networks, including quantification of main stream sinuosity (Nikora, 1991; Hjeimfeit, 1988), quantification of how the network fills space (La Barbera and Rosso, 1989; Takayasu, 1990; Foufoula-Georgiou and Sapozhnikov, 2001; Gangodagamage et al., 2011, 2014), and simultaneous quantification of both features (Tarboton et al., 1988; Rosso et al., 1991; Tarboton, 1996; Veltri et al., 1996). River basins have also been analysed with fractal analysis. For instance, Bendjoudi and Hubert (2002) showed that the perimeters of the Danube (eastern Europe) and Seine (France) river basins are too tortuous to be scale independent. Rainfall occurrence patterns also appear to exhibit fractal features (Lovejoy and Mandelbrot, 1985; Lovejoy and Schertzer, 1985; Olsson et al., 1993; Hubert et al., 1995). In extensions including the use of multifractal tools, i.e. for fields and not simply geometrical shapes, such tools have also been used to study river discharges and rainfall time series (see Tessier et al., 1996, or Pandey et al., 1998, for examples combining both). Such analysis was also carried out on simulated discharged in urban context (Gires et al., 2012).

Some authors relied on the same concept of fractal dimension for characterizing land use cover in various contexts. For example Chen et al. (2001) computed a fractal dimension for various land use classes and used it to analyse land use change between two areal pictures taken 20 years apart over a $4\,\mathrm{km}^2$ mountainous catchment. Darrel and Wu (2001) computed fractal dimensions of three land use classes – desert, agriculture, and urban – and used it to analyse their evolution during a century over a $69\,\mathrm{km} \times 89\,\mathrm{km}$ area around Phoenix (Arizona, USA). This allowed for investigating the effect of urbanization over landscape and was used to develop a model to reproduce observed features. Similarly, Tannier et al. (2011) used this concept to identify the morphological boundary of urban areas in a scale-invariant way. Iverson (1988) estimated fractal dimensions for numerous land use types to study the evolution of landscape over 160 years in Illinois (USA). Soil features have also been studied with fractal analysis. For instance Wang et al. (2006) analysed particle size distribution with fractal concepts. A feature emphasized by many authors is the relationship between fractal features and power-law decay (i.e. non-Gaussian behaviour) of various fields such as river portion length, rainfall event duration, particle size distribution, or distance between

buildings (Mandelbrot, 1983; Lavergnat and Golé, 1998; Tarboton, 1996; Wang et al., 2006; Tannier et al., 2011). This implies that up- and downscaling of meteorological and hydrological parameters needs to account for this non-Gaussian behaviour. For hydrological analysis it means that hydrological models are likely to be sensitive to scale differences between rainfall input and catchment characterization (Ogden and Julien, 1994).

Despite this wide range of applications, fractal analysis has seldom been used to specifically address the topic of urban hydrology. Initial attempts to characterize urban drainage networks (Sarkis, 2008; Gires et al., 2014) or imperviousness (Gires et al., 2014) have been carried out on limited areas. In this paper we go a step further and implement fractal analysis on 10 urban catchments with different characteristics located across five European countries. The investigation includes analysis of the sewer network geometry and distribution of imperviousness derived from available GIS data, including the way in which it is represented in operational semi-distributed hydrodynamic urban drainage models. In order to be able to use the same technique to analyse both sewer networks and maps of distributed imperviousness, we use fractal tools on them, and not multifractal ones such as the one found in De Bartolo et al. (2004, 2006) for river networks. Multifractals will be used in the characterization of the representation of imperviousness in models. This multi-catchment investigation allows for obtaining robust results that are representative of a range of hydrological characteristics. The opportunity to carry out this multi-catchment investigation arose from the Interreg north-west Europe (NWE) project RainGain, which focuses on improving rainfall estimation and pluvial flood modelling and management in urban areas across NWE.

The paper is organized as follows. In Sect. 2 the available dataset over the 10 pilot catchments is described. The concept of fractal dimension and the methodology used to compute it are explained in Sect. 3. Results are presented and discussed in Sect. 4. In Sect. 5, the main conclusions are presented and future work is discussed.

2 Experimental sites and datasets

In total, 10 urban catchments, with areas in the range of 2–$8\,\mathrm{km}^2$ and located in five European countries (UK, France, the Netherlands, Belgium, and Portugal) were adopted as pilot sites in this study. The general location of the pilot catchments is shown in Fig. 1 and their main characteristics are summarized in Table 1.

For each pilot catchment, three types of data are analysed in this paper and Fig. 2 displays them for all the catchments:

i. The sewer system is considered as a network of linear pipes (left column in Fig. 2). The level of precision of available data is not the same for all the catchments. Indeed for the Morée-Sausset and Torquay catchments,

Table 1. General characteristics of the pilot urban catchments and their semi-distributed urban drainage models.

	Catchment characteristics					Model characteristics		
	Area [ha]	Length[a] [km]	Slope[b] [m m^{-1}]	Land use[c]	Pop. density [per/ha]	Total pipe length [km]	Num. of SC[d]	Mean/SD SC size [ha]
Cranbrook, UK	865	6.10	0.0093	R&C	48	98	1765	0.49/0.71
Torquay (town centre), UK	570	5.35	0.0262	R&C	60	41	492	1.16/1.09
Morée-Sausset, FR	560	5.28	0.0029	R&C	70	15	47	11.92/10.34
Sucy-en-Brie, FR	269	4.02	0.0062	R&C	95	4	9	29.89/27.47
Herent, BE	511	8.16	0.0083	R	20	67	683	0.71/1.27
Jouy-en-Josas, FR	302	2.47	0.037	R	15	–	–	–
Ghent, BE	649	4.74	0.0001	R	24	83	1424	0.46/0.89
Rotterdam-Kralingen, NL	670	~2[e]	0.0003	R&C	154	143	2435	0.12/0.13
Rotterdam-Centrum, NL	340	~1[e]	0.0001	R&C	88	140	2832	0.0769/0.0737
Coimbra, PT	158	4.21	0.0333	R&C	116	34.75	911	0.17/0.28

[a] Length of longest flow path (through sewers) to catchment outfall.

[b] Catchment slope is difference in ground elevation between point most upstream and outlet/catchment length. This simplistic indictor is used to estimate whether the catchment exhibits strong slopes on average (Ochoa-Rodriguez et al., 2015). Other types of studies, such as those on of surface runoff, would indeed require more refined analysis of the topography but they are outside the scope of this paper and refined digital elevation models were not available for all studied areas.

[c] Predominant land use types: R is residential; C is commercial.

[d] SC is sub-catchments.

[e] The definition (a) is not straightforward due to the loopedness of the catchment

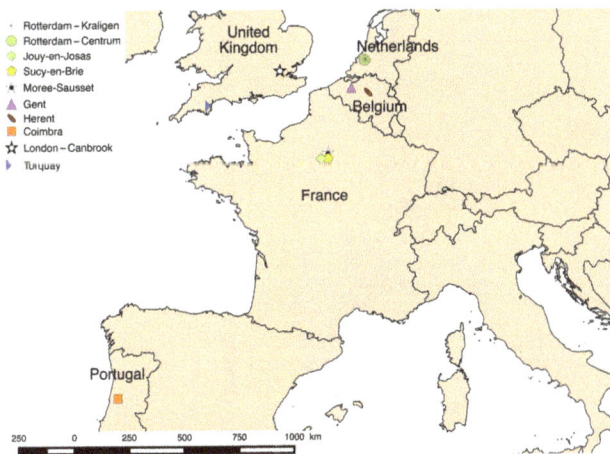

Figure 1. Location of the pilot urban catchments.

only the main pipes are taken into account, whereas for the others all pipes down to street level (not the connections from building or houses to the network) are available.

ii. An imperviousness map at a resolution of $2\,m \times 2\,m$ generated with the help of QGIS (http://www.qgis.org) is based on data derived mainly from Open Street Map (http://www.openstreetmap.org/) (middle column

in Fig. 2). More precisely, for each catchment the road layer (of polyline type) was retrieved from the Open Street Map platform and a 4 m buffer (adopted based on normal width of roads in urban and peri-urban catchments) was set on both sides of this polyline layer. The building layer was retrieved either from the same platform or from local building register datasets. These two datasets were rasterized in a map with pixels of size $2\,m \times 2\,m$. An imperviousness map was then derived in which a pixel containing roads or buildings is marked as impervious and other pixels are marked as pervious.

iii. A map of imperviousness is derived from catchment representation in semi-distributed hydrodynamic models (right column in Fig. 2). A validated operational semi-distributed hydrodynamic model was available for each of the pilot catchments, except for Jouy-en-Josas. In this type of model, the whole catchment is split into a number of sub-catchment, an independent hydrological block corresponding to a portion of the full catchment. The models are not the same for all the pilot sites but they all function with the same underlying principles. Each sub-catchment contains a mix of pervious and impervious surfaces, whose runoff drains to a common outlet point, which could be either a node of the drainage network or another sub-catchment (Rossman, 2010). Each sub-catchment is characterized by a num-

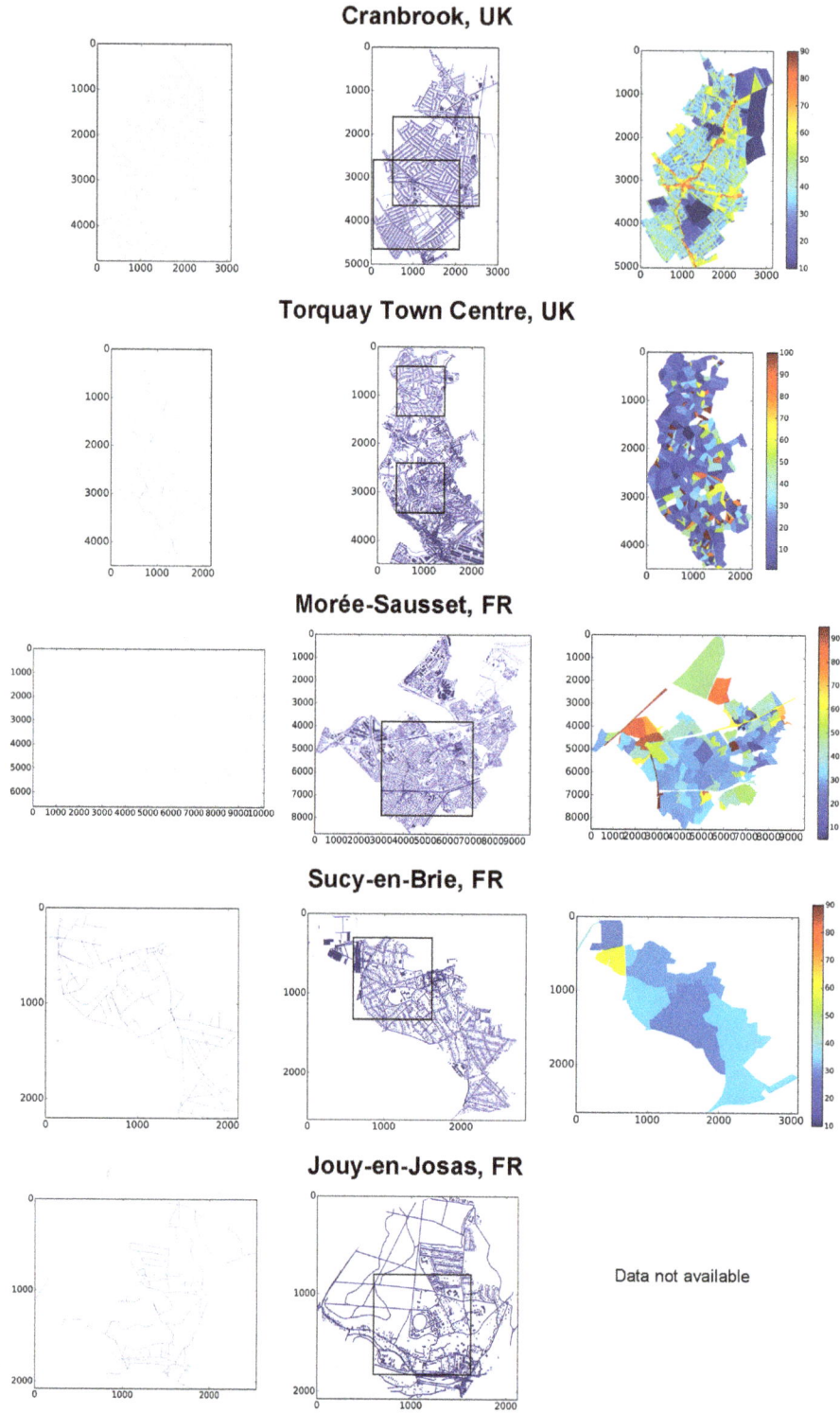

Cranbrook, UK

Torquay Town Centre, UK

Morée-Sausset, FR

Sucy-en-Brie, FR

Jouy-en-Josas, FR

Data not available

Figure 2.

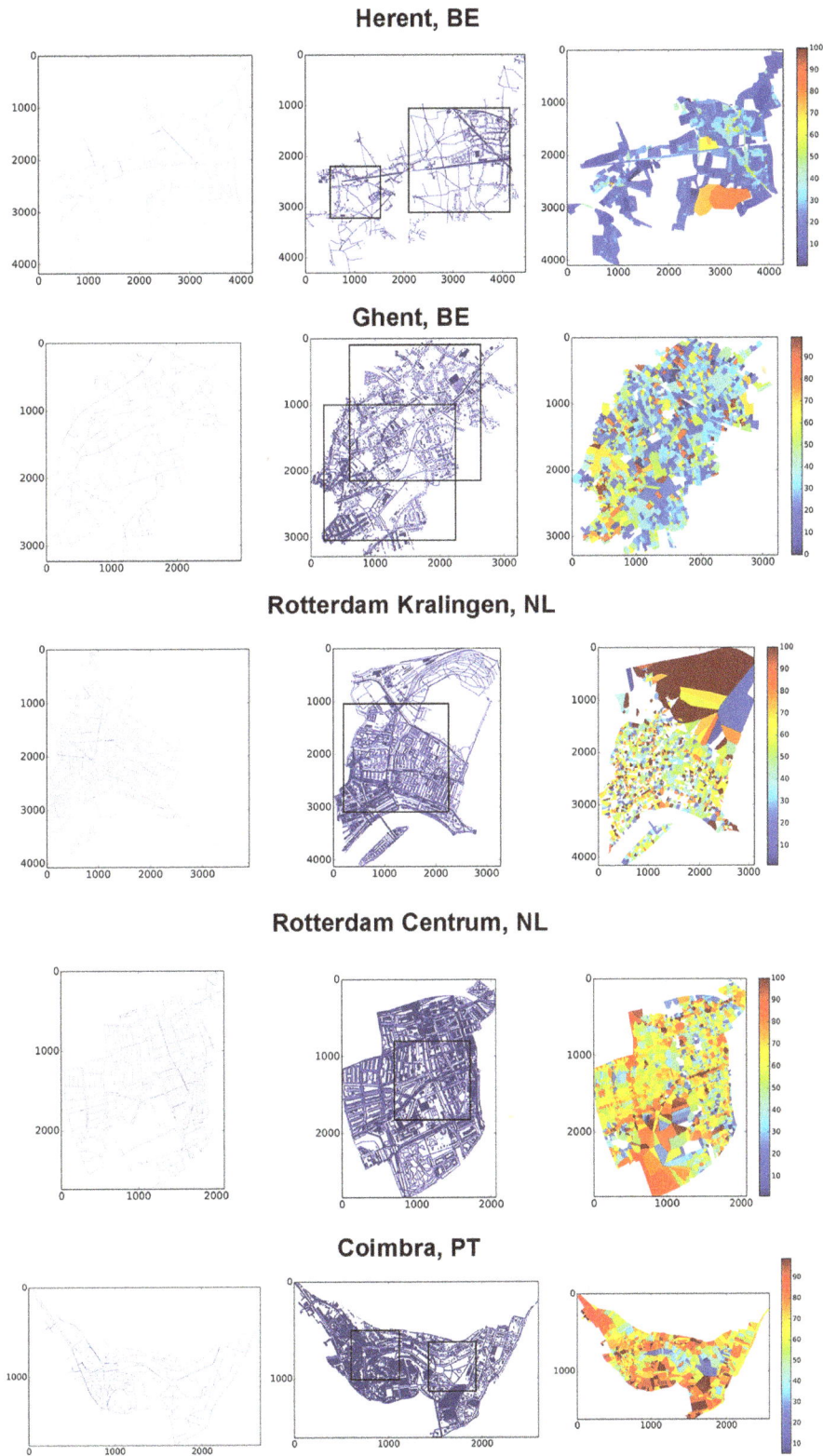

Figure 2. Sewer system (left), distributed imperviousness map with pixels a size of 2 m (middle), and maps of the imperviousness (%) as assigned to each sub-catchment in the semi-distributed models (right) of the pilot catchments. The axes correspond to metres (m). The black squares (visible in the middle column) correspond to the studied areas in the fractal analysis.

ber of parameters, including total area, length, slope, proportion of each land use, and soil type characteristics. Rainfall is inputted as homogeneous in space within each sub-catchment, and based on the sub-catchment's characteristics, the total runoff is estimated with the help of a lumped model and routed to the outlet point. The flow in pipes is then represented with the help if numerical approximation of one-dimensional (1-D) shallow-water equations. The size and distribution of sub-catchments depend on the modeller's choices according to the local features, the available data, and desired level of precision. Based on the percentage of impervious areas assigned to each sub-catchment within each pilot catchment, a raster map with pixels of size 2 m × 2 m was generated for each pilot site. The distribution of sub-catchments is visible in Fig. 2 because the values of imperviousness are uniform over them. The average size of sub-catchment elements varies greatly according to the studied area (see Table 1). For instance, it is much greater in Sucy-en-Brie than in Rotterdam-Kralingen. The purpose of the paper is not to evaluate the performance of those models all previously validated and used operationally by practitioners but to characterize their inputs, notably in comparison with more refined impervious data maps. Discussions on outputs of these models can be found in Ochoa-Rodriguez et al. (2015).

3   Methodology

As explained in Sect. 1, the concept of fractal dimension was used in this paper to characterize various geometrical sets (namely the sewer network and imperviousness), embedded in a 2-D space. Let us consider such a bounded set A of outer scale l_0. The first step consists in changing its resolution, i.e. modifying its observation scale l. The resolution λ is defined as the ratio between the outer scale and the observation scale ($\lambda = \frac{l_0}{l}$). This is achieved by representing it with the help of non-overlapping pixels of size l. At a given scale the set A is represented by all pixels overlaying the geometrical set. A range of values is tested for l. In this study, the analysis started at the smallest pixel size available, i.e. 2 m. The pixel size is then multiplied by two at each step, i.e. four adjacent pixels are merged, up to a maximum pixel size that covers as much of the total catchment area as possible. An illustration of this process for the sewer system of the Herent case is displayed in Fig. 3. Limited differences are visible when changing the observation scale from 2 to 4 m (some details are lost in the intersections, and close pipes merged), and they are much more pronounced with observation scales equal to 16 and 64 m (merging of numerous pipes). These observations are actually consistent with the scale break at 64 m, which will be identified and discussed in Sect. 4.1.

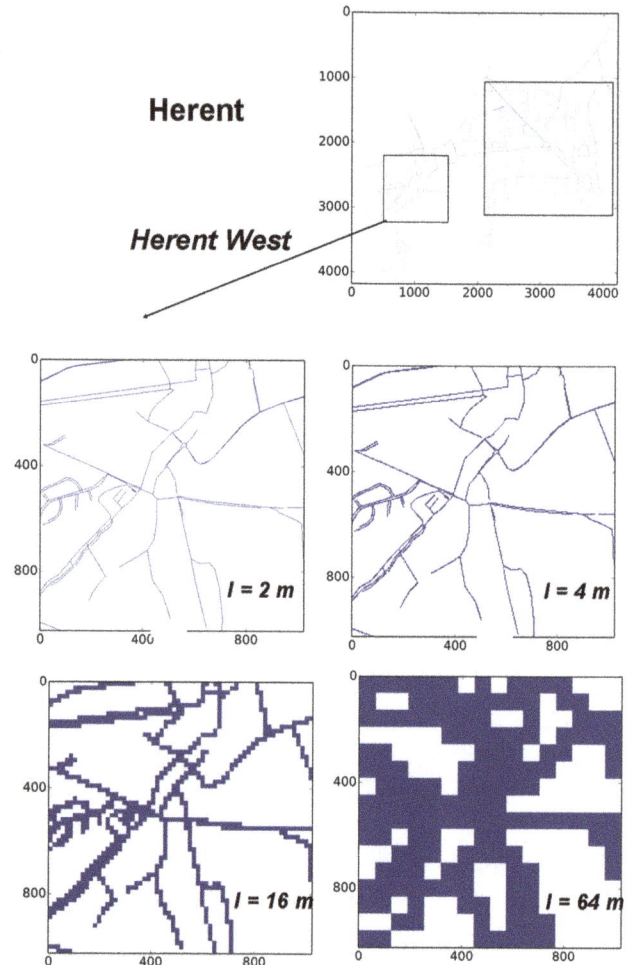

Figure 3. The sewer network of the Herent west study area observed with the help of pixels of various sizes. The axes correspond to metres (m).

This means that the outer scale of the studied set will necessarily be the original pixel size multiplied by a power of 2, closest to the maximum catchment scale (pixels are merged 4 by 4 in order to maximize the number of points in the following linear regression; less reliable results would be obtained with by merging pixels 9 by 9 or 25 by 25). As a consequence, square areas are extracted from the studied catchments to be analysed with the help of fractal analysis. Their size is chosen as a balance between achieving the greatest possible coverage (which increases the range of available scales) and limiting the portion of the square extending outside the catchment boundary (given that the artificial zeros in these portions might bias the analysis due to side effects). The studied areas within each catchment are shown in Fig. 2 for all catchments. In four catchments (Cranbrook, Ghent, Herent and Torquay) two areas are studied, sometimes slightly overlapping (Cranbrook and Ghent).

Now that the methodology to change the resolution of the dataset has been explained, it is possible to describe the computation of its fractal dimension with the help of the box-counting method (Hentschel and Proccacia, 1983; Lovejoy et al., 1987). Let $N_{\lambda,A}$ be the number of non-overlapping pixels of size l necessary to cover the set A. For a fractal object, this number and the resolution are power-law related in the high-resolution limit ($\lambda \to +\infty$), with an exponent equal to the fractal dimension (D_F) of the set; i.e. we have

$$N_{\lambda,A} \approx \lambda^{D_F}. \tag{1}$$

A standard technique to estimate a fractal dimension is the box-counting one, which relies on the previous equation. To implement this technique, one defines non-overlapping pixels of size l as explained in the previous paragraph and plots Eq. (1) on a log–log scale. For a fractal set, the points will be along a straight line, whose slope is equal to D_F. The quality of the scaling is assessed with the help of the coefficient of determination r^2 of the linear regression. It is an imperfect indicator, especially given the limited number of points available, and should be completed by visual inspection, The fractal dimension quantifies the sparseness of the set A; i.e. how much space it fills across scales.

The notion of fractal dimension is well suited for studying binary fields such as a sewer network or map of imperviousness. However, when the field can have more than two states, as it is the case in this paper for the maps of representation of imperviousness inputted in semi-distributed hydrodynamics models, multifractal tools might be needed. Intuitively such fields are characterized with the help of various fractal dimensions; i.e. for each threshold, the geometrical set of the areas where the field exceeds it exhibits a different fractal dimension. More rigorously the notion of threshold, which is scale dependent, is replaced by the scale-invariant one of singularity, γ. Then and the portions of a multifractal field ε_λ where it exceeds the threshold λ^γ at a given resolution λ are studied. Their probability scales as

$$\Pr\left(\varepsilon_\lambda > \lambda^\gamma\right) \approx \lambda^{-c(\gamma)}, \tag{2}$$

where $c(\gamma)$ is the co-dimension function, which fully characterizes the variability not only at a single scale but also across scales of ε (see Schertzer and Lovejoy, 2011, and references therein for a recent review of this formalism). $c(\gamma)$ corresponds to the fractal co-dimension (equal to the embedding Euclidian dimension − 2 here − minus the fractal dimension) of the geometrical set where ε_λ exceeds λ^γ. In the specific framework of universal multifractals (Schertzer and Lovejoy, 1987, 1997), the co-dimension function only depends on three parameters that have a physical interpretation: H the non-conservation parameter that measures the scaling behaviour of the mean of the studied field ($\langle\varepsilon_\lambda\rangle \approx \lambda^H$, $H = 0$ for a conservative field); C_1 the mean intermittency that measures the clustering of the average intensity (mathematically it is $c(\gamma_1)$, where γ_1 is the singularity corresponding to the mean; $C_1 = 0$ for an homogenous field); and α

the multifractality, which measures how the mean intermittency evolves when considering singularities slightly different from γ_1 ($\alpha = 0$ for a fractal field). These parameters are estimated with the help of the double trace moment (DTM) technique (Lavallée et al., 1993).

4   Results and discussion

4.1   Sewer network and distributed land use

Figure 4 shows a log–log plot of $N(\lambda)$ vs. λ (Eq. 1) for the Torquay north case study. A single scaling behaviour over the whole range of available scales is not retrieved. Indeed, the plot exhibits a scale break at roughly 64 m pixel scale, separating two distinct scaling regimes. Over each regime, the scaling is robust with r^2 all above 0.99, and visible straight lines. Similar qualitative features, i.e. two distinct well-defined scaling regimes separated by a break, are retrieved for the other studied areas and not displayed. Numerical values of the computed fractal dimensions and the values of scale break for all studied area are reported in Table 2.

For the scaling regime associated with small scales (i.e. right portion of the graph), a fractal dimension basically equal to 1 is found for all the study areas. This does not contain information on the network's features but simply reflects the linear structure of the pipes at these scales. It also means that the maximum resolution of the available data (2 m pixels here) is not critical to the analysis and does not introduce a potential bias. Indeed, increasing or decreasing it would simply yield to extending or shrinking the widths of the scale range of this regime but will not affect the values at larger scales discussed below. The break is located at roughly 64 m for most of the areas, which is consistent with the distance between two streets. It is at 32 m in Coimbra and Rotterdam-Centrum, which correspond to densely urbanized city centres. The break at 128 m for the Morée-Sausset sewer is due to the fact that only major sewer pipes are available and included in the numerical network model meaning small-scale features simply extend over wider range of scales. Including more pipes would likely lead to shifting the scale break to smaller scales. It appears that for all the catchments the break is observed at roughly the approximate inter-pipe distance of the portion of network taken into account. For the large-scales regime ($\sim$ 64 to 2048 m), an actual fractal dimension between 1 and 2 characterizing the space filled by the network is retrieved. According to the catchment, we find D_F ranging from 1.69 to 1.94. With smaller scales, this regime is expected to continue until the physical scales of structures is reached below which a fractal dimension of 2 would obviously be found. It, is in any case, smaller than 2 meaning that the network does not completely fill the 2-D space. An interpretation of these values is that these are representative of the level of urbanization of the areas. For instance, we find the greater fractal dimensions in the Rotterdam districts and

Table 2. Estimated fractal dimensions of the sewer system and impervious areas for all the studied areas.

| | Outer scale (m) | Sewer system | | | Distributed imperviousness | | $\%_{diff}{}^{*}$ | UM parameters for imperviousness map for semi-distributed models | |
		D_F for large scales	D_F for small scales	Scale of the break	D_F for all scales	% of impervious pixels		α	C_1
Rotterdam-Centrum	1024	1.94	1.07	32	1.93	61	−9	1.29	0.017
Rotterdam-Kralingen	2048	1.94	1.17	64	1.89	46	−3	0.71	0.064
Cranbrook north	2048	1.94	0.97	64	1.83	29	14	1.36	0.018
Cranbrook south	2048	1.90	0.97	64	1.81	26	17	1.25	0.025
Coimbra west	512	1.90	1.03	32	1.96	75	−18	1.37	0.009
Ghent north	2048	1.86	1.06	64	1.80	24	14	1.10	0.057
Ghent south	2048	1.85	1.06	64	1.82	27	16	1.01	0.054
Herent west	1024	1.82	1.06	64	1.71	19	−1	1.28	0.074
Herent east	2048	1.81	1.08	64	1.72	16	16	0.87	0.083
Sucy-en-Brie	1024	1.80	1.00	64	1.79	26	11	1.60	0.013
Coimbra east	512	1.79	0.97	32	1.86	45	13	1.71	0.20
Jouy-en-Josas	1024	1.79	1.79	64	1.75	22	x	x	x
Torquay south	1024	1.77	1.77	64	1.86	38	-16	1.45	0.062
Torquay north	1024	1.71	1.71	64	1.82	29	−6	1.44	0.084
Morée-Sausset	4096	1.69	1.69	128	1.88	34	−1	1.64	0.023

* See explanations in last paragraph of Sect. 4.2 and Fig. 9

smaller ones in less-urbanized Jouy-en-Josas and Torquay. This will need to be confirmed with the analysis of imperviousness maps.

These results are consistent with values found in similar studies for drainage networks. Sarkis (2008) found a fractal dimension equal to 1.67 for the pluvial drainage network of the Val-de-Marne County (south-east of Paris), based on an analysis at scales of 290 m to 18 km, only considering the main pipe network. Typical values for natural river network fractal dimensions (computed with the box-counting technique) are usually smaller than those found here for urban catchments. For instance Takayasu (1990) found D_F for the Amazon and Nile rivers equal to 1.85 and 1.4 respectively.

Figure 5 displays the impervious pixels (in blue), along with the computation of the fractal dimension of the corresponding geometrical set for the Torquay north area. It appears that a unique scaling regime on the whole range of available scales is identified (single straight line), resulting in

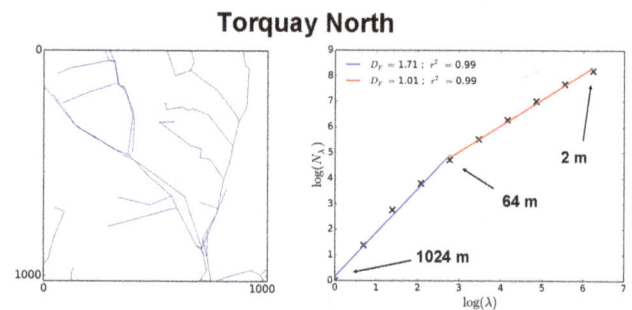

Torquay North

Figure 4. Sewer system (left) and computation of the corresponding fractal dimension, i.e. Eq. (1) in log–log plot (right), for the Torquay north study area. For the left figure, the axes correspond to metres (m).

fractal dimension 1.81. Unique scale regimes are also found for impervious surface distributions in all the other studied areas. The scaling regime is robust with visible straight lines as in Fig. 5 (right) and r^2 is always greater than 0.995. The uniqueness of the regime also means that results are not sensitive to the initial pixel size of 2 m as for the sewer system analysis (but for a different reason). Increasing this size would simply reduce the width of the range of scales available to compute the fractal dimension but not change its value. Numerical values of these fractal dimensions are reported in Table 2. Despite the fact that the impervious pixels do not represent the majority of the pixels at a 2 m resolution, their fractal dimension is rather elevated meaning that the impervious areas fill the space in urban areas. As expected less-urbanized areas exhibit lower fractal dimension.

For a given catchment, numerical values of fractal dimension for distributed imperviousness are similar to the ones found at large scales in the sewer system analysis. Discrepancies are usually smaller than 0.1; smaller than the differences between the various catchments. Areas of similar urban density have similar fractal dimensions and lower density urban areas are consistently characterized by lower fractal dimensions. These numerical similarities are worth noting and actually one of the main finding of this analysis, confirmed on a wide set of study areas. Indeed it suggests that the scaling behaviours observed on sewer networks and distributed land use have the same physical basis and reflect a unique underlying level of urbanization. The only difference being that it stops at the inter-pipe distance for the sewer network, whereas it expands down to 2 m scale for the imperviousness. Contrary to other formalisms, such as the use of a single percentage of imperviousness defined with data at an arbitrary scale, this fractal dimension is quantity valid across scales and furthermore based on the characterization of two aspects related to urbanization (namely the sewer network and the distributed imperviousness), which makes it robust.

4.2 Representation of imperviousness in semi-distributed models

After having investigated the fractal behaviour of sewer system and imperviousness with the help of distributed data, the imperviousness distribution used in operational semi-distributed hydrodynamic models is studied in this section. A given threshold T is selected and fractal features of the geometrical sub-set made up of the sub-catchments with imperviousness greater than the threshold T, representing different degrees of imperviousness in this case, are analysed. Figure 6 illustrates the corresponding sub-sets and computation of the fractal dimensions for T equal to 20, 50, and 80 % for the Torquay north study area. Figure 7 displays r^2 (coefficient of determination of the linear regressions defining D_F) vs. T (top) and D_F vs. T (bottom) for all pilot areas.

As expected, at higher thresholds, the remaining impervious areas are smaller and the associated fractal dimensions are also smaller. It should be noted that the quality of the scaling also tends to diminish for increasing imperviousness thresholds. This effect is significant for some areas such as

Torquay North

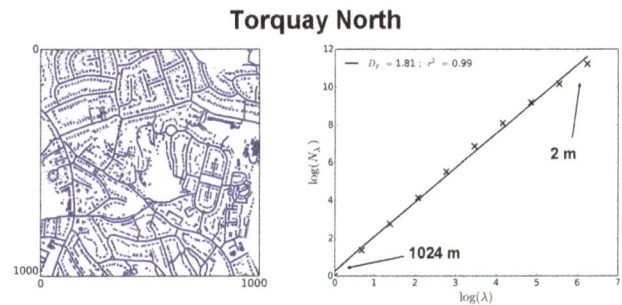

Figure 5. Impervious pixels at a 2 m resolution (left) and computation of the fractal dimension of the corresponding geometrical set, i.e. Eq. (1) in log–log plot, (right) for the Torquay north study area. For the left figure, the axes correspond to metres (m).

Moree-Sausset, Herent, and Sucy-en-Brie and hence limits the possible interpretation of this analysis. In these cases, there is a very limited (one or sometimes even zero) number of remaining sub-catchments at high imperviousness thresholds, which is likely to bias the analysis. This phenomenon is due the smaller number of sub-catchment in these cases. The most critical one is that of Sucy-en-Brie, for which the model consists of only eight sub-catchments (see Fig. 2). Such low spatial resolution hampers implementation of fractal analysis and this is reflected in the low r^2 for thresholds greater than 40 % (no data for $T > 60$ %). Computations on larger areas, which would include more sub-catchments or a higher model resolution (smaller sub-catchment size and greater number of sub-catchments as was done in other study areas) with a high degree of imperviousness (as it is the case for the Rotterdam-Centrum study area), would be needed to confirm this interpretation. This issue illustrates the need for models with a number of sub-catchment enabling one to fully represent the variability of imperviousness. The use of fully distributed models is a way to improve this representation. For hydrological purposes the use of more distributed model also enable one to better account for the spatio-temporal rainfall variability, which is known to have a significant impact on simulated outputs (Gires et al., 2014).

Interestingly, the fractal dimension estimates are in overall agreement with the level of urbanization discussed in the previous section, i.e. the most urbanized areas exhibit the greatest fractal dimension for all thresholds. This is especially true for thresholds lower than 60 %. For greater ones, whose estimates are less reliable, more differences are noted. For instance D_F with $T > 60$ % for London-Cranbrook are much smaller than for Ghent, whereas the estimates from the distributed data are rather close (Table 2). This reflects different choices by the modellers in the representation of the urban catchment. Indeed, imperviousness is one of the main "tuning" variables used in the calibration of urban drainage models. The differences in imperviousness observed between semi-distributed models and distributed datasets may be caused by "lumping" of catchment characteristics in the

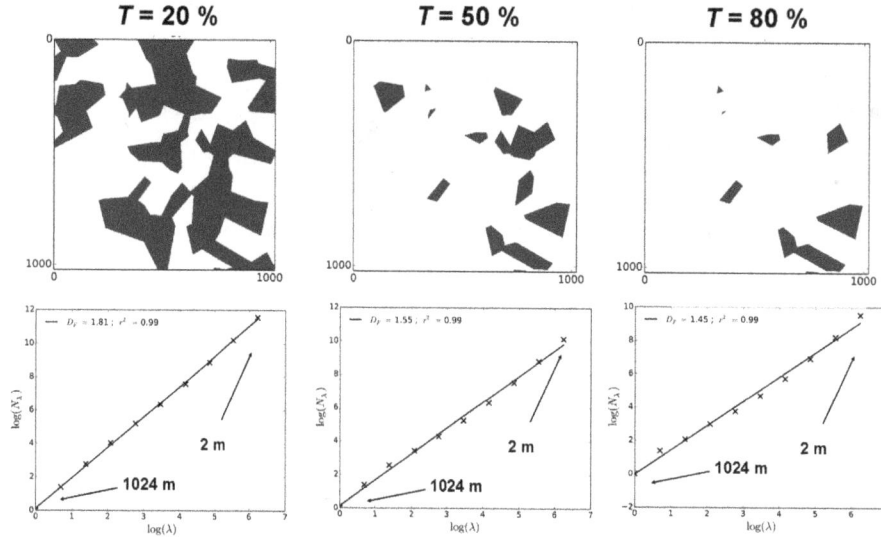

Figure 6. Illustration of the computation of the fractal dimension of the area covered by the sub-catchments, whose imperviousness is greater than a threshold T for T equal to 20 % (left), 50 % (middle), and 80 % (right) for the Torquay north study area: corresponding geometrical set (top) and Eq. (1) in log–log plot (bottom). For the upper figures, the axes correspond to metres (m).

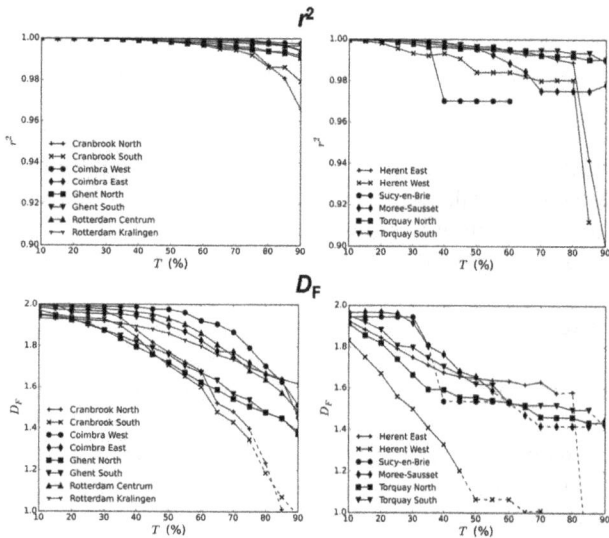

Figure 7. Fractal dimension analysis of the area covered by the subcatchments with imperviousness greater than a threshold T for various values of T: r^2 vs. T (top) and D_F vs. T (bottom). On the bottom curves the dash portions correspond to thresholds for which $r^2 < 0.99$ meaning the estimates are less reliably robust (poorer quality of the scaling). Fractal dimension are computed on the whole range of available scales (i.e. between 2 m and 512–4096 m according the study area).

would need to be further confirmed by analysis on a larger number of datasets, is simply that the spatial structure of the highly impervious areas could exhibit a clear difference with regards to less-urbanized ones (see also multifractal analysis). It should be mentioned that similar to the findings of the previous section, estimates obtained for various areas within a given catchment are rather similar, except for Herent. In Herent the impervious areas fill a greater space in the east study area than in the west one, which was not the case for the imperviousness from the distributed data. This is explained by different modelling choices with respect to the level of detail in catchment representation. Models could also have been calibrated long time before the GIS data were obtained. For Coimbra the differences, especially for low thresholds, are smaller than the ones observed on the sewer system and the distributed imperviousness.

Given that we found that the fractal dimension of subcatchments' imperviousness of semi-distributed models was dependent on the threshold used to define it, we naturally investigated the possibility of using a multifractal framework to analyse this dependency. This is achieved by checking the adequacy of the empirical co-dimension function $c(\gamma)$ with its theoretical expected shape. More precisely, at the maximum resolution Λ, for each studied threshold T, the corresponding singularity γ_T is estimated as $\log_\Lambda \frac{T}{\langle T \rangle}$, where $\langle T \rangle$ is the average of the studied thresholds and equal to 50 here. The empirical value of $c(\gamma_T)$ is then simply given by the fractal co-dimension $(2 - D_F)$. Finally, $2 - D_F$ is plotted as a function of γ_T, along with the theoretical shape of $c(\gamma)$. This technique is known as functional box counting in the literature (Lovejoy et al., 1987). The UM parameters α and C_1 used are those retrieved from DTM analysis and reported in Table 2.

models and errors in the model and/or in the distributed datasets. This effect also partially explains the fact that disparities between the catchments tend to strengthen with increasing thresholds, which are likely to be more affected by modellers' choices. Another possible explanation, which

Torquay North Herent West Cranbrook North Coimbra West

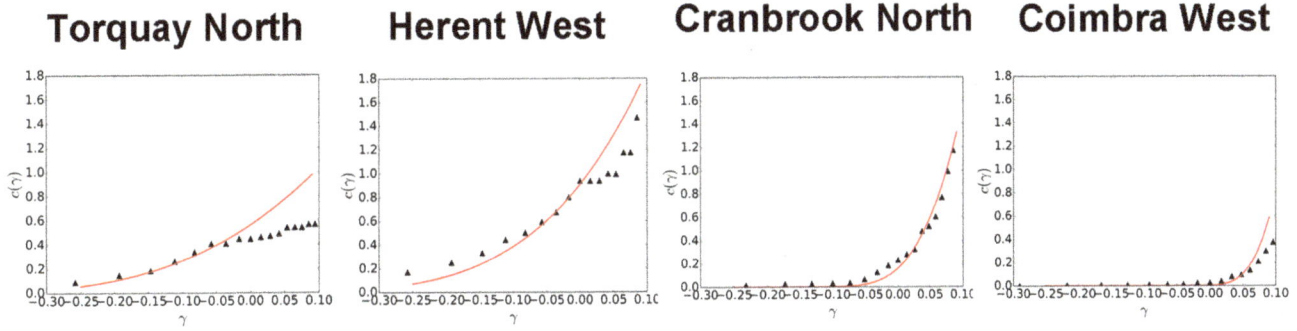

Figure 8. Functional box-counting analysis of the map of sub-catchments imperviousness for four selected catchments. Triangles: for each threshold $2 - D_F$ (Fig. 7) vs. the corresponding singularity γ_T is estimated as $\log_\Lambda \frac{T}{\langle T \rangle}$ (where $\langle T \rangle$ is the average of the studied thresholds and equal to 50 here). Solid line: theoretical shape of $c(\gamma)$ with UM parameters estimated with the help of DTM technique (Table 2).

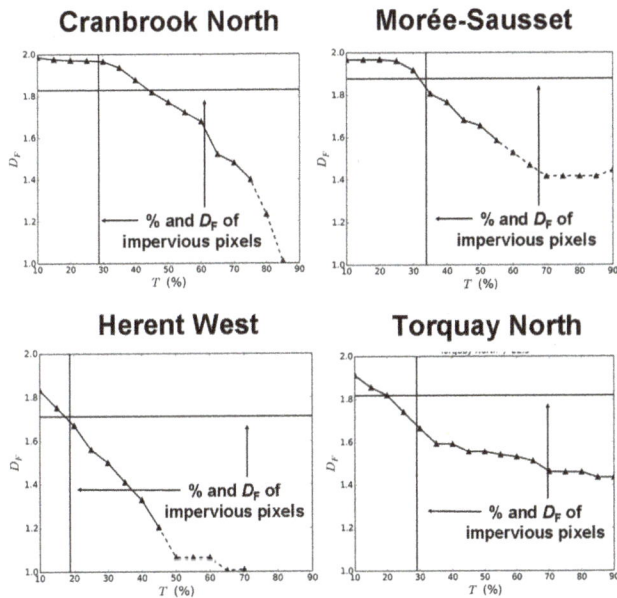

Figure 9. For four study areas: D_F vs. T for the map of sub-catchments imperviousness in model is plotted (same as in Fig. 7), fractal dimension from the distributed data (horizontal line), and percentage of impervious pixel at the two metre resolution (vertical line).

They are generally in the range 1.2–1.6 for α and 0.01–0.09 for C_1. The quality of the scaling related to α and C_1 is low with the coefficient of determination in the linear regressions of the order of 0.8–0.9, meaning that their reliability is not very high. Figure 8 displays these curves for four representative cases. It should be mentioned that the theoretical curve of $c(\gamma)$ was shifted horizontally "manually" to better fit the empirical points. This mimics the effect of H, with which it was not possible to estimate robustly with this dataset. It appears that the agreement between the empirical points and theoretical expectations is good in most of the cases (Herent west, Cranbrook and Torquay on Fig. 8), and it remains valid on a large range of $c(\gamma)$. In other cases such as Coim-

bra west, it is less good and some discrepancies are visible. These results should only be taken as preliminary ones that should be confirmed by further analysis on extended datasets given the limitations of this study: small range of available scales, low quality of the data, which is not actual physical data but a representation with different resolution in models, and manual fitting of H. In some cases, such as Torquay north and to a smaller extent Herent west in Fig. 8, there seems to be a linear behaviour for empirical points associated with large singularities. This is the signature of a multifractal phase transition, which reflects the large-scale influence of small-scale variability. Such behaviour is commonly found in geophysical fields. It is associated with a power-law tail for the probability distribution of the pixels' imperviousness. Results are not reliable enough to get definitive conclusions, but they are encouraging and should be a first step before a more in-depth analysis of the notion of imperviousness and its characterization in a scaling framework. A possible useful application would be the possibility to easily and realistically fill gaps of missing data in imperviousness maps.

Finally, fractal dimensions of the imperviousness computed for the semi-distributed models were compared to those derived from fully distributed GIS data (Sect. 4.1). This is done in Fig. 9 for three studied areas. D_F vs. T for the model is plotted (same as in Fig. 7 bottom) along with the fractal dimension from the distributed data (horizontal line) and the percentage of impervious pixels with 2 m size pixels (vertical line). If the spatial distribution of the average catchment imperviousness is realistically represented in the model, the intersection of these two straight lines should be located on the D_F vs. T curve. This is clearly visible in Fig. 8 for Morée-Sausset and Herent west; much less for Cranbrook. The location of the intersection of the two straight lines below the curve indicates that the Cranbrook model overestimates space filled by the areas with imperviousness greater than the average. In order to quantify this effect, the difference (denoted $\%_{diff}$) between the value of T at the intersection of the D_F vs. T curve with the horizontal line and the percentage of impervious pixels is reported in Table 2.

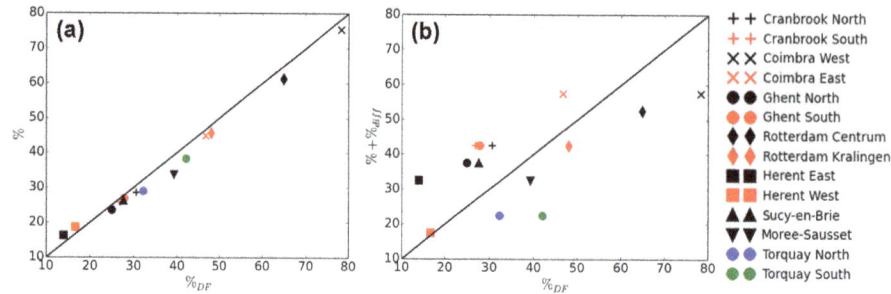

Figure 10. The percentages of distributed imperviousness (%) at the highest data resolution **(a)** and of the imperviousness of semi-distributed models (%+%$_{\text{diff}}$) **(b)** as a function of the percentages of imperviousness resulting from the fractal dimension estimates (%$_{D_F}$). The black line indicates the first bisector.

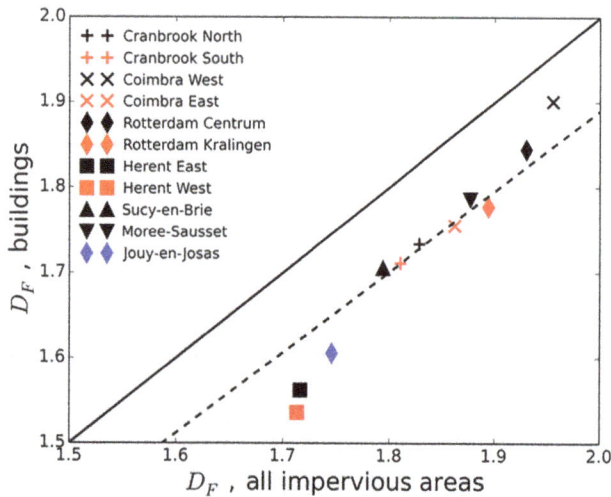

Figure 11. Empirical relation between the fractal dimensions of the total impervious area and of buildings only. The continuous line indicates the first bisector, while dotted line is given by $D_{F_{\text{build}}} = 0.945 D_{F_{\text{all}}}$.

The absolute value of this difference is always smaller than 18 % and smaller than 10 % in five cases. There is no obvious relation between the numerical value of this quantity and the level of resolution of the hydrodynamic model.

The percentages of distributed imperviousness (%) at the highest resolution Λ and of the imperviousness of semi-distributed models (%+%$_{\text{diff}}$) could be compared to the percentages of imperviousness resulting from the fractal dimension estimates: %$_{D_F} = 100\Lambda^{D_F-2}$. Figure 10 displays the results of such a comparison. First of all, this figure (Fig. 10a) demonstrates that for several catchments, uncertainties in scaling estimates result in visible discrepancies between (%) and (%$_{D_F}$) that are expected to be identical in the case of a "perfect" scaling. The difference of these two estimates is based on the fact that the percentages of distributed imperviousness (%) is computed at the highest resolution Λ only, whereas the fractal dimension estimates are computed across all the scales and hence result in a multiscale characteristic

for each catchment. Then, the adjusted percentage of the imperviousness of semi-distributed models, in general, diverges even stronger with regard to the one resulting from the fractal dimension estimates. The only two improvements were observed for the Rotterdam-Kralingen and Herent west catchments (see Fig. 10b).

Such analysis could support validation of the representation of catchments in semi-distributed models; the smaller the difference, the better catchment imperviousness is represented by the model. It should be mentioned that this interpretation assumes that data available for analysing distributed imperviousness are accurate and complete, which is generally supported by the scaling behaviour of the data.

4.3   Representation of imperviousness of buildings

In this sub-section we discuss the results of the comparison of fractal dimensions computed on two different geometrical sets: the total imperviousness areas as roads and buildings ($D_{F_{\text{all}}}$) and the buildings only ($D_{F_{\text{build}}}$). Obtained results show that for each catchment the geometrical set of buildings alone behaves as a fractal set. Indeed as for the analysis carried out in Sect. 4.1 (total imperviousness) straight lines are found in the linear regression of Eq. (1) in log–log plot (not shown) with r^2 remaining greater than 0.99, meaning that numerical values of fractal dimensions are robust. Obviously $D_{F_{\text{build}}}$ could not be greater than $D_{F_{\text{all}}}$, since the building areas are embedded within a larger fractal set of all impervious areas, and we have $N_{\text{build}} = \lambda^{D_{F_{\text{build}}}} = \lambda^{aD_{F_{\text{all}}}}$. The empirical results displayed in Fig. 11 suggest that a common value $a = 0.945$ remains suitable for the majority of the catchments. Such a small coefficient may influence the scaling at the smallest scales only. The changes seem to increase with smaller values either meaning that the network of road has a greater importance in these cases, or is simply due to a slight decline of scaling. Indeed, by comparing Figs. 10 and 11, one may note a slight amplification of scaling issues compared to those observed for the percentages of distributed imperviousness.

This analysis was made to investigate the relationships between the fractality of building distributions, as a source for potential green roofs implementation for water flow management, within fractality of the whole imperviousness areas. Indeed green roofs are one of the available tools that can be used to optimize (if needed) water flows in urban and peri-urban areas, hence the need to better understand their potential distribution. More precisely, to increase the functionality of green roofs over the full range of catchment scales (Versini et al., 2016), an optimization of green roof locations could be made to increase their fractal dimension up to the fractal dimension of the total imperviousness area. The fractal tools could also be used to evaluate the potential impact of green roofs.

5 Conclusions

In this paper we implemented (multi-)fractal analysis in the context of urban hydrology on 10 catchments located in five European countries. The results have consequences both in terms of urban catchment characterization and representation in urban hydrological models.

First, it appears that the fractal dimension of either the sewer network or the impervious pixels (roads or houses) on a 2 m pixels map can be used to characterize the level of urbanization of a given area. In fact, for a given area similar estimates are obtained for both geometrical sets. The main difference is that the scale invariance is valid from one or few kilometres down to only approximately inter-pipe distance for the sewer network, whereas it extends down to 2 m for imperviousness, which matches with the spatial resolution of the imperviousness datasets. This tool is innovative in the context of urban hydrology, because it provides a quantitative estimate of a level of urbanization, which is valid across scales and not only at the scale at which it is defined as for other tools. These findings open new practical perspectives that should be explored in future work. An example is the possibility of identifying consistent – across scales – areas that should be modelled separately. Another one is the possibility of relying on the scale-invariance features to fill gaps of missing data in a realistic way. This issue is increasingly visible as one goes toward higher-resolution model. It is furthermore an acknowledgment of the complexity of the notion of imperviousness which is usually simplified in state-of-the-art urban hydrological models in which it is often represented as a mere percentage, thus neglecting without taking into account its heterogeneous distribution. Using scale-invariant concepts able to handle more appropriately these features is a lead that should used to innovatively improve distributed hydrological models.

Second, the representation of imperviousness in operational semi-distributed models was analysed. It appears that, by analysing the geometrical set made of sub-catchments with imperviousness greater than a given threshold, it is possible to retrieve urbanization patterns. In this study, it was found that fractal dimension values decrease from 1.9–2.0 for imperviousness degrees above 10 % down to 1.4–1.6 for imperviousness degrees above 90 %. Results for higher imperviousness degrees were subject to larger uncertainty as a result of data scarcity; findings should be verified in studies based on larger datasets.

It was also shown that comparing fractal dimension values related to modelled imperviousness to imperviousness represented in high-resolution GIS datasets allows one to quantify how well imperviousness is represented in urban hydrological models. These results open perspectives for the development of tools to verify whether a hydrological model properly represents the degree of imperviousness in a catchment and also to study urbanization patterns emerging at different degrees of imperviousness. Such insights could latter be used in support of hydrological analysis as well as other urban development analyses.

Competing interests. The authors declare that they have no conflict of interest.

Acknowledgements. The authors greatly acknowledge partial financial support from European Union INTERREG IV NWE RainGain Project (http://www.raingain.eu). Authors acknowledge Julien Richard for his help in preparing the data and Fig. 1.

Edited by: C. Onof

References

Bendjoudi, H. and Hubert, P.: Le coefficient de compacite de Gravelius: analyse critique d'un indice de forme des bassins versants, Hydrological Sciences Journal-Journal Des Sciences Hydrologiques, 47, 921–930, 2002.

Berne, A., Delrieu, G., Creutin, J.-D., and Obled, C.: Temporal and spatial resolution of rainfall measurements required for urban hydrology, J. Hydrol., 299, 166–179, 2004.

Chen, L., Wang, J., Fu, F., and Qiu, Y.: Land-use change in a small catchment of northern Loess Plateau, China. Agriculture, Ecosyst. Environ., 86, 163–172, doi:10.1016/S0167-8809(00)00271-1, 2001.

Darrel, J. and Wu, J.: Analysis and simulation of land-use change in the central Arizona – Phoenix region, USA, Land. Ecol., 16, doi:10.1023/A:1013170528551, 2001.

De Bartolo, S. G., Gaudio, R., and Gabriele, S.: Multifractal analysis of river networks: Sandbox approach, Water Resour. Res., 40, W02201, doi:10.1029/2003WR002760, 2004.

De Bartolo, S. G., Primavera, L., Gaudio, R., D'Ippolito, A., and Veltri, M.: Fixed-mass multifractal analysis of river networks and braided channels, Phys. Rev. E, 74, doi:10.1103/PhysRevE.74.026101, 2006.

Foufoula-Georgiou, E. and Sapozhnikov, V.: Scale invariances in the morphology and evolution of braided rivers, Math. Geol., 33, 273–291, doi:10.1023/A:1007682005786, 2001.

Gangodagamage, C., Belmont, P., and Foufoula-Georgiou, E.: Revisiting scaling laws in river basins: new considerations across hillslope and fluvial regimes, Water Resour. Res., 47, W07508, doi:10.1029/2010WR009252, 2011.

Gangodagamage, C., Foufoula-Georgiou, E., and Belmont, P.: River basin organization around the mainstem: scale invariance in tributary branching and the incremental area function, J. Geophys. Res.-Earth Surf., 119, 2174–2193, doi:10.1002/2014JF003304, 2014.

Gires, A., Tchiguirinskaia, I., Schertzer, D., and Lovejoy, S.: Multifractal analysis of an urban hydrological model on a Seine-Saint-Denis study case, Urban Water J., 10, 195–208, 2012.

Gires, A., Giangola-Murzyn, A., Abbes, J. B., Tchiguirinskaia, I., Schertzer, D., and Lovejoy, S.: Impacts of small scale rainfall variability in urban areas: a case study with 1D and 1D/2D hydrological models in a multifractal framework, Urban Water J., 12, 607–617, doi:10.1080/1573062X.2014.923917, 2014.

Gires, A., Tchiguirinskaia, I., Schertzer, D., Ochoa-Rodriguez, S., Willems, P., Ichiba, A., ten Veldhuis, M.-C., Wang, L.-P., Pina, R., Van Assel, J., Bruni, G., Murla Tuyls, D., and ten Veldhuis, M.-C.: Data for "Fractal analysis of urban catchments and their representation in semi-distributed models: imperviousness and sewer system" [Data set], Zenodo, doi:10.5281/zenodo.571181, 2017.

Hentschel, H. E. and Proccacia, I.: The infinite number of generalized dimensions of fractals and strange attractors, Physica, 8D, 435–444, 1983.

Hjelmfelt, A.: Fractals and the river-length catchment-area ratio, Water Resour. Bull., 24, 455–459, 1988.

Hubert, P., Friggit, F., and Carbonnel, J. P.: Multifractal structure of rainfall occurrence in west Africa, in: New Uncertainty Concepts in Hydrology and Water Resources, edited by: Kundzewicz, Z. W., Cambridge University Press, Cambridge, 109–113, 1995.

Iverson, L.: Land-use changes in Illinois, USA: The influence of landscape attributes on current and historic land use, Land. Ecol., 2, 45–61, doi:10.1007/BF00138907, 1988.

La Barbera, P. and Rosso, R.: On the fractal dimension of stream networks, Water Resour. Res., 25, 735–741, doi:10.1029/WR025i004p00735, 1989.

Lavallée, D., Lovejoy, S., and Ladoy, P.: Nonlinear variability and landscape topography: analysis and simulation, in: Fractals in geography, edited by: De Cola, L. and Lam, N., New York, Prentice-Hall, 158–192, 1993.

Lavergnat, J. and Golé, P.: A stochastic rainfrop time distribution model, J. Appl. Met, 37, 805–818, 1998.

Lovejoy, S. and Mandelbrot, B.: Fractal properties of rain and a fractal model, Tellus, 37A, 209–232, 1985.

Lovejoy, S. and Schertzer, D.: Generalized scale-invariance in the atmosphere and fractal models of rain, Water Resour. Res., 21, 1233–1250, 1985.

Lovejoy, S., Schertzer, D., and Tsonis, A. A.: Functional box-counting and multiple elliptical dimensions in rain, Science, 235, 1036–1038, 1987.

Mandelbrot, B. B.: The Fractal Geometry of Nature, W.H. Freeman and Company, New York, 468 pp., 1983.

Nikora, V.: Fractal structures of river plan forms, Water Resour. Res., 27, 1327–1333, doi:10.1029/91WR00095, 1991.

Ochoa-Rodriguez, S., Wang, L.-P., Gires, A., Pina, R., Reinoso-Rondinel, R., Bruni, G., Ichiba, A., Gaitan, S., Cristiano, E., van Assel, J., Kroll, S., Murlà-Tuyls, D., Tisserand, B.,

Schertzer, D., Tchiguirinskaia, I., Onof, C., Willems, P., and ten Veldhuis, M.-C.: Impact of spatial and temporal resolution of rainfall inputs on urban hydrodynamic modelling outputs: A multi-catchment investigation, J. Hydrol., 531, 389–407, doi:10.1016/j.jhydrol.2015.05.035, 2015.

Ogden, F. L. and Julien, P. Y.: Runoff model sensitivity to radar rainfall resolution, J. Hydrol., 158, 1–18, 1994.

Olsson, J., Niemczynowicz, J., and Berndtsson, R.: Fractal analysis of high-resolution rainfall time series, J. Geophys. Res., 98, 23265–23274, 1993.

Pandey, G., Lovejoy, S., and Schertzer, D.: Multifractal analysis including extremes of daily river flow series for basis five to two million square kilometres, one day to 75 years, J. Hydrol., 208, 62–81, 1998.

Rossman, L. A.: Storm Water Management Model User's Manual Version 5.0, Cincinnati, Ohio, 2010.

Rosso, R., Bacchi, B., and La Barbera, P.: Fractal relation of mainstream length to catchment area in river networks, Water Resour. Res., 27, 381–387, doi:10.1029/90WR02404, 1991.

Sarkis, B.: Etude multi-échelle des réseaux d'assainissement, MSc Thesis, Ecole des Ponts ParisTech, 2008.

Schertzer, D. and Lovejoy, S.: Physical modelling and analysis of rain and clouds by anisotropic scaling and multiplicative processes, J. Geophys. Res., 92, 9693–9714, 1987.

Schertzer, D. and Lovejoy, S.: Universal Multifractals do Exist!: Comments on "A statistical analysis of mesoscale rainfall as a random cascade", J. Appl. Meteor., 36, 1296–1303, 1997.

Schertzer, D. and Lovejoy, S.: Multifractals, Generalized Scale Invariance and Complexity in Geophysics, International Journal of Bifurcation and Chaos, 21, 3417–3456, 2011.

Takayasu, H.: Fractals in the physical sciences, Manchester University Press, Manchester, 1990.

Tannier, C., Thomas, I., Vuidel, G., and Frankhauser, P.: A Fractal Approach to Identifying Urban Boundaries, Geogr. Anal., 43, 211–227, doi:10.1111/j.1538-4632.2011.00814.x, 2011.

Tarboton, D., Bras, R., and Rodriguez-Iturbe, I.: The fractal nature of river networks, Water Resour. Res., 24, 1317–1322, doi:10.1029/WR024i008p01317, 1988.

Tarboton, D.: Fractal river networks, Horton's laws and Tokunaga cyclicity, J. Hydrol., 187, 105–117, doi:10.1016/S0022-1694(96)03089-2, 1996.

Tessier, Y., Lovejoy, S., Hubert, P., Schertzer, D., and Pecknold, S.: Multifractal analysis and modeling of rainfall and river flows and scaling, causal transfer functions, J. Geophys. Res., 31D, 26427–26440, 1996.

Veltri, M., Veltri, P., and Maiolo, M.: On the fractal description of natural channel networks, 187, 137–144, doi:10.1016/S0022-1694(96)03091-0, 1996.

Versini, P. A., Gires, A., Tchinguirinskaia, I., and Schertzer, D.: Toward an operational tool to simulate green roof hydrological impact at the basin scale: a new version of the distributed rainfall–runoff model Multi-Hydro, Water Sci. Technol., 74, 1845–1854, doi:10.2166/wst.2016.310, 2016.

Wang, X., Li, M. H., Liu, S., and Liu, G.: Fractal characteristics of soils under different land-use patterns in the arid and semiarid regions of the Tibetan Plateau, China. Geoderma, 134, 56–61, doi:10.1016/j.geoderma.2005.08.014, 2006.

Flow pathways and nutrient transport mechanisms drive hydrochemical sensitivity to climate change across catchments with different geology and topography

J. Crossman[1,2], M. N. Futter[3], P. G. Whitehead[2], E. Stainsby[4], H. M. Baulch[5], L. Jin[6], S. K. Oni[7], R. L. Wilby[8], and P. J. Dillon[1]

[1]Chemical Sciences, Trent University, Peterborough, ON, Canada
[2]Oxford University Centre for the Environment, Oxford University, Oxford, UK
[3]Department of Aquatic Sciences and Assessment, Swedish University of Agricultural Sciences, Uppsala, Sweden
[4]Ontario Ministry of Environment, Etobicoke, ON, Canada
[5]School of Environment and Sustainability and Global Institute for Water Security, University of Saskatchewan, Saskatoon, SK, Canada
[6]Department of Geology, State University of New York College at Cortland, Cortland, NY, USA
[7]Department of Forest Ecology and Management, Swedish University of Agricultural Science, Umeå, Sweden
[8]Department of Geography, Loughborough University, Leicestershire, UK

Correspondence to: J. Crossman (jillcrossman@trentu.ca)

Abstract. Hydrological processes determine the transport of nutrients and passage of diffuse pollution. Consequently, catchments are likely to exhibit individual hydrochemical responses (sensitivities) to climate change, which are expected to alter the timing and amount of runoff, and to impact in-stream water quality. In developing robust catchment management strategies and quantifying plausible future hydrochemical conditions it is therefore equally important to consider the potential for spatial variability in, and causal factors of, catchment sensitivity, as it is to explore future changes in climatic pressures. This study seeks to identify those factors which influence hydrochemical sensitivity to climate change. A perturbed physics ensemble (PPE), derived from a series of global climate model (GCM) variants with specific climate sensitivities was used to project future climate change and uncertainty. Using the INtegrated CAtchment model of Phosphorus dynamics (INCA-P), we quantified potential hydrochemical responses in four neighbouring catchments (with similar land use but varying topographic and geological characteristics) in southern Ontario, Canada. Responses were assessed by comparing a 30 year baseline (1968–1997) to two future periods: 2020–2049 and 2060–2089. Although projected climate change and uncertainties were similar across these catchments, hydrochemical responses (sensitivities) were highly varied. Sensitivity was governed by quaternary geology (influencing flow pathways) and nutrient transport mechanisms. Clay-rich catchments were most sensitive, with total phosphorus (TP) being rapidly transported to rivers via overland flow. In these catchments large annual reductions in TP loads were projected. Sensitivity in the other two catchments, dominated by sandy loams, was lower due to a larger proportion of soil matrix flow, longer soil water residence times and seasonal variability in soil-P saturation. Here smaller changes in TP loads, predominantly increases, were projected. These results suggest that the clay content of soils could be a good indicator of the sensitivity of catchments to climatic input, and reinforces calls for catchment-specific management plans.

1 Introduction

Phosphorus (P) is an essential nutrient in riverine and lotic ecosystems (Jarvie et al., 2002), however when present in concentrations surplus to requirements it can result in eutrophication, where death and decay of excess aquatic plant matter leads to reduction in stream oxygen concentrations (Nicholls, 1995; Jarvie et al., 2006) which affects fish spawning and survival (Evans, 2006). High P loads can also bring about alterations in algal species composition, leading to dominance of blue-green bacteria, some of which produce toxic compounds (Chorus and Bartram, 1999), and blooms can lead to increases in water quality treatment costs (Smith, 2003). Eutrophication of many water bodies around the world is now a pressing concern making nutrient monitoring and management increasingly important.

There are numerous sources and pathways of P to rivers, which may travel either in the dissolved (DP) or particulate (PP) form. As some of the PP may be rapidly converted to bioavailable DP, the movement of both forms (total phosphorus; TP) is generally monitored. Delivery to streams may be direct through sewage effluent, or more diffuse through soils (overland flow and leaching). Leaching can be a significant P transport mechanism where the ability of soil to adsorb P is low (e.g. shortly after fertiliser applications), and where leachate rarely comes into contact with adsorption sites (e.g. through macropore flow) (Haygarth et al., 1998; Hooda et al., 1999; Borling, 2003).

Nutrient transport pathways are influenced by hydrology, which is in turn driven by topography, geology and climate. Changes in climate, therefore, have far reaching implications for the future of catchments, with increasing temperatures and altered precipitation patterns affecting the timing and magnitude of runoff and soil moisture, changing lake levels, groundwater availability and river discharge regimes (Gleik, 1989; Bates et al., 2008). It follows, therefore, that individual catchments are likely to respond to climate change in different ways (van Roosmalen et al., 2007). The extent to which the hydrology and nutrient concentrations of a catchment respond to alterations in climate drivers can be termed its "sensitivity" to climate change (Bates et al., 2008). In developing robust management strategies and quantifying plausible future ranges of hydrology and water quality it is important to explore possible changes in climatic and non-climatic pressures (Whitehead et al., 2009), but equally so to consider the potential for spatial variability in, and causal factors of, catchment sensitivity.

A complicating factor when determining future conditions is climatic uncertainty. This stems from a wide range of sources, including natural variability, the inability to predict future emissions of greenhouse gases, and an imprecise understanding of climate systems (modelling uncertainty) (Goddard and Baethgen, 2009). The incomplete knowledge of the physics of climate systems introduces a significant source of uncertainty into global climate models (GCMs),

known as parameter uncertainty (Wilby and Harris, 2006; Meehl and Stocker, 2007), whereby GCMs developed by different scientists may predict various outcomes under similar forcing conditions. It is therefore important to explore a range of different plausible futures, to provide a better understanding of the vulnerability of catchments (McSweeney and Jones, 2010), including possible threshold effects. Only recently have credible techniques become available to more fully quantify the effect of physical or parametric uncertainty on future projections. These methods include multi-model- and perturbed physics ensembles (PPEs). The former explores uncertainties stemming from differences in model structure by combining outputs from multiple GCMs (Collins et al., 2006). The latter alters the physical parameterisation of a single GCM, within scientifically accepted ranges, to create an array of model variants with specific climate sensitivities (UKCP, 2012). Using a PPE, alterations are made to the GCM in a systematic way, and the effects of different sensitivities on climate projections can be directly quantified (Collins et al., 2006); therefore a wider range of physically plausible future climates can be explored (McSweeney and Jones, 2010).

This study explores the underlying determinants of catchment hydrochemical sensitivity to climate change, examining if any particular catchment characteristics might be used as an indicator of that sensitivity. A comparison of hydrochemical sensitivity is made across four major subcatchments of a large lake in southern Ontario, Canada (Lake Simcoe), by applying five variants of the regionally downscaled Met Office Hadley Centre PPE to a range of catchments with varying characteristics, using process-based hydrochemical models. Objectives include (a) comparing the hydrological and water quality sensitivity (to climate uncertainty) across catchments with different quaternary geology and topography and (b) generating likelihood estimates of future water quality, having accounted for uncertainty in GCM parameters. A greater understanding of catchment sensitivity is intended to facilitate adaptive management strategies, which could look to reduce the extent of a catchment's hydrochemical response to climatic change. Such an approach would lessen reliance upon the current "project-then-act" paradigm, which relies heavily upon the accuracy of individual climate projections. The Simcoe region was chosen for study as 50 years of research had previously been undertaken into the sources and pathways of phosphorous. Agricultural practices, urbanisation, wetland drainage and sewage effluent have all been linked to increasing P levels in the lake (Evans et al., 1996; LSRCA, 2009, Jin et al., 2013), and associated with depleted oxygen concentrations and decreasing populations of trout, herring and whitefish (LSEMS, 1995).

Table 1. Land use cover of the four study catchments within the Lake Simcoe catchment reported to 1 decimal place (1 d.p.).

Land type		% Land cover			
		Holland	Pefferlaw	Beaver	Whites
Agriculture	Intensive	34.9	23.4	28.1	20.4
	Non-intensive	17.8	25.3	37.0	45.7
Urban		19.6	10.6	5.1	1.7
Wetland		10.8	18.2	19.3	21.2
Forest		16.9	22.5	10.5	11.0

2 Methods

2.1 Site description

Lake Simcoe is situated in southern Ontario, approximately 45 km north of Toronto, Canada. It drains north into the Severn River, and ultimately into Georgian Bay (Fig. 1). The total catchment area is 2914 km^2 (with a lake area of 720 km^2), comprised of 21 subcatchments. The four focus catchments (referred to in this study as Holland, Pefferlaw, Beaver and Whites) are situated in the south, and contribute some of the most significant P-inputs to the Lake (Winter et al., 2002, 2007). The combined east and west branches of the Holland form the largest catchment, with an area of 614 km^2. The Pefferlaw and Beaver are a similar size at 444 and 325 km^2; and the Whites is the smallest at 85 km^2. All catchments are characterised by a dominance of agricultural land use (Table 1), though the Beaver and Whites have the highest coverage at over 65 %. Major crops in the area are alfalfa, corn (for grain), soybeans and winter wheat (Statistics Canada, 2011). The Holland has the greatest degree of urbanisation at $\sim$20 %. There are three sewage treatment works (STWs) in the Holland, two in the Beaver, and one in the Pefferlaw. Being much smaller, there are no STWs within the boundaries of the Whites catchment.

A proportion of the headwaters of each catchment are located on the Oak Ridges Moraine, an important area of groundwater recharge (Johnson, 1997), characterised by steep slopes, high infiltration capacity and relatively low surface overland flow (LSRCA, 2012a). The lower reaches of each catchment flows through the Peterborough drumlin fields, Algonquin Plains and Rolling Plains (Johnson, 1997). These are characterised by low slopes, and clayey-till soils (LSRCA, 2012a). This transition in topography and quaternary geological permeability creates areas of waterlogging and marshlands proximal to the Lake (Miles, 2012).

Differences in catchment dominant quaternary geology leads to variations in surface overland flow and soil water residence times (Table 2), and the catchments can be split into two distinct typologies; those with a high proportion of runoff, and rapid throughflow (the Beaver and the Whites) (Miles, 2012); and those transporting water more slowly through the catchment (the Holland and the Peffer-

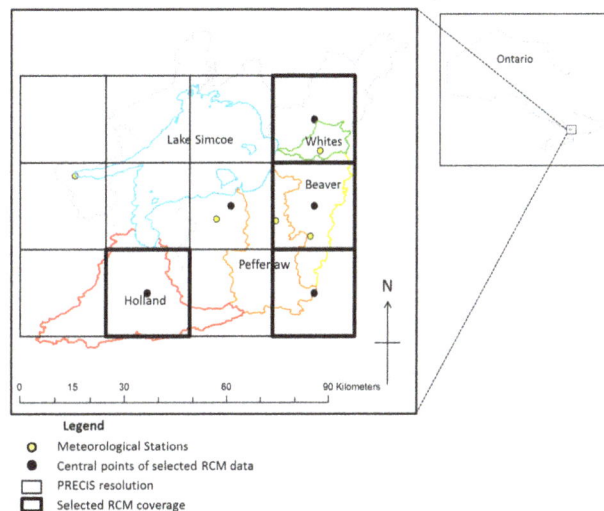

Figure 1. Map of 4 study catchments within the Simcoe Basin, southern Ontario, demonstrating PRECIS (Providing Regional Climates for Impacts Studies) RCM (regional climate model), available meteorological stations, and catchment boundaries.

law) (LSRCA, 2010b), with less runoff, and slow rates of soil matrix flow. The Beaver and Whites have a high coverage of low-permeability clayey soils (Soil Landscapes of Canada Working Group, 2010), and a high density of tile drains; characteristics which are often associated high runoff rates and macropore flow (Huang, 1995; Burt and Pinay, 2005) respectively. The saturation excess threshold of these soils is low, particularly in the northern end of the catchment, where wetlands dominate (Miles, 2012). Slower water transport in the Holland and Pefferlaw might be attributed to a higher soil storage capacity, due to the lower coverage of clayey soils, greater dominance of sandy loams and, in the Pefferlaw, low density of tile drains and a low percentage of impervious surfaces (LSRCA, 2012b). Additionally, in the Holland, extensive seasonal drainage of wetlands is performed (LSRCA, 2010b). Wetland drainage increases soil bulk density, which results in greater soil water retention capacity, and reduction of hydraulic conductivity and of overland flow (Burke, 1967, 1975; Schlotzhauer and Price, 1999). Over 50 % of the Holland's former marshlands are now drained in spring for

Table 2. Details of key parameters and data sources used in model calibration (1 d.p.).

Parameter	Data description	Study site				Data Source
		Holland	Pefferlaw	Beaver	Whites	
Catchment characteristics	Catchment area (km²)	613.8	444.4	325.3	85.0	Modelled using a 2 m vertical resolution DEM (Global Land Cover Facility, 2002)
	Number of subcatchments	39	41	30	23	
	Quaternary Geology (% clay composition)	8.3	12.6	27.5	41.9	Catchment average calculations from GIS maps derived from Agriculture and Agri-Food Canada (2010)
Hydrological characteristics	Temperature (°C) daily average data	7.4	7.3	7.3	7.3	Measurements from local Environment Canada meteorological stations
	Precipitation daily total data (mm d⁻¹)	2.5	2.5	2.5	2.0	
	SMD (mm d⁻¹)	42.8	48.0	47.2	22.1	Derived from HBV model
	HER (mm d⁻¹)	0.5	0.9	1.1	0.7	
	Soil water residence time (days)	3.0	2.7	2.4	1.4	Calculations from flow hydrographs derived from field monitoring data (Dave Woods, personal communication, 2013; LSRCA, 2010a; LSEMS, 2013; PWQMN, 2009)
	Saturation excess threshold (m³ s⁻¹)	0.6	3E-2	0.5	0.5	
	Flow-velocity relationship (flow a and b parameters)	a: 6.0×10^{-2} b: 0.70	a: 5.0×10^{-2} b: 0.20	a: 8.0×10^{-2} b: 0.67	a: 6.0×10^{-2} b: 0.7	Derived from monitoring data (Dave Woods, personal communication, 2012; LSRCA, 2010a; LSEMS, 2013; PWQMN, 2009)
P budget in non-intensive agriculture	Fertiliser inputs from grazing animals and applications to crops (kg ha⁻¹ yr⁻¹)	33.2	32.4	30.0	25.8	Calculations using data from Statistics Canada (2011); OMAFRA (2009); Bangay (1976) using methods of Wade et al. (2007b)
	Septic tank inputs (kg ha⁻¹ yr⁻¹)	1.2	0.8	0.2	0.2	Calculations based on data from Statistics Canada (2011) and Scott et al. (2006); using methods of Paterson et al. (2006) and Stephens (2007)
	Maximum plant uptake (kg ha⁻¹ yr⁻¹)	100	100	100	100	Based on previous INCA applications to the Simcoe region by Jin et al. (2013)
P budget in intensive agriculture	Fertiliser inputs from grazing animals and applications to crops (kg ha⁻¹ yr⁻¹)	21.2	24.0	32.5	32.4	Calculations using data from Statistics Canada (2011); OMAFRA (2009); Bangay (1976) using methods of Wade et al. (2007b)
	Maximum plant uptake (kg ha⁻¹ yr⁻¹)	100	100	100	100	Based on previous INCA applications to the Simcoe region by Jin et al. (2013)
P inputs to whole catchments	Atmospheric deposition: regional values (kg ha⁻¹ yr⁻¹)	0.21	0.21	0.21	0.21	Regional monitoring data (Ramwekellan et al., 2009)
	Groundwater TDP concentration (mg L⁻¹)	7.0E-3	6.0E-3	1.0E-1	7.0E-7	Based on information from the Provincial Groundwater Monitoring Network (PGWMN, 2012)
Sewage (STW) P inputs to river reaches	Number of STWs	3	1	2	N/A	XCG consultants Ltd (2010) and KMK consultants (2004)
	Average Inputs (kg P yr⁻¹)	103.6	102.9	67.2	N/A	

agricultural purposes, resulting in extensive modification to the catchment hydrology (Nicholls and MacCrimmon, 1975; LSRCA, 2010b). Urbanisation has had little influence on surface overland flow (LSRCA, 2010b), and has been attributed to township locations in the more permeable sediments of the Oak Ridges Moraine, where high infiltration capacities counteract the impact of impervious surfaces (Oni et al., 2014).

The average 30 year climate of the study regions (1968–1997) has a distinct seasonal cycle with average peak temperatures in the summer (June–August) of between 17 to 20 °C. Minimum temperatures occur in winter (December–February) of between −4 to −6 °C. Precipitation patterns are highly localised (Smith, 2010), but are generally greatest in summer (June–August) and in autumn (September–November) with between 2.9 and 3.4 mm of precipitation per day, and least in spring (March–May) with between 2.0 and 2.3 mm per day. There is a strong seasonal input of snow to the Simcoe catchment, with local meteorological stations recording an average of 92.3 cm of snow falling during winter, and 26.5 cm during spring. Annual average in-stream TP concentrations range from 0.03 mg L^{-1} in the Beaver to 0.14 mg L^{-1} in the Holland.

2.2 Dynamic modelling of hydrology and phosphorus using INCA-P

2.2.1 Model description

A process based, spatially distributed model was chosen for this study as it was considered more suitable for studying impacts of climate change on hydrochemistry than empirical alternatives (Adams et al., 2013). Empirical models are developed using correlative relationships, based upon a mechanistic understanding, and thus reflect current relationships between dependent and independent variables. Their applicability under future conditions is therefore questionable (Leavesley, 1994). Process-based models, however, have a greater focus on representing the underlying physical processes that describe system behaviour (Adams et al., 2013). As model parameters have a specific physical meaning, which can be defined for future altered catchment states (Bathurst and O'Connell, 1992)these models can be applied with more confidence outside the range of data under which they were developed, and hence have a greater ability to deliver credible projections under a changing climate (Leavesley, 1994).

The dynamic, process-based INCA-P model (INtegrated CAtchment model of Phosphorus dynamics) has been applied to over 40 catchments across Europe and North America (Wade et al., 2002a; Whitehead et al., 2011, 2014; Jin et al., 2013; Baulch et al., 2013 inter alia). It uses a semi-distributed approach to simulate the flow of water and nutrients through the terrestrial system to river reaches, differentiated by land use type. The model can simulate fully branched river networks, with unlimited numbers of tributaries and stream orders. Information flows through the model from the individual process equations, via the subcatchment comprised of up to six land use types, to a network of multiple reaches and tributaries. A full mass balance is imposed at each level (Supplement SI 1).

The flow of water and phosphorus through INCA is modelled through four storage zones (SI 2), using a series of detailed process equations, solved using numerical integration routines (Wade et al., 2002a, b, 2007a). By analysing phosphorus inputs from diffuse sources, sewage treatment works (STWs), sediment interactions and biological processes (Wade et al., 2002a) the model estimates daily values for a range of variables both in the terrestrial and aquatic phase. Terrestrial model outputs used in this study include nutrient export coefficients, and concentrations of both soil water total dissolved phosphorus (TDP) and labile P. Aquatic outputs include flow, and concentrations of in-stream total phosphorus (TP), TDP, and particulate phosphorus (PP).

INCA-P requires a daily input time series of precipitation, temperature, hydrologically effective rainfall (HER) and soil moisture deficit (SMD). Precipitation and temperature are obtained from local meteorological stations, and HER and SMD from the rainfall-runoff model Nordic HBV (Hydrologiska Byråns Vattenbalansavdelningen) (Saelthun, 1995). There are five main storage components within HBV, and the physical processes operating within and between these zones are represented through simplified mathematical expressions which can be adjusted through a series of parameters to within recommended ranges, to attain a best fit compared with observed river flow records (Oni et al., 2011). An individual HBV model set-up was used for each of the study catchments.

2.2.2 Model calibration

Calibration periods were selected that maximise available spatial resolution and temporal longevity of observed data, and as a result, the calibration period for each catchment varied according to availability of data (SI 3). Data was not reserved for independent validation as these are open, dynamic and natural systems, and no amount of independent data can account for a system's potential to change in unanticipated ways (Oreskes et al., 1994); i.e. validation by independent data does not prove accuracy of future predictions any more than calibration. Again, the use of a process-based model (as opposed to empirical) gives higher confidence that models will perform adequately under future conditions (Leavesly, 1994; Adams et al., 2013). Lengthening the calibration period has been shown to enhance model performance and has proven the most effective method in applying water quality models to predictive studies (Larssen et al., 2007); hence the chosen method was to calibrate models using the widest possible range of conditions.

Whilst the Holland, Pefferlaw and Beaver were all calibrated for over a decade (20, 17 and 12 years respectively), the short observed record available for the Whites (2010–2012) limited calibration to three years. Intensive monitoring over 17 sites within the Whites, however, enabled a detailed calibration across the whole catchment. Given the highly localised nature of rainfall patterns within the Simcoe region (Smith, 2010), catchment-specific observed daily temperature, precipitation and flow data was used for each HBV model (PWQMN, 2009; LSEMS, 2013; LSRCA, 2010a; Environment Canada, 2013; D. Woods, personal communication, 2013), and the derived HER and SMD were used to complete the daily time series required for INCA-P. For each catchment, the most proximal meteorological station data was used (Fig. 1). Where station record length was insufficient, the nearest available records were used and adjusted to the local baseline conditions using bias correction techniques of Futter et al. (2009).

Each of the four study catchments was subdivided into their constituent subcatchment network using ArcHydro GIS software, and a hydrological network developed from a 2 m vertical resolution digital elevation model (DEM) (Global LandCover Facility, 2002) (SI3). Land uses were grouped into five cover classes; urban, intensive agricultural, non-intensive agricultural, wetlands and forest, derived from Ecological Land Classification of Ontario data (Ontario Ministry of Natural Resources, 2007) (Table 1). Parameter inputs for model calibration were then calculated. As a distributed model, INCA can be calibrated individually to each land use, subcatchment and river reach within the catchment, giving extensive flexibility where conditions vary significantly within local areas. As there are over 107 parameters within the INCA model, plausibility of calibration values was assessed through a variety of methods including field measurements, GIS and digital elevation assessments, literature values, expert judgement and model performance statistics. For those model parameters which have the greatest impact on model output, it is especially important to obtain accurate calibration values. Within INCA-P these parameters include stream velocity, Freundlich coefficients, and soil P concentrations (Crossman et al., 2013c; Lepistö et al., 2013). Details of methods used to derive values for these (and other) parameters are given below, with average values for each catchment given in Table 2.

P-inputs from fertilisers (kg P day^{-1}) were calculated using the methods of Wade et al. (2007b) and Baulch et al. (2013), from local recommended P-application rates (OMAFRA, 2009) and crop types (Statistics Canada, 2011). Inputs of P from livestock waste were calculated using current livestock assessment numbers (Statistics Canada, 2011) combined with Ontario livestock phosphorus nutrient coefficients (Bangay, 1976). Based on OMAFRA recommendations, model fertiliser inputs were extended over a 60 day period to intensively cropped areas, and a 120 day period to non-intensive agricultural lands beginning in late April

(Baulch et al., 2013; Whitehead et al., 2011). Plant uptake in these agricultural areas was set to a maximum of 100 kg ha^{-1} yr^{-1}, based on previous INCA-P applications to the Simcoe region (Jin et al., 2013). In addition to effluent from STWs (Table 2), a large proportion of households in rural areas contribute P loads through septic systems. Input loads were determined using methods of Scott and Winter (2006) and Paterson et al. (2006), calculated as a function of annual P input per person, combined with septic system usage. Data used included GIS information on households connected to septic systems (Louis Berger Group Inc., 2010), household population data (Statistics Canada., 2011), and the average TP load from excretion (2.6 g TP person^{-1} day^{-1}; Stephens, 2007). As the efficiency of P removal is estimated at 57 % (Whitehead et al., 2011), only 43 % of the calculated septic TP load was input to the model.

Initial soil P concentrations were based on values measured by Fournier et al. (1994), and soil equilibrium coefficients based on laboratory-derived equilibrium phosphorus concentration (EPC$_0$) and Freundlich isotherm values for different land use and soil types (Peltouvuori, 2006; Väänänen, 2008; Koski-Vähälä, 2001). These values were applied to the dominant soil types of each land use category within INCA (Olding et al., 1950; Soil Landscapes of Canada Working Group, 2010). Average catchment values for EPC$_0$ ranged from 0.01 in the Beaver to 0.08 in the Holland, and are consistent with those used by Baulch et al. (2013) and Whitehead et al. (2011). An overview of key input variables and data sources is given in Table 2, with model structures provided in SI 3.

2.3 Climate modelling using perturbed physics ensemble

The PPE used in this study is designed to quantify uncertainty in model predictions (QUMP) (McSweeney and Jones, 2010), and was developed from the HADCM3 global climate model (GCM), under the SRES-A1B scenario. Whilst ideally the widest range of ensemble member climate sensitivities would be explored, it is important that the models used continue to represent regional climatic behaviours (Collins et al., 2006). Therefore, as suggested by McSweeney and Jones (2010), a subset of five of the available 17 members were chosen, which cover the widest range of climate sensitivities, but which also accurately reproduce baseline temperature and precipitation patterns (Table 3). In the Simcoe basin, ensemble members with higher sensitivities were less representative of regional behaviours; and a maximum appropriate sensitivity of 4.0 was determined (Table 3).

The five members were dynamically downscaled by the Institute for Energy, Environment and Sustainable Communities (IEESC, 2012) to 25 km^2 grid cells using the regional climate model PRECIS (Providing Regional Climates for Impacts Studies). Complete details of the model components within PRECIS and of its application to southern Ontario

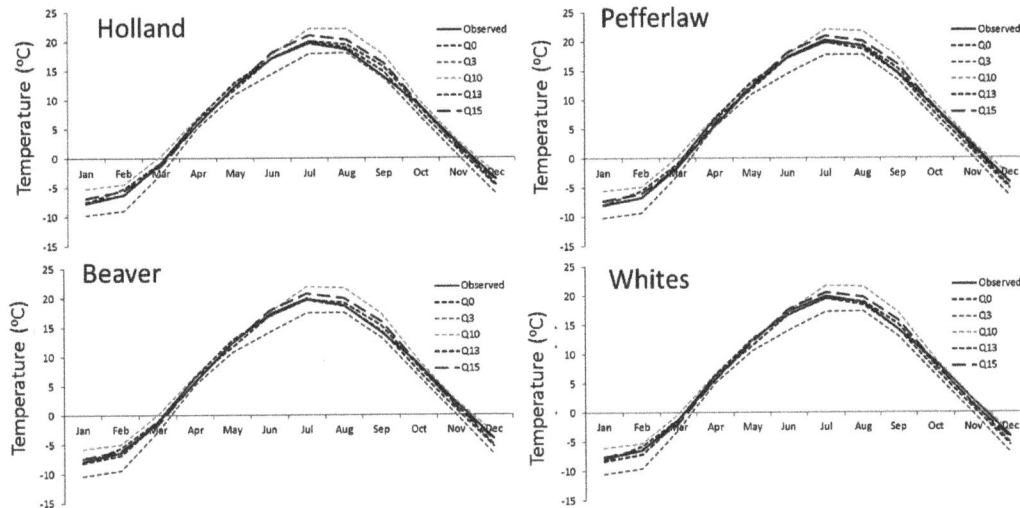

Figure 2. Comparison of monthly temperature data from 5 ensemble members (Q0–15), regionally downscaled using PRECIS, with observed meteorological station data over 30 year baseline period (1968–1997).

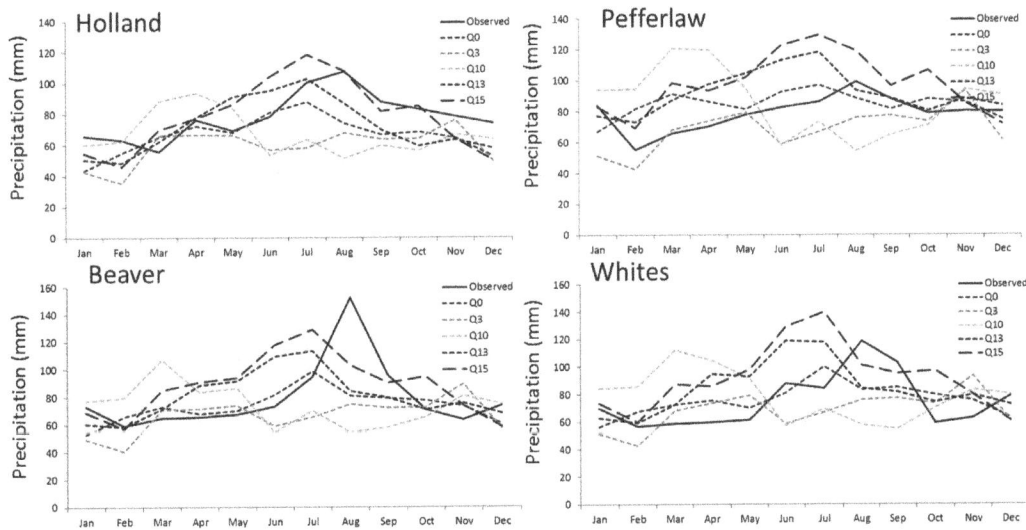

Figure 3. Comparison of monthly precipitation data from 5 ensemble members (Q0–15), regionally downscaled using PRECIS, with observed meteorological station data over 30 year baseline period (1968–1997).

are given in Jones et al. (2004), and IEESC (2012). PRECIS was run from 1968 to 2100, and daily outputs of mean temperature and total precipitation obtained for each ensemble member. Output grid cells from PRECIS were selected for analysis which covered the respective Simcoe catchments. At this resolution most catchments fall within a single grid cell; where two or more PRECIS outputs were available for a single catchment, that which was geographically closest to the catchment's observed monitoring station was selected. Individual grid cells were used for each catchment to reflect the highly localised nature of Simcoe precipitation patterns.

Whilst GCM outputs are not intended to be weather forecasts it is important that each ensemble member provides plausible climate information (Futter et al., 2009). The

standard climatic baseline for southern Ontario has been established as the 30-year period of 1968–1997 (IEESC, 2012). Therefore to establish the suitability of each scenario, monthly average comparisons for both temperature and precipitation were made between QUMP outputs and observed data over this baseline period. Although reproduction of temperature by QUMP members was highly successful (Fig. 2), with an average monthly R^2 value of 0.99, precipitation was less accurate with an R^2 of only 0.23 (Fig. 3). The distinctive seasonal patterns of minimum winter, and maximum summer precipitation values are well represented by ensemble members Q0, Q13 and Q15; whereas ensemble members Q3 and Q10 present inverse seasonal patterns. Whilst removing these less accurate members from the study would restrict

Table 3. Quantifying Uncertainty in Model Predictions (QUMP) members listed in order of climate sensitivity under a doubling of CO_2 concentrations (from Collins et al., 2006). Bold members indicate those selected for use in the study.

QUMP member	Climate sensitivity
Q5	2.2
Q9	2.2
Q1	2.6
Q15	**2.9**
Q16	3.0
Q2	3.1
Q14	3.1
Q13	**3.2**
Q11	3.2
Q7	3.3
Q	3.4
Q0	**3.5**
Q6	3.8
Q3	**3.8**
Q10	**3.9**
Q4	4.6
Q8	4.9
Q12	4.9

the range of future conditions applied in the analysis, use of many of the available bias correction techniques might also be considered questionable for a PPE study. Linear scaling, local intensity scaling, variance scaling and power transformations, for example, all perturb future regional climate model (RCM) simulations to account for biases detected during the baseline period (Lenderink et al., 2007) and in doing so would alter the future change of each ensemble member to different extents. This would render the study incapable of directly assessing impacts of climate variability (originating from internal GCM parameter uncertainty) upon catchment responses.

In contrast to the above, however, the Delta Change (Δ Change) method perturbs a time series of observed meteorological data to match projected future changes. Where there are discrepancies between baseline observed and baseline GCM data, this method assumes that the RCM is better able to simulate relative change than it is to provide absolute values (Hay et al., 2000). In the case of the PPE it also preserves the original relationship between QUMP ensemble members. Δ Change is an established technique for directly assessing impacts of climate change, and was the primary method used to generate future scenarios in the U.S National Assessment (Hay et al., 2000). It was chosen here in order to ensure plausible projections for the Simcoe region under all ensemble members, whilst incorporating the maximum range of conditions in the study. Whilst all bias correction methods have their limitations and assumptions, it should be noted that delta change specifically does not account for changes

in future event frequency (Teutschbein and Seibert, 2012). The same, however, is true of the linear-scaling method. In truth, all correction algorithms assume that the same correction factor will apply under future climate conditions. Despite their various limitations, however, simulation performance is greatly improved following the correction for bias (Teutschbein and Seibert, 2012).

Δ Change was applied individually to each study catchment, using the associated gridded PRECIS outputs and observed meteorological data. Monthly Δ Change was calculated as the average monthly difference between each ensemble baseline (1968–1997) and a future 30-year period; future periods selected for comparison were 2020–2049 (the 2030s), and 2060–2089 (the 2070s). These differences were applied to the observed meteorological conditions of the baseline period, so that each month in the observed data set underwent the same degree of change in climate as was originally demonstrated by the particular ensemble member. The standard formulae were used to determine the Δ Change for future daily temperature (T_f, daily) and future daily precipitation (P_f, daily) (Akhtar et al., 2008), where additive change factors are used for temperature, and multiplicative for precipitation:

$$T_{f,daily} = T_{o,daily} + \left(T_{f,monthly} - T_{b, monthly}\right) \quad (1)$$

$$P_{t,daily} = P_{o,daily} \times \frac{P_{f,monthly}}{P_{b,monthly}} \quad (2)$$

where $T_{o,daily}$ and $P_{o,daily}$ are the observed daily temperature and precipitation; $T_{f,monthly}$ and $P_{f,monthly}$ are the mean monthly QUMP simulated future temperature and precipitation; $T_{b,monthly}$ and $P_{b,monthly}$ are the mean monthly QUMP simulated baseline temperature and precipitation. Monthly Δ change values of precipitation (%) and temperature (°C) are provided in SI 4 and SI 5 for reference. The new Δ change-adjusted temperature and precipitation time series were then passed through respective HBV catchment models to derive associated SMD and HER. Eleven time series were generated for each catchment; five ensemble members over two different future 30 year time periods, and one baseline 30-year period of observed meteorological data. Finally, these time series of temperature, precipitation, SMD and HER were used as input for the INCA-P models to generate future values of water quality (TP) and flow.

2.4 Data analysis

Future change in variables (TP concentrations, TP loads and hydrology) were assessed as a percentage difference between 30 year baseline and 30 year future time periods. In this way, the impact of potential INCA-P model deficiencies during the calibration period were minimised, if model performance accuracy is presumed to be similar under both current and future scenarios. The ensemble cumulative distribution function (CDF) of monthly changes in each variable (temperature, precipitation, flow, and TP) was calculated to

Table 4. Model performance statistics presented for locations of highest calibration accuracy and at downstream extent (for which corresponding observed flow, TP and loads data were available). Results are reported to 2 d.p. For individual performance statistics of each tributary, see Fig. 6.

| Catchment | Maximum model coefficient | | | | | Model coefficient at downstream extent | | | | |
| | Flow | | TP concentration | | TP loads** | Flow | | TP concentration | | TP loads** |
	R^2	Model error (MAE) (%)	R^2	MAE (%)	R^2	R^2	MAE (%)	R^2	MAE (%)	R^2
Holland	0.95	−24.78	0.72	−4.36	0.74	0.95	−24.78	0.52*	−20.12*	0.60*
Pefferlaw	0.91	+14.06	0.37	−23.4	0.71	0.91	+46.28	0.34	−23.41	0.61
Beaver	0.98	−15.30	0.79	+4.40	0.82	0.85	33.50	0.05	−58.37	0.72
Whites	0.92	+3.03	0.47	−26.34	0.94	0.92	+3.03	0.29	−26.34	0.77

* Due to lack of observed data available for model performance assessment in the convergence zone of the East and West branch of the Holland, reported outflow performance statistics are an average (mean) of statistics from the two individual branches. ** Due to the different temporal resolution of sampling vs. modelled data "observed loads" are not directly comparable with modelled loads. Both are calculated using monthly means of flow and total phosphorus data, however, although degree of load variance explained by the model can be determined (R^2), a direct quantification of over or underestimation (model error) cannot be calculated.

identify the likelihood of a change *being less than a certain amount*, having accounted for the projections from all ensemble members. Uncertainty was calculated as the range of values projected by ensemble members at a specific likelihood level (Fig. 4), and "average uncertainty" was calculated as the mean value of uncertainty across all likelihood levels. The term "likelihood" is used in contrast to probability, where likelihood estimates give the odds of an event given specific data, and probability requires all outcomes to be accounted for. Due to the random and open character of natural systems, it is impossible to account for all possible outcomes, and therefore probability cannot be used in the statistical sense. Whilst the term "subjective probability" would also be appropriate (UKCP, 2010), likelihood is more commonly used.

The full spectrum of catchment responses projected (both hydrological and chemical) was assessed by looking at the range of catchment results (i.e. the maximum catchment value minus the minimum catchment value). In addition, catchment sensitivity metrics were derived to assess catchment responses independent of between-catchment differences in climate drivers. The ensemble average catchment response variable (monthly change in hydrology – $m^3 s^{-1}$ – and phosphorus concentration – $mg L^{-1}$) was divided by monthly change in precipitation (mm), giving an output of "catchment response per unit change in precipitation". A similar metric was not established for unit change in temperature due to insignificant differences in this driving variable between catchments. Sensitivities were compared across catchments to provide an indicator as to which underlying factors might contribute to their responses.

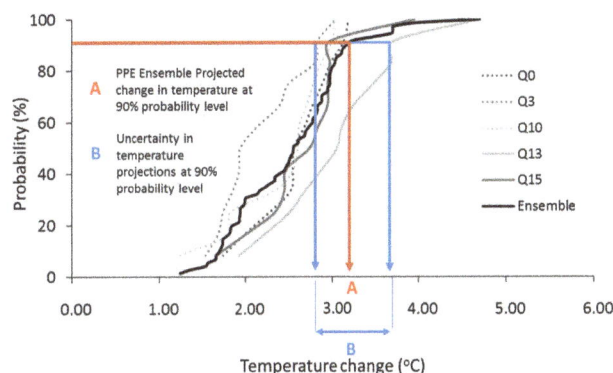

Figure 4. Calculations of (**a**) monthly change in temperature and (**b**) projected uncertainty from cumulative distribution functions of perturbed physics ensemble temperature data of the Holland River in 2030.

3 Results

3.1 INCA-P model calibration

Model outputs were compared to observed flow data (PWQMN, 2009; LSEMS, 2013; LSRCA, 2010a Environment Canada, 2013; D. Woods, personal communication, 2013), and grab samples of TP. Although long term TP records were collated from a number of different sources (Table 2), all providers used similar methods of water quality analysis (colorimetric analysis following a digestion on the whole water sample). A summary of model performance statistics is provided in Table 4, an example of daily model output is given in Fig. 5, and the spatial variability of model fit throughout the catchment is provided in Fig. 6. All model performance statistics represent monthly averages, where model average error (MAE) is calculated as in Crossman et

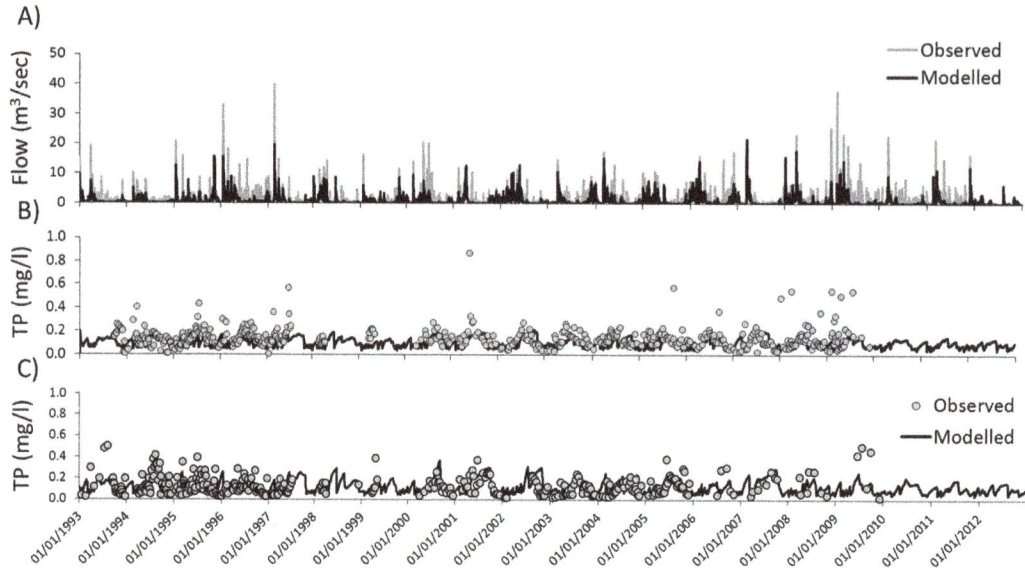

Figure 5. Example of INCA model output. Calibration results from Holland River comparing modelled and observed data of (**a**) flow at East Branch outflow, (**b**) total phosphorus (TP) concentrations at East Branch outflow, (**c**) TP concentration at West Branch outflow.

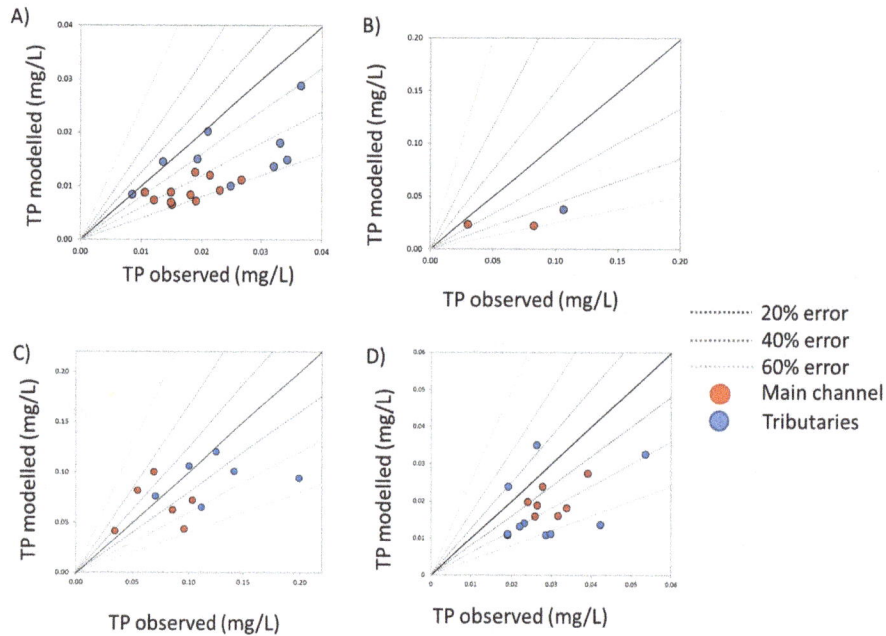

Figure 6. Spatial variability in model accuracy of total phosphorus (TP) concentrations throughout INCA models calibrated across multiple reaches, illustrating relative model error in (**a**) Holland, (**b**) Pefferlaw, (**c**) Beaver and (**d**) Whites catchments. Data represents average TP concentrations over the calibration period.

al. (2013a):

$$\text{MAE} = \left(\sum \frac{\left(A^t o - A^t m \right)}{\sum A^t o} \right) \times 100 \qquad (3)$$

where $A^t o$ is the modelled value at time step t, and $A^t m$ is the observed value at time step t.

During calibration periods there was considerable spatial variability in TP concentration model performance statistics within each catchment (Fig. 6). In general, model accuracy was greatest in catchment outflows and lower in tributaries. In the Holland the greatest flow accuracy was achieved near the outflow, with $R^2 = 0.95$, and MAE $= -24.8\%$. The accuracy of TP concentrations was also highest in the main chan-

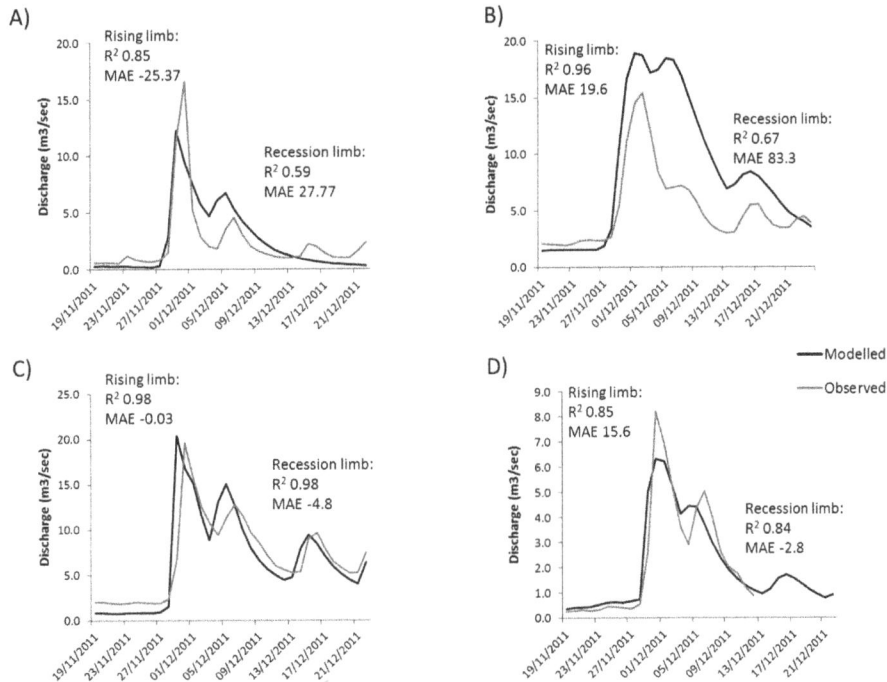

Figure 7. Analysis of INCA model fit to event data at outflow of (**a**) Holland, (**b**) Pefferlaw, (**c**) Beaver and (**d**) Whites catchments. Rising and recession limb at analysed separately for R^2 and model average error (MAE) to determine model performance under precipitation events. All events were analysed during the same four day period of intense precipitation in November 2011. Although event curves can be used to confirm model representation of reality, it should be noted they should not be used to compare response times between catchments.

nel, with an $R^2 = 0.72$, and MAE $= -4.4\%$. TP loads are well represented at the downstream extent where $R^2 = 0.74$. In the Pefferlaw, model performance within the main channel ranged from $R^2 = 0.91$ to 0.65 for flow, and MAE from 14.1 to 46.3 %. Near the outflow, explained variance in annual TP concentrations is fair ($R^2 = 0.34$) with an error of -23.4%. This need not be interpreted as a poor model fit as accuracy varied significantly with season, where MAE is as low as -13.0% in winter, but as high as -36.0% in spring and can be attributed to difficulty in representing occasional peaks in TP during March and April. Model accuracy in reproducing annual loads varies from good to excellent with an R^2 of 0.61 to 0.71.

In the Beaver, the greatest model flow accuracy was found in the upstream tributaries, where flow $R^2 = 0.98$, and MAE was as low as -15.3%. R^2 of TP concentrations $= 0.79$ and MAE was $+4.4\%$. The explained variance for TP load was also high (up to 0.82). At the downstream extent, flow variability was simulated well, with $R^2 = 0.85$, though MAE was $+33.5\%$. Although the fit to TP concentrations at the outflow was poor ($R^2 = 0.05$ and MAE $= -58.4\%$) that of TP loads remained good ($R^2 = 0.72$). Despite the shorter duration of observed data available for the Whites, the explained variance at the outflow for TP concentrations, flow and TP loads were 0.92, 0.29 and 0.77, respectively. Average annual flow error at this site was only 3.0 %. The model does underpredict TP concentrations with an outflow MAE of -26.3%,

which is lowest during autumn and spring (-6.6 and 7.6 %, respectively) and highest during summer (-36.2%). In headwaters and tributaries, R^2 of flow and TP loads of up to 0.90 and 0.94 were achieved. Success in modelling TP concentrations was highly spatially variable (Fig. 6), with a maximum $R^2 = 0.47$.

In addition to the monthly model performance statistics, model responses to precipitation and seasonal snowmelt events were assessed at catchment outflows (Figs. 7 and 8). High R^2 and low model errors during daily analysis of both the rising and recession limbs of hydrographs in each study catchment demonstrate the accuracy of, and give confidence in, the runoff simulations performed by the HBV-INCA model chain. Similarly, there was high seasonal accuracy of snowmelt responses (Fig. 8), increasing support for the models' ability to simulate flow pathways.

TP soil export coefficients ($kg\,TP\,km^{-2}\,yr^{-1}$) were calculated and compared with previous studies to assess the plausibility of simulated nutrient transport pathways, to determine the relative contributions of TP from each land use type, and to identify catchments with noteworthy P export. INCA source apportionment outputs ($kg\,TP\,km^{-2}$) were divided over the years of model runs ($kg\,TP\,km^{-2}\,yr^{-1}$). Values were highly variable between sites, with catchment averages ranging from $3.4\,kg\,km^{-2}\,yr^{-1}$ in the Beaver, to $13.1\,kg\,km^{-2}\,yr^{-1}$ in the Holland (Table 5). These findings are consistent with previous studies of these catchments

Table 5. Export coefficients (kg km^{-2} yr^{-1}) of total phosphorus from diffuse and point sources, and contributions of each source as a percentage of total phosphorus export (1 d.p.).

	Source	Coefficients (kg P km^{-2} yr^{-1})				Source (% contribution)			
		East/West Holland	Pefferlaw	Beaver	Whites	East/West Holland	Pefferlaw	Beaver	Whites
In-stream	STW	0.8	0.3	0.5	0.00	5.5	3.7	13.3	0.00
Soils	Urban	9.9	4.8	2.4	2.1	14.0	6.1	3.1	0.4
	Intensive agriculture	16.1	8.3	3.7	8.7	40.8	23.0	26.3	21.2
	Non-intensive agriculture	16.0	13.8	4.2	8.9	20.6	41.3	38.9	48.7
	Wetlands	14.4	5.1	2.5	8.1	11.3	11.0	12.1	20.4
	Forest	6.4	5.6	2.4	7.1	7.8	14.9	6.3	9.3
	Average soils export	*13.1*	*8.1*	*3.4*	*8.5*	*100*	*100*	*100*	*100*

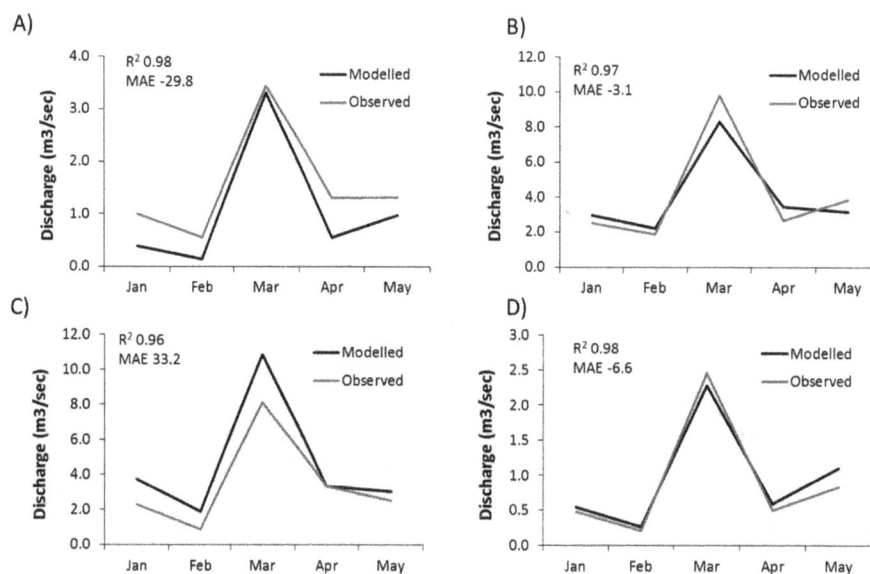

Figure 8. Analysis of model response to observed spring melt in (**a**) Holland, (**b**) Pefferlaw, (**c**) Beaver and (**d**) Whites catchments. Results analysed during spring melt of 2010.

(Thomas and Sevean, 1985; Winter et al., 2007; Baulch et al., 2013), although higher exports have previously been reported for the Holland (Winter et al., 2007). This could be attributed to the larger spatial domain used in this study (covering both East and West branches of the river), plus a recent decline in TP exports due to implementation of best management practices.

In each catchment, the highest exports of TP originated from areas of agricultural land use, ranging from 3.7 kg km^{-2} yr^{-1} (Beaver) to 16.1 kg km^{-2} yr^{-1} (Holland). The lowest exports of TP originated from sewage treatment works, with the lowest outputs from the Whites (no output) and the greatest from the Holland (0.8 kg km^{-2} yr^{-1}). It is notable that in the Holland exceptionally high export coefficients were found in wetlands (14.4 kg km^2 yr^{-1}). This is consistent with drainage of former wetlands and use for intensive agriculture. The percentage contribution of each

source to TP output from the catchment was calculated by multiplying each coefficient by land use area, and (in the case of STWs) by calculating the load differences resulting from models run with and without STW inputs (Table 5). In the Whites, Beaver and Pefferlaw, agriculture was by far the major source of TP (ranging from 64.3 to 69.9 %). In the Holland, although agricultural sources dominate, there is also a large contribution from urban areas and STW (14.0 and 5.5 %, respectively).

Terrestrial model outputs indicate significant seasonal variability in soil TDP concentrations within the agricultural areas of each catchment, with large increases in summer following fertiliser additions (SI 6), and associated increases in the soil labile pool, followed by reductions in winter and spring. These simulated processes are consistent with those described by Haygarth et al. (1998) and Borling (2003). The consistency of modelled nutrient and hydrological pro-

Table 6. A measure of average uncertainty of temperature, precipitation, flow, TP concentrations and loads across the perturbed physics ensemble, where "average uncertainty" is calculated as the mean value of uncertainty across all probability levels within the cumulative distribution function (1 d.p.).

Date	Catchment	Average uncertainty				
		Temperature (°C)	Precipitation (%)	Flow (%)	TP (%)	Loads (%)
	Holland	0.9	19.6	34.0	5.9	39.3
	Pefferlaw	0.9	16.2	23.3	7.8	27.9
2030	Beaver	0.9	16.9	29.0	11.3	36.6
	Whites	0.9	19.7	43.2	20.8	58.9
	Range	3.0×10^{-2}	3.5	19.9	15.0	30.9
	Holland	1.9	23.7	37.3	7.1	43.0
	Pefferlaw	1.9	23.1	27.6	11.1	33.4
2070	Beaver	1.9	23.2	35.1	11.9	38.7
	Whites	2.0	22.60	40.1	23.6	53.8
	Range	0.1	1.1	12.5	16.5	20.4

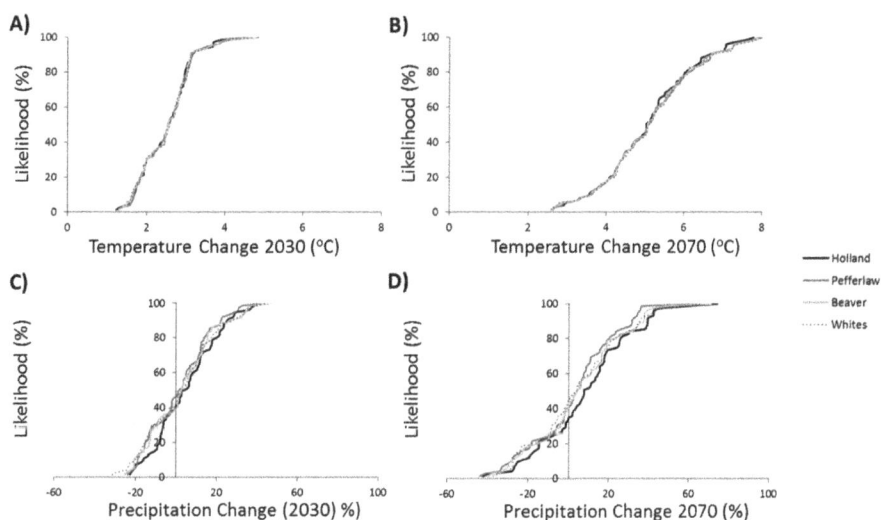

Figure 9. Between-catchment comparison of the CDFs from the perturbed physics ensemble of (**a**) monthly change in temperature by 2030, (**b**) monthly change in temperature by 2070, (**c**) monthly change in precipitation by 2030, (**d**) monthly change in precipitation by 2070 (CDF: cumulative distribution function).

cesses with existing system understanding gives confidence that catchment nutrient transport pathways are accurately represented.

3.2 Climate change

3.2.1 Likelihood of change

Projected changes in temperature and precipitation for 10, 50 and 90 % likelihood levels indicate that changes will become progressively more extreme and more likely between the 2030s and 2070s (SI 7). Notably, there is very little difference between catchments in projected climatic drivers; by 2070 at the 50 % likelihood level there is a cross catch-

ment range of only 0.1 °C (temperature) and 4.1 % (precipitation). The average temperature uncertainty within all catchments was low (Table 6), at < 1 °C in 2030, and < 2 °C in 2070. Average precipitation uncertainty was higher, at < 19 % in 2030 and < 23 % in 2070. Of note is the similarity in uncertainty between catchments (Fig. 9a–d), with a cross-catchment range in average temperature uncertainty of only 3.0×10^{-2} (2030) and 0.1 (2070) (Table 6), and cross-catchment range in average precipitation uncertainty of only 3.5 % (2030) and 1.1 % (2070) (Table 6).

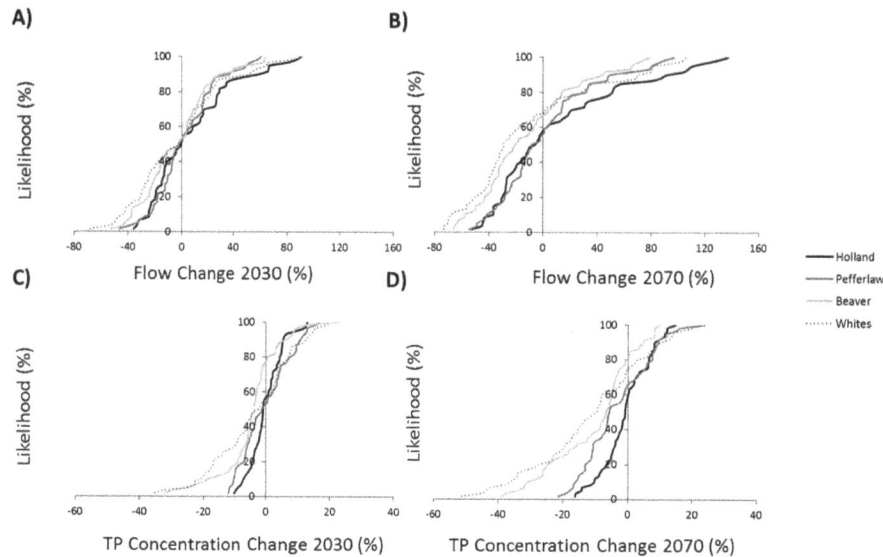

Figure 10. Between-catchment comparison of the CDFs from the perturbed physics ensemble of **(a)** monthly change in flow by 2030, **(b)** monthly change in flow by 2070, **(c)** monthly change in total phosphorus (TP) concentration by 2030, **(d)** change in TP concentration by 2070 (CDF: cumulative distribution function).

3.2.2 Seasonality of change

Seasonal changes in climate drivers (SI 8, SI 9) demonstrate that the largest increases in temperature generally occur during winter. Average seasonal changes in temperature were similar between catchments, with a maximum cross-catchment range occurring during winter (0.2 °C in 2070). The Pefferlaw was, however, consistently projected to experience the greatest winter increases (+3.0 °C in 2030 and +5.9 °C in 2070), and the Whites the least (+2.9 °C in 2030 and 5.8 °C in 2070). The largest increases in precipitation were projected to occur during winter and spring, with lower increases and modest reductions during summer and autumn. Seasonal precipitation changes varied more markedly between catchments than did temperature, with a maximum cross-catchment range during winter of 10.9 % (2070). The largest changes in winter precipitation are projected to occur in the Holland (12.2 %, 2030 and 26.2 %, 2070), and the lowest in the Pefferlaw (5.8 %, 2030 and 15.3 %, 2070). The largest reductions in summer precipitation are projected for the Whites (−0.7 % in 2030 and −9.2 % in 2070), and the lowest for the Holland (+2.8 % in 2030 and −2.4 % in 2070).

3.3 Hydrologically effective rainfall and river flow

3.3.1 Likelihood of change

Similar to temperature and precipitation, the likelihood and extent of flow changes increased between the 2030s and 2070s (SI 10). There were, however, markedly greater differences in projected flow changes between catchments than for corresponding climatic drivers. At the 50 % likelihood level

there is a cross catchment range of 23.4 % by 2070, where projected changes are significantly smaller in the Holland and Pefferlaw (−5.8 and −8.2 %) compared to the Whites and Beaver (−29.2 and −20.4 %) respectively (Fig. 10). The average flow uncertainty within each catchment was also generally higher than that of temperature and precipitation, reaching up to 43.2 % (2030) and 40.1 % (2070) (Table 6). In contrast to climatic drivers, average uncertainty in projected flow varied considerably between catchments, with a cross-catchment range of 19.9 % (2030) and 12.5 % (2070), and was consistently largest in the Whites and lowest in the Pefferlaw.

3.3.2 Seasonality of change

Seasonal changes in HER and flow (SI 11, SI 12) demonstrate that, similar to precipitation, the largest increases occur during winter, with reductions occurring during summer and autumn. In spring, however, in contrast to the increases projected for precipitation, large reductions are projected for both HER and flow. Similar to precipitation, the maximum cross-catchment range occurred in winter, though seasonal variability in HER and flow between catchments was notably greater than that of climatic drivers, with a winter HER range of 39.0 % (2030) and 68.1 % (2070) and flow range of 21.7 % (2030) and 46.8 % (2070). Similar to precipitation, the Holland projects the largest mean seasonal increases in winter HER and flow (of up to 96.1 % HER and 84.6 % flow), whilst the Pefferlaw and Beaver project the smallest, with projected change as low as 28.0 % (HER) and 38.0 % (flow). The largest mean seasonal reductions in summer flow and

Table 7. Calculation of quantity of hydrologically effective rainfall (HER) (mm) generated per mm of precipitation (1 d.p.), and comparison with catchment characteristics (hydrological and quaternary geology).

Catchment	Change in HER per mm change in precipitation			% surface runoff		% soil matrix flow		Soil water residence	Clay content
	2030	2070	Average	2030	2070	2030	2070	times (days)	(%)
Holland	4.9×10^{-1}	5.5×10^{-1}	5.2×10^{-1}	6.1	5.7	93.9	94.3	3.0	8.3
Pefferlaw	8.0×10^{-1}	9.3×10^{-1}	8.7×10^{-1}	12.3	11.2	87.7	88.8	2.7	12.6
Beaver	8.9×10^{-1}	9.9×10^{-1}	9.4×10^{-1}	16.1	31.8	83.9	68.2	2.4	27.5
Whites	8.6×10^{-1}	1.0	9.4×10^{-1}	22.6	31.6	77.4	68.4	1.4	41.9

HER occur in the Whites, and the smallest in the Holland and Pefferlaw.

Changes in soil moisture deficit (SMD) were also analysed. Annually, SMD increased in all catchments, though increases in the Whites and Beaver (up to 45.8 %) were up to four times greater than those in the Holland and Pefferlaw (up to 11.0 %). Seasonally, the largest SMD increases were projected during spring, and were consistently higher in the Beaver and Whites than those in the Holland and Pefferlaw.

3.3.3 Sensitivity to change

To determine the sensitivity of catchment HER to precipitation input, the HER generated per unit change in precipitation was calculated (Table 7). On average, the Holland and Pefferlaw yield the least HER in response to changes in precipitation, whilst the Beaver and Whites generate the most. In both the 2030s and 2070s a greater proportion of HER flowed as surface runoff in the Beaver and Whites than in the Holland and Pefferlaw (Table 7), where soil matrix flow contributions were particularly high (up to 94.3 % in the Holland). The difference between catchments increased between 2030 and 2070, where proportions of surface runoff decreased in the Holland and Pefferlaw, and increased in the Beaver and Whites. Proportions of overland and subsurface flow varied by land use type, with highest runoff proportions in saturated wetlands and urban areas, and lowest in forests and intensively farmed (and drained) agricultural lands.

3.4 Water quality

Ensemble average projections (accounting for system uncertainty) of total annual TP loads for the 2070's were projected to decrease in the Beaver and Whites by −14.7 and −13.1 %, respectively, but lower reductions, and even some increases were projected for the Pefferlaw and Holland, of −1.0 and 14.7 %, respectively (Table 8). Projections varied considerably between catchments with a range of 14.6 % (2030) to 29.4 % (2070).

Table 8. Annual change in total phosphorus loads between baseline and future periods, for all ensemble members of the perturbed physics ensemble. Includes ensemble average (EA) change (1 d.p.).

		Mean annual % change in TP loads					
		Q0	Q3	Q10	Q13	Q15	EA
2030	Holland	22.6	11.7	−8.0	10.8	3.7	8.2
	Pefferlaw	13.5	10.5	−2.9	−3.6	−6.6	2.2
	Beaver	4.5	0.2	−12.8	−7.6	−16.4	−6.4
	Whites	13.5	2.4	−9.0	−4.2	−25.1	−4.5
	Range						14.6
2070	Holland	29.6	17.6	−2.0	20.9	7.3	14.7
	Pefferlaw	13.2	9.1	−14.2	−5.7	−7.5	−1.0
	Beaver	0.3	−7.2	−26.3	−16.5	−23.8	−14.7
	Whites	7.0	−3.2	−30.5	−14.8	−24.1	−13.1
	Range						29.4

3.4.1 Likelihood of change

Again, projected changes in monthly TP concentrations and loads at the 10, 50 and 90 % likelihood levels amplify between 2030 and 2070 (SI 13) and indicate that larger changes are more likely by 2070. The cross-catchment range of TP loads is considerable, and at 32.7 % (2070 at the 50 % likelihood level) is wider than that of all other variables. Similarly there is a larger difference between catchments in TP concentration changes than for corresponding climatic drivers (8.9 % by 2070). Reductions in TP concentration are projected at the 50 % likelihood level, and are more extreme in the Whites and Beaver than in the Holland and Pefferlaw. Similarly, at the 50 % likelihood level markedly higher monthly reductions in TP loads are projected for the Whites and Beaver (−39.7 % by 2070) compared to the Holland and Pefferlaw (−11.2 by 2070).

In some catchments (Holland and Pefferlaw) the average uncertainty for TP concentrations was lower than that of the corresponding climatic drivers. However, similar to flow, uncertainty varied significantly between catchments (Table 6), with a cross-catchment range of 15.0 % (2030) and 16.5 % (2070). In contrast, the average uncertainty for TP

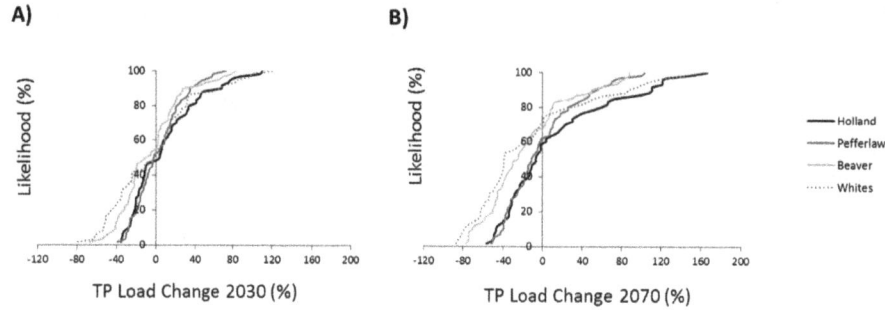

Figure 11. Between-catchment comparison of the CDFs from the perturbed physics ensemble of (**a**) change in total phosphorus loads (TP) by 2030, (**b**) change in TP loads by 2070 (CDF: cumulative distribution function).

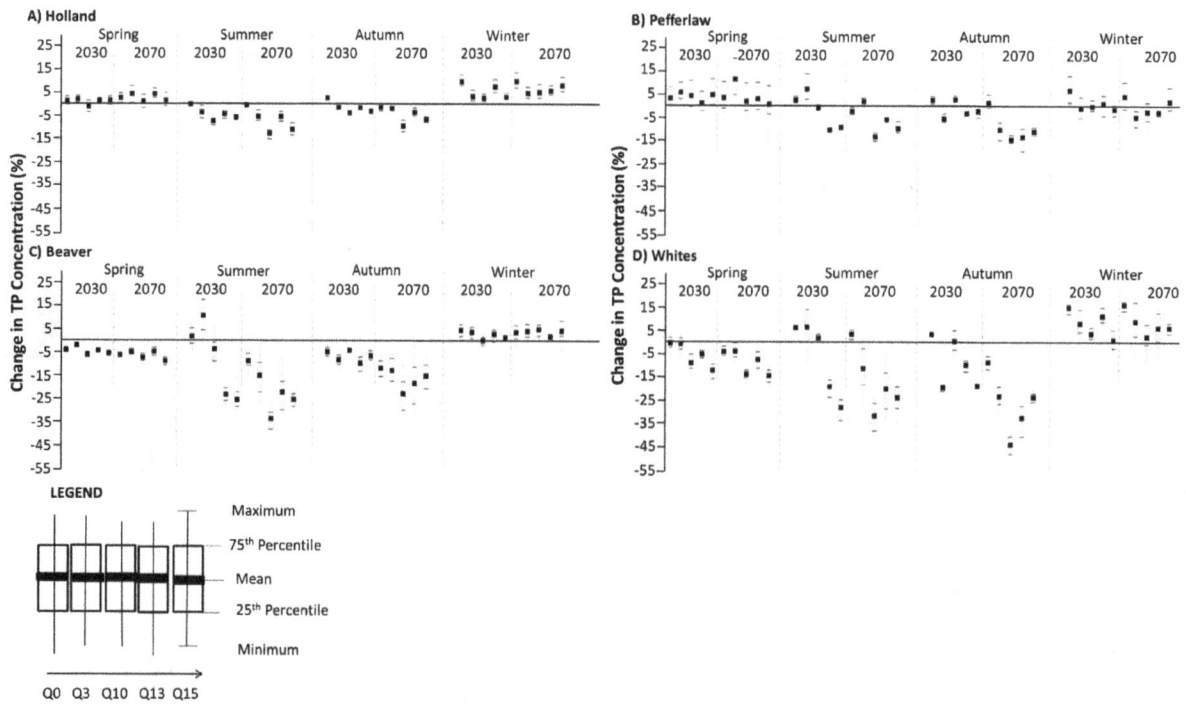

Figure 12. Box plot of seasonal changes in total phosphorus concentrations across all perturbed physics ensemble members (one box per ensemble member) in (**a**) Holland, (**b**) Pefferlaw, (**c**) Beaver and (**d**) Whites.

loads within each catchment was markedly higher than corresponding catchment drivers, flow and TP concentrations, at 30.9 % in 2030, and 20.4 % in 2070. Average uncertainty of TP concentrations was consistently highest in the Whites and Beaver (Table 6). Similarly, uncertainty in TP loads is highest in the Whites and lowest in the Pefferlaw (Fig. 11).

3.4.2 Seasonality of change

Seasonal changes in TP concentrations demonstrate that the largest increases are projected during winter (up to +7.6 %) and the largest reductions during summer and autumn (up to −26.6 %) (Fig. 12), and during this period are positively associated with changes in HER and flow. In spring, however, only in the Beaver and Whites did projected TP reductions

(up to −9.0 %) correspond with projected reductions in HER and flow. In contrast, increases in spring TP concentrations are projected for the Holland and Pefferlaw (up to 3.9 %), and are negatively associated with projected hydrological changes (flow reductions). Accordingly, the spring cross-catchment range of projections was high (9.3 % in 2030 and 12.9 % in 2070). Ensemble average seasonal changes in TP concentration varied to a greater extent between catchments than did corresponding seasonal climatic drivers with a maximum cross-catchment range in autumn of 22.3 % (2070).

Seasonal changes in TP loads demonstrate that, similar to HER and flow, increases are projected during winter and reductions projected during spring, summer and autumn. Similar to hydrological and climatic variables, the maximum

Table 9. Estimation of average catchment sensitivity of water quality to uncertainty in precipitation: projected change in total phosphorus ($mg\,L^{-1}$) per mm change in precipitation averaged over 2030's and 2070's (1 d.p.) highlighting (in bold) the seasons of greatest sensitivity.

Catchment	Season	Projected change in TP ($\mu g\,L$) per mm of change in precipitation	Proportion of annual change (%)	Clay content (%)
Holland	Spring	5.9	6.7	
	Summer	24.4	27.6	
	Autumn	54.0	**61.1**	
	Winter	4.1	4.6	
	Annual average	22.1	100	8.3
Pefferlaw	Spring	5.0	11.7	
	Summer	23.8	**56.1**	
	Autumn	4.0	9.5	
	Winter	9.7	22.7	
	Annual average	10.6	100	12.6
Beaver	**Spring**	201.0	**86.2**	
	Summer	4.9	2.1	
	Autumn	13.5	5.8	
	Winter	13.8	5.9	
	Annual average	58.3	100	27.5
Whites	Spring	28.3	5.0	
	Summer	5.5	1.0	
	Autumn	18.2	3.2	
	Winter	518.8	**90.9**	
	Annual average	142.7	100	41.9

cross-catchment range occurs during winter (26.4 % in 2030, and 53.2 % in 2070), and minimum in spring (12.4 % in 2030, and 19.1 % in 2070). The degree of difference in TP loads between catchments is, however, notably higher than for all other variables, suggesting additional factors are influencing TP loads. In winter, corresponding with climatic and hydrological changes, markedly higher increases in TP load are projected for the Holland (56.2 % in 2030, and 97.0 % in 2070), than for the Beaver (29.8 % in 2030 and 43.8 % in 2070). In summer, greater reductions in TP loads are projected for the Beaver and Whites, (9.6 and −14.8 % in 2030, respectively; −40.0 and −45.9 % in 2070), than for the Pefferlaw and Holland (−2.2 and −6.6 % in 2030, respectively; −15.1 and −20.9 % in 2070). This large cross-catchment difference in seasonal balances likely accounts for a significant portion of the difference in projections of annual loads between the Pefferlaw and Holland (increases) and Beaver and Whites (reductions) (Table 8).

3.4.3 Sensitivity to change

An analysis of change in TP ($mg\,L^{-1}$) per unit change in precipitation was undertaken, to determine catchment sensitivity to climate drivers (Table 9). The Beaver and Whites

had a higher sensitivity to changes in precipitation throughout the year (58.3 and $142.7\,\mu g\,L^{-1}$ TP generated per unit of change in precipitation), compared to the Holland and Pefferlaw (22.1 and $10.6\,\mu g\,L^{-1}$ TP per unit change in precipitation). In the Beaver and Whites the majority of changes in TP export occurred during spring and winter, in contrast to the Holland and Pefferlaw where greater changes occurred during summer and autumn.

4 Discussion

4.1 Likelihood of future changes in hydrochemistry

Climate change and variability present long term water management challenges both now, and persisting into the future. This study takes that uncertainty in our knowledge of climate systems into consideration, and results indicate that changes in climatic drivers (temperature and precipitation) become progressively more likely and extreme between the 2030s and 2070s, increasingly perturbing catchment hydrology and water quality from present conditions. Notably, although the changes in climatic drivers were similar between four neighbouring study catchments, responses in HER and flow varied considerably between sites.

With a significant seasonal snowmelt influence, catchments such as those within the Lake Simcoe basin may be very sensitive to climatic inputs (Barnett et al., 2005), and even small changes in climate could have large implications for catchment hydrology (Hamlet et al., 2005; Barnett et al., 2005), specifically with respect to the magnitude and timing of snowfall and snowmelt. The PPE projected greatest increases both in temperature and in precipitation to occur during winter, where larger quantities of precipitation previously falling as snow could in future fall as rain, likely resulting in an earlier snowmelt of reduced magnitude (Regonda et al., 2005; Crossman et al., 2013a, b). Projected soil moisture deficits (SMD) suggest that whilst in winter these higher inputs of rain (as opposed to snow) and an earlier snowmelt will limit rises in SMD, in spring a marked SMD increase is to be expected, due in part to the earlier depletion of frozen water stores.

Seasonally, the direction of change in projected HER and flow did not always match those of climatic drivers. Although each indicated that changes will be more extreme and likely between the 2030s and 2070s, spring reductions in HER and flow contrast markedly to projected precipitation increases. HER and flow are affected by a number of variables, including evapotranspiration and preceding soil moisture conditions. It is likely, therefore, that spring reductions in HER and flow are due to the increases in spring SMD, the magnitude of which was not offset by increases in spring rainfall.

The response of water quality (TP concentrations and TP loads) to climatic drivers also varied between catchments. Although, similar to climate drivers and hydrology, larger changes were more likely between 2030 and 2070, the direction of change was not consistent between catchments. Seasonally, in the Beaver and Whites the direction of change in TP concentration was consistently positively associated with that of HER and flow; whereas in the Holland and Pefferlaw a negative association was demonstrated in spring. These different hydrological and water quality responses to very similar climatic drivers are significant in terms of annual TP loads, where large annual reductions were projected for the Beaver and Whites, compared to much smaller reductions, and even some increases, projected in the Holland and Pefferlaw. Differences in hydrological flow pathways and mechanisms of nutrient delivery are a possible explanation for these different responses.

4.2 Catchment characteristics driving seasonal differences in hydrochemical responses

Model calibrations demonstrated that in every catchment TDP soil water concentrations increased in summer following fertiliser applications, which is consistent with observations of Haygarth et al. (1998) where fertiliser additions and plant decomposition enriched near-surface soil stores.

Research has shown that when TDP is present in excess of plant requirements, it can be transported from these surface pools to increase in-stream concentrations via overland and subsurface flow of soil water in highly fertilised organic soils, with low phosphorus sorption capacities i.e. shortly after the application of fertilisers in areas of shallow water tables e.g. wetlands (Borlin, 2003). By late autumn, the surface pools of TDP may be exhausted, and lower concentrations of TDP are delivered through subsurface flows, though soils have the capacity to release some P in an attempt to re-establish an equilibrium. Generally, soil water that leaches into streams in spring and winter will be less P-enriched than it was during the summer months; the manner in which this process is represented by the models is shown in Fig. SI6 (in the Supplement).

Where catchments, such as the Beaver and Whites, have significant overland and macropore flow contributions, or tile drainage (where rapid flow mobilises PP, and bypasses P-binding sites in the soils; Hooda et al., 1999), TP can be delivered directly to the streams by erosion in spring (Heathwaite and Dils, 2000), and a positive relationship between HER and TP is maintained. Under future climates, spring flow and TP concentrations are therefore reduced simultaneously. However, where catchment pathways are dominated by soil matrix flow (Table 7), the dilute leachate enhanced by spring meltwater has a significant impact on in-stream TP concentrations (Borling, 2003), and a negative relationship between HER and TP is created. When future spring HER is reduced (and less dilute leachate delivered to streams), the in-stream TP concentrations increase. This effect may be further enhanced in the Holland and Pefferlaw catchments due to dilution of sewage treatment work contributions (point sources).

The large reductions in annual TP loads of the Whites and Beaver are projected due to large scale reductions in flow and TP concentrations in spring, summer and autumn, combined with only small scale increases in winter flow and TP concentrations. In contrast, spring flow reductions in the Holland and Pefferlaw enhance TP concentrations, which in combination with larger scale winter flow and TP concentrations (and smaller scale summer flow reductions), leads to projections of annual increases in TP loads.

The earlier thawing of frozen soils in spring under a changing climate may also influence nutrient flow pathways. In the Holland and Pefferlaw, earlier thawing was projected to lead to an increased proportion of subsurface flow, enhancing interactions with the soil matrix, and may be an additional causal factor behind the dilution of the spring and winter leachate. In the Beaver and Whites, however, with large macropores and tile drainage, soil freeze–thaw is projected to have less impact on infiltration (as observed by Harris, 1972). In addition, as water movement through the soil profile in the Beaver and Whites is rapid, any changes in subsurface flow generated during an earlier spring melt would have a limited effect on soil water concentrations reaching the stream. A final consideration is the drainage of wetlands (polders) in the Holland River during spring (Burke, 1967, 1975), where artificial drainage of soils can significantly increase levels of P lost through subsurface leaching (McDowell et al., 2001). This practice might therefore contribute to the negative HER associated with TP during spring, where the addition of artificial drainage waters with high TP concentrations will in future be less diluted by flows with a lower spring melt contribution.

4.3 Impact of catchment sensitivity compared to future climate uncertainty

In all catchments, uncertainty in temperature is low (across the range of QUMP scenarios generated by the PPE), and similar between catchments. As is common amongst climate change studies, uncertainty in precipitation is higher (Hawkins and Sutton, 2011; Maskey et al., 2004), though remains relatively consistent between sites. Importantly, hydrological (HER and flow) uncertainty is considerably higher than corresponding climatic drivers, and varies more significantly between catchments. The catchment-specific response can be defined by the quantity of HER generated per unit of precipitation input (Table 7). This relative index of HER sensitivity to climate change normalises the catchment response from differences in original driving forces (precipitation). On average, the Beaver and Whites catchments demonstrated a greater HER sensitivity to precipitation change (> 0.9 mm) than the Holland and Pefferlaw (< 0.6 mm). This sensitivity might be attributed to shorter soil water residence times in the Beaver and Whites, whereby more rapid soil drainage through macropores and tile drains results in prompt responses of soil moisture content to changes in precipitation; and the dominance of overland flow results in greater exposure to evaporation and rising temperatures. As a result, reductions in summer and autumn precipitation have a greater impact upon SMD in the Beaver and Whites (as indicated by model results) and thus upon HER and flow. Similarly, reductions in spring-melt waters within these catchments have a greater impact on SMD, as soils with a dominant overland flow component would previously have responded most rapidly and significantly to snowmelt input (Dunne and Black, 1971).

In contrast, in the Holland and Pefferlaw projected changes in SMD are lower, and the impact of a changing climate upon hydrology less extreme, due to a higher soil storage capacity, longer soil water residence times, and a greater proportion of water directed through the soil matrix. Slow passage of waters through soils in the Holland and Pefferlaw reduces its response rate to precipitation changes and meltwater reductions, and its exposure to evaporation (Dunne and Black, 1971; van den Hurk et al., 2004), meaning soil moisture deficits can be recharged with a lesser affect upon HER and stream flow. The hydrological sensitivity of the Pefferlaw is higher than that of the Holland, but is consistent with

the difference in soil water residence times and clay content between the two catchments (Tables 2 and 7).

The sensitivity of TP concentrations to changing climate also varies considerably between catchments, with a greater response of TP per unit change in precipitation in the Whites and Beaver (up to $142.7 \mu g \, L^{-1}$) than in the Holland and Pefferlaw (up to $22.1 \mu g \, L^{-1}$). In Simcoe this variance between catchments appears to be associated with flow pathways and timings of nutrient export which may act as a buffer to uncertainty within some catchments. For instance, in the Holland and Pefferlaw, the majority of changes in TP export are associated with changes in future summer and autumn precipitation (Table 9), despite there being smaller absolute changes in water volumes during these periods. This is likely due to the high mobility of soil TP in the Holland and Pefferlaw at this time of fertiliser applications, with high initial P soil concentrations, and limited P sorption capacities. As the uncertainty of future climate change is lowest during this period, these are the catchments for which the lowest TP uncertainty is calculated. During the periods of high climate uncertainty (winter and spring), TP mobility is low, and the effects of the climate change uncertainty upon water quality are buffered. In the Beaver and Whites, a much larger proportion of the changes in TP export occurs during winter and spring. This is likely due to the high runoff rates and macropore flow, which facilitates movement of TP directly to the stream during all seasons. Climate uncertainty is highest during the large projected spring and winter changes, therefore uncertainty in future TP concentrations are higher in these catchments. Given that nutrient loads are directly derived from a combination of flow and TP concentrations, it follows that the uncertainty from each propagates up into the load estimations. As such, uncertainty of future TP loads is also greatest in the Whites.

It is evident then that the future certainty in both hydrology and water quality depends not only upon climatic inputs (between which there is little difference in these catchments), but also largely upon the relative sensitivity of a catchment to changing drivers. Results from this study suggest that catchment sensitivity is strongly influenced by quaternary geology, seasonal P-inputs and P-saturation thresholds. Sensitivity to climatic drivers was highest in clayey catchments, dominated by overland flow, with little influence of P-saturation (Beaver and Whites); it was lowest in sites with greater contributions from soil matrix flow, and where seasonal variability in soil P-saturation had a marked influence on water quality (Holland and Pefferlaw). It is likely that these characteristic drivers of sensitivity will be applicable to other catchments. For instance, van den Hurk (2004) determined that future inter-annual variability in wintertime runoff in the Rhine was smallest in catchments with high storage capacities, whilst van Roosmalen et al. (2007) determined from a range of sites in Denmark with differing quaternary geology, that sites with clayey soils responded to precipitation change with the highest increases in stream discharge and overland flow proportions. The consistency of driving forces behind catchment sensitivity to climate change requires further investigation, but indicates that the clay content of soils might be used as an indicator of catchment hydrochemical sensitivity to climate change.

4.4   Study limitations and further research

By analysing outputs from a range of physical structures of a single global climate model (a perturbed physics ensemble), this study explores the sensitivity of catchments to uncertainty in climate system behaviour and parameter values of global climate models. Whilst this enables an examination of a wider range of plausible future projections than could be investigated through a multi-model ensemble approach, the PPE used is derived only from the HADCM3 model. A wider range of plausible futures could be generated by perturbing the physics of additional GCMs (Collins et al., 2006). Currently, however, PPEs are available for only a limited number of GCMs.

As discussed previously, the choice of GCM bias correction method impacts both climatic and hydrochemical outputs. The Δ Change method was selected as it preserves the impact of perturbing the physical structure of the climate model (i.e. it maintains the difference between PPE ensemble members). This method, however, does not account for potential changes in distribution of wet and dry days, and outputs from this study should not be used in assessments of future flood or drought frequency.

Finally, parametric uncertainty within both INCA-P and the rainfall-runoff model HBV might also influence model outputs. In this study, care was taken to accurately calibrate the most sensitive parameters. Within HBV these have previously been identified as parameters associated with partitioning precipitation into runoff and soil moisture (Seibert, 1997; Uhlenbrook et al., 1999). The assessment of hydrological responses to individual storm events and seasonal snowmelt events (Figs. 7 and 8) enhance confidence in runoff partitioning within these model applications. Within INCA-P, sensitive parameters have been identified as Freudlich soil coefficients and stream flow velocities (Crossman et al., 2013c; Lepistö et al., 2013). Laboratory and field values specific to each catchment were used for these parameters. The similarity of model process responses (including soil water TDP responses to fertiliser inputs) to those in the literature, and of model soil TP export coefficients to those calculated by Winter et al. (2007), gives confidence in model representation of reality. While models were calibrated using these measured and calculated values, there are over 107 parameters within INCA, and parametric uncertainty is an issue that should be considered (Wade et al., 2001; Starrfelt and Kaste, 2014). Although beyond the scope of this study, future work will explore the extent to which parametric uncertainty might influence assessments of catchment sensitivity.

Whilst there are a range of uncertainties associated with hydrochemical modelling, process based modelling does provide the best method of understanding catchment responses to possible future conditions (Jin et al., 2013). These results provide a useful basis for the development of catchment-specific management approaches. The overall outcome of this study indicates that management strategies might focus on internal processes dynamics that define the extent of hydrochemical responses to changes in climatic input, and as such reinforces the call for catchment-specific management plans. In the Holland and Pefferlaw, where the majority of P export appears to be derived from leachate during summer and autumn, a reduction in labile P concentrations and an increase in soil sorption capacity would be beneficial. It has been shown that soil P levels can take over 20 years to fully recover from extensive periods of excess fertiliser applications (McCollum, 1991; McLauchlan, 2006), but may gradually be achieved through methods such as avoidance of nutrient applications in excess of plant requirements (Whitehead et al., 2011), and establishing nutrient vulnerable zones where use is no longer required. In the Beaver and Whites, where the majority of P export appears to be derived from surface runoff and soil erosion, a more appropriate strategy might include efforts to reduce rates of surface runoff from fertilised fields, through injecting fertiliser directly to the root zone, constructing vegetated riparian areas (Sharpley et al., 2001), restricting fertiliser applications to periods of higher soil moisture deficit (Westerman and Overcash, 1980; Sharpley et al., 2001), and limiting tile drainage.

5   Conclusions

Accounting for climate modelling uncertainty, this study suggests that in the Beaver and Whites, large scale reductions in annual TP loads could result from reductions in flow and TP concentrations in spring, summer and autumn, combined with minor increases in winter. In contrast, much smaller reductions and even some increases in annual TP loads are expected in the Holland and Pefferlaw, due to much smaller scale reductions in summer and autumn flow and TP concentrations, combined with larger scale increases in winter, and enhanced spring TP concentrations. These findings indicate that all changes will become more extreme and more likely between 2030 and 2070.

Uncertainty in future projections of climate change had different impacts on uncertainty of future hydrological and water quality projections within each catchment. This impact depended upon (a) catchment characteristics which determine the sensitivity of a catchment to climate drivers and (b) the degree of projected climate change within each catchment. Influential catchment characteristics include quaternary geology, seasonal P-inputs and P-saturation thresholds. Catchment sensitivity to climatic drivers was relatively high where catchments had large proportions of overland flow and a direct association between TP concentrations, loads and

HER was established. Catchment sensitivity was lower in catchments with larger proportions of soil matrix flow, and where seasonal variability in soil P-saturation influenced water quality; here internal catchment processes had a greater impact on hydrology and water quality.

The different ranges of uncertainty demonstrated by the catchments indicate that in water quality management it is important to consider every catchment as a distinct hydrological unit. Effectiveness of management strategies will vary depending on the dominant nutrient sources and P transport mechanisms in each area. Significantly, although the hydrology and water quality of the Whites and Beaver Catchments are more sensitive to changes in climate, the majority of projections under the PPE indicate reductions in TP concentrations and loads. Although the Holland and Pefferlaw are less sensitive to change, the dominant response is an increase in annual TP loads; as such the latter catchments might be seen as priority target areas for nutrient management.

Acknowledgements. The authors would like to thank Jennifer Winter at the Ontario Ministry of Environment, for her invaluable scientific inputs and support throughout the project. Thanks also to the Hadley Centre, for generating the original PPE data sets, and to Gordon Huang and associates at the University of Regina, for generating the regionally downscaled PRECIS data sets of southern Ontario (IEESC, 2012). Finally, thanks to Dave Woods and Eavan O'Connor, for collection of water quality and flow monitoring data in the Lake Simcoe catchments. Funding for this study was provided by the Ontario Ministry of Environment (grant number 55-58313).

Edited by: A. D. Reeves

References

Adams, H. D., Williams, A. P., Chonggang, X., Rauscher, S. A., Jiang, X., and McDowell, N. G.: Empirical and process-based approaches to climate-induced forest mortality loads, Front. Plant. Sci., doi:10.3389/fpls.2013.00438, in press, 2013.

Akhtar, M., Ahmad, N., and Booij, M. J.: The impact of climate change on the water resources of Hindukush-Karakorum-Himalaya region under different glacier coverage scenarios, J. Hydrol., 355, 148–163, 2008.

Bangay, G. E.: Livestock and poultry wastes in the Great Lakes Basin. Environmental Concerns and Management Issues, Social Science Series No. 15, Environment Canada, Inland Waters Directorate, Ontario Region, Water planning and Management Branch, Burlington, Ontario, 1976.

Barnett, T. P., Adam, J. C., and Lettenmaier, D. P.: Potential impacts of a warming climate on water availability in snow-dominated regions, Nature, 438, 303–309, doi:10.1038/nature04141, 2005.

Bates, B. C., Kundzewicz, Z. W., Wu, S., and Palutikof, J. P. (Eds.): Climate Change and Water, Technical Paper VI of the Intergovernmental Panel on Climate Change, IPCC Secretariat, Geneva, Switzerland, 2008.

Bathurst, J. C. and O'Connell, P. E.: Future of distributed modelling: the Systeme Hydrologique Europeen, Hydrol. Process., 6, 265–277, 1992.

Baulch, H. M., Futter, M. N., Whitehead, P. G., Woods, D. T., Dillon, P. J., Butterfield, D. A., Oni, S. K., Aspden, L. P., O'Conner, E. M., and Crossman, J.: Phosphorus dynamics across intensively monitored subcatchments in the Beaver River, J. Inland Wat., 1, 187–206, 2013.

Borling, K.: Effects of long-term inorganic fertilisation of cultivated soils, Doctoral Thesis, Swedish University of Agricultural Sciences, Uppsala, 2003.

Burke, W.: Principles of drainage with special reference to peat, Irish Forestry, 24, 1–7, 1967.

Burke, W.: Aspects of the hydrology of blanket peat in Ireland. Hydrology of marsh-ridden areas. Proceedings of the Minsk symposium, June 1972, IAHS Studies and Reports in Hydrology 19, Unesco Press, Paris, 171–182, 1975.

Burt, T. P. and Pinay, G.: Linking hydrology and biogeochemistry in complex landscapes, Prog. Phys. Geogr., 29, 297–316, 2005.

Chorus, I. and Bartram, J. (Eds.): Toxic cyanobacteria in water – A guide to their public health consequences, E and FN Spon, London, England, 1999.

Collins, M., Booth, B. B. B., Harris, G. R., Murphy, J. M., Sexton, D. M. H., and Webb, M. J.: Towards quantifying uncertainty in transient climate change, Clim. Dynam., 27, 127–147, 2006.

Crossman, J., Futter, M. N., Oni, S. K., Whitehead, P. G., Jin, L., Butterfield, D., Baulch, H. M., and Dillon, P. J.: Impacts of climate change on hydrology and water quality: future proofing management strategies in the Lake Simcoe watershed, Canada, J. Great Lakes Res., 39, 19–32, 2013a.

Crossman, J., Futter, M. N., and Whitehead, P. G.: The significance of shifts in precipitation patterns: modelling the impacts of climate change and glacier retreat on extreme flood events in Denali National Park, Alaska, PLOS ONE, 8, e74054, doi:10.1371/journal.pone.0074054, 2013b.

Crossman, J., Whitehead, P. G., Futter, M. N., Jin, L., Shahgedanova, M., Castellazzi, M., and Wade, A. J.: The interactive responses of water quality and hydrology to changes in multiple stressors, and implications for the long-term effective management of phosphorus, Sci. Total Environ., 454–455, 230–244, 2013c.

Dunne, T. and Black, R. D.: Runoff processes during snowmelt, Water Resour. Res., 7, 1160–1172, 1971.

Environment Canada: Water survey of Canada, www.wsc.ec.gc.ca (last access: January 2014), 2013.

Evans, D. O.: Effects of hypoxia on scope-for-activity of lake trout: defining a new dissolved oxygen criterion for protection of lake trout habitat, Technical Report 2005-01, Aquatic Research and Development Section, Applied Research and Development Branch, Ministry of Natural Resources, Peterborough, Ontario, 2006.

Evans, D. O., Nicholls, K. H., Allen, Y. C., and McMurtry, M. J.: Historical land use, phosphorus loading, and loss of fish habitat in Lake Simcoe, Canada, Can. J. Fish Aquat. Sci., 53, 194–218, 1996.

Fournier, R. E., Morrison, I. K., and Hopkin, A. A.: Short range variability of soil chemistry in three acidic soils in Ontario, Canada, Commun. Soil Sci. Plan., 25, 3069–3082, 1994.

Futter, M. N., Forsius, M., Homberg, M., and Starr, M.: A long-term simulation of the effects of acidic deposition and climate change on surface water dissolved organic carbon concentrations in a boreal catchment, Hydrol. Res., 402, 291–305, 2009.

Gleik, P. H.: Climate change, hydrology, and water resources, Rev. Geophys., 27, 329–344, 1989.

Global LandCover Facility: Digital Elevation Model, Earth Science Data Interface, available at: http://www.landcover.org/ (last access: 1 March 2013), 2002.

Goddard, L. and Baethgen, W.: Better estimating uncertainty with realistic models, EOS, 90, 343–344, 2009.

Hamlet, A. F., Mote, P. W., Clark, M. P., and Lettenmaier, D. P.: Effects of temperature and precipitation variability on snowpack trends in the Western United States, J. Climate, 18, 4545–4561, 2005.

Harris, A. R.: Infiltration rate as affected by soil freezing under three cover types, Am. Soc. Agron., 36, 489–492, 1972.

Hawkins, E. and Sutton, R.: The potential to narrow uncertainty in projections of regional precipitation change, Clim. Dynam., 37, 407–418, 2011.

Hay, L .E., Wilby, R. L., and Leavesley, G. H.: A comparison of delta change and downscaled GCM scenarios for the three mountainous basins in the United States, J. Am. Water Resour. Assoc., 36, 387–397, 2000.

Haygarth, P. M., Hepworth, L., and Jarvis, S. C.: Forms of phosphorus transfer in hydrological pathways from soil under grazed grassland, Eur. J. Soil Sci., 49, 65–72, 1998.

Heathwaite, A. L. and Dils, R. M.: Characterising phosphorus loss in surface and subsurface hydrological pathways, Sci. Total Environ., 25, 523–538, 2000.

Hooda, P. S., Moynaghm, M., Svoboda, F. I., Edwards, A. C., Anderson, H. A., and Sym, G.: Phosphorus loss in drainflow from intensively managed grassland soils, Am. Soc. Agron., 28, 1235–1242, 1999.

Huang, M.: Old and subsurface water contributions to storm runoff generation in flat, fractured, clayey terrain, Ph.D. Thesis, Department of Civil and Environmental Engineering, University of Windsor, Windsor, 1995.

IEESC – Institute for Energy, Environment and Sustainable Communities: Producing High-Resolution (25 km by 25 km) Probabilistic Climate Change Projections over Ontario Using UK PRECIS, Report submitted to the Ontario Ministry of Environment, edited by: Huang, G. H., Wang, X. Q., Lin, Q. G., Yao, Y., Cheng, G. H., Fan, Y. R., Li, Z., Lv, Y., Han, J. C., Wang, S., Suo, M. Q., Dong, C., Chen, J. P., Chen, X. J., and Zhou, X., data available online at: http://env.uregina.ca/moe/ (last access: 25 July 2013), 2012.

Jarvie, H. P., Neal, C., Williams, R. J., Neal, M., Wickham, H., Hill, L. K., Wade, A. J., Warwick, A., and White, J.: Phosphorus sources, speciation and dynamics in a lowland eutrophic chalk river, Sci. Total Environ., 282–283, 203, 2002.

Jarvie, H. P., Neal, C., and Withers, P. J. A.: Sewage-effluent phosphorus: a greater risk to river eutrophication than agricultural phosphorus?, Sci. Total Environ., 282–283, 175–203, 2006.

Jin, L., Whitehead, P. G., Baulch, H. M., Dillon, P. J., Butterfield, D., Oni, S. K., Futter, M. N., Crossman, J., and O'Connor, E. M.: Modelling phosphorus in Lake Simcoe and its subcatchments: scenario analysis to assess alternative management strategies, J. Inland Wat., 3, 207–220, 2013.

Johnson, F. M.: The Landscape Ecology of the Lake Simcoe Basin, Lake Reserv. Manage., 13, 226–239, doi:10.1080/07438149709354313, 1997.

Jones, R. G., Noguer, M., Hassel, D. C., Hudson, D., Wilson, S. S., Jenkins, G. J., and Mitcehll, J. F. B.: Generating high resolution climate change scenarios using PRECIS, Met Office Hadley Centre, Exeter, UK, 1–40, 2004.

KMK Consultants: Schomberg Water Pollution Control Plant Class Environmental Assessment, Environmental Study Report, prepared for The Regional Municipality of York, Ontario, 2004.

Koski-Vähälä, J.: Role of resuspension and silicate in internal phosphorus loading. Dissertation in Limnology, Department of Limnology and Environmental Protection, Department of Applied Chemistry and Microbiology, University of Helskini, Helsinki, 2001.

Larssen, T., Høgåsen, T., Cosby, B. J.: Impact of time series data on calibration and prediction uncertainty for a deterministic hydrogeochemical model, Ecol Model., 207, 22–33, 2007.

Leavesley, G. H.: Modelling the effects of climate change on water resources – a review, Climatic Change, 28, 159–173, 1994.

Lenderink, G., Buishand, A., and van Deursen, W.: Estimates of future discharges of the river Rhine using two scenario methodologies: direct versus delta approach, Hydrol. Earth Syst. Sci., 11, 1145–1159, doi:10.5194/hess-11-1145-2007, 2007.

Lepistö, A., Etheridge, J. R., Granlund, K., Kotamäki, N., Maulve, O., Rankinen, K., and Varjopuro, R.: Adaptive strategies to mitigate the impacts of climate change on European Freshwater Ecosystems, Deliverable 5.5 Report on the biophysical catchment-scale modelling of Yläneenjoki-Pyhäjärvi demonstration site, REFRESH EU Seventh Framework Programme, retrieved from: http://www.refresh.ucl.ac.uk/webfm_send/2161 (last access: 7 January 2014), 2013.

LSEMS – Lake Simcoe Environmental Management Strategy: Lake Simcoe: Our Waters, Our Heritage, in: Lake Simcoe environmental management strategy implementation program summary of phase I progress and recommendations for phase II, edited by: Heathcote, I., published by Lake Simcoe Region Conservation Authority, Ministry of Natural Resources, Ministry of Environment and Energy and Ministry of Agriculture, Food and Rural Affairs, available online at: http://agrienvarchive.ca/download/L-simcoe_our_waters_heritage95.pdf (last access: 25 June 2014), 1995.

LSEMS – Lake Simcoe Environmental Management Strategy: Flow and nutrient monitoring data supplied by Lake Simcoe Region Conservation Authority, available to download from: http://www.lsrca.on.ca/programs/watershed_monitoring/systemintro.php (last access: 15 December 2013), 2010.

Louis Berger Group Inc.: Estimation of the Phosphorus Loadings to Lake Simcoe, Final Report Submitted to Lake Simcoe Region Conservation Authority, available for download at: http://www.lsrca.on.ca/pdf/reports/phosphorus_estimates_2010.pdf (last access: 11 July 2014), 2010.

LSRCA – Lake Simcoe Region Conservation Authority: Report on phosphorus loads to Lake Simcoe 2004–2007, Lake Simcoe Region Conservation Authority, Ontario, 2009.

LSRCA – Lake Simcoe Region Conservation Authority: Routine Monitoring Data Strategy O'Connor, E. Personal Communication, 2010a.

LSRCA – Lake Simcoe Region Conservation Authority, Holland River Subwatershed Plan, Durham Region, available for download at: http://www.lsrca.on.ca/pdf/reports/east_holland_subwatershed_2010.pdf (last access: 7 July 2014), 2010b.

LSRCA – Lake Simcoe Region Conservation Authority: Beaver River Subwatershed Plan, Durham Region, available for download at: http://www.lsrca.on.ca/pdf/reports/beaver_river_subwatershed_plan_2012.pdf (last access: 7 July 2014), 2012a.

LSRCA – Lake Simcoe Region Conservation Authority: Pefferlaw River Subwatershed Plan, Durham Region, available to download at: http://www.lsrca.on.ca/pdf/reports/pefferlaw_river_subwatershed_plan_2012.pdf (last access: 7 July 2014), 2012b.

Maskey, S., Guinot, V., and Prices, R. K.: Treatment of precipitation uncertainty in rainfall-runoff modelling: a fuzzy set approach, Adv. Water Resour., 27, 889–898, 2004.

McCollum, R. E.: Buildup and decline in soil phosphorus: 30-year trends on a Typic Umprabuult, Agron J., 83, 77–85, 1991.

McDowell, R. W., Sharpley, A. N., Condron, L. M., Haygarth, P. M., and Brookes, P. C.: Processes controlling soil phosphorus release to runoff and implications for agricultural management, Nut. Cycl. Agroecosys., 59, 269–284, 2001.

McLauchlan, K.: The nature and longevity of agricultural impacts on soil carbon and nutrients: a review, Ecosystems, 9, 1364–1382, 2006.

McSweeney, C. and Jones, R.: Selecting members of the 'QUMP' perturbed-physics ensemble for use with PRECIS, Met Office Hadley Centre, 2010.

Meehl, G. A. and Stocker, T. F.: Global climate projections in Climate Change: The Physical Science basis, in: Contribution of Working Group 1 to the Fourth Assessment Report of the Intergovernmental Panel on Climate Change, edited by: Solomon, S., Quin, D., Manning, M., Chen, Z., Marquis, M., Averyt, K. B., Tignor, M., and Miller, H. L., Cambridge University Press, New York, 2007.

Miles, J.: The relationship between land use and forms of phosphorus in the Beaver River watershed of Lake Simcoe, Ontario, MSc Thesis in Environmental and Life Sciences Graduate Program, Trent University, Trent, 2012.

Nicholls, K. H.: Some Recent water quality trends in Lake Simcoe, Ontario: implications for basin planning and limnological research, Can. Water Resour. J., 20, 213–226, 1995.

Nicholls, K. H. and MacCrimmon, H. R.: Nutrient Loading to Cook Bay of Lake Simcoe from the Holland River Watershed, Int. Revue Ges. Hydrobiol., 60, 159–193, 1975.

Olding, A. B., Wicklund, R. E., and Richards, N. R.: Soil Survey of Ontario County, Report No. 23 of the Ontario Soil Survey, Experimental Farms Service, Canada Department of Agriculture, Ottawa, and the Ontario Agricultural College, Toronto, 1950.

OMAFRA – Ontario Ministry of Agriculture, Food and Rural Affairs: Agronomy Guide for Field Crops, available from: http://www.omafra.gov.on.ca/english/crops/pub811/1toc.htm (last access: 3 January 2014), 2009.

Oni, S. K., Futter, M. N., and Dillon, P. J.: Landscape-scale control of carbon budget of Lake Simcoe: A process based modelling approach, J. Great Lake Res., 37, 160–165, 2011.

Oni, S. K., Futter, M. N., Molot, L. A., and Dillon, P. J.: Adjacent catchments with similar patterns of land use and climate have

markedly different dissolved organic carbon concentration and runoff dynamics, Hydrol. Process., 28, 1436–1449, 2014.

Ontario Ministry of Natural Resources: Ecological Land Classification of Ontario, available for download at: https://www.javacoeapp.lrc.gov.on.ca/geonetwork/srv/en/main.home (last access: 15 January 2014), 2007.

Oreskes, N., Shrader-Frechette, K., and Belitz, K.: Verification, Validation, and Confirmation of Numerical Models in the Earth Sciences, Science, 263, 641–646, 1994.

Paterson, A. M., Dillon, P. J., Hutchinson, N. J., Futter, M. N., Clark, B. J., Mills, R. B., Reid, R. A., and Scheider, W. A.: A review of the components, coefficients and technical assumptions of Ontario's Lakeshore Capacity Model, Lake Reserv. Manage., 22, 7–18, 2006.

Peltouvouri, T.: Phosphorus in agricultural soils of Finland – characterization of reserves and retention in mineral soil profiles, Pro Terra No. 26, Academic dissertation, University of Helsinki, Helsinki, 2006.

PGWMN – Provincial Groundwater Monitoring Network Program: Groundwater level data, groundwater chemistry data, and precipitation data, Ministry of Environment, available from the world wide web: https://www.javacoeapp.lrc.gov.on.ca/geonetwork/srv/en/metadata.show?id=13677 (last access: 15 January 2014), 2012.

PWQMN – Provincial Water Quality Monitoring Network: Ministry of Environment, available for download at: http://www.metoffice.gov.uk/media/pdf/e/3/SelectingCGMsToDownscale.pdf (last access: 10 June 2014), 2009.

Ramwekellan, J., Gharabaghi, B., and Winter, J. G.: Application of weather radar in estimation of bulk atmospheric deposition of total phosphorus over Lake Simcoe, Can. Water Resour. J., 34, 37–60, 2009.

Regonda, S., Rajagopalan, B., Clark, M., and Pitlick, J.: Seasonal cycle shifts in hydroclimatology over the western United States, J. Climate, 18, 372–384, doi:10.1175/JCLI-3272.1, 2005.

Saelthun, N.: Nordic HBV Model, Norwegian Water Resources and Energy, Administration, Oslo, Norway, 1995.

Schlotzhauer, S. M. and Price, J. S.: Soil water flow dynamics in a managed cutover peat field, Quebec: Field and laboratory investigations, Water Resour. Res., 35, 3675–3686, 1999.

Scott, L. D., Winter, J. G., and Girard, R. E.: Annual water balances, total phosphorus budgets and total nitrogen and chloride loads for Lake Simcoe (1998–2004), Technical Report No. Imp. A. 6, Lake Simcoe Environmental Management Strategy Implementation Phase III, Ontario, Canada, 2006.

Seibert, J.: Esimation of parameter uncertainty in the HBV model, Nord. Hydrol., 28, 246–262, 1997.

Sharpley, A. N., Kleinman, P., and McDowell, R.: Innovative management of agricultural phosphorus to protect soil and water resources, Commun. Soil Sci. Plan., 32, 1071–1100, 2001.

Smith, G. J.: Deriving spatial patterns of severe rainfall in Southern Ontario from rain gauge and radar data, A thesis presented to the University of Waterloo, Waterloo, Ontario, 2010.

Smith, V. H.: Eutrophication of freshwater and coastal marine ecosystems, Environ. Sci. Poll. Res., 10, 126–139, 2003.

Soil Landscapes of Canada Working Group: Soil Lanscapes of Canada v.3.2, Agriculture and Agri-Food Canada, 2010.

Starrfelt, S. and Kaste, O.: Bayesian uncertainty assessment of a semi-distributed integrated catchment model of phosphorus transport, Environ. Sci. Process. Imp., 16, 1578–1587, 2014.

Statistics Canada: Farm and Operator Data, Census of Agriculture, 2011.

Stephens, S. L. S.: Optimizing agricultural and urban pollution remediation measures using watershed modeling: Review, calibration, validation and applications of the CANWET model in the Lake Simcoe watershed, Ontario, Canada, MS thesis, Trent University, Trent, 2007.

Teutschbein, C. and Seibert, J.: Bias correction of regional climate model simulations for hydrological climate-change impact studies: review and evaluation of different models, J. Hydrol., 456–457, 12–29, 2012.

Thomas, R. L. and Sevean, G.: Leaching of phosphorus from the organic soils of the Holland Marsh, A report to the Ministry of Environment, Archive of Agri-Environmental Programs in Ontario, available for download at: http://agrienvarchive.ca/download/P_leaching_holland_marsh85.pdf (last access: 1 December 2014), 1985.

Uhlenbrook, S., Seibert, J., Leibundgut, C., and Rodhe, A.: Prediction uncertainty of conceptual rainfall-runoff models caused by problems in identifying model parameters and structure, Hydrolog. Sci. J., 44, 779–797, 1999.

UKCP – UK Climate Projections: Probability in UKCP09, available from: http://ukclimateprojections.metoffice.gov.uk/21680 (last access: 24 October 2014), 2010.

UKCP – UK Climate Projections: Perturbed Physics Ensembles, Coupled Model Intercomparison Project, available from: http://www.ukclimateprojections.metoffice.gov.uk/23251 (last access: 3 January 2014), 2012.

Väänänen, R.: Phosphorus retention in forest soils and the functioning of buffer zones used in forestry, Dissertationes Forestales 60, Department of Forest Ecology, University of Helsinki, Helsinki, p. 42, 2008.

van den Hurk, B., Hirshci, M., Shcär, C., Lenderink, G., van Meijgaard, E., van Ulden, A., Rockel, B., Hagemann, S., Graham, P., Kjellström, E., and Jones, R.: Soil control on runoff response to climate change in regional climate model simulations, J. Climate, 18, 3536–3551, 2004.

van Roosmalen, L., Christensen, B. S. B., and Sonnenborg, T. O.: Regional differences in climate change impacts on groundwater and stream discharge in Denmark, Vadose Zone J., 6, 554–571, 2007.

Wade, A. J., Hornberger, G. M., Whitehead, P. G., Jarvie, H. P., and Flynn, N.: On modelling the mechanisms that control in-stream phosphorus, macrophyte, and epiphyte dynamics: an assessmemt of a new model using general sensitivity analysis, Water Resour. Res., 37, 2777–2792, 2001.

Wade, A. J., Whitehead, P. G., and Butterfield, D.: The Integrated Catchments model of Phosphorus dynamics (INCA-P), a new approach for multiple source assessment in heterogeneous river systems: model structure and equations, Hydrol. Earth Syst. Sci., 6, 583–606, doi:10.5194/hess-6-583-2002, 2002a.

Wade, A. J., Whitehead, P. G., and O'Shea, L. C. M.: The prediction and management of aquatic nitrogen pollution across Europe: an introduction to the Integrated Nitrogen in European Catchments project (INCA), Hydrol. Earth Syst. Sci., 6, 299–313, doi:10.5194/hess-6-299-2002, 2002b.

Wade, A. J., Butterfield, D., Lawrence, D. S., Bärlund, I., Durand, P., Lazar, A., and Kaste, O.: The integrated catchment model of phosphorus dynamics (INCA-P), a new structure to simulate particulate and soluble phosphorus transport in European catchments, in: Diffuse Phosphorus Loss: Risk Assessment, Mitigation options and Ecological Effects in River Basins, 5th Int. Phosphorus Workshop (IPW5), 3–7 September 2007, Silkeborg, Denmark, Aarhus Universite, Aarhus, Heckrath, 2007a.

Wade, A. J., Butterfield, D., Griffiths, T., and Whitehead, P. G.: Eutrophication control in river-systems: an application of INCA-P to the River Lugg, Hydrol. Earth Syst. Sci., 11, 584–600, doi:10.5194/hess-11-584-2007, 2007b.

Westerman, P. W. and Overcash, M. R.: Short-term attenuation of runoff pollution potential for land-applied swine and poultry manure, in: Livestock waste – a renewable resource, Proceedings of the th International Symposium on Livestock Wastes, Am. Soc. Agric. Eng., St. Joseph, MI, 1980.

Whitehead, P. G., Wilby, R. L., Battarbee, R., Kernan, M., and Wade, A.: A review of the potential impacts of climate change on surface water quality, Hydrolog. Sci. J., 54, 101–123, 2009.

Whitehead, P. G., Jin, L., Baulch, H. M., Butterfield, D., Oni, S. K., Dillon, P. J., Futter, M., Wade, A. J., North, R., O'Connor, E. M., and Jarvie, H. P.: Modelling phosphorus dynamics in multi-branch river systems: a study of the Black River, Lake Simcoe, Ontario, Canada, Sci. Total Environ., 412–413, 315–323, 2011.

Whitehead, P. G., Jin, L., Crossman, J., Comber, S., Johnes, P. J., Daldorph, P., Flynn, N., Collins, A. L., Butterfield, D., Mistry, R., Bardon, R., Pope, L., and Willows, R.: Distributed and dynamic modelling of hydrology, phosphorus and ecology in the Hampshire Avon and Blashford Lakes: evaluating alternative strategies to meet WFD standards, Sci. Total Environ., 481, 157–166, 2014.

Wilby, R. L. and Harris, I.: A framework for assessing uncertainties in climate change impacts: Low-flow scenarios for the River Thames UK, Water Resour. Res., 42, 1–10, doi:10.1029/2005WR004065, 2006.

Winter, J. G., Dillon, P. J., Futter, M. N., Nicholls, K. H., Wolfgang, A. S., and Scott, L. D.: Total phosphorus budgets and nitrogen loads: Lake Simcoe, Ontario (1990–1998), J. Great Lakes Res., 28, 301–314, 2002.

Winter, J. G., Eimers, M. C., Dillon, P. J., Scott, L. D., Scheider, W. A., and Campbell, C. W.: Phosphorus inputs to Lake Simcoe from 1990 to 2003: Declines in tributary loads and observations on Lake Water Quality, J. Great Lakes Res., 333, 381–396, 2007.

XCG Consultants: Review of Phosphorus Removal at Municipal Sewage Treatment Plants Discharging to the Lake Simcoe Watershed, Prepared for Water Environment Association of Ontario Lake Simcoe Clean-Up Fund, Environment Canada, Milton, Ontario, 2010.

7

Mean transit times in headwater catchments: insights from the Otway Ranges, Australia

William Howcroft[1], **Ian Cartwright**[1,2], **and Uwe Morgenstern**[3]

[1]School of Earth, Atmosphere and Environment, 9 Rainforest Walk, Monash University, Clayton, VIC 3800, Australia
[2]National Centre for Groundwater Research and Training, G.P.O. Box 2100, Flinders University, Adelaide, SA 5001, Australia
[3]GNS Science, 1 Fairway Drive, Avalon, P.O. Box 368, Lower Hutt 5040, New Zealand

Correspondence: William Howcroft (billhowcroft@gmail.com)

Abstract. Understanding the timescales of water flow through catchments and the sources of stream water at different flow conditions is critical for understanding catchment behaviour and managing water resources. Here, tritium (^{3}H) activities, major ion geochemistry and streamflow data were used in conjunction with lumped parameter models (LPMs) to investigate mean transit times (MTTs) and the stores of water in six headwater catchments in the Otway Ranges of southeastern Australia. ^{3}H activities of stream water ranged from 0.20 to 2.14 TU, which are significantly lower than the annual average ^{3}H activity of modern local rainfall, which is between 2.4 and 3.2 TU. The ^{3}H activities of the stream water are lowest during low summer flows and increase with increasing streamflow. The concentrations of most major ions vary little with streamflow, which together with the low ^{3}H activities imply that there is no significant direct input of recent rainfall at the streamflows sampled in this study. Instead, shallow younger water stores in the soils and regolith are most likely mobilised during the wetter months.

MTTs vary from approximately 7 to 230 years. Despite uncertainties of several years in the MTTs that arise from having to assume an appropriate LPM, macroscopic mixing, and uncertainties in the ^{3}H activities of rainfall, the conclusion that they range from years to decades is robust. Additionally, the relative differences in MTTs at different streamflows in the same catchment are estimated with more certainty. The MTTs in these and similar headwater catchments in southeastern Australia are longer than in many catchments globally. These differences may reflect the relatively low rainfall and high evapotranspiration rates in southeastern Australia compared with headwater catchments elsewhere.

The long MTTs imply that there is a long-lived store of water in these catchments that can sustain the streams over drought periods lasting several years. However, the catchments are likely to be vulnerable to decadal changes in land use or climate. Additionally, there may be considerable delay in contaminants reaching the stream. An increase in nitrate and sulfate concentrations in several catchments at high streamflows may represent the input of contaminants through the shallow groundwater that contributes to streamflow during the wetter months. Poor correlations between ^{3}H activities and catchment area, drainage density, land use, and average slope imply that the MTTs are not controlled by a single parameter but a variety of factors, including catchment geomorphology and the hydraulic properties of the soils and aquifers.

1 Introduction

Determining the timescales over which precipitation is transmitted from a recharge area through a catchment to where it discharges into rivers or streams (the transit time) is important for understanding catchment behaviour and is of inherent interest to resource managers. Streams with long MTTs are connected to relatively large stores of water in the underlying aquifers (Maloszewski and Zuber, 1982; Morgenstern et al., 2010) that may sustain streamflow during droughts that last up to a few years. However, longer-term changes, such as deforestation, agricultural development, climate change, and/or landscape change following bushfires, are likely to affect both the quality and the quantity of river flows.

Headwater streams are important as they commonly support diverse ecosystems, provide recreational opportunities and in many catchments contribute a significant proportion of the total river flow (Freeman et al., 2007). Headwater streams also differ from lowland rivers in terms of their potential water inputs. Unlike lowland rivers, which typically receive groundwater inflows from regional aquifers or near-river floodplain sediments, the sources of water in headwater streams are far less well understood. Headwater streams are commonly developed at elevations well above those of the regional water tables and/or occur on relatively impermeable bedrock. Yet such streams continue to flow even during prolonged dry periods. There are several potential water stores that could contribute to stream flow, including the soil zone, weathered or fractured basement rocks, and/or perched aquifers at the soil–bedrock interface (e.g. Sklash and Farvolden, 1979; Kennedy et al., 1986; Swistock et al., 1989; Bazemore et al., 1994; Fenicia et al., 2006; Jensco and McGlynn, 2011).

Estimates of MTTs in headwater catchments range from a few months to several decades (e.g. Soulsby et al., 2000; McGuire and McDonnell, 2006; Hrachowitz et al., 2009; McDonnell et al., 2010; Stewart and Fahey, 2010; Stewart et al., 2010; Mueller et al., 2013; Stockinger et al., 2014; Atkinson, 2014; Cartwright and Morgenstern, 2015, 2016a, b; Duvert et al., 2016). However, in many regions globally the range of MTTs in headwater catchments is not well known. Additionally, it is not always clear why MTTs vary between different areas. This lack of knowledge limits our abilities to protect and manage headwater catchments.

1.1 Estimating mean transit times (MTTs)

Groundwater follows a myriad of flow paths between the recharge areas to where it discharges into streams or rivers. Consequently, groundwater discharge does not have a discrete age but rather has a distribution of transit times. MTTs are commonly estimated using lumped parameter models (LPMs) that describe the distribution of water with different ages or tracer concentrations in simplified aquifer geometries (Maloszewski and Zuber, 1982, 1996; Maloszewski et al., 1983; Cook and Bohlke, 2000; Maloszewski, 2000; Zuber et al., 2005). LPMs represent a viable and commonly used alternative to estimating MTTs using numerical groundwater models that rely upon hydraulic parameters that are seldom known with certainty and which vary spatially. However, the LPMs are only approximations of actual flow systems and the MTTs may be broad estimates rather than specific values.

The LPMs may be utilised with stable (O, H) isotopes or major ions if the concentrations vary seasonally in rainfall (e.g. Soulsby et al., 2000; McGuire and McDonnell, 2006; Tetzlaff et al., 2007, 2009; Hrachowitz et al., 2009, 2010;

Kirchner et al., 2010). Determining MTTs from stable isotope ratios or major ion concentrations relies on tracking the delay and dampening of the seasonal variations between precipitation and discharge. However, use of these tracers typically requires sub-weekly sampling over time periods equal to or exceeding that of the transit times (Timbe et al., 2015). In addition, these tracers become ineffective when transit times exceed 4 to 5 years as the initial variations in rainfall are progressively dampened to below the point at which they can be detected (Stewart et al., 2010).

Gaseous tracers (e.g. ^{3}He, chlorofluorocarbons, or SF$_6$) are effective in determining residence times of groundwater (Cook and Bohlke, 2000) but are difficult to apply to surface water due to gas exchange. With a half-life of 12.32 years, tritium (^{3}H) has been used to estimate MTTs of up to 150 years (e.g. Morgenstern et al., 2010; Stewart et al., 2010). Unlike other radioactive tracers (e.g. ^{14}C), ^{3}H is part of the water molecule and its activities are affected only by radioactive decay and dispersion and not by geochemical or biogeochemical reactions in the soils or aquifers. Because ^{3}H activities are not affected by processes in the unsaturated zone, the MTTs reflect both recharge through the unsaturated zone and flow in the groundwater system.

Utilisation of ^{3}H as a tracer is facilitated by the fact that the ^{3}H activities of rainfall have been measured globally for several decades (International Atomic Energy Agency, 2016). Due to atmospheric nuclear testing, ^{3}H activities of rainfall peaked during the 1950s and 1960s (the "bomb pulse"). The bomb-pulse ^{3}H activities in the Southern Hemisphere were much lower than in the Northern Hemisphere (Tadros et al., 2014) and have now largely declined to below those of modern rainfall (Morgenstern et al., 2010). As a consequence, MTTs can generally be determined from single ^{3}H measurements (Morgenstern et al., 2010; Morgenstern and Daughney, 2012) in an analogous manner to how other radioactive isotopes (e.g. ^{14}C or ^{36}Cl) are used in regional groundwater systems. This also allows MTTs at different streamflows to be estimated (Morgenstern et al., 2010; Duvert et al., 2016; Cartwright and Morgenstern, 2015, 2016a, b).

Using LPMs to estimate MTTs has a number of uncertainties. Due to the attenuation of the ^{3}H bomb pulse in the Southern Hemisphere, the suitability of the LPM can no longer be evaluated by time-series ^{3}H measurements (Cartwright and Morgenstern, 2016a) as is still possible in the Northern Hemisphere (e.g. Blavoux et al., 2013). Hence, LPMs must be assigned based upon knowledge of the geometry of the flow system and/or information from previous time-series studies in similar catchments. While not being able to assess the form of the LPM results in uncertainties in the calculated MTTs, the MTTs are less sensitive to the choice of LPM than is the case in the Northern Hemisphere (e.g. Blavoux et al., 2013).

Rivers can receive water from numerous stores, including groundwater, tributaries, soil water, and perched aquifers, each of which may have different MTTs. The mixing of water from different flow systems potentially produces water samples with a residence time distribution that does not correspond to those in the LPMs, and calculated MTTs are lower than actual MTTs. This is known as the aggregation error (Kirchner, 2016; Stewart et al., 2017) and it increases as the difference between the transit times of the individual endmembers increases. For transit times estimated from single ^{3}H activities, the aggregation error decreases with an increasing number of endmembers as the mixing of numerous aliquots water with different transit times is similar to what is represented by the LPMs (Cartwright and Morgenstern, 2016a).

Despite the uncertainties in calculating MTTs, because the ^{3}H activities of the remnant bomb-pulse waters have largely decayed, Southern Hemisphere waters with low ^{3}H activities have longer MTTs than waters with high ^{3}H activities. This permits relative mean transit times to be readily assessed. Because ^{3}H is radioactive, there is no requirement for flow in the catchment to be time-invariant as long as the flow path geometry remains relatively constant.

1.2 Predicting mean transit times

Fundamentally, MTTs are a function of the recharge rate, length of groundwater flow paths, and rates of groundwater flow, and parameters that control those factors will control the MTTs. Large catchments may have some long groundwater flow paths and consequently have long MTTs (e.g. McGlynn et al., 2003; Hrachowitz et al., 2010). Catchments with higher drainage densities (i.e. higher total stream length per unit area) may contain numerous short groundwater flow paths and consequently have short MTTs (e.g. Hrachowitz et al., 2009). Large groundwater storage volumes will likely also result in long MTTs (e.g. Ma and Yamanaka, 2016). Groundwater flow is likely to be more rapid through steeper catchments due to the higher hydraulic gradients, resulting in shorter MTTs (e.g. McGuire et al., 2005). Forested catchments may have higher evapotranspiration and lower recharge rates than cleared catchments (Allison et al., 1990), and the degree of forest cover exerts a control on MTTs (e.g. Tetzlaff et al., 2007). The hydraulic conductivities of the bedrock and soils are also important in controlling the timescales of water movement through catchments (e.g. Tetzlaff et al., 2009; Hale and McDonnell, 2016).

Identifying the controls on MTTs is important for understanding catchment functioning. It also potentially allows first-order estimates of MTTs to be made in similar catchments for which detailed geochemical tracer data do not exist. In some catchments, correlations between ^{3}H activities and major ion geochemistry or the runoff coefficient (the proportion of rainfall exported from the catchment by the stream) also allow first-order estimates of MTTs to be made (Morgenstern et al., 2010; Cartwright and Morgenstern, 2015, 2016a).

1.3 Objectives

This study evaluates the range of and controls on MTTs in headwater streams from the upper Gellibrand catchment of the Otway Ranges in southeastern Australia. Specifically, we test the following hypotheses. Firstly that, in common with headwater catchments elsewhere in southeastern Australia, the MTTs are several years to decades. Secondly, that the MTTs are most likely controlled by catchment attributes such as land cover, slope, or drainage density. Lastly, that shallower water stores within the catchment become progressively mobilised during higher rainfall periods contribute to streamflow at those times. We also use this study to evaluate whether there are geochemical proxies that could be used to make first-order predictions of MTTs at times when no ^{3}H data are available. Documenting MTTs is critical to understanding and protecting headwater catchments and, while this study is based on a specific area, the results have relevance to catchments globally. There is not a complete understanding of the range of MTTs in headwater catchments, nor what controls these. Thus, these are important gaps in our understanding of headwater catchments.

2 Study area

The Otway Ranges are located in southern Victoria, Australia, approximately 150 km southwest of Melbourne (Fig. 1). The region has a temperate climate, with average rainfall varying from approximately 1000 mm yr^{-1} at Gellibrand and Forrest to approximately 1600 mm yr^{-1} at Mount Sabine (Department of Environment, Land, Water and Planning, 2017) (Fig. 1) with the majority of rainfall occurring during the austral winter (July to September). Average potential evapotranspiration is 1000 to 1100 mm yr^{-1} and exceeds precipitation during the summer months (Bureau of Meteorology, 2016). The Otway Ranges occur within the Great Otway National Park, and have ecological, cultural, historical, and recreational significance. Much of the area is dominated by eucalyptus forest but also includes some commercial forestry, much of which is also eucalyptus.

The geology of the study area is described by Tickell et al. (1991). The basement comprises the Early Cretaceous Otway Group, which consists primarily of volcanogenic sandstone and mudstone with minor amounts of shale, siltstone, and coal. The Otway Group is considered to be a poor aquifer and crops out across most of the Lardners Creek and Gellibrand river catchments, as well as within the higher elevation areas of the Yahoo Creek and Ten Mile Creek catchments (Fig. 1).

The Otway Group is uncomformably overlain by Tertiary sediments of the Eastern View Formation, Demons Bluff Formation, Clifton Formation and Gellibrand Marl. The Eastern View Formation is composed of three sand and gravel units that collectively form the Lower Tertiary Aquifer. These sed-

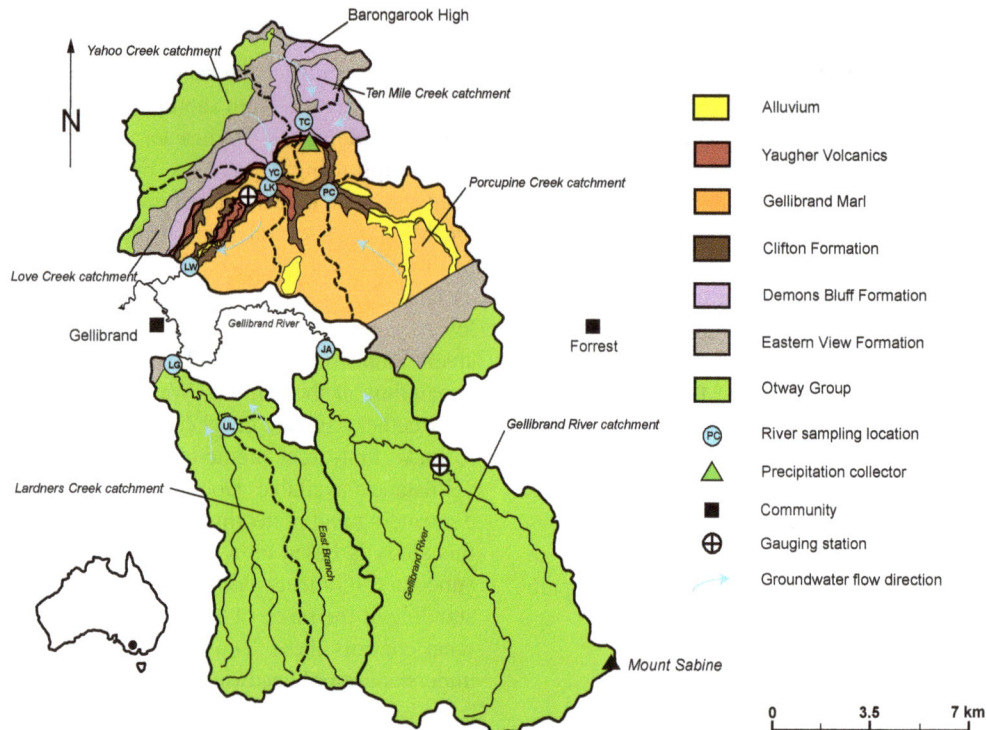

Figure 1. Map of study area showing catchments, sampling locations and bedrock geology. Inset map shows location of study area in Australia. Source: DataSearch Victoria (2015). LG = Lardners Gauge, UL = Upper Lardners, JA = Gellibrand River at James Access, PC = Porcupine Creek, TC = Ten Mile Creek, YC = Yahoo Creek, LK = Love Creek Kawarren, and LW = Love Creek Wonga. Current or discontinued gauging stations exist at all sites except for Upper Lardners.

iments crop out at various locations across the study area including at the Barongarook High (Fig. 1), which is the primary recharge area for the aquifer (Stanley, 1991; Petrides and Cartwright, 2006). The Eastern View Formation is overlain by the Demons Bluff Formation, which is a calcareous silt with negligible permeability. The formation crops out sparsely within the study area, mainly along Yahoo and Ten Mile creeks. Overlying this unit is the Clifton Formation, which is a limonitic sand and gravel aquifer. This unit crops out along Porcupine, Ten Mile, Yahoo, and Love creeks. The Clifton Formation is overlain by the Gellibrand Marl, which consists of approximately 200 to 300 m of calcareous silt. The Gellibrand Marl crops out extensively within the Love Creek and Porcupine Creek catchments and acts as a regional aquitard. Along Love Creek and parts of the Gellibrand River, the Tertiary units have been intruded by the Yaugher Volcanics, which consist primarily of basalt, tuff, and volcanic breccia. Deposits of alluvium are present along most of the stream courses, particularly Porcupine Creek and Love Creek.

Regional groundwater flows from the recharge area in the Barongarook High to the south and southwest (Leonard et al., 1981; Stanley, 1991; Atkinson et al., 2014). Additionally, localised recharge may occur elsewhere across the study area (Atkinson et al., 2014), particularly where the Eastern View

Formation crops out. Regional groundwater discharges into the Gellibrand River, Love Creek, Porcupine Creek, Ten Mile Creek and Yahoo Creek (Hebblethwaite and James, 1990; Atkinson et al., 2013; Costelloe et al., 2015). In the higher elevations of the study area, including the upper reaches of Lardners Creek, the regional water table is likely to be below the base of the streambed (Costelloe et al., 2015). Based upon ^{14}C and ^{3}H activities, residence times of the regional groundwater are between 100 and 10000 years (Petrides and Cartwright, 2006; Atkinson et al., 2014).

The Gellibrand River (Fig. 1) flows west-southwest for approximately 100 km from its highest point in the Otway Ranges before discharging into the Southern Ocean. This study focuses on six headwater catchments of the upper Gellibrand River: Lardners Creek, Love Creek, Porcupine Creek, Ten Mile Creek, Yahoo Creek, and the Gellibrand River upstream of James Access (Fig. 1). The Lardners Creek catchment includes the whole catchment (Lardners Gauge) and a smaller upper sub-catchment (Upper Lardners) (Fig. 1). Similarly, Love Creek includes the whole catchment (Love Creek Wonga) and a smaller portion of the upper catchment (Love Creek Kawarren). Porcupine Creek, Ten Mile Creek and Yahoo Creek are also tributaries to Love Creek. Love Creek and Lardners Creek flow into the Gellibrand River near Gellibrand (Fig. 1). These headwater streams contribute a sig-

Table 1. Summary of the attributes of the upper Gellibrand River catchments.

Catchment (Fig. 1)	Drainage area (km^2)	Drainage density (m m^{-2})	Forest cover (%)	Average slope (°)	Runoff coefficient (%)
Upper Lardners (UL)	20.0	1.0×10^{-3}	92	11.0	nc*
Lardners Gauge (LG)	51.6	1.1×10^{-3}	91	11.0	33.0
Gellibrand River at James Access(JA)	81.0	9.2×10^{-4}	95	11.3	39.0
Porcupine Creek (PC)	33.6	9.5×10^{-4}	88	5.9	11.4
Ten Mile Creek (TC)	9.6	8.8×10^{-4}	88	5.7	12.0
Yahoo Creek (YC)	16.6	8.7×10^{-4}	95	8.6	10.5
Love Creek Kawarren (LK)	74.4	9.3×10^{-4}	82	6.4	10.6
Love Creek Wonga (LW)	91.7	9.2×10^{-4}	78	6.7	8.6

* Not calculated.

nificant portion of flow to the Gellibrand River, which in turn provides water for several towns, supports important aquatic and terrestrial fauna, and provides water for agriculture. Current land use in the upper Gellibrand catchment, including the cleared agricultural land which replaced the native eucalyptus forest, has been established for several decades. Despite their significance, the headwater catchments of the Otway Ranges face a number of threats, including urbanisation, further clearing of native vegetation, drought, and bushfire, all of which have the potential to impact the quantity and quality of water within the streams.

The six catchments have areas ranging from 9.6 km^2 (Porcupine Creek) to 91.7 km^2 (Love Creek Wonga) (Table 1). Drainage densities are relatively similar and range from 8.7×10^{-4} m m^{-2} at Yahoo Creek to 1×10^{-3} m m^{-2} at Lardners Gauge and Upper Lardners (Table 1). Forest cover is lowest in the Love Creek Wonga (78 %) and Love Creek Kawarren (82 %) catchments. Forest cover in the other catchments is 88 % in the Porcupine Creek and Ten Mile Creek catchments, 91 to 92 % in the Lardners Gauge and Upper Lardners catchments, and 95 % in the Gellibrand River and Yahoo Creek catchments. Average slopes range from 5.7° (Ten Mile Creek) to 11.3° (at James Access).

3 Methods

3.1 Sampling and streamflow

River water samples were collected from eight locations in the catchments (Fig. 1). Lardners Creek was sampled at an active gauging station (Lardners Gauge) that is maintained by the Department of Environment, Land, Water and Planning (DELWP) (site 235210) and from the Lardners Creek East Branch (Upper Lardners), approximately 3.5 km upstream from Lardners Gauge. Love Creek was sampled at Kawarren (Love Creek Kawarren), approximately 1 km upstream of DELWP gauging station 235234 and at the Wonga Road crossing (Love Creek Wonga), approximately 4.5 km downstream of Kawarren. River water samples were col-

lected from the Gellibrand River, Porcupine Creek, Ten Mile Creek, and Yahoo Creek at the sites of former DELWP gauging stations (sites 235235, 235241, 235239, and 235240, respectively).

Streamflow at the time of sampling was determined for each of the eight locations with the exception of Upper Lardners, which is ungauged. Sub-daily streamflow is currently measured at Lardners Gauge (site 235210) and at Love Creek (site 235234) (Department of Environment, Land, Water and Planning, 2017) (Fig. 1). Streamflow at James Access on the Gellibrand River was estimated using a correlation ($R^2 = 0.97$, p value $= 10^{-8}$) between streamflow at the former gauging station at this location and that at the existing Upper Gellibrand River gauging station (site 235202), approximately 7 km upstream (Fig. 1). Likewise, streamflow at the Porcupine Creek, Ten Mile Creek, and Yahoo Creek sampling sites was estimated using correlations ($R^2 = 0.95$, 0.77, and 0.84, respectively, with p values $< 10^{-6}$) between streamflow at the former gauging stations at these locations and the Love Creek gauging station.

River water samples were collected from each site in July 2014, September 2014, March 2015, and September 2015 (Supplement). An additional round of river water samples was collected from Lardners Gauge, Porcupine Creek, Ten Mile Creek, and Love Creek Kawarren in November 2015. The water samples were collected from close to the centre of the streams using a polyethylene container fixed to an extendable pole. Additional data for James Access are from Atkinson (2014). A single precipitation sample was collected from Birnam in the Otway Ranges near Ten Mile Creek (Fig. 1) in September 2014 using a rainfall collector. The collector consisted of a polyethylene storage container equipped with a funnel positioned approximately 0.5 m above ground level. Prior to collection of the precipitation sample, the collector had been in the field for 78 days, during which time approximately 198 mm of rainfall was recorded at Forrest while 431 mm of rainfall was recorded at Mount Sabine (Department of Environment, Land, Water and Planning, 2017).

3.2 Geochemical analyses

The electrical conductivity (EC) and pH of the river water and precipitation samples was measured in the field using a calibrated TPS® hand-held water quality meter and probes. The EC measurements have a precision of $1\,\mu S\,cm^{-1}$. Cation concentrations were measured at Monash University using a Thermo Fischer ICP-OES on samples that had been filtered through $0.45\,\mu m$ cellulose nitrate filters and acidified to a pH < 2 using double-distilled 16 M HNO_3. Anion concentrations were measured at Monash University on filtered, unacidified samples using a Metrohm ion chromatograph. The precision of the cation and anion analyses, based upon replicate sample analysis, is $\pm 2\,\%$ while accuracy based on analysis of certified water standards is $\pm 5\,\%$. HCO_3 concentrations were measured by colorimetric titration with H_2SO_4 using a Hach digital titrator and reagents and are precise to $\pm 5\,\%$. Concentrations of total dissolved solids (TDSs) were determined by summing the concentrations of cations and anions. Geochemical data are presented in the Supplement.

^{3}H analysis was conducted at the GNS Water Dating Laboratory in Lower Hutt, New Zealand. The samples were vacuum distilled and electrolytically enriched prior to analysis by liquid scintillation counting, as described by Morgenstern and Taylor (2009). Following further improvements the sensitivity is now further increased to a lower detection limit of 0.02 TU (tritium units) via tritium enrichment by a factor of 95, and reproducibility of tritium enrichment of 1 % is achieved via deuterium calibration for every sample. ^{3}H activities are expressed as absolute values in tritium units where 1 TU represents a ^{3}H $/ ^1$H ratio of 1×10^{-18}. The precision (1σ) is $\sim 1.8\,\%$ at 2 TU.

3.3 Catchment attributes

Catchment attributes (Table 1) were determined using ArcGIS 10.2 (ESRI, 2013) and datasets from DataSearch Victoria (2015). The hydrology modelling tools in ArcGIS were used to generate the stream network from a 20 m digital elevation model. A threshold catchment area of 50 Ha reproduces the observed perennial stream network of the area. Catchment areas upstream of each sampling site and drainage densities were determined using the watershed tool. Mean slopes were calculated using the spatial analysis tools. Vector-based land use datasets were converted to raster formats and reclassified. Land use was assigned as forest (native vegetation and plantations) and cleared land, which includes urban and agricultural regions. Runoff coefficients were calculated using streamflow data for each of the catchments (except Upper Lardners) for March 1986 to July 1990 (Department of Environment, Land, Water, and Planning, 2017), the only interval for which continuous streamflow data are available for each catchment. The runoff coefficient calculations assumed a uniform average annual rainfall of 1.3 m for each catchment (Bureau of Meteorology, 2016). Correlations between catchment attributes and other parameters are considered to be strong where $R^2 \geq 0.7$.

3.4 Calculating mean transit times

The lumped parameter models implemented in the TracerLPM Excel workbook (Jurgens et al., 2012) were used to estimate MTTs. The ^{3}H activity of water sampled from a stream at time $t(C_0(t))$ is related to the input (C_i) of ^{3}H via the convolution integral:

$$C_0(t) = \int_0^\infty C_i\,(t - T)\,g(T)e^{-\lambda T}\mathrm{d}T, \qquad (1)$$

where T is the transit time, $t - T$ is the time that the groundwater entered the flow system, λ is the decay constant ($0.0563\,yr^{-1}$ for ^{3}H), and $g(T)$ is the exit age distribution function, for which closed-form analytical solutions have been derived (e.g. Maloszewski and Zuber, 1982, 1992, 1996; Kinzelbach et al., 2002). MTTs were estimated by matching the predicted ^{3}H activities from the LPMs to the observed ^{3}H activities of the samples.

As discussed earlier, the use of single ^{3}H activities to estimate MTTs requires that an LPM be assigned. Here two LPMs were utilised: the exponential piston-flow model (EPM) and the dispersion model (DM), which are among the most commonly used LPMs (McGuire and McDonnell, 2006; Stewart et al., 2010). The EPM describes flow in aquifers with both exponential and piston-flow portions. This model may be applied to unconfined aquifers where recharge through the unsaturated zone resembles piston flow and flow within the aquifer resembles exponential flow (Morgenstern et al., 2010). TracerLPM defines an EPM ratio, which represents the relative contribution of exponential and piston flow (Jurgens et al., 2012). The EPM ratio is $1/f - 1$, where f is the proportion of aquifer volume exhibiting exponential flow.

The dispersion model is based on the one-dimensional advection–dispersion equation for a semi-infinite medium (Jurgens et al., 2012). While this model can be applied to a wide variety of aquifer configurations, conceptually it is probably less realistic than other LPMs. Nonetheless, it has been successfully used to predict tracer concentrations over time in a number of flow systems (e.g. Maloszewski, 2000). Utilisation of this model requires defining a dispersion parameter, D_p, which represents the ratio of dispersion to advection.

The average annual ^{3}H activities of modern rainfall in central and southeastern Australia are predicted to vary between 2.4 and 3.2 TU (Tadros et al., 2014). ^{3}H activities of 9- to 17-month rainfall samples from elsewhere in Victoria are between 2.72 and 2.99 TU (Atkinson, 2014; Cartwright and Morgenstern, 2015; Cartwright et al., 2018) and fall within the range of predicted ^{3}H activities for their locations. Interpolating the data from that study suggests that modern rainfall in the Otway Ranges has an annual average ^{3}H activity of

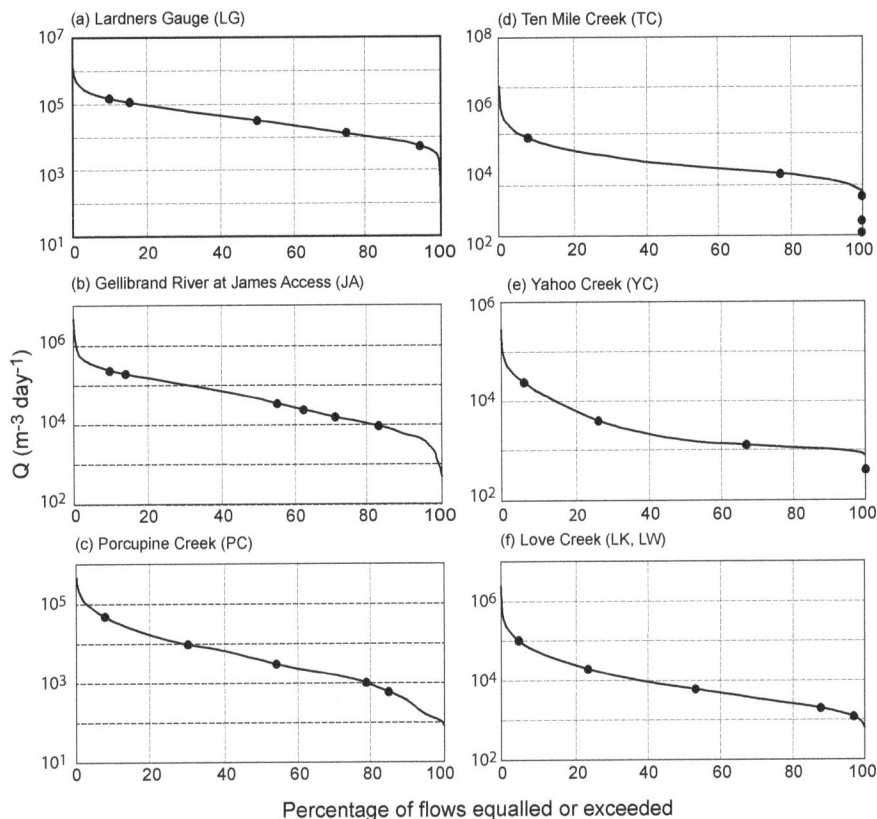

Figure 2. Streamflows at which samples were collected relative to flow duration curves for Lardners Gauge (**a**), Gellibrand River at James Access (**b**) – additional data (black circles) from Atkinson (2014), Porcupine Creek (**c**), Ten Mile Creek (**d**), Yahoo Creek (**e**) and Love Creek (**f**). Streamflow data from Department of Environment, Land, Water and Planning (2017).

~ 2.8 TU (which is slightly lower than the ~ 3.0 TU recorded at Melbourne ~ 150 km to the east of the study area). A value of 2.8 TU was used as the average annual ^{3}H activity of modern (2010 to 2016) rainfall as well as for the years prior to the atmospheric nuclear tests (pre-1951). The ^{3}H input in the intervening years is based on the ^{3}H activities of rainfall in Melbourne (International Atomic Energy Agency, 2016; Tadros et al., 2014). These were decreased by 6.7 % to account for the expected difference in ^{3}H activities in the rainfall between the Otway Ranges and Melbourne.

There are several uncertainties in the MTT calculations. The analytical uncertainty ranges between 0.02 and 0.04 TU (Supplement). To assess the effect of uncertainties in rainfall ^{3}H activities, MTTs were recalculated assuming that modern and pre-1950 rainfall had an average ^{3}H activity of either 2.4 or 3.2 TU, with the ^{3}H activities of the intervening years adjusted proportionally. As this range encompasses the estimated annual ^{3}H activities of rainfall over most of central and southeastern Australia, it allows a conservative estimate of uncertainties to be made.

The aggregation or macroscopic mixing of waters also introduces uncertainties (Kirchner, 2016; Stewart et al., 2017). Consider a stream fed by several tributaries. The expected

MTT (MTT$_e$) can be calculated using the streamflow data, ^{3}H activities, and MTTs of each tributary via

$$\text{MTT}_e = a\text{MTT}_1 + b\text{MTT}_2 + c\text{MTT}_3 + \ldots \quad (2)$$

(Stewart et al., 2017). In Eq. (2), a, b, and c, represent the fraction of total flow contributed by tributaries 1, 2, and 3. If the aggregation is minimal, MTT$_e$ will be similar to that estimated from the measured ^{3}H activity via the LPM. The successful application of Eq. (2) relies on the MTTs of the different tributaries being defined by their ^{3}H activities (which in itself may not be straightforward due to aggregation within those sub-catchments). Nevertheless, it provides a broad estimate of the error due to macroscopic mixing that is otherwise difficult to assess.

3.5 Groundwater volumes

The volume (V in m^3) of groundwater stored within an aquifer that interacts with the stream (sometimes referred to as the turnover volume) is related to the MTT by

$$V = Q \cdot \text{MTT}, \quad (3)$$

where Q is streamflow (m^3 yr^{-1}) (Maloszewski and Zuber, 1982, 1992; Morgenstern et al., 2010).

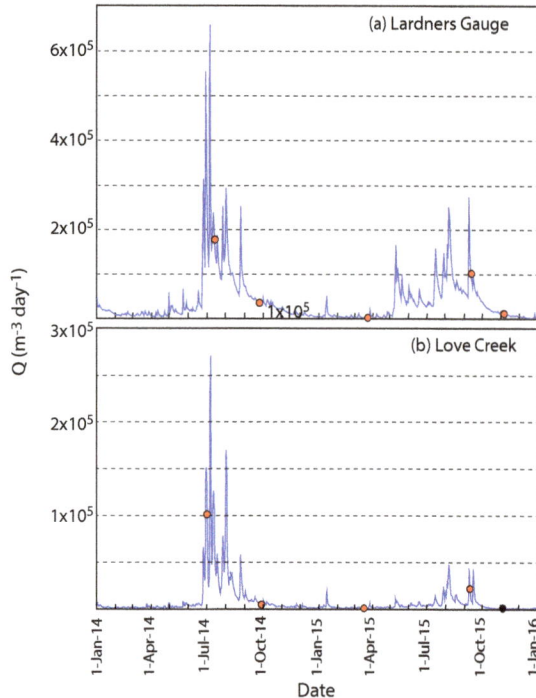

Figure 3. Hydrographs for Lardners Gauge **(a)** and Love Creek **(b)** together with the timing of sample collection. Data from Department of Environment, Land, Water and Planning (2017).

4 Results

4.1 Streamflow

Streamflow was highest during July 2014 (Supplement), ranging from $8.6 \times 10^3 \, m^3 \, day^{-1}$ at Ten Mile Creek to $255 \times 10^3 \, m^3 \, day^{-1}$ at James Access. Discharge was lowest during March and November 2015, ranging from $0.1 \times 10^3 \, m^3 \, day^{-1}$ at Ten Mile Creek to $8.8 \times 10^3 \, m^3 \, day^{-1}$ at James Access. Figure 2 illustrates the streamflows for the sampling rounds relative to the flow duration curves for the catchments. Samples were generally collected between the 10th and 100th percentiles of streamflow, which encompasses a wide range of flow conditions. Samples were collected during the recession periods after high-flow events that follow rainfall or during baseflow conditions (Fig. 3). Overland flow was not observed during any of the sampling events and small ephemeral tributaries in the catchments were dry.

Runoff coefficients range from 33 and 39 % at Lardners Gauge and James Access, respectively, to between 9 and 12 % at Porcupine Creek, Ten Mile Creek, Yahoo Creek Wonga, and Love Creek Kawarren (Table 1). The higher runoff coefficients at Lardners Gauge and James Access relative to the other catchments may be due to the fact that these rivers drain steeper catchments and are underlain almost entirely by low hydraulic conductivity Otway Group basement rocks (Fig. 1).

4.2 Tritium activities

As discussed above, the annual average ^3H activities of modern rainfall in much of central and southeastern Australia are between 2.4 and 3.2 TU (Tadros et al., 2014). The 78-day precipitation sample collected from near Ten Mile Creek in September 2014 had a tritium activity of 2.45 TU. This is lower than both the expected ^3H activities for the Otway Ranges (~ 2.8 TU: Tadros et al., 2014) and those of 9- to 12-month rainfall samples elsewhere in Victoria (2.72 to 2.99 TU: Atkinson, 2014; Cartwright and Morgenstern, 2015, 2016a; Cartwright et al., 2018). However, the Ten Mile Creek sample reflects rainfall over only part of the year and may not be representative.

Tritium activities of the rivers are < 2.14 TU, which are lower than the average annual ^3H activities of modern rainfall and indeed the Ten Mile Creek rainfall sample. The ^3H activities vary from 0.20 TU at Porcupine Creek in March 2015 to 2.14 TU at Yahoo Creek in July 2014 (Fig. 4). The higher ^3H activities in the rivers are within the range of ^3H activities of 1.80 to 2.25 TU for soil pipe water in higher elevations in the Gellibrand Catchment (Atkinson, 2014) (Fig. 4). In general, ^3H activities were highest at high streamflow (July 2014) and lowest at low streamflow (March and November 2015).

The ^3H activities of Love Creek at the upstream (Love Creek Kawarren) and downstream (Love Creek Wonga) locations in individual events varied by < 0.1 TU. The ^3H activities in Lardners Creek between Upper Lardners and Lardners Gauge were slightly more variable (up to 0.17 TU). The range of ^3H activities between the events was most variable at Porcupine Creek (0.20 to 1.97 TU), followed by Yahoo Creek (0.43 to 2.14 TU), Love Creek Kawarren (0.48 to 1.91 TU), Love Creek Wonga (0.55 to 1.88 TU), Ten Mile Creek (0.44 to 1.74 TU), Upper Lardners (1.54 to 1.99 TU), James Access (1.73 to 2.08 TU), and Lardners Gauge (1.64 to 1.97 TU) (Fig. 4). Overall, the highest ^3H activities were similar across all catchments but the lower ^3H activities varied considerably. The ^3H activities increase with increasing streamflow up to approximately $10^4 \, m^3 \, day^{-1}$, above which ^3H activities do not increase appreciably (Fig. 4). Despite differences in catchment size, slope, geology, and land use, there is a strong correlation between ^3H activities and streamflow across the catchments ($^3H = 0.2613 \ln(Q) + 0.8973$; $R^2 = 0.75$, p value $= 0.15$).

4.3 Major ion geochemistry

River water geochemistry is similar across all catchments and is dominated by Na, Cl, and HCO_3 (Supplement). TDS concentrations are generally less than $100 \, mg \, L^{-1}$ at Lardners Gauge, Upper Lardners, and James Access but typically exceed $200 \, mg \, L^{-1}$ in Love Creek Wonga, Love Creek Kawarren, Porcupine Creek, Ten Mile Creek, and Yahoo Creek. TDS concentrations increase downstream in Lardners

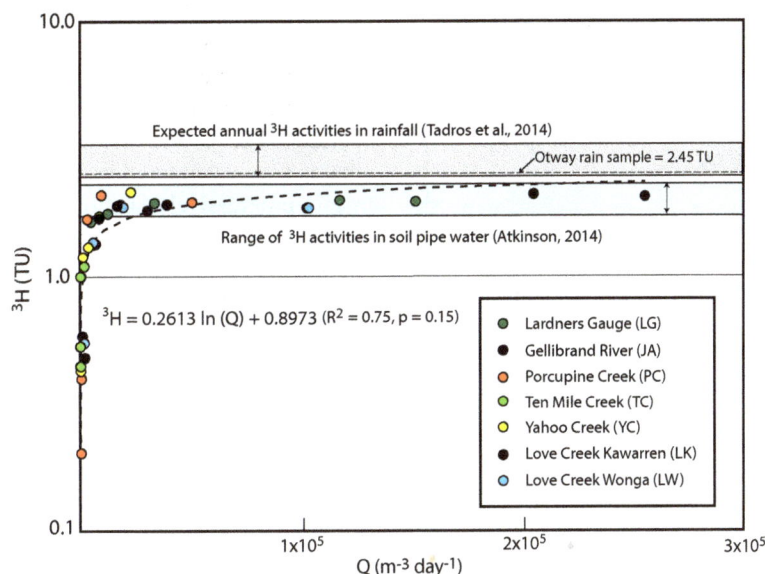

Figure 4. ^{3}H activities of stream water as a function of streamflow for all catchments except Upper Lardners which is ungauged. ^{3}H data from Supplement, streamflow data from Department of Environment, Land, Water and Planning (2017) or calculated as discussed in the text. Shaded boxes show the expected annual average of rainfall ^{3}H activities from Tadros et al. (2014) and soil waters from Atkinson (2014).

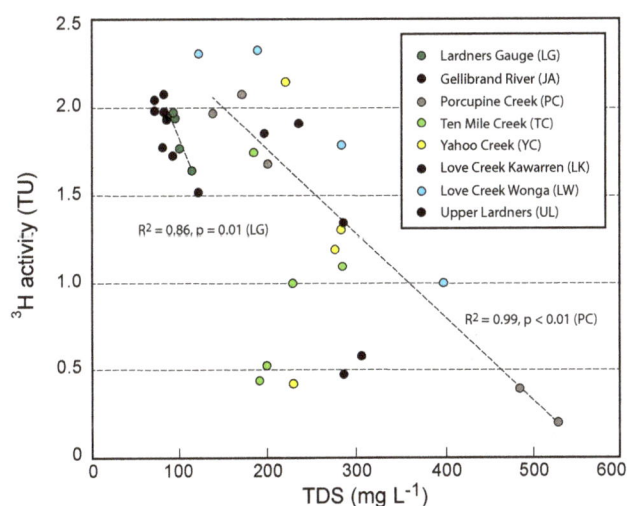

Figure 5. ^{3}H activities as a function of TDS for all catchments (data from Supplement). Strong inverse correlations between ^{3}H activities and TDS exist for Lardiners Gauge and Porcupine Creek.

and Love creeks and are inversely correlated with streamflow in all catchments.

At Love Creek, Ten Mile Creek, Yahoo Creek, and Upper Lardners, there is no correlation between ^{3}H activities and EC, TDS, or major ion concentrations (Fig. 5). However, at Porcupine Creek, there is a strong correlation ($R^2 > 0.95$, p value < 0.01) between ^{3}H activities and EC, TDS, and all

major ion concentrations with the exception of chloride, nitrate, and sulfate. In addition, there is a strong correlation ($R^2 = 0.86$, p value $= 0.01$) between ^{3}H activities and TDS at Lardners Gauge (Fig. 5).

At Upper Lardners, James Access, and Ten Mile Creek, there is a strong correlation ($R^2 > 0.8$, p values < 0.11) between nitrate concentration and ^{3}H activities (Fig. 6a). The range of nitrate concentrations (0.08 to 2.0 mg L^{-1}) were relatively similar during each sampling event across all catchments except for in July 2014, when nitrate concentrations exceeded 3 mg L^{-1} at Love Creek Kawarren and Love Creek Wonga. A similar correlation exists between sulfate concentrations and ^{3}H activities at James Access and at Upper Lardners, but not at Ten Mile Creek (Fig. 6b). However, sulfate concentrations at these locations are lower than they are in the other catchments.

5 Discussion

The combination of streamflow, ^{3}H activities, major ion geochemistry, and catchment attributes allows aspects of the behaviour of the upper Gellibrand catchments to be understood. This section addresses the changing stores of water in the catchments, the range and uncertainties of MTTs, and whether MTTs can be predicted from catchment attributes or geochemical data.

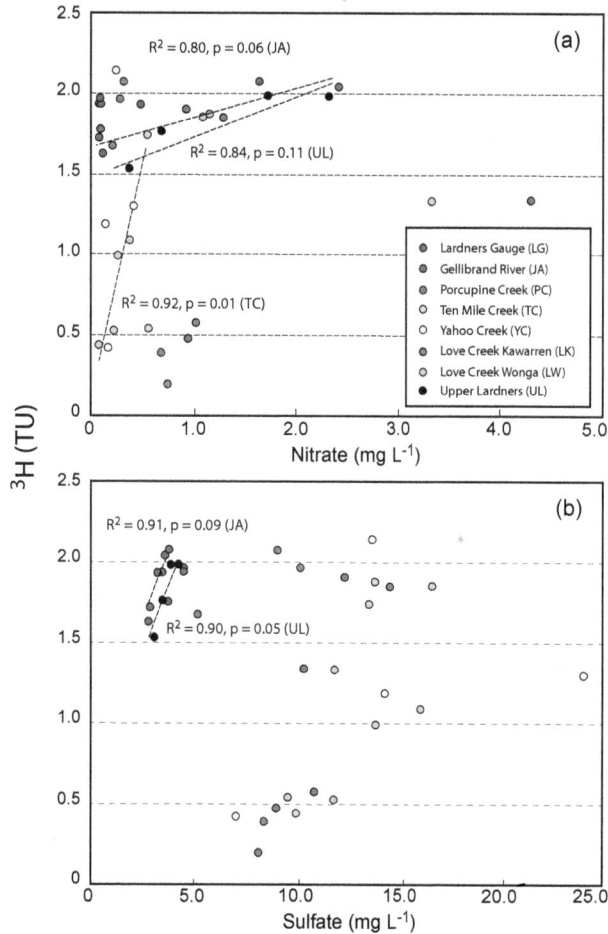

Figure 6. ^{3}H activities as function of nitrate concentrations **(a)** and sulfate concentrations **(b)**. Data from Supplement. Strong ($R^2 > 0.7$) correlations indicated.

and Farvolden, 1979; Kennedy et al., 1986; Jensco and McGlynn, 2011; Cartwright and Morgenstern, 2015).

Together, these observations suggest that there is no significant direct input of recent rainfall during the sampling periods. The flow system is concluded to be a continuum that is dominated by older groundwater inflows at low flows while progressively shallower and younger stores of water (such as soil water or perched groundwater) are mobilised during wetter periods. The observations that nitrate and sulfate concentrations in several of the catchments are higher at high streamflows (Fig. 6) may reflect the input of contaminants from recent agricultural activities to the streams. This observation agrees with the conceptualisation that shallower stores of water in the catchment, which are more likely to be impacted by contamination, are mobilised during the wetter periods of the year.

5.2 Mean transit times

If the conceptualisation of the flow system is correct, MTTs may be calculated using a single LPM. If there were some dilution by recent rainfall, using a single LPM yields the minimum MTT of the baseflow component (Morgenstern et al., 2010). MTTs in the headwaters catchments were estimated using the EPM and the DM. For the EPM, EPM ratios of 0.33 (75 % exponential flow), 1.0 (50 % exponential flow) and 3.0 (25 % exponential flow) were adopted. The EPM model accords with the expected geometry of flow in the catchment (vertical recharge through the unsaturated zone followed by flow along flow paths of varying length), and EPM models with these EPM ratios have reproduced the ^{3}H time series in headwater catchments with similar geometries elsewhere (Maloszewski and Zuber, 1982; Morgenstern and Daughney, 2012; Blavoux et al., 2013; Morgenstern et al., 2010). For the DM, D_p values of 0.05 and 0.5 were adopted, which are appropriate for kilometre-scale flow systems (Zuber and Maloszewski, 2001; Gelhar et al., 1992). Utilisation of a variety of LPMs allows the impact of the assumed model on the MTTs to be assessed.

Calculated MTTs ranged from approximately 7 years at Yahoo Creek in July 2014 to 230 years at Porcupine Creek in March 2015 (Table 3). In general, the lowest MTTs were estimated from the EPM with EPM ratio = 3.0 while the highest MTTs were estimated using the DM with $D_p = 0.5$. Because of the remnant bomb-pulse ^{3}H, a few samples with ^{3}H activities between 1.2 and 1.7 TU yield MTTs that are non-unique for models with high piston flow components (i.e. the EPM with EPM ratio = 3.0 and the DM with $D_p = 0.05$; Table 3, Fig. 7). The choice of the LPM has little impact on MTTs for ^{3}H activities greater than 1 TU (Fig. 7). However, as ^{3}H activities decrease, the relative difference between the MTTs from the different LPMs increases. At the lowest ^{3}H activity of 0.20 TU, the difference between the MTT estimates is approximately 164 years.

5.1 Sources of river inflows

It is important to determine how the water stores that contribute to streamflow change between high and low flows. Groundwater inflows are most probably the dominant source of water during the summer months. However, at times of higher streamflow there may be mobilisation of younger shallower water stores (e.g. water from the soils or the regolith) as the catchment wets up (e.g. Hrachowitz et al., 2013; Cartwright and Morgenstern, 2015, 2016a) or mixing between baseflow and recent rainfall (e.g. Morgenstern et al., 2010). The river water samples were collected during baseflow conditions or during recession periods after high streamflows that follow rainfall (Fig. 3) when recent rainfall is less likely to directly contribute to streamflows. That the major ion geochemistry varies little with streamflow also suggests that there is not significant dilution of groundwater inflows with recent rainfall during the sampling periods (e.g. Sklash

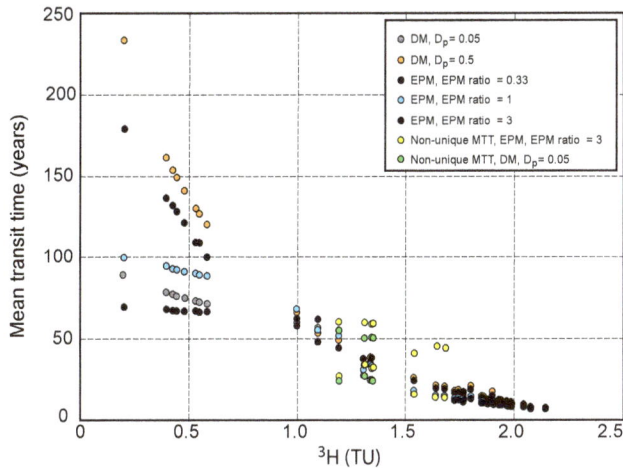

Figure 7. Estimated MTTs vs. ^{3}H activities in the stream waters calculated using the exponential piston-flow model (EPM) with EPM ratios of 0.33, 1.0, and 3.0 and the dispersion model (DM) with D_p values of 0.05 and 0.5. Data from Supplement and Table 3.

MTTs for Lardners Gauge, Upper Lardners, and James Access were similar, and are between 7 and 26 years. In contrast, MTTs for Porcupine Creek ranged from approximately 7 to 230 years, while those for Ten Mile Creek, Yahoo Creek, Love Creek Wonga, and Love Creek Kawarren ranged from approximately 13 to 150, 7 to 15, and 10 to 140 years, respectively. In all catchments, the longest MTTs are recorded at the lowest streamflows (March 2015) while the shortest MTTs occur at the highest streamflows (July 2014 and September 2015) (Fig. 8). At Lardners Gauge, James Access, Porcupine Creek, and Love Creek, the samples collected at the highest flow rates have MTTs that are slightly longer than those of the samples collected at the second highest streamflow (Fig. 8). Whether this reflects changes to the flow system or is due to uncertainties in the MTT estimates is not certain.

The volume of water in the aquifers that contributes to the streamflow may be estimated from Eq. (3). Both the Lardners Gauge and the Love Creek Wonga catchments have active streamflow monitoring, and the calculations are carried out for these catchments. Using the relationships between MTT and streamflow (Fig. 8) and streamflow data for 2014 and 2015 (Department of Environment, Land, Water, and Planning, 2017), the average MTT for the two catchments is estimated to be 29.7 years (Love Creek Wonga) and 10.8 years (Lardners Gauge). For the average annual streamflow over those 2 years, the turnover volumes are 2.6×10^5 m^3 (Love Creek Wonga) and 4.5×10^5 m^3 (Lardners Gauge). These volumes are small relative to the likely volumes of water stored in the catchments. For the catchment areas (Table 1) and a porosity of 0.1 to 0.3, which is appropriate for most soils and aquifers, this volume of water could be stored in a layer that is 0.01 to 0.1 m thick.

5.3 Uncertainties in MTT estimates

The uncertainties in the MTTs arising from the analytical uncertainties (Supplement) range from ± 0.9 years for the sample with the highest ^{3}H activity to ± 10 years for the sample with the lowest ^{3}H activity. These equate to relative uncertainties of $\sim \pm 10$ %. Having to assume an LPM reflects a major uncertainty for calculating the MTTs, especially for waters with ^{3}H activities < 1 TU (Fig. 7). For a water with a ^{3}H activity of 2 TU, the uncertainty in MTTs is ± 1.2 years (± 13 %), while for waters with ^{3}H activities of 1 and 0.5 TU they are ± 5 years (± 8 %) and ± 31 years (± 30 %), respectively. The EPM with an EPM ratio of 3.0 and the DM with a D_p value of 0.05 have a large component of piston flow and are possibly less realistic representations of the flow systems; however, the differences between the MTTs estimated using the other LPMs are still considerable.

The influence of uncertainties in the ^{3}H input was assessed by varying the modern and pre-bomb-pulse ^{3}H activities between 2.4 and 3.2 TU and adjusting the ^{3}H activities in the intervening years accordingly. As discussed above, this encompasses the predicted range of average annual ^{3}H activities in most of central and southeastern Australia. These calculations used the EPM with an EPM ratio of 1.0 but the effect is similar in the other models. The relative difference between MTTs is generally highest when ^{3}H activities exceed 1 TU (Fig. 9). For ^{3}H activities of 2 TU, the uncertainty in MTTs is ± 5 years (± 54 %), while for waters with ^{3}H activities of 1 and 0.5 TU they are ± 10 years (± 15 %) and ± 5 years (± 5 %), respectively.

^{3}H activities in rainfall can vary seasonally. Catchments with MTTs in excess of a few years do not preserve seasonal variations in stable isotope ratios or major ion concentrations (Stewart et al., 2010). In a similar way, the seasonal variation in rainfall ^{3}H activities are unlikely to be preserved in the catchment waters (Morgenstern et al., 2010). Thus, using annual ^{3}H activities as the input is appropriate. However, if recharge has a strong seasonality, its ^{3}H activities may be different from those of annual rainfall. Rainfall in the Otway Ranges is distributed throughout the year and it is likely that some recharge occurs throughout the year. Less recharge probably occurs during summer due to some rainfall being lost to evapotranspiration. However, as is the case elsewhere in the Southern Hemisphere (Morgenstern et al., 2010), the ^{3}H activities in summer rainfall are closely similar to the average annual ^{3}H activities (Tadros et al., 2014; International Atomic Energy Agency, 2017). The observation that the ^{3}H activities of summer (December to February) rainfall at Mount Buffalo in northeastern Victoria were similar (2.86 TU) to those of two annual rainfall samples (2.99 and 2.85 TU) support this assertion (Cartwright and Morgenstern, 2015). With such a seasonal distribution of ^{3}H activities, the uncertainties in MTTs resulting from using the average annual ^{3}H activities are less than those that arise from the general uncertainty in the ^{3}H input function.

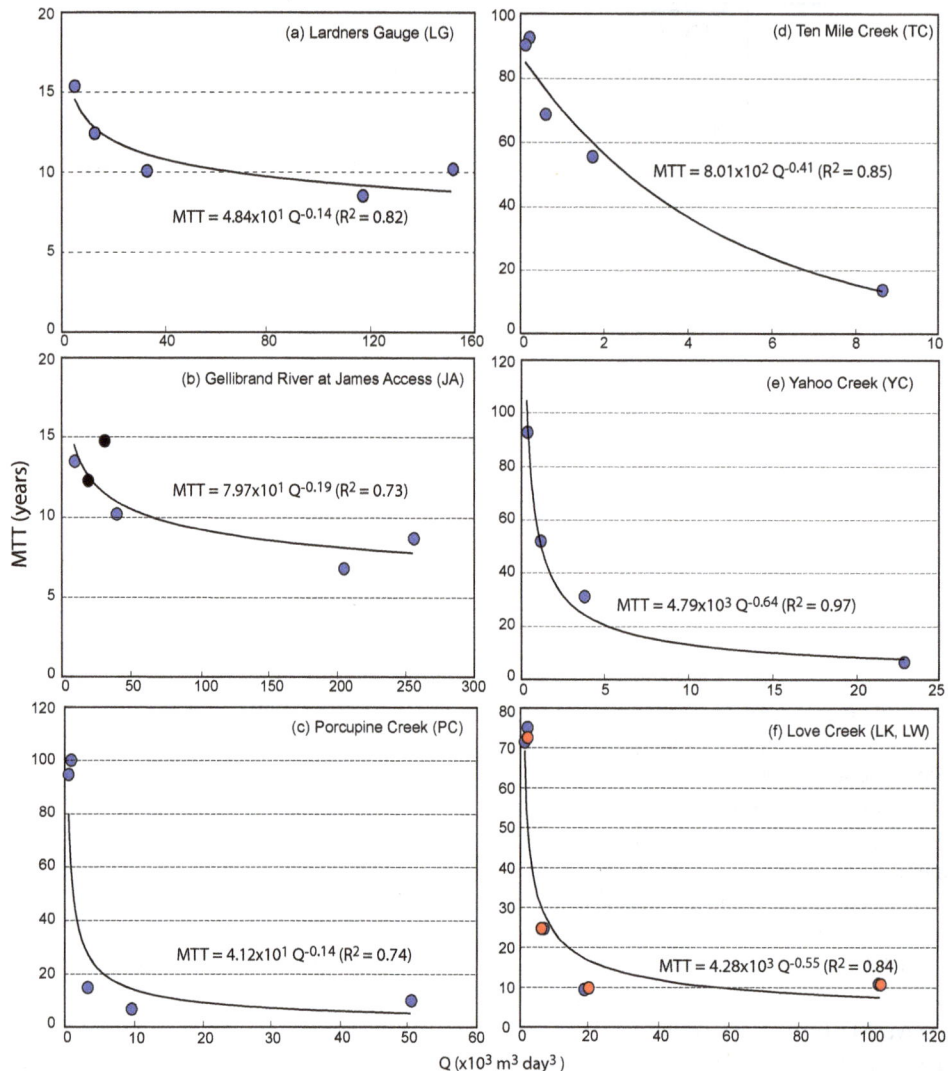

Figure 8. MTTs calculated using the EPM model with an EPM ratio of 1.0 (Table 3) as a function of streamflow (Q) for Lardners Gauge **(a)**, Gellibrand River at James Access **(b)** – black circles are data from Atkinson (2014), Porcupine Creek **(c)**, Ten Mile Creek **(d)**, Yahoo Creek **(e)**, and Love Creek **(f)** – blue circles are Love Creek Kawarren and red circles Love Creek Wonga. Curves are exponential trend lines. Streamflow data from Department of Environment, Land, Water and Planning (2017) or calculated as discussed in the text.

The impact of macroscopic mixing was estimated using Eq. (2) and the streamflow data and MTTs for Porcupine, Ten Mile, and Yahoo creeks that flow into Love Creek upstream of Love Creek Kawarren (Fig. 1). The analysis used the EPM with an EPM ratio of 1.0 (Table 3), but again similar results were obtained with the other LPMs. Based on the streamflow data, these three streams contribute 77 to 82 % of total stream flow at Love Creek Kawarren (Table 3). The remaining portion of flow in Love Creek is assumed to be contributed by undefined inputs such as groundwater inflow and inputs from smaller tributaries. It was assumed that there was one unidentified input, the ^{3}H activity of which was estimated by the difference between the weighted ^{3}H activities of Porcupine, Ten Mile, and Yahoo creeks and the ^{3}H ac-

tivity at Love Creek Kawarren. The MTT of this input was determined from the ^{3}H activity using the EPM.

In March 2015, the estimated MTT calculated using the LPM at Love Creek Kawarren was higher than MTT_e calculated using Eq. (2) by 3.7 years or 4 % (Table 4). At other times, the differences were 3.9 to 7.4 years (18 to 37 %). These calculations may not truly address aggregation as there may be more than one unidentified additional store of water and there may be aggregation within the individual subcatchments (which impacts their estimated MTTs). Nevertheless, they do indicate that the potential uncertainties in MTTs due to aggregation are potentially several years (as discussed by Stewart et al., 2017). For waters with similar ^{3}H activities, Cartwright and Morgenstern (2016a) estimated

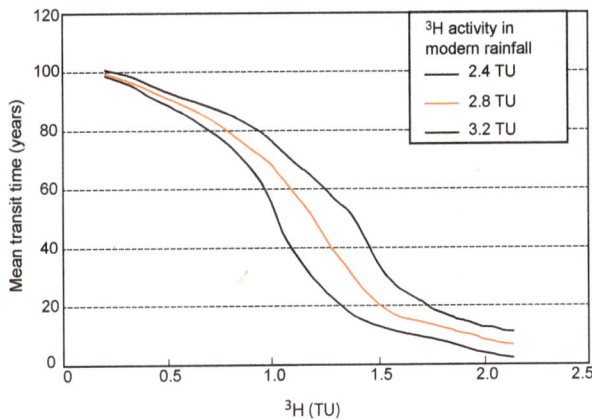

Figure 9. Impact of varying rainfall ^{3}H inputs on MTTs calculated using the EPM model with an EPM ratio of 1.0. The three rainfall inputs modern and pre bomb-pulse ^{3}H activities of 2.4, 2.8, and 3.2 TU and the ^{3}H activity of the bomb-pulse rainfall was varied by a similar proportion as discussed in the text.

Table 2. Correlation between catchment attributes and ^{3}H activities for the upper Gellibrand River catchments.

Catchment attribute	Sampling date	R^2
Area	Jul 2014	0.01
	Sep 2014	0.26
	Mar 2015	0.06
	Sep 2015	0.57
Drainage density	Jul 2014	0.00
	Sep 2014	0.58
	Mar 2015	0.40
	Sep 2015	0.40
Runoff coefficient	Jul 2014	0.10
	Sep 2014	0.66
	Mar 2015	0.94
	Sep 2015	0.19
Forest cover	Jul 2014	0.51
	Sep 2014	0.15
	Mar 2015	0.24
	Sep 2015	0.01
Slope	Jul 2014	0.39
	Sep 2014	0.55
	Mar 2015	0.87
	Sep 2015	0.15

that the aggregation error may be up to 20 % where two waters with MTTs of 10 and 50 years or 1 and 5 years mixed but noted that this error became progressively lower if more stores of water with a similar range of MTTs mixed.

If the uncertainties are uncorrelated, the overall uncertainty is given by the square root of the sum of the squares of the individual uncertainties. The analysis assumes that uncertainties due to analytical uncertainties and aggregation are uniformly 10 and 20 %, respectively, and the uncertainties from the range of LPMs and the ^{3}H input of rainfall are as discussed above. For a water with a ^{3}H activity of 2 TU, the overall uncertainty in MTTs are approximately $\pm 60\%$ (± 5.4 years), whereas for waters with ^{3}H activities of 1 and 0.5 TU they are $\pm 28\%$ (± 17 years) and $\pm 38\%$ (± 35 years), respectively.

While these uncertainties are considerable, the observation that the ^{3}H activities of the streams are locally 10 % of those of modern rainfall (and far less than the rainfall ^{3}H activities at the peak of the bomb pulse) necessitates that the MTTs must be several decades. Because the aggregation error, which is probably the most difficult to assess, results in MTTs being underestimated (Kirchner et al., 2016; Stewart et al., 2017) some MTTs may be longer than calculated. Relative differences in MTTs between and within catchments may be estimated with more certainty. Because the catchments are located in a relatively small area, the ^{3}H inputs are likely to be closely similar. Thus, uncertainties in the ^{3}H input are thus less likely to impact the comparison of MTTs between catchments. Additionally, as the geometry of the flow system in each catchment is unlikely to vary substantially at different streamflows, not being able to assess the suitability of the LPM has less impact on the relative differences in MTTs at different streamflows in the same catchment.

5.4 Predicting mean transit times

There are weak ($R^2 \leq 0.7$) or no correlations between ^{3}H activities and catchment area, drainage density, or forest cover (Table 2). There is a strong correlation between ^{3}H activities and average slope ($R^2 = 0.87$, p value 0.01) during March 2015, when streamflow was lowest but not at other times. The variability of MTTs from James Access, Lardners Gauge, and Upper Lardners (which occur on the Otway Group: Fig. 1) and from Porcupine Creek, Yahoo Creek, Love Creek, and Ten Mile Creek (which have similar lithologies in their catchments: Fig. 1) indicates the MTTs are not simply related to the geology. A combination of the catchment properties together with the hydraulic properties of the soils and aquifers or evapotranspiration rates likely control the MTTs. The hydraulic properties and evapotranspiration rates are probably spatially variable and are difficult to estimate, which makes it difficult to assess their influence. The observation that relationships between ^{3}H activities and streamflow in all the catchments are similar (Fig. 4) suggests that the MTTs at high flows reflect the inflow of water from the shallower water stores which will be largely independent of the catchment attributes.

There are strong positive correlations between ^{3}H activities and the runoff coefficient ($R^2 = 0.94$, p value $= 0.27$) (Fig. 10). This may be due to both the runoff coefficient and MTTs being controlled by the rates of recharge and ground-

Table 3. Summary of calculated mean transit times (MTTs) for the upper Gellibrand River catchments.

Location (Fig. 1)	Date (dd/mm/yyyy)	Q^a $10^3\,m^3\,day^{-1}$	^3H (TU)	EPM[b] 0.33	1.0	3.0	DM[c] 0.05	0.5
Upper Lardners (UL)	10/07/2014	–	1.99	9.9	9.6	8.8	9.0	11.2
	28/09/2014	–	1.77	15.7	12.9	11.8	12.2	17.6
	20/03/2015	–	1.54	24.2	18.5	(16.2, 41.4)[d]	16.3	26.2
	10/09/2015	–	1.99	8.8	8.2	8.6	8.3	9.9
Lardners Gauge (LG)	10/07/2014	151.3	1.94	10.8	10.2	9.3	9.6	12.3
	28/09/2014	32.8	1.94	10.6	10.1	9.2	9.5	12.1
	20/03/2015	5.0	1.64	19.8	15.4	(14.1, 45.7)	14.2	21.6
	10/09/2015	116.6	1.97	9.1	8.5	8.7	8.6	10.2
	04/11/2015	12.7	1.77	13.8	12.4	11.2	11.6	15.8
Gellibrand River at James Access (JA)	13/03/2012	18.5	1.90	15.5	12.3	11.8	11.7	17.7
	26/04/2012	30.4	1.80	19.2	14.8	13.1	13.4	21.4
	10/07/2014	255.2	2.04	8.7	8.7	8.1	8.2	9.7
	28/09/2014	39.1	1.93	10.8	10.2	9.4	9.7	12.4
	20/03/2015	8.8	1.73	16.2	13.5	12.2	12.6	18.2
	10/09/2015	204.4	2.08	7.3	6.8	7.7	7.0	8.1
Porcupine Creek (PC)	10/07/2014	50.4	1.97	10.3	9.8	9.0	9.2	11.7
	27/09/2014	3.3	1.68	19.3	14.9	(13.9, 44.7)	13.8	21.0
	20/03/2015	1.0	0.20	179	100	69.5	89.6	234
	10/09/2015	9.7	2.08	7.3	6.8	7.7	7.0	8.1
	04/11/2015	0.6	0.40	137	94.8	68.4	78.7	162
Ten Mile Creek (TC)	10/07/2014	8.6	1.74	17.1	13.6	12.5	12.7	18.8
	27/09/2014	0.6	1.00	58.3	68.5	62.5	60.1	66.3
	20/03/2015	0.2	0.44	128	92.5	67.2	76.4	150
	10/09/2015	1.7	1.09	48.3	55.5	62.0	57.0	53.5
	04/11/2015	0.1	0.53	109	90.3	67.2	73.3	130
Yahoo Creek (YC)	11/07/2014	23.0	2.14	6.9	6.8	7.2	7.0	7.6
	28/09/2014	1.2	1.19	44.7	52.0	(60.6, 27.4)	(55.3, 24.8)	49.2
	20/03/2015	0.4	0.43	132	93.1	67.4	77.2	154
	10/09/2015	3.9	1.30	34.8	31.3	(34.3, 60.0)	(27.6, 50.7)	37.9
Love Creek Kawarren (LK)	10/07/2014	102.9	1.85	13.3	11.5	10.5	10.9	15.0
	27/09/2014	6.7	1.34	35.3	33.5	(32.3, 59.2)	(24.8, 51.2)	38.4
	20/03/2015	2.0	0.48	121	91.2	67.0	75.1	141
	10/09/2015	18.6	1.91	10.4	9.8	9.5	9.5	11.9
	04/11/2015	1.2	0.58	100	88.6	66.8	71.5	120
Love Creek Wonga (LW)	10/07/2014	103.5	1.86	13.1	11.4	10.4	10.8	14.8
	28/09/2014	6.0	1.34	35.7	34.2	(32.1, 59.3)	(24.8, 51.4)	38.8
	20/03/2015	2.0	0.55	109	89.4	66.4	72.6	127
	10/09/2015	19.6	1.88	11.0	10.4	9.8	9.9	12.6

[a] Discharge. [b] Exponential piston-flow model with EPM parameter of 0.33, 1, and 3. [c] Dispersion model with dispersion parameter of 0.05 and 0.5. [d] Non-unique MTTs.

water flow. The Lardners Gauge and James Access sites have much higher runoff coefficients than the other catchments, and the correlation with ^3H activities may reflect the difference between the two groups of catchments. If this is the case, the runoff coefficient may be useful in determining gross rather than subtle differences in MTTs.

EC and streamflow were measured on a monthly basis at the gauging station on Porcupine Creek (site 235241) between January 1990 and January 1994 (Department of Environment, Land, Water and Planning, 2017). A strong correlation between MTTs and EC at this location (MTT $= 1.362\,e^{0.0061 \cdot EC}$: $R^2 = 0.96$, p value $= 10^{-8}$) allows MTTs at this site to be estimated over this 4-year period (Fig. 11). The estimated MTTs range from 3 to 50 years, with the longest MTTs corresponding to low summer flows and the shortest MTTs during high winter flows. Although based upon a limited number of samples, these results demonstrate

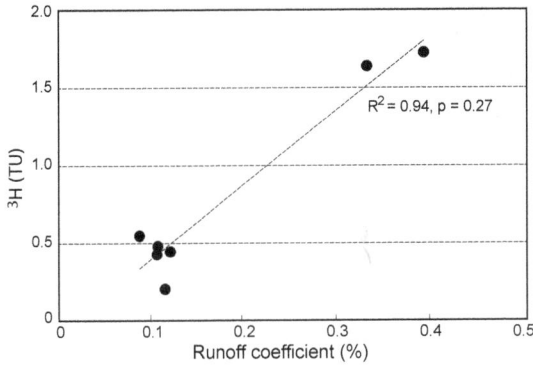

Figure 10. ^{3}H activities vs. runoff coefficients for the March 2015 samples (data from Table 1 and Supplement). Although a strong correlation ($R^2 = 0.94$) exists, it may be a result of the grouping of the samples.

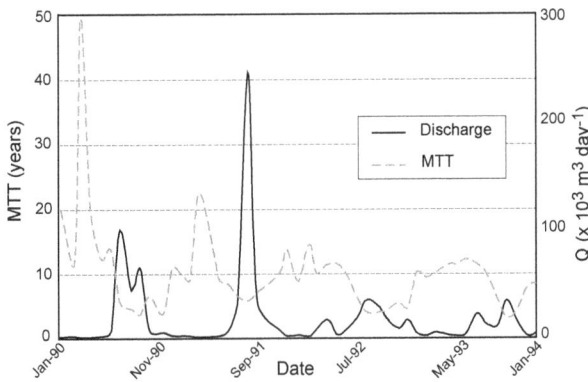

Figure 11. Variation in MTT as a function of streamflow at Porcupine Creek for January 1990 to January 1994 calculated using the relationship between EC and ^{3}H activity (Supplement) and monthly EC data from the Department of Environment, Land, Water and Planning (2017). Streamflow data also from Department of Environment, Land, Water and Planning (2017).

the high variability of transit times within the catchment and the value of finding proxies for ^{3}H.

6 Summary and conclusions

The calculated MTTs in the six headwater catchments in the Upper Gellibrand catchment of Otway Ranges vary from approximately 7 to 230 years, verifying the hypothesis that these streams are sustained by relatively old water. While there are significant uncertainties in the MTT estimates, the conclusion that they range from years to several decades and are longer at low streamflows is robust. Similar MTTs are recorded in other catchments in southeastern Australia (e.g. Cartwright and Morgenstern, 2015, 2016a, b). Especially at low streamflows, the MTTs are far longer than in most headwater catchments worldwide (e.g. Stewart et al., 2010) and

Table 4. Estimates of the difference between calculated mean transit times (MTTs) and that estimated from the mixing of waters from different tributaries at Love Creek Kawarren.

Sample date	MTT (years)	
10/07/2014	MTT$_e$[a]	15.4
	Sample MTT[b]	11.5
	Difference (years)	3.9
	Difference (%)	25.5
27/09/2014	MTT$_e$	40.9
	Sample MTT	33.5
	Difference (years)	7.4
	Difference (%)	18.1
20/03/2015	MTT$_e$	87.4
	Sample MTT	91.2
	Difference (years)	3.8
	Difference (%)	4.4
10/09/2015	MTT$_e$	15.5
	Sample MTT (years)	9.8
	Difference (years)	5.7
	Difference (%)	36.7

[a] Estimated from the tributary inputs (Eq. 2). [b] Estimated using the EPM (1.0) lumped parameter model (Table 3).

are some of the longest yet recorded. The average MTT of 15 ± 22 years calculated by Stewart et al. (2010) was for MTTs based on ^{3}H activities, which makes it directly comparable with MTTs from the south Australian catchments.

Understanding the reasons for the difference in MTTs between catchments is important for understanding catchment behaviour. The catchments in southeastern Australia have similar dimensions, slopes, and stream densities to those elsewhere, making it unlikely that the differences in MTTs result from catchment geomorphology. The Gellibrand catchments have only thin near-river alluvial sediments, thus diminishing the likelihood of bank storage and return flows of young waters during the recession from the high streamflows. However, many headwater catchments globally lack extensive alluvial sediments. The hydraulic properties of the soils and aquifers may also result in slow recharge rates and long MTTs. These are very poorly known and it is difficult to assess their influence.

Due to the high transpiration rates of eucalyptus forests, recharge rates in Australian catchments are generally lower than elsewhere globally (Allison et al., 1990). However, the observation that there is no correlation between the percentage of forest cover and MTTs in the upper Gellibrand catchments where land clearing occurred several decades ago is problematic for proposing this as a simple control. Despite being in the more temperate region of southeastern Australia, the average rainfall in the Otway Ranges of 1000 to 1600 mm yr^{-1} is modest compared with upland areas in many parts of the world and the average evapotranspiration rate of 1000 to 1100 mm yr^{-1} includes a sizeable component of evaporation (which is more prevalent on the cleared land) (Bureau of Meteorology, 2016). The long MTTs in the catchments from southeastern Australia may, therefore, reflect the low rainfall and high evaporation and/or transpiration rates that limit recharge.

The long MTTs are significant for understanding and managing the catchments. Firstly, there are likely to be long-lived stores of water in these catchments that can sustain the streams during droughts that last up to a few years, although longer-term changes (such as land use change or climate change) may eventually affect the streamflows. The long MTTs also imply that any contaminants in groundwater are likely to be released into the streams over years to decades (e.g. Morgenstern and Daughney, 2012). The locally higher nitrate and sulfate concentrations at high streamflows may reflect the input of contaminants from recent agricultural activities to the streams via the younger groundwater that is mobilised at those times.

Even at baseflow conditions, it was not possible to simply predict the MTTs across the catchments from catchment attributes or the geochemistry, although local correlations exist (this refutes one of the hypotheses). The MTTs are most likely controlled by a combination of catchment attributes and also soil properties, hydraulic conductivities, and evapotranspiration rates. This is in keeping with the observation that previous studies have identified correlations between a range of parameters and MTTs (i.e. no single attribute appears to provide the dominant control on MTTs across different regions). Characterising hydraulic properties and evapotranspiration rates on a catchment-wide scale is difficult, which limits the ability to predict MTTs. The runoff coefficient that is a reasonable indicator of MTTs elsewhere in southeastern Australia (Cartwright and Morgenstern, 2015) was the best predictor of MTTs. This may reflect the fact that both the runoff coefficient and MTTs are controlled by recharge and groundwater flow rates.

This study illustrates that, while broad ranges of MTTs may be estimated using ^{3}H, precise determination of MTTs is difficult. Additionally, it highlights the challenge in understanding the reasons for the long MTTs in the Australian catchments compared with headwater catchments elsewhere. The potential controls on MTTs is catchments are numerous, and more studies in catchments with different climate, land use, geomorphology, and geology are needed if the desire to be able to predict catchment behaviour regionally or globally is to be realised.

Author contributions. WH undertook the sampling program and oversaw the analysis of the geochemical parameters and the MTT calculations. UM was responsible for the ^{3}H analysis. The paper was prepared by WH, IC, and UM.

Acknowledgements. Field work and laboratory analyses were conducted with the help of Massimo Raveggi, Rachelle Pearson, Wang Dong, Kwadwo Osei-Bonsu, and Lei Chu. Funding for this project was provided by Monash University and the National Centre for Groundwater Research and Training (NCGRT). NCGRT is an Australian Government initiative supported by the Australian Research Council and the National Water Commission via Special Research Initiative SR0800001. We also thank two anonymous reviewers and the editor Markus Hrachowitz for their perceptive and helpful comments.

Edited by: Markus Hrachowitz

References

Allison, G. B., Cook, P. G., Barnett, S. R., Walker, G. R., Jolly, I. D., and Hughes, M. W.: Land clearance and river salinisation in the western Murray Basin, Australia, J. Hydrol., 119, 1–20, https://doi.org/10.1016/0022-1694(90)90030-2, 1990.

Atkinson, A. P.: Surface water – groundwater interactions in an upland catchment (Gellibrand River, Otway Ranges, Victoria, Australia), PhD Thesis, Monash University, Australia, 2014.

Atkinson, A. P., Cartwright, I., Gilfedder, B. S., Hoffman, H., Unland, N. P., Cendon, D. I., and Chisari, R.: A multi-tracer approach to quantifying groundwater inflows to an upland river; assessing the influence of variable groundwater chemistry, Hydrol. Process., 24, 1–12, https://doi.org/10.1002/hyp.10122, 2013.

Atkinson, A. P., Cartwright, I., Gilfedder, B. S., Cendón, D. I., Unland, N. P., and Hofmann, H.: Using ^{14}C and ^{3}H to understand groundwater flow and recharge in an aquifer window, Hydrol. Earth Syst. Sci., 18, 4951–4964, https://doi.org/10.5194/hess-18-4951-2014, 2014.

Bazemore, D. E., Eshleman, K. N., and Hollenbeck, K. J.: The role of soil water in storm flow generation in a forested headwater catchment: Synthesis of natural tracer and hydrometric evidence, J. Hydrol., 162, 47–75, https://doi.org/10.1016/0022-1694(94)90004-3, 1994.

Blavoux, B., Lachassagne, P., Henriot, A., Ladouche, B., Marc, V., Beley, J.-J., Nicoud, G., and Olive, P.: A fifty-year chronicle of tritium data for characterising the functioning of the Evian and Thonon (France) glacial aquifers, J. Hydrol., 494, 116–133, https://doi.org/10.1016/j.jhydrol.2013.04.029, 2013

Bureau of Meteorology: Comonwealth of Australia, Bureau of Meteorology, available at: http://www.bom.gov.au, last access: 21 January 2016.

Cartwright, I. and Morgenstern, U.: Transit times from rainfall to baseflow in headwater catchments estimated using tritium: the Ovens River, Australia, Hydrol. Earth Syst. Sci., 19, 3771–3785, https://doi.org/10.5194/hess-19-3771-2015, 2015.

Cartwright, I. and Morgenstern, U.: Contrasting transit times of water from peatlands and eucalypt forests in the Australian Alps determined by tritium: implications for vulnerability and the source of water in upland catchments, Hydrol. Earth Syst. Sci., 20, 4757–4773, https://doi.org/10.5194/hess-20-4757-2016, 2016a.

Cartwright, I. and Morgenstern, U.: Using tritium to document the mean transit time and source of water contributing to a chain-of-ponds river system: Implications for resource protection, Appl. Geochem., 75, 9–19, https://doi.org/10.1016/j.apgeochem.2016.10.007, 2016b.

Cartwright, I., Irvine, D., Burton, C., and Morgenstern, U.: Assessing the controls and uncertainties on mean transit times in contrasting headwater catchments, J. Hydrol., 557, 16–29, https://doi.org/10.1016/j.jhydrol.2017.12.007, 2018.

Cook, P. G. and Bohlke, J. K.: Determining timescales for groundwater flow and solute transport, in: Environmental Tracers in Subsurface Hydrology, Kluwer, Boston, USA, 1–30, 2000.

Costelloe, J. F., Peterson, T. J., Halbert, K., Western, A. W., and McDonnell, J. J.: Groundwater surface mapping informs sources of catchment baseflow, Hydrol. Earth Syst. Sci., 19, 1599–1613, https://doi.org/10.5194/hess-19-1599-2015, 2015.

DataSearch Victoria: Victoria Department of Sustainability and Environment Spatial Warehouse, available at: http://services.land. vic.gov.au/SpatialDatamart/index.jsp, last access: 10 June 2015.

Department of Environment, Land, Water and Planning: Water Measurement Information System, available at: http://data.water. vic.gov.au/monitoring.htm, last access: 10 February 2017.

Duvert, C., Stewart, M. K., Cendón, D. I., and Raiber, M.: Time series of tritium, stable isotopes and chloride reveal short-term variations in groundwater contribution to a stream, Hydrol. Earth Syst. Sci., 20, 257–277, https://doi.org/10.5194/hess-20-257-2016, 2016.

ESRI: ArcGis Desktop: Release 10.2, Redlands, CA: Environmental Systems Research Institute, 2013.

Fenicia, F., Savenije, H. H. G., Matgen, P., and Pfister, L.: Is the groundwater reservoir linear? Learning from data in hydrological modelling, Hydrol. Earth Syst. Sci., 10, 139–150, https://doi.org/10.5194/hess-10-139-2006, 2006.

Freeman, M. C., Pringle C. M., and Jackson, C. R.: Hydrologic connectivity and the contribution of stream headwaters to ecologic integrity at regional scales, J. Am. Water Resour. As., 43, 5–14, https://doi.org/10.1111/j.1752-1688.2007.00002.x, 2007.

Gelhar, L. W., Welty, C., and Rehfeldt, K. R.: A critical review of data on field-scale dispersion in aquifers, Water Resour. Res., 28, 1955–1974, https://doi.org/10.1029/92WR00607, 1992.

Hale, V. C. and McDonnell, J. J.: Effect of bedrock permeability on stream base flow mean transit time scaling relationships: 1. A multiscale catchment intercomparison, Water Resour. Res., 52, 1358–1374, https://doi.org/10.1002/2014WR016124, 2016.

Hebblethwaite, D. and James, B.: Review of surface water data, Kawarren, Hydrology & Surface Water Resources Section, Investigations Branch, Technical Services Division, Rural Water Commission of Victoria, Investigations Report No. 1990/45, 1990.

Hrachowitz, M., Soulsby, C., Tetzlaff, D., Dawson, J. J. C., and Malcom, I. A.: Regionalization of transit time estimates in montane catchments by integrating landscape controls. Water Resour. Res., 45, WR05421, https://doi.org/10.1029/2008WR007496, 2009.

Hrachowitz, M., Soulsby, C., Tetzlaff, D., and Speed, M.: Catchment transit times and landscape controls – does scale matter?, Hydrol. Process., 24, 117–125, https://doi.org/10.1002/hyp.7510, 2010.

Hrachowitz, M., Savenije, H., Bogaard, T. A., Tetzlaff, D., and Soulsby, C.: What can flux tracking teach us about water age distribution patterns and their temporal dynamics?, Hydrol. Earth Syst. Sci., 17, 533–564, https://doi.org/10.5194/hess-17-533-2013, 2013.

International Atomic Energy Agency: Global Network of Isotopes in Precipitation, available at: http://www.iaea.org/water, last access: March 2016.

Jensco, K. G. and McGlynn, B.: Hierarchical controls on runoff generation: Topographically driven hydrologic connectivity, geology and vegetation, Water Resour. Res., 47, W11527, https://doi.org/10.1029/2011WR010666, 2011.

Jurgens, B. C., Bohlke, J. K., and Eberts, S. M.: TracerLPM (Version 1): An Excel® workbook for interpreting groundwater age distributions from environmental tracer data, US Geol. Surv., Techniques and Methods Report 4-F3, US Geological Survey, Reston, USA, 60 pp., 2012.

Kennedy, V. C., Kendall, C., Zellweger, G. W., Wyerman, T. A., and Avanzino, R. J.: Determination of the components of storm flow using water chemistry and environmental isotopes, Mattole River Basin, California, J. Hydrol., 84, 107–140, https://doi.org/10.1016/0022-1694(86)90047-8, 1986.

Kinzelbach, W., Aeschbach, W., Alberich, C., Goni, I. B., Beyerle, U., Brunner, P., Chiang, W. H., Rueedi, J., and Zoellmann, K.: A survey of methods for groundwater recharge in arid and semiarid regions: early warning and assessment report series: Nairobi, Kenya, Division of Early Warning and Assessment, United Nations Environment Programme, 101 pp., 2002.

Kirchner, J. W.: Aggregation in environmental systems – Part 1: Seasonal tracer cycles quantify young water fractions, but not mean transit times, in spatially heterogeneous catchments, Hydrol. Earth Syst. Sci., 20, 279–297, https://doi.org/10.5194/hess-20-279-2016, 2016.

Kirchner, J. W., Tetzlaff, D., and Soulsby, C.: Comparing chloride and water isotopes as hydrologic tracers in two Scottish catchments, Hydrol. Process., 24, 1631–1645, https://doi.org/10.1002/hyp.7676, 2010.

Leonard, J., Lakey, R., and Cumming, S.: Gellibrand groundwater investigation interim report, December 1981, Geological Survey of Victoria, Department of Minerals and Energy, Unpublished Report 1981/132, 1981.

Ma, W. and Yamanaka, T.: Factors controlling inter-catchment variation of mean transit time with consideration of temporal variability, J. Hydrol., 534, 193–204, https://doi.org/10.1016/j.jhydrol.2015.12.061, 2016.

Maloszewski, P.: Lumped-parameter models as a tool for determining the hydrologic parameters of some groundwater systems based on isotope data, IAHS-AISH Publication 262, Vienna, Austria, 271–276, 2000.

Maloszewski, P. and Zuber, A.: Determining the turnover time of groundwater systems with the aid of environmental tracers: 1. Models and their applicability, J. Hydrol., 57, 207–231, https://doi.org/10.1016/0022-1694(82)90147-0, 1982.

Maloszewski, P. and Zuber, A.: On the calibration and validation of mathematical models for the interpretation of tracer experiments in groundwater, Adv. Water Resour., 15, 47–62, https://doi.org/10.1016/0309-1708(92)90031-V, 1992.

Maloszewski, P. and Zuber, A.: Lumped parameter models for the interpretation of environmental tracer data, in: International Atomic Energy Agency, Manual on mathematical models in isotope hydrogeology, TECDOC-910: Vienna, Austria, International Atomic Energy Agency Publishing Section, 9–58, 1996.

Maloszewski, P., Rauert, W., Stichler, W., and Herrmann, A.: Application of flow models in an alpine catchment area using tritium and deuterium data, J. Hydrol., 66, 319–330, https://doi.org/10.1016/0022-1694(83)90193-2, 1983.

McDonnell, J. J., McGuire, K., Aggarwal, P., Beven, J., Biondi, D., Destouni, G., Dunn, S., James, A., Kirchner, J., Kraft, P., Lyon, S., Maloszewski, P., Newman, B., Pfister, L., Rinaldo, A., Rodhe, A., Sayama, T., Seibert, J., Solomon, K., Soulsby, C., Stewart, M., Tetzlaff, D., Tobin, C., Troch, P., Weiler, M., Western, A., Worman, A., and Wrede, S.: How old is streamwater? Open questions in catchment transit time conceptualization, modelling and analysis, Hydrol. Process., 24, 1745–1754, https://doi.org/10.1002/hyp.7796, 2010.

McGlynn, B., McDonnell, J., Stewart, M., and Seibert, J.: On the relationship between catchment scale and streamwater residence time, Hydrol. Process., 17, 175–181, https://doi.org/10.1002/hyp.5085, 2003.

McGuire, K. J. and McDonnell, J. J.: A review and evaluation of catchment transit time modelling, J. Hydrol., 330, 543–563, https://doi.org/10.1016/j.jhydrol.2006.04.020, 2006.

McGuire, K. J., McDonnell, J. J., Weiler, M., Kendall, C., McGlynn, B. L., Welker, J. M., and Seibert, J.: The role of topography on catchment-scale water residence time, Water Resour. Res., 41, W05002, https://doi.org/10.1029/2004WR003657, 2005.

Morgenstern, U. and Daughney, C. J.: Groundwater age for identification of baseline groundwater quality and impacts of land-use intensification – The National Groundwater Monitoring Programme of New Zealand, J. Hydrol., 456–457, 79–93, https://doi.org/10.1016/j.jhydrol.2012.06.010, 2012.

Morgenstern, U. and Taylor, C. B.: Ultra low-level tritium measurement using electrolytic enrichment and LSC, Isot. Environ. Healt. S., 45, 96–117, https://doi.org/10.1080/10256010902931194, 2009.

Morgenstern, U., Stewart, M. K., and Stenger, R.: Dating of streamwater using tritium in a post nuclear bomb pulse world: continuous variation of mean transit time with streamflow, Hydrol. Earth Syst. Sci., 14, 2289–2301, https://doi.org/10.5194/hess-14-2289-2010, 2010.

Mueller, M. H., Weingartner, R., and Alewell, C.: Importance of vegetation, topography and flow paths for water transit times of base flow in alpine headwater catchments, Hydrol. Earth Syst. Sci., 17, 1661–1679, https://doi.org/10.5194/hess-17-1661-2013, 2013.

Petrides, B. and Cartwright, I.: The hydrogeology and hydrogeochemistry of the Barwon Downs Graben aquifer, southwestern Victoria, Australia, Hydrogeol. J., 14, 809–826, https://doi.org/10.1007/s10040-005-0018-8, 2006.

Sklash, M. G. and Farvolden, R. N.: The role of groundwater in storm runoff, J. Hydrol., 43, 45–65, https://doi.org/10.1016/0022-1694(79)90164-1, 1979.

Soulsby, C., Malcolm, R., Helliwell, R., Ferrier, R. C., and Jenkins, A.: Isotope hydrology of the Allt a' Mharcaidh catchment, Cairngorms, Scotland: implications for hydrologic pathways and residence time, Hydrol. Process., 14, 747–762, https://doi.org/10.1002/(SICI)1099-1085(200003)14:4<747::AID-HYP970>3.0.CO;2-0, 2000.

Stanley, D. R.: Resource evaluation of the Kawarren Sub-Region on the Barwon Downs Graben, RWC Investigations Branch Un-

published Report 1991/36, Rural Water Commission of Victoria, 1991.

Stewart, M. K. and Fahey, B. D.: Runoff generating processes in adjacent tussock grassland and pine plantation catchments as indicated by mean transit time estimation using tritium, Hydrol. Earth Syst. Sci., 14, 1021–1032, https://doi.org/10.5194/hess-14-1021-2010, 2010.

Stewart, M. K., Morgenstern, U., and McDonnell, J. J.: Truncation of stream residence times: how the use of stable isotopes has skewed our concept of streamwater age and origin, Hydrol. Process., 24, 1646–1659, https://doi.org/10.1002/hyp.7576, 2010.

Stewart, M. K., Morgenstern, U., Gusyev, M. A., and Maloszewski, P.: Aggregation effects on tritium-based mean transit times and young water fractions in spatially heterogeneous catchments and groundwater systems, Hydrol. Earth Syst. Sci., 21, 4615–4627, https://doi.org/10.5194/hess-21-4615-2017, 2017.

Stockinger, M. P., Bogena, H. R., Lucke, A., Diekkruger, B., Weiler, M., and Vereecken, V.: Seasonal soil moisture patterns: Controlling transit time distributions in a forested headwater catchment, Water Resour. Res., 50, 5270–5289, https://doi.org/10.1002/2013WR014815, 2014.

Swistock, B. R., DeWalle, D. R., and Sharp, W. E.: Sources of acidic storm flow in an Appalachian headwater stream, Water Resour. Res., 25, 2139–2147, https://doi.org/10.1029/WR025i010p02139, 1989.

Tadros, C. V., Hughes, C. E., Crawford, J., Hollins, S. E., and Chisari, R.: Tritium in Australian precipitation: a 50 year record, J. Hydrol., 513, 262–273, https://doi.org/10.1016/j.jhydrol.2014.03.031, 2014.

Tetzlaff, D., Malcolm, I. A., and Soulsby, C.: Influence of forestry, environmental change and climatic variability on the hydrology, hydrochemistry and residence times of upland catchments, J. Hydrol., 346, 93–111, https://doi.org/10.1016/j.jhydrol.2007.08.016, 2007.

Tetzlaff, D., Seibert, J., and Soulsby, C.: Inter-catchment comparison to assess the influence of topography and soils on catchment transit times in a geomorphic province; the Cairngorm mountains, Scotland, Hydrol. Process., 23, 1874, https://doi.org/10.1002/hyp.7318, 2009.

Tickell, S. J., Cummings, S., Leonard, J. G., and Withers, J. A.: Colac: 1 : 50 000 Map Geological Report, Geological Survey Report No. 89, Victoria Department of Manufacturing and Industry Development, 1991.

Timbe, E., Windhorst, D., Celleri, R., Timbe, L., Crespo, P., Frede, H.-G., Feyen, J., and Breuer, L.: Sampling frequency trade-offs in the assessment of mean transit times of tropical montane catchment waters under semi-steady-state conditions, Hydrol. Earth Syst. Sci., 19, 1153–1168, https://doi.org/10.5194/hess-19-1153-2015, 2015.

Zuber, A. and Maloszewski, P.: Lumped Parameter Models, Environmental Isotopes in Hydrological Cycle, Vol. VI, Modelling, IAEA/UNESCO, Paris, Technical Document, 2001.

Zuber, A., Witczak, S., Rozanski, K., Sliwka, I., Opoka, M., Mochalski, P., Kuc, T., Karlikowska, J., Kania, J., Jackowicz-Korczynski, M., and Dulinski, M.: Groundwater dating with ^3H and SF_6 in relation to mixing patterns, transport modelling and hydrochemistry, Hydrol. Process., 19, 2247–2275, https://doi.org/10.1002/hyp.5669, 2005.

8

The potential of urban rainfall monitoring with crowdsourced automatic weather stations in Amsterdam

Lotte de Vos[1,2], **Hidde Leijnse**[2], **Aart Overeem**[1,2], and **Remko Uijlenhoet**[1]

[1]Hydrology and Quantitative Water Management Group, Department of Environmental Sciences, Wageningen University, 6708 PB Wageningen, the Netherlands
[2]Research and Development Observations and Data Technology, Royal Netherlands Meteorological Institute, 3732 GK De Bilt, the Netherlands

Correspondence to: Lotte de Vos (lotte.devos@wur.nl)

Abstract. The high density of built-up areas and resulting imperviousness of the land surface makes urban areas vulnerable to extreme rainfall, which can lead to considerable damage. In order to design and manage cities to be able to deal with the growing number of extreme rainfall events, rainfall data are required at higher temporal and spatial resolutions than those needed for rural catchments. However, the density of operational rainfall monitoring networks managed by local or national authorities is typically low in urban areas. A growing number of automatic personal weather stations (PWSs) link rainfall measurements to online platforms. Here, we examine the potential of such crowdsourced datasets for obtaining the desired resolution and quality of rainfall measurements for the capital of the Netherlands. Data from 63 stations in Amsterdam ($\sim 575\,\text{km}^2$) that measure rainfall over at least 4 months in a 17-month period are evaluated. In addition, a detailed assessment is made of three Netatmo stations, the largest contributor to this dataset, in an experimental setup. The sensor performance in the experimental setup and the density of the PWS network are promising. However, features in the online platforms, like rounding and thresholds, cause changes from the original time series, resulting in considerable errors in the datasets obtained. These errors are especially large during low-intensity rainfall, although they can be reduced by accumulating rainfall over longer intervals. Accumulation improves the correlation coefficient with gauge-adjusted radar data from 0.48 at 5 min intervals to 0.60 at hourly intervals. Spatial rainfall correlation functions derived from PWS data show much more small-scale variability than those based on gauge-adjusted radar data and those found in similar research using dedicated rain gauge networks. This can largely be attributed to the noise in the PWS data resulting from both the measurement setup and the processes occurring in the data transfer to the online PWS platform. A double mass comparison with gauge-adjusted radar data shows that the median of the stations resembles the rainfall reference better than the real-time (unadjusted) radar product. Averaging nearby raw PWS measurements further improves the match with gauge-adjusted radar data in that area. These results confirm that the growing number of internet-connected PWSs could successfully be used for urban rainfall monitoring.

1 Introduction

Urban catchments are characterized by a high proportion of impervious surfaces, leading to a large fraction of rainfall producing direct runoff and a fast hydrological response. This makes cities especially vulnerable to flooding. The temporal and spatial resolutions of rainfall data required for urban applications exceed those needed for rural catchments (Schilling, 1991). The rainfall information at spatial and temporal resolutions of typically 1 km by 1 km and 5 min generated by most operational weather radars is considered valuable for urban hydrological analysis and forecasting (Liguori et al., 2012). However, radar has significant limitations; rainfall is determined indirectly, over an atmospheric volume with a size depending on the distance from the radar station, which may not be representative for rainfall at ground level

(Einfalt et al., 2004; Peleg et al., 2013). Errors in rainfall estimates from radar due to sampling uncertainties can be significant. In addition, there is an optimum spatial resolution corresponding to a given temporal resolution (Fabry et al., 1994; Bell and Moore, 2000). Rain gauges, if well maintained, provide accurate ground-based measurements, although they are limited in their spatial representation; Villarini et al. (2008) showed that approximations of true spatial rainfall fields with rain gauges requires a dense network and/or large temporal measurement intervals.

Hydrological models, designed to deal with high-resolution input, provide the best simulation results not just when the temporal resolution or the spatial resolution is high, but particularly when the combination thereof is optimal. The required spatiotemporal resolutions for urban applications have been studied extensively. Berne et al. (2004) determined a relation between the space–time resolution required for hydrological applications as a function of the catchment size for Mediterranean conditions. It was found that for urban catchments in the order of $10\,km^2$, rainfall data are needed at a temporal resolution of 5 min and a spatial resolution of 3 km. For urban catchments of $1\,km^2$, these resolutions were 3 min and 2 km, respectively. The space–time scales of four types of rainfall are evaluated by Emmanuel et al. (2012). With the use of variograms of 24 storm events, the spatial resolutions required to capture these types of rainfall at urban scales range from 0.8 to 3 km for instantaneous monitoring and from 2.5 to 8 km for 30 min intervals.

Gires et al. (2012) found an outflow uncertainty of up to 20 % in an urban catchment of $9\,km^2$ due to rainfall variability at scales smaller than the typical C-band radar resolution of 1 km by 1 km and 5 min. Bruni et al. (2015) addressed the loss in urban hydrodynamic model accuracy due to smoothing and smearing. Radar data of four storm events in 1 min temporal resolution were aggregated to various spatial and temporal resolutions (highest range resolution of 30 m) and used as precipitation input in a $3.4\,km^2$ Dutch urban catchment. Smoothing occurs when the ratio of radar resolution over catchment size becomes larger than 0.2 and storms that move near the catchment boundary are averaged partly out of the catchment. Smearing becomes significant when the ratio of the spatial resolution of radar measurements over the rainfall correlation length exceeds 0.9, leading to averaging of rainfall over the coarse spatial grid and resulting in underestimation of rainfall rates in areas within the storm cells and overestimation in the surrounding areas. Also, a runoff peak time shift of up to 6 min was found due to temporal aggregation (from 1 min to 5 and 10 min) of rainfall input.

Ochoa-Rodriguez et al. (2015) evaluated the required spatial and temporal resolutions of rainfall in a simple spatiotemporal scaling framework. A spatial resolution of 1 km, typically found in radar, was found to give good hydrodynamic model results, although some extremes were missed. Temporal resolutions should ideally be below the 5 min intervals currently available in most operational weather radar-

products. Nevertheless, the accuracy of 5 min radar data can be improved with the use of an accumulation procedure that assumes constant velocity of the rainfall field and rainfall intensity to vary linearly in time (Fabry et al., 1994). Coarsening temporal resolution has more impact on the accuracy than coarsening spatial resolution. Initial results from an ongoing study by the authors indicate that this impact is reduced when temporal resolutions are coarsened through aggregation (i.e., similar to rain gauges) instead of sampling. Lobligeois et al. (2014) evaluated the circumstances where hydrological model performance is enhanced by higher spatial resolution of rainfall. They did so by comparing lumped and semi-distributed models with subcatchment sizes of 64, 16 and $4\,km^2$. From comparisons between the various model outputs and observations in 181 catchments in France, it was found that model accuracy improvement depends on scale, catchment and event characteristics, and that the spatial representation of rainfall can be a highly important factor in the model performance.

From these works it becomes evident that an increase of the number of measurements would yield a higher accuracy of rainfall fields and would improve hydrological applications. Adding sensors (rain gauges or others) to a network is costly, although there are alternatives. For instance, rain maps can be produced from received signal strength in cellular communication networks, as the microwave signals propagating over the link paths are attenuated by rainfall (Overeem et al., 2016). Weather data can also be provided directly by crowdsourcing measurements from amateurs in various ways (Muller et al., 2015). A growing number of weather enthusiasts measure their local weather with automatic personal weather stations (PWSs). PWS accuracy on measuring temperature, relative humidity, radiation, pressure, rainfall, wind speed and direction has been evaluated for popular high-end expensive weather stations (Jenkins, 2014; Bell et al., 2015), as well as for the cheaper, user-friendly Netatmo type (temperature only) (Meier et al., 2015), which have grown rapidly in number over the past years. So far, weather stations have been used to obtain air temperature data to examine the urban heat island effect (Steeneveld et al., 2011; Wolters and Brandsma, 2012), although other meteorological variables, such as rainfall, are measured by some of these stations as well.

A large number of PWSs share data on online platforms, both on the owner's own initiative (Gharesifard and Wehn, 2016) or automatically as an intrinsic software feature of the product (i.e., for Netatmo). Netatmo has its own online platform collecting and visualizing data from all operational Netatmo stations. The WunderMap of company Weather Underground is a similar online platform. Data from Netatmo stations are automatically linked to the WunderMap. Owners of other PWS types can actively transmit their measurements to this platform as well. A growing number of automatic weather stations are linked to these platforms; in May 2016 there were 258 personal weather stations linked to Wun-

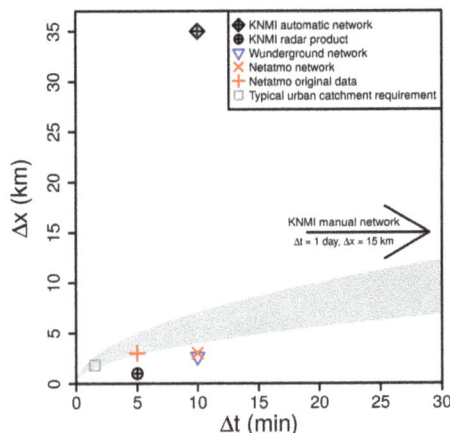

Figure 1. Temporal and spatial resolution of unfiltered rainfall measurements in the Netherlands with PWS network obtained via Netatmo API, WunderMap API and the potential availability of Netatmo measurements, as well as the resolution of KNMI's automatic and manual rainfall measurement network and radar product. The curve represents a relation between the temporal and the spatial resolution of rainfall measurement required for urban hydrology as determined by Berne et al. (2004) for Mediterranean climate, where the square represents the value for an urban catchment with surface area of $0.1\,\text{km}^2$.

derMap in the Amsterdam metropolitan area ($\sim 575\,\text{km}^2$) alone (239 of type Netatmo), of which 83 stations measured rainfall (64 of type Netatmo). By contrast, the official national automatic weather station network in the Netherlands ($\sim 35\,000\,\text{km}^2$) consists of 31 stations, and these are, as a rule, always located outside urban areas. Figure 1 shows the relative resolutions in the Netherlands of networks discussed in this paper. At many locations, the density of PWS stations collecting rainfall data far exceeds that of any realistic operational network implemented by national weather services or local authorities beyond experimental campaigns. As the online platforms collecting and sharing PWS weather data are not nation-bound, global rainfall measurements have become easily available, with especially high densities in western Europe, USA and Japan.

Although rainfall data availability with PWS networks is cause for optimism for urban hydrological applications, errors are expected to be larger than those in traditional measurements. PWSs come in many types, a large fraction of which are low cost with expected low sensor quality. In most cases, there is no information available on the PWS type, the installation setup, maintenance of the sensor or data postprocessing while transferring measurements to the online platform. Bell et al. (2013) examine the potential improvement on the UK's observational network with the real-time and local weather measurements of air temperature, relative humidity and pressure collected from WunderMap. The most critical issue was found to be the estimation of data quality. Validation procedures like range tests (i.e., a check whether

the measurement is within predefined extremes limits) and internal consistency tests should be applied to precipitation data from automatic weather stations (Estévez et al., 2011). The integration of crowdsourced data with variable temporal resolutions in hydrological discharge modeling by accounting for different uncertainties for data of various sources has been addressed in recent research (Mazzoleni et al., 2015).

It becomes clear that urban applications would benefit from high-resolution rainfall measurements. The potential of crowdsourced PWS rainfall data for this purpose has not previously been explored. Using the existing PWS network requires minimal financial investment, and would therefore be an economically reasonable alternative to conventional techniques to increase measurement resolutions. This study aims to determine the added value of crowdsourcing automatic weather stations for urban rainfall monitoring. For this purpose, the most common PWS is tested in an experimental setup with a high-quality rain gauge reference. Additionally, a dataset of 63 crowdsourced PWS stations in Amsterdam is validated with a gridded dataset based on radar data, a manual network and a WMO-certified automatic rain gauge network. These combined results provide insight on the rainfall measurement accuracy of the most commonly used PWS, as well as any issues that occur in operational crowdsourcing of PWS rain measurements. Following this introduction is the Methods section, where Sect. 2.1 describes the data and Sect. 2.2 gives an outline to determine the achieved measurement scales and quality of PWS, respectively. The results of an experimental PWS setup, a comparison of a larger dataset in Amsterdam with gauge-adjusted radar data and an analysis on inter-gauge spatial correlation of this dataset are given in Sect. 3. Finally, a discussion on the state and future role of PWS networks in (urban) hydrological applications and conclusions are given in Sects. 4 and 5, respectively.

2 Methods

2.1 Data collection

2.1.1 Personal weather stations

From the WunderMap website, a dataset of 63 automatic weather stations located in the Amsterdam area ($\sim 575\,\text{km}^2$) has been retrieved. Stations were selected based on the availability of rainfall measurements, which should cover at least 4 months between December 2014 and April 2016. Of these stations, 49 are of brand Netatmo, 7 are of brand Davis and 7 are of other unspecified brands. No details on the devices are given. According to the product specifications provided by the manufacturer, the Netatmo rain gauges have a measurement range of 0.2–$150\,\text{mm}\,\text{h}^{-1}$ with an accuracy of $1\,\text{mm}\,\text{h}^{-1}$. The plastic tipping buckets have a volume of $0.1\,\text{mm}$ and a collecting funnel with a diameter of $13\,\text{cm}$. The rain gauge module communicates in a wireless manner

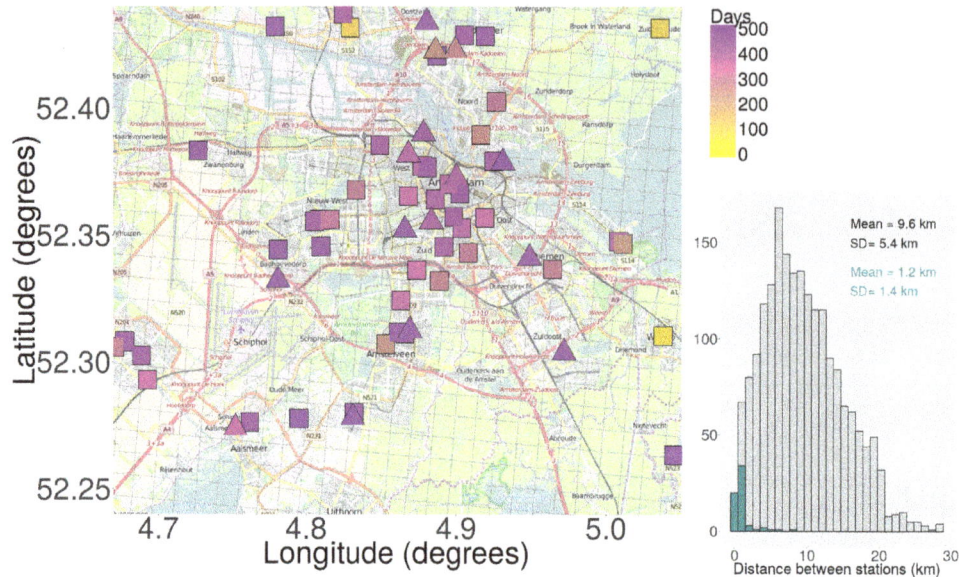

Figure 2. Locations and operational days (i.e., days with measurements) of Netatmo (squares) and other types (triangles) of PWSs, with the radar pixel grid in the Amsterdam metropolitan area. Inter-station distances are represented in the histogram, colored green for nearest neighbor distances. The background map is taken from ©OpenStreetMap (www.openstreetmap.org).

to the Netatmo indoor module over distances up to 100 m. The number of tips in the previous interval is communicated every ~ 5 min from the indoor module to the online dashboard via a WiFi connection, where it can be monitored by the weather station owner. Simultaneously, the measurement is linked to the Netatmo weather map from which it is sent every ~ 10 min to the WunderMap. The WunderMap stations that contribute to the dataset are visualized in Fig. 2. The WunderMap platform collects the rainfall measurements and rewrites them into rainfall over the past hour and cumulative rainfall for that day. Daily rainfall only becomes non-zero once the 0.3 mm threshold is reached and subsequent rainfall is only reported if the rounded daily rainfall increases by at least 0.2 mm.

While Netatmo hardware can store measurements for a period of time in the event of bad connectivity with the server, only real-time data are automatically transferred to the WunderMap. This causes gaps in the WunderMap datasets where there may be none in the original Netatmo data, which are only accessible to the weather station owner. WunderMap time series are characterized by (large) gaps in the dataset and irregular measurement frequencies, though often 5, 10 or 15 min. Also, the locations of Netatmo weather stations on the WunderMap are obtained from the settings at the Netatmo platform without notice to or confirmation from the PWS owner. Relocations of the station that are communicated to the Netatmo platform are not simultaneously adjusted on the WunderMap, potentially leading to large errors in sensor location.

We process the data obtained via WunderMap by calculating the difference in cumulative daily rainfall compared with the previous time step. Since these time steps are not fixed, this results in rainfall accumulations over time intervals of varying lengths. In order to obtain compatible time series, the rainfall is interpolated on a fixed timeline with constant steps, where constant rainfall within the original intervals is assumed. Original intervals longer than 20 min are discarded. Faulty values in precipitation data from automatic weather stations can be identified with range tests and internal consistency tests (Estévez et al., 2011). As a first quality check, values of the interpolated time series are compared with the median rainfall of all stations for each time interval. Values exceeding this median by more than $50\,\mathrm{mm\,h^{-1}}$ are excluded. Dry periods in the dataset are identified as periods of at least 24 h where the median of all PWS measurements indicate zero rainfall. If a PWS reports continuous zero rainfall for at least 12 h outside of this dry reference, the measurements in this dry period are considered as faulty zero rainfall measurements and are discarded. Finally, inter-gauge correlations are determined. If a low correlation (i.e., average and median < 0.21) is found between a station and all other stations, the entire time series for that station is excluded. Visual comparison with corresponding radar rainfall time series showed that a filter based on these criteria was suitable in excluding obviously incorrect data from the datasets. This filter could be applied in real time, although for operational uses beyond this dataset, adjustments will be required.

2.1.2 Radar

As rainfall reference, we use gauge-adjusted radar data from a climatological rainfall dataset by the Royal Netherlands

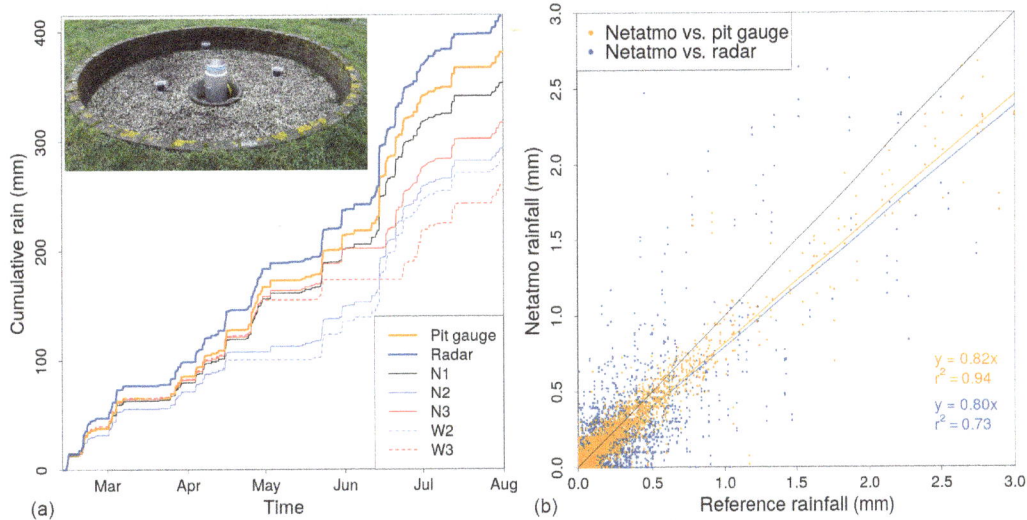

Figure 3. (a) Cumulative rainfall according to reference pit gauge, gauge-adjusted radar, Netatmo stations (N1, N2 and N3) and Netatmo stations obtained via WunderMap (W2 and W3). N2 (and, as a consequence, W2) was offline between 20 April and 1 May. The photo shows the experimental setup of the rain gauges in the pit gauge configuration. **(b)** Scatter plots of 10 min rainfall and linear fits of rainfall according to N1, N2 and N3 as compared to reference pit gauge (orange) and gauge-adjusted radar (blue).

Meteorological Institute (KNMI) (Overeem et al., 2009a, b, 2011), freely available as "Radar precipitation climatology" via http://climate4impact.eu. This dataset is based on data from two C-band Doppler weather radars in De Bilt and Den Helder and has a temporal resolution of 5 min and a spatial resolution of $0.92 \, \text{km}^2$, covering the entire land surface of the Netherlands. This radar makes volumetric scans in all directions, measuring instantaneous rainfall at a location every 5 min. In this product, radar composite images have been adjusted with rainfall measurements from the KNMI rain gauge networks (31 automatic and 325 manual gauges). For details on the method of adjusting, we refer to Overeem et al. (2009a, b, 2011). It should be noted that, due to their different representativeness, there can be significant differences between radar pixel areal rainfall and point rainfall (Schilling, 1991; Einfalt et al., 2004; Villarini et al., 2008; Peleg et al., 2013). Using a radar product that is adjusted with ground measurements will likely reduce this difference.

2.1.3 Netatmo experimental setup

As the majority of the weather stations linked to the WunderMap is of type Netatmo, we examine the quality of Netatmo rain gauges in a dedicated experimental setup; see Fig. 3, photo inset. As reference, we use a high-quality KNMI pit gauge at the Cabauw Experimental Site for Atmospheric Research (CESAR) (Leijnse et al., 2010), that measures cumulative rainfall in intervals of 12 s. This electronic rain gauge is placed in a so-called pit gauge configuration: a small hill of diameter 6.2 m with a circular pit with diameter 3 m and a depth of 40 cm in the middle. Precipitation is collected in the instrument (collecting funnel with a diameter

of 16 cm, i.e., $200 \, \text{cm}^2$) and in the event of solid precipitation melted by a heating element in the funnel. The amount of liquid water is measured by the position of a floating unit connected to a potentiometer. Rainfall is measured every 12 s within the range of 0–0.7 mm with a resolution of 0.1 mm and an accuracy of 0.2 mm. The Netatmo sensors are placed at $\sim 40 \, \text{cm}$ around the electronic sensor in the center of the pit in such a way that the top of each sensor is level with the rim of the pit. The period considered is from 12 February to 25 May 2016. The datasets, as collected directly from the Netatmo personal account in millimeters of rainfall per interval of typically 5 min, as well as via the WunderMap platform, are compared to the pit gauge reference. One of the stations was offline between 20 April and 1 May, and one station could not be accessed via Weather Underground.

2.2 Analysis

2.2.1 Station measurement density

As mentioned previously, the original PWS data temporal resolution from WunderMap is quite irregular. The number of stations containing rainfall measurements for time series per 5 and 10 min shows that the data availability is quite variable; see Fig. 4. Moreover, the fraction of the measurements over the period that is filtered out does not seem to vary significantly in time. It is not straightforward how to attribute a certain measurement resolution to a network that has highly irregular measurement frequencies and station locations at irregular distances from one another. When the Amsterdam area is divided into grid cells, or pixels, of a certain size, the number of pixels that contain at least one measurement is

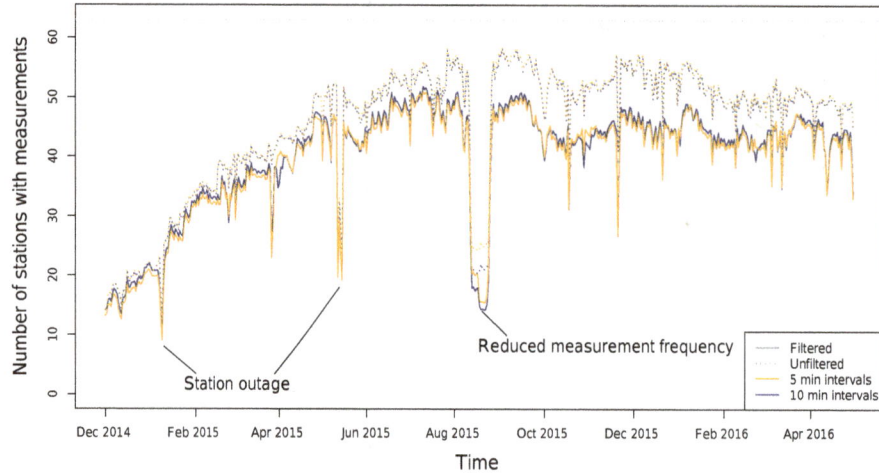

Figure 4. Number of stations with rainfall data from the PWS dataset, before and after applying filter, for every 5 and 10 min interval over the entire period, smoothed per day. The two indicated dips correspond to complete outage of stations, the third with a longer period of fewer measurements in all stations.

an indication of the network resolution. The fraction of total pixels that contain at least one measurement has been calculated for all time steps over the entire period, for various combinations of pixel sizes and time step lengths in the scale range relevant for urban applications. It is found that for the Amsterdam dataset (before filter has been applied), the fraction of pixels containing at least one measurement is more limited by the number of stations than the measurement frequency; see Fig. 5. Only when dividing the period in time steps shorter than 10 min, an increase of measurement frequency will result in a higher fraction. This is unsurprising as most stations in the dataset link their measurements to the Weather Underground platform approximately every 10 min. Adding stations will result in an increase in fraction at all time step sizes in this range. The PWS network consists of more stations than the number examined in this dataset and continues to grow, which will have a positive effect on the PWS network measurement resolution.

2.2.2 Station measurement quality

With the Netatmo experimental setup, the performance of this type of PWS and the consequences of transferring its data to the online platform are examined. The measurements are compared to the high-resolution pit gauge as well as to the radar rainfall at the corresponding pixel. These two comparisons should give an indication of the differences due to sensor performance and those due to differences in representativeness of radar and rain gauges.

Rainfall measurements of the PWS dataset in Amsterdam are compared with the radar rainfall measurement at their corresponding radar pixels. When comparing station data with gauge-adjusted radar data, the coefficient of variation of the residuals (CV) is calculated. The standard deviation of the differences between the datasets is divided by the mean

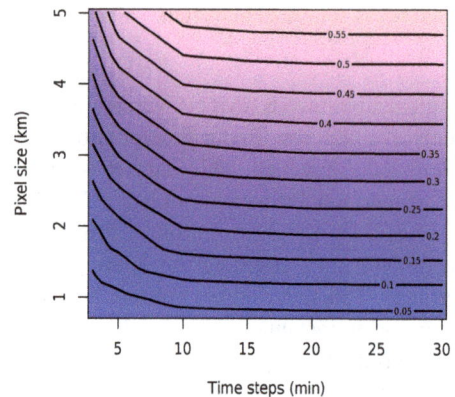

Figure 5. Indicated with curves as well as colors are the fractions of pixels containing at least one measurement of the unfiltered dataset for combinations of time step length and pixel grid size over the Amsterdam area between December 2014 and April 2016.

of the gauge-adjusted radar data. A low value of CV indicates a good match between the datasets. Additionally, spatial correlations between stations are estimated with the use of Pearson's product–moment correlation coefficient (r):

$$r = \frac{E[XY] - E[X]E[Y]}{\sqrt{(E[X^2] - E[X]^2) \cdot (E[Y^2] - E[Y]^2)}}, \qquad (1)$$

where $E[\cdot]$ is the expectation (estimated as the arithmetic mean) and (X, Y) are corresponding time series of rainfall measurements. Because of the spatial and temporal variability of rainfall, the correlation of two point locations decreases with distance between these points. A three-parameter exponential function is suggested by Habib et al. (2001) to describe this spatial dependency relation between inter-station correlation (r) and distance (d):

$$r = r_0 \exp\left[-\left(\frac{d}{X_0}\right)^{S_0}\right],\qquad(2)$$

where r_0 is the nugget parameter, X_0 is the correlation distance and S_0 is the shape factor. The nugget parameter r_0 is a measure of small-scale variability and/or measurement error and is equal to 1 for perfect zero-distance correlation. Correlation distance X_0 indicates the distance at which the rainfall decorrelates (i.e., the distance beyond which the correlation drops below e^{-1}), which should be interpreted with caution when it exceeds the investigated spatial extent.

The relationship in Eq. (2) is sensitive to rainfall extremes (Habib et al., 2001), climatic regimes (Krajewski et al., 2003) and seasonality (van de Beek et al., 2011; Tokay and Öztürk, 2012) as well as strongly dependent on time interval (Krajewski et al., 2003; Ciach and Krajewski, 2006; van de Beek et al., 2011; Tokay and Öztürk, 2012; van de Beek et al., 2012; Peleg et al., 2013). For the PWS dataset in Amsterdam, correlograms are constructed and compared with spatial dependencies found in literature. Special consideration is given to the correlations between Netatmo stations as compared to the other types of rain gauges.

3 Results

3.1 Netatmo comparison with pit gauge

The original data of three Netatmo stations (measurement frequency of ~5 min) are compared with pit gauge data (measurement frequency of 12 s) and gauge-adjusted radar data (measurement frequency of 5 min), over the period February–May 2016. Over this period, the cumulative rainfall of station 2 was lower than that of the others; see Fig. 3a. This was the result of station outage. In general, the Netatmo stations measure less rainfall than the pit gauge and radar reference over this period. The scatter plots in Fig. 3b do not include the intervals where one or both of the time series contain no measurements (in the event of station outage), and show a good r^2 of 0.94 between Netatmo measurements and the pit gauge reference. Even though this r^2 suggests a small measurement error in Netatmo, the comparison with radar shows significant scatter away from the perfect fit. This is inherent to comparisons between point locations and pixel averages, and the scatter plot resembles those reported in Peleg et al. (2013), though the radar value used there was an average value of 12 pixels instead of 1.

The correlation between Netatmo and the pit gauge is calculated for a multitude of accumulation intervals; see Fig. 6. This correlation reflects small-scale rainfall variability and thus is closely related to the nugget parameter in Eq. (2). As expected, an increase of correlation is found for larger accumulation intervals. However, the correlations of data from the same devices obtained via WunderMap with the same pit

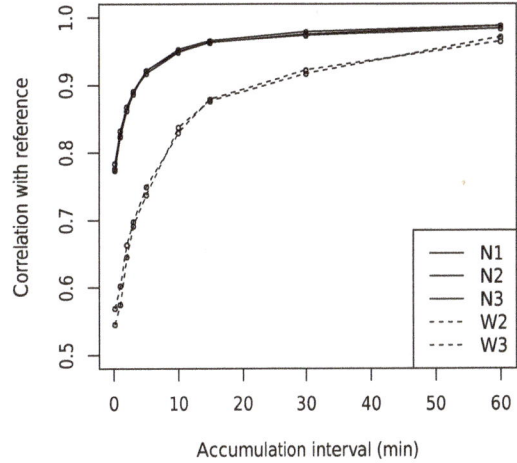

Figure 6. Correlation between rainfall measurements by Netatmo stations as obtained via personal dashboard (N1, N2 and N3), as well as those obtained via WunderMap (W2 and W3), and the pit gauge reference for various accumulation time steps.

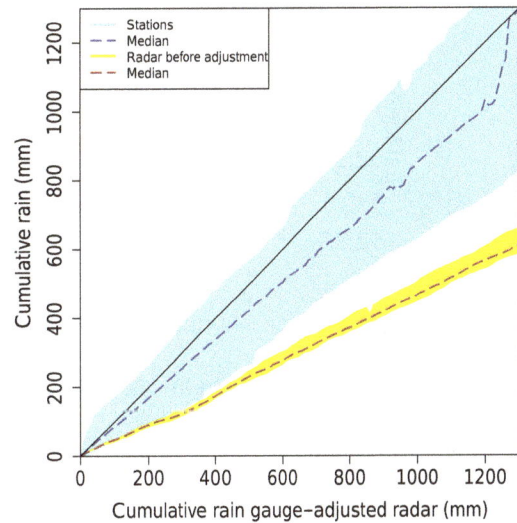

Figure 7. Double mass plots of station filtered rainfall measurements and real-time radar data with gauge-adjusted radar rainfall at the corresponding location in the period between December 2014 and March 2016. Only intervals where both radar and station contain measurements are taken into account. Colored regions indicate the range between double mass plot of stations with minimum and maximum steepness and dashed lines represent the median of the combined datasets.

gauge reference show far lower values; see Fig. 6. The original Netatmo data have typical time steps of 5 min against 10 min for the WunderMap data. If this was the only difference between the time series, the correlation graphs should overlap for accumulation intervals above 10 min. As they only approach one another for hourly accumulations, it can be concluded that besides this effect, additional information is lost in the transfer of data between platforms.

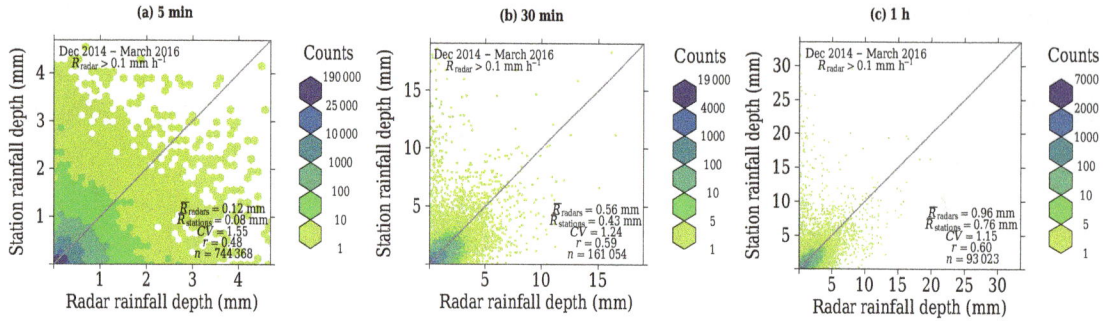

Figure 8. Scatter density plots of all station rainfall measurements against the gauge-adjusted radar rainfall data in the corresponding radar pixel when radar reported non-zero rainfall (> 0.1 mm). The $\overline{R}_{\text{radars}}$, $\overline{R}_{\text{stations}}$, CV, r and n values in the panels represent the average rainfall according to the gauge-adjusted radar data, the average rainfall according to the stations, the coefficient of variation of the residuals, the correlation and the number of intervals, respectively. Graphs are made for 5 min, 30 min and hourly accumulation intervals.

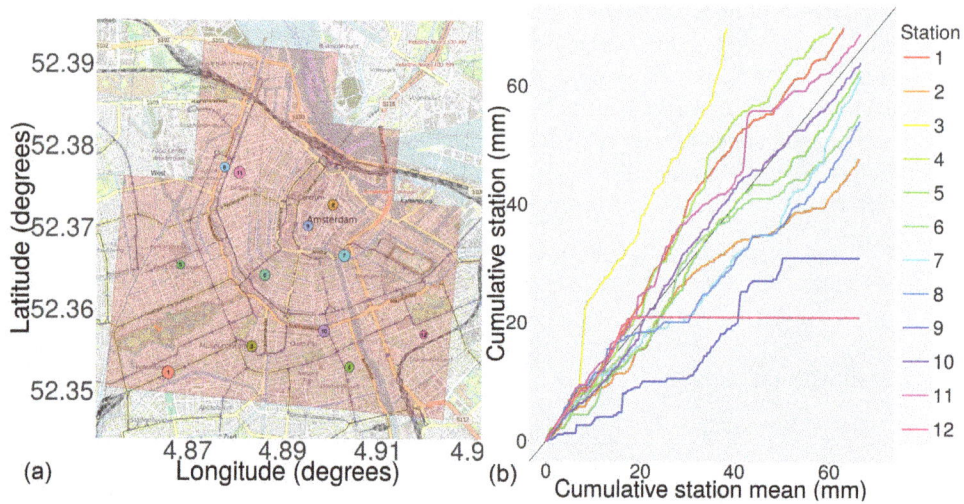

Figure 9. (a) Locations of 12 stations and 20 radar pixels in the city center of Amsterdam, where symbol size represents the number of unique days with measurements by the station (range of 371–514 days). The background map is taken from ©OpenStreetMap (www.openstreetmap.org). **(b)** Double mass plots of the station measurements as compared to the mean of all 12 stations over the intervals where all stations contain measurements.

Besides the Netatmo dashboard (available to the station owner) and WunderMap, Netatmo data are also accessible from the Netatmo weather map platform. In this research, real-time measurements from the three stations in the experimental setup were obtained with from this platform with an application programming interface (API). It was found that rainfall measurements from this dataset were attributed with a time stamp of the moment the data were collected, instead of the time stamp of the measurement itself. In the event of sensor outage, the last available measurement was collected repeatedly. These artifacts resulted in faulty interval attribution of rainfall and negatively affected the correlations with the original dataset as well as with gauge-adjusted radar data. An API containing such processing errors will result in datasets that contain considerable errors, though these errors are easily overlooked without the original data. Fortunately, the original data can also be obtained from the Netatmo plat-

form. These time series are identical to the data from the Netatmo dashboard (N1, N2 and N3 from the experimental setup) and can be obtained in real time.

3.2 Amsterdam weather station comparison with radar

Figure 7 shows the double mass plot of the filtered PWS dataset in Amsterdam, as well as the unadjusted (real-time) radar with the gauge-adjusted radar reference at the same locations. The only intervals considered are those where both time series contain measurements. Even though individual stations often do not follow the diagonal line representing a perfect match, the median of all available stations only shows a slight underestimation as compared to the gauge-adjusted radar rainfall data. This underestimation is far greater in the real-time radar product. Though large deviations occur, the median of the stations resembles the reference quite well.

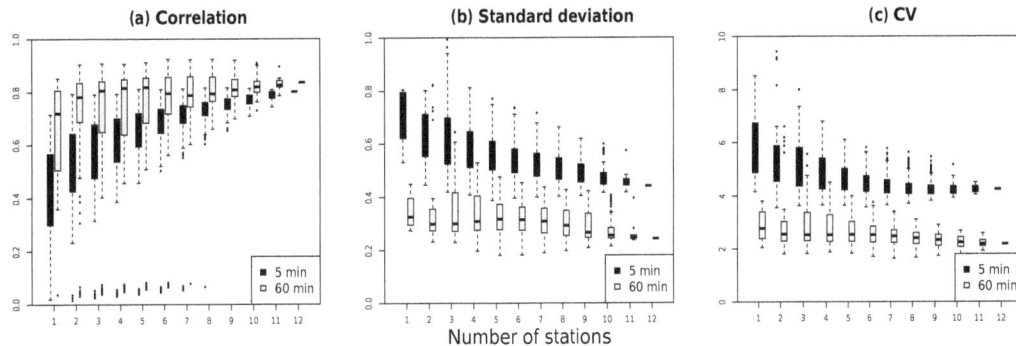

Figure 10. Box plots of correlation, standard deviation and coefficient of variation of residuals (CV) of averaged rainfall intensity time series. The box plots contain the outcomes for all possible subsets within the 12 stations in the Amsterdam city center, as compared to the gauge-adjusted radar rainfall intensity 20-pixel mean for interpolated 5 min and hourly time series.

When comparing station rainfall against corresponding gauge-adjusted radar rainfall data over the entire period with the condition that the radar measures non-zero rainfall, a better correspondence is found for longer time steps; see Fig. 8. A similar scatter as in Fig. 8 is found by Peleg et al. (2013). At longer accumulation intervals, the averages resemble each other more, the CV decreases and the r increases, indicating a better resemblance between gauge-adjusted radar and station datasets.

3.3 Amsterdam center average comparison

In order to investigate whether the generally poor quality of individual PWS measurements can (partly) be compensated by the generally high quantity of measurements, averages of unfiltered PWS measurements are compared with radar pixel averages over a small area in Amsterdam. The selected area is the region with highest parking rates: the densely populated and touristic area of the city center and Museum Square, as floods in this area will heavily impact residents, businesses and tourism alike. This region of $\sim 20\,\mathrm{km}^2$ is shown in Fig. 9, where the cumulative rainfall of each station relative to the mean of the 12 stations is shown. From Fig. 9, the variation between station measurements becomes evident. Some stations measure highly unlikely values considering the measurements of their nearby stations, such as stations 3, 9 and 12.

The means of all possible subsets of the 12 PWSs are compared with the average of the 20 radar pixels over the selected Amsterdam center region. For each subset, the correlation, standard deviation and CV of the residuals of rainfall intensity is calculated over all intervals where each station contains measurements. The resulting outcomes of each subset are represented with box plots in Fig. 10 per number of stations contributing to the PWS mean. The correlation increases and the standard deviation and CV decrease when averaging multiple stations, even when some of the station time series consist of obviously faulty measurements; see Fig. 10. By averaging the unfiltered measurements of a

dozen stations, crowdsourced measurements seem to be able to describe rainfall in the city center. As expected, the values based on 60 min rainfall intensities show a better correspondence with gauge-adjusted radar data than 5 min rainfall intensities.

3.4 Amsterdam weather station spatial correlations

Rainfall variability is often described with correlograms; see Sect. 2.2.2, describing Pearson's product–moment correlation between station pairs as a function of distance. Correlograms of PWS data at longer accumulation intervals show higher inter-station correlations and the decrease with distance is not as steep; see Fig. 11. This is similar to the results reported by Villarini et al. (2008), Peleg et al. (2013) and Tokay and Öztürk (2012). Especially in winter (see upper panels of Fig. 11) and for short accumulation intervals, the non-Netatmo pairs show higher correlation with one another. However, the goodness of fit of the correlograms differs significantly from those found by Villarini et al. (2008), Peleg et al. (2013) and Tokay and Öztürk (2012).

The correlations of all station pairs in the dataset are fitted with the relation in Eq. (2). Fitting was done by determining the nonlinear (weighted) least-squares estimates of the parameters with the Gauss–Newton algorithm. The resulting parameters for the total dataset, as well as winter and summer individually, are given in Fig. 12. The graphs for winter show the most deviating response, suggesting irregularities in this subset in particular. The nugget parameter r_0 of the total dataset varies between 0.50 and 0.67 for this accumulation interval range. Villarini et al. (2008) found a similar nugget parameter of 0.51 for 1 min accumulations, though with far larger values at higher accumulation intervals. The nuggets found by Krajewski et al. (2003) (0.95–0.97 for 15 min and longer), Ciach and Krajewski (2006) (0.995 and higher for 1 min and longer), Tokay and Öztürk (2012) (0.97 and higher for 5 min and longer) and Peleg et al. (2013) (0.92 and higher for 1 min and longer), are all considerably higher than the nugget parameters found here. This is not surprising as the

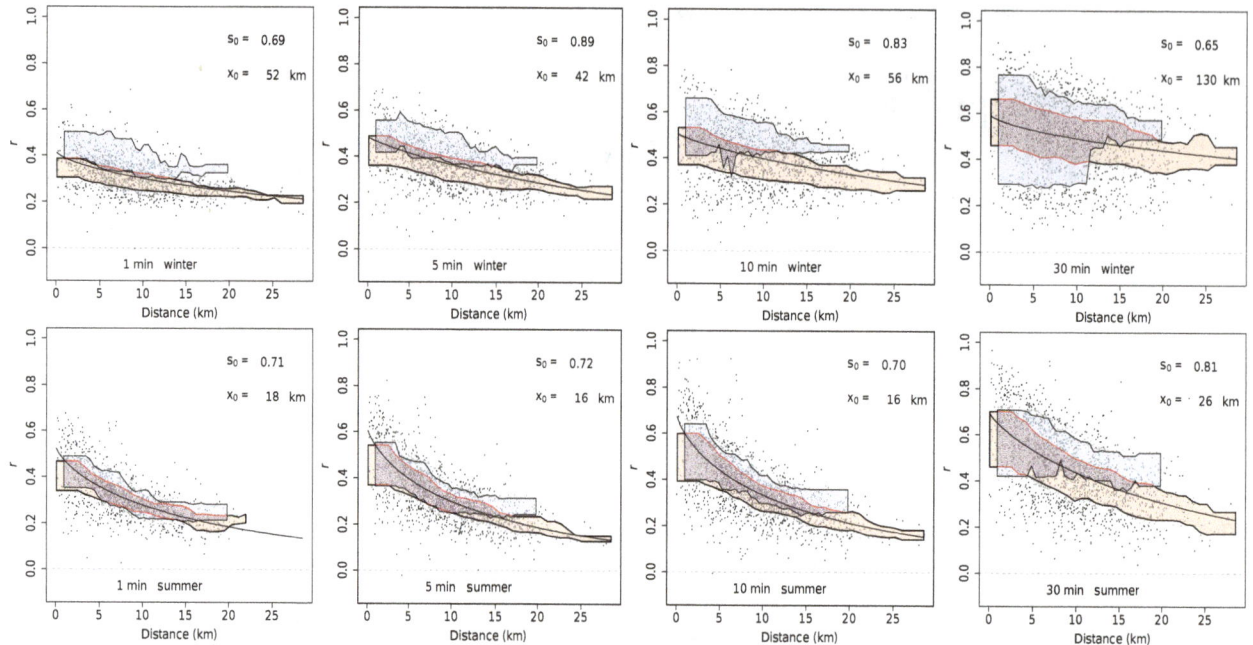

Figure 11. Correlograms of all stations after filtering at various accumulation intervals for winter (top panels) and summer (bottom panels). The red and blue areas represent the interquartile range of the Netatmo stations and non-Netatmo stations, respectively. The areas are constructed with a moving window of width 5 km. The scatter plots are fitted with the exponential relation of Eq. (2), the parameters of which are given in the panels.

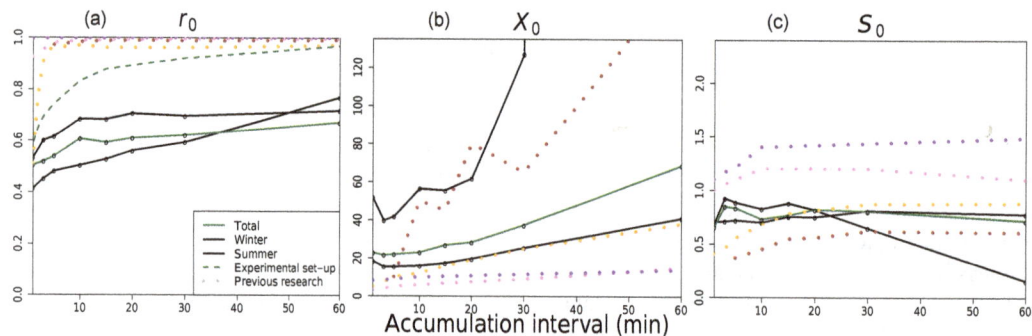

Figure 12. Timescale dependency of nugget (**a**), correlation distance (**b**) and shape factor (**c**) parameters from fit described in Eq. (2), for the total PWS dataset, as well as winter and summer only. Dotted lines represent values found in previous research by Peleg et al. (2013) (violet), Villarini et al. (2008) (orange), Tokay and Öztürk (2012) (brown) and Ciach and Krajewski (2006) (purple), where the dashed line in the first panel shows the timescale dependency of the Netatmo station nugget found in the experimental setup as previously shown in Fig. 6.

gauges in the networks evaluated in those papers are carefully controlled and of higher sensor quality than typical PWSs.

The correlation distance of the total PWS dataset increases with interval size in a similar manner as in previous research; see Fig. 12. The erratic response of the winter graphs suggests a poor fit resulting from other factors than rainfall variability. Likely the correlation distance of stratiform winter rainfall is larger than the spatial scale examined here. The shape parameters do not seem to follow an obvious dependence, similar to Peleg et al. (2013), though other research found this parameter to increase with interval size (Krajew-

ski et al., 2003; Ciach and Krajewski, 2006; Villarini et al., 2008; Tokay and Öztürk, 2012).

4   Discussion

In the experimental setup in Cabauw, the immediate overlying radar pixel that was first considered as reference turned out to show a significant bias as compared to gauge-adjusted radar rainfall data in all neighboring pixels. The next nearest pixel to the setup was then used as reference instead. The distance between radar pixel center and experimental setup thereby increased slightly from 428.9 to 473.5 m. Faulty measurements can occur in the gauge-adjusted radar dataset, which should be kept in mind when it is used as a reference. When comparing the Amsterdam area radar pixels used in this research to their combined mean value over the 17-month period, individual time series showed up to 10 % consistent higher or lower values. Biases in gauge-adjusted radar could result in a larger spread in Fig. 8, although they have a far smaller influence on the results found in Fig. 10 as, in that case, the values are averaged.

Each aspect of this research, i.e., the Netatmo experimental setup, the analysis of the station data obtained with the Netatmo API and the Amsterdam PWS dataset from WunderMap, concerned time series over a different, though partly overlapping, time period. As the shorter time series were examined with the purpose of identifying artifacts in the data, those conclusions can be carried over to the longer, more robust analyses. The results on PWS data availability (see Figs. 1 and 5) do not take measurement quality into account. Because of the faulty attribution of rainfall to measurement intervals due to rounding in the data transfer, the measurements in the current form should be accumulated to larger intervals to reduce errors, although this reduces the temporal resolution appreciably. It would be more desirable to address the collection method of the PWS data in the platforms in order to maintain the quality of the original PWS rainfall measurements before data transfer.

The filter applied on the PWS dataset in this paper was based on all stations in the dataset. For operational purposes, the median value that is used as a selection criterion should be based on nearby stations only. Large rainfall values were excluded based on a limit on maximum rainfall of $50\,\mathrm{mm\,h^{-1}}$ above the median rainfall at all PWSs at that interval. This potentially excludes rainfall with plausible return times: we take the example of a 10 min interval during which the median rain intensity of the stations is $4\,\mathrm{mm\,h^{-1}}$. A measurement of $54\,\mathrm{mm\,h^{-1}}$ and higher would then be excluded, though this corresponds to an event that would occur statistically every 1.5 years (Buishand and Wijngaard, 2007). Because of the small spatial scales and the lack of extremely heavy precipitation in this dataset, the current filter was applicable, as confirmed by visual comparison with gauge-adjusted radar data.

Although a large fraction of the PWS networks consists of Netatmo stations, this does not imply similar performance of these datasets. Factors like placement and maintenance are unknown and not necessarily equally interfering with the measurements. Even less metadata is available on the other PWS types in the dataset, since information on data transfer and the sensors used is not provided for those PWSs. It is expected that there is a positive correlation between the purchase costs of the PWS and the importance of maintenance and high-quality measurements to its owner, although this assumption could not be examined with our dataset. Furthermore, the location of the station is based on the setting provided by the PWS owner, although these may be faulty due to inaccurate localization, rounding of the longitude and latitude or relocation of the station at a later time. Even when relocations of PWSs are accurately provided to the Netatmo platform, this is not automatically communicated to WunderMap, resulting in inaccurate time series for that location. This issue is found to arise in the PWS dataset, though the filter criterion regarding minimum correlation with the other stations excludes time series of those stations entirely.

Different spatial correlation parameters between studies are to be expected due to different climates, rainfall types, gauge network density and quality. However, the nugget-parameter r_0 (1 for perfect correlation between time series) found here is significantly lower than in other studies. Additionally, the nugget values of the Amsterdam dataset are significantly lower than the correlation found between the Netatmo datasets with the electronic rain gauge reference in the experimental setup when the data were obtained via the WunderMap platform; see also Fig. 6. This suggests the interference of additional factors besides sensor measurement errors and data transfer rounding when rainfall measurements are gathered in a less controlled manner. Such factors could be measurement errors due to station placement and poor maintenance.

It is important to note that, even though gauge-adjusted radar rainfall is used as a rainfall reference, differences with point measurements are to be expected because of representativeness errors. Ideally, a high-density gauge network could be used to improve this rainfall product in the future. A non-identical match should therefore not directly be interpreted as negative. However, as the nugget parameter from the station analyses was considerably lower than could be explained by rainfall variability alone, differences with gauge-adjusted radar data here are likely mainly caused by errors in the PWS dataset. Besides data transfer errors that heavily influence the nugget parameter, the installation errors (e.g., due to shielding), that are minimized in the experimental setup, further decrease the nugget in the Amsterdam dataset. When comparing nuggets from the experimental setup and the Amsterdam dataset in the left panel in Fig. 12, the correlations found in the former do indeed reach higher values than those influenced by installation errors in the latter.

5 Conclusions

The resolution and quality of crowdsourced PWS rainfall measurements from the platform with the most dense PWS network were analyzed to establish whether this data source allows urban hydrological applications. Although the required resolutions (as described by Schilling, 1991, Berne et al., 2004, Emmanuel et al., 2012 and Ochoa-Rodriguez et al., 2015) are not yet achieved by the current PWS networks, the density of these networks is expected to increase. As the resolution of the current network in Amsterdam is more limited in the spatial resolution than the temporal resolution, the expected continued growth of PWSs that share rainfall measurements via online platforms will yield a network approaching the desired resolutions. This offers a vast contrast compared to KNMI's automatic rain gauge network which, in the Amsterdam metropolitan area, only measures rainfall at one location outside of the city (at Schiphol airport).

From comparisons between Netatmo rainfall time series in an experimental setup that reduces the errors due to faulty installation to a minimum, the measurements closely resemble those from the high-resolution electronic rain gauge. Larger differences are found with radar rainfall, likely due to differences in representativeness between pixels and point measurements. Although the sensor performance of this largest contributor of data in the PWS network considered in this research looks promising, there is a significant loss in accuracy due to transfer of data to the online platform WunderMap. In this study, the daily cumulative rainfall values as obtained from WunderMap are rewritten as the difference in rainfall as compared to the previous time step. WunderMap cumulative daily rainfall can only become non-zero when at least 0.3 mm rainfall has been collected, and later increases are only registered if they amount to at least 0.2 mm. Especially in the event of light rain, rainfall could occur for a longer period than the interval length in which the daily cumulative rainfall increases. The rainfall is then attributed to a single interval instead of all previous intervals in which it may have been raining as well. This causes significant errors at small timescales. These errors result in inter-station correlations that were considerably poorer than those found in literature, especially in winter and at short accumulation intervals.

The median rainfall of the Amsterdam PWS dataset shows less systematic bias than the real-time available radar product. Averaging PWS time series further improves correlation, standard deviation and coefficient of variation with the averaged gauge-adjusted radar rainfall in a certain region ($\sim 20\,\mathrm{km}^2$). Provided that the degree and likelihood of overestimation of rainfall by PWSs is similar to the degree and likelihood of rainfall underestimation, as was the case in our Amsterdam city center dataset, a dense subset of PWSs can provide good rainfall estimation over a small area, even for intervals of 5 min and without applying a quality filter.

The largest obstacles for the use of crowdsourced PWS datasets are the errors resulting from data transfer, errors due to poor maintenance and faulty installations (i.e., at shielded locations). The rounding of cumulative daily rainfall measurements occurring in the WunderMap platform and the time stamp uncertainty of measurements obtained from platforms with faulty APIs can lead to considerable errors in the time series, which are only reduced at large accumulation intervals. For the purpose of a high-quality rainfall measurement network with PWS data, these issues need to be addressed first. Processing errors can be avoided by obtaining raw data from the Netatmo weather map platform, though

the station density is slightly lower than that of the network linked to the WunderMap. When the processing of data is no longer interfering with the quality of the datasets, the potential of PWS platforms becomes significant. It provides rainfall measurements from all over the world that are easy to collect, located in rural areas as well as in cities, with station densities and coverage exceeding those from national weather services, and growing towards a level matching the reported resolutions that are required for urban hydrological applications.

6 Data availability

The gauge-adjusted radar data used in this research are freely available as "radar precipitation climatology" via http://climate4impact.eu. Measurements from personal weather stations can be accessed via the online platforms to which they are linked: https://www.wunderground.com/wundermap and https://weathermap.netatmo.com.

Competing interests. The authors declare that they have no conflict of interest.

Acknowledgements. This research was performed as part of the RainSense project, funded by the Amsterdam Institute for Advanced Metropolitan Solutions (AMS) and the SMART city project (project no. 13760) funded by Netherlands Technology Foundation (STW). The data were made available by Weather Underground, and subsequently the weather enthusiasts sharing their weather data with the online community at the online platform WunderMap. The authors would like to thank Marcel Brinkenberg of KNMI for his assistance with the experimental setup of the weather stations at the Cabauw Experimental Site for Atmospheric Research (CESAR). Thanks is also due to Tom de Ruijter from MeteoGroup for providing data and insight on weather measurements obtained with the Netatmo API.

Edited by: K. Arnbjerg-Nielsen

References

Bell, S., Cornford, D., and Bastin, L.: The state of automated amateur weather observations, Weather, 68, 36–41, 2013.

Bell, S., Cornford, D., and Bastin, L.: How good are citizen weather stations? Addressing a biased opinion, Weather, 70, 75–84, 2015.

Bell, V. A. and Moore, R. J.: The sensitivity of catchment runoff models to rainfall data at different spatial scales, Hydrol. Earth Syst. Sci., 4, 653–667, doi:10.5194/hess-4-653-2000, 2000.

Berne, A., Delrieu, G., Creutin, J.-D., and Obled, C.: Temporal and spatial resolution of rainfall measurements required for urban hydrology, J. Hydrol., 299, 166–179, 2004.

Bruni, G., Reinoso, R., van de Giesen, N. C., Clemens, F. H. L. R., and ten Veldhuis, J. A. E.: On the sensitivity of urban hydrodynamic modelling to rainfall spatial and temporal resolution, Hydrol. Earth Syst. Sci., 19, 691–709, doi:10.5194/hess-19-691-2015, 2015.

Buishand, T. A. and Wijngaard, J.: Statistiek van extreme neerslag voor korte neerslagduren [Statistics of extreme rainfall for short durations], Royal Netherlands Meteorologic Institute, 2007.

Ciach, G. J. and Krajewski, W. F.: Analysis and modeling of spatial correlation structure in small-scale rainfall in Central Oklahoma, Adv. Water Resour., 29, 1450–1463, 2006.

Einfalt, T., Arnbjerg-Nielsen, K., Golz, C., Jensen, N.-E., Quirmbach, M., Vaes, G., and Vieux, B.: Towards a roadmap for use of radar rainfall data in urban drainage, J. Hydrol., 299, 186–202, 2004.

Emmanuel, I., Andrieu, H., Leblois, E., and Flahaut, B.: Temporal and spatial variability of rainfall at the urban hydrological scale, J. Hydrol., 430, 162–172, 2012.

Estévez, J., Gavilán, P., and Giráldez, J. V.: Guidelines on validation procedures for meteorological data from automatic weather stations, J. Hydrol., 402, 144–154, 2011.

Fabry, F., Bellon, A., Duncan, M. R., and Austin, G. L.: High resolution rainfall measurements by radar for very small basins: the sampling problem reexamined, J. Hydrol., 161, 415–428, 1994.

Gharesifard, M. and Wehn, U.: To share or not to share: Drivers and barriers for sharing data via online amateur weather networks, J. Hydrol., 535, 181–190, 2016.

Gires, A., Onof, C., Maksimovic, C., Schertzer, D., Tchiguirinskaia, I., and Simoes, N.: Quantifying the impact of small scale unmeasured rainfall variability on urban runoff through multifractal downscaling: A case study, J. Hydrol., 442, 117–128, 2012.

Habib, E., Krajewski, W. F., and Ciach, G. J.: Estimation of rainfall interstation correlation, J. Hydrometeorol., 2, 621–629, 2001.

Jenkins, G.: A comparison between two types of widely used weather stations, Weather, 69, 105–110, 2014.

Krajewski, W. F., Ciach, G. J., and Habib, E.: An analysis of small-scale rainfall variability in different climatic regimes, Hydrolog. Sci. J., 48, 151–162, 2003.

Leijnse, H., Uijlenhoet, R., van de Beek, C. Z., Overeem, A., Otto, T., Unal, C. M. H., Dufournet, Y., Russchenberg, H. W. J., Figueras i Ventura, J., Klein Baltink, H., and Holleman, I.: Precipitation measurement at CESAR, the Netherlands, J. Hydrometeorol., 11, 1322–1329, doi:10.1175/2010JHM1245.1, 2010.

Liguori, S., Rico-Ramirez, M. A., Schellart, A. N. A., and Saul, A. J.: Using probabilistic radar rainfall nowcasts and NWP forecasts for flow prediction in urban catchments, Atmos. Res., 103, 80–95, 2012.

Lobligeois, F., Andréassian, V., Perrin, C., Tabary, P., and Loumagne, C.: When does higher spatial resolution rainfall information improve streamflow simulation? An evaluation using 3620 flood events, Hydrol. Earth Syst. Sci., 18, 575–594, doi:10.5194/hess-18-575-2014, 2014.

Mazzoleni, M., Verlaan, M., Alfonso, L., Monego, M., Norbiato, D., Ferri, M., and Solomatine, D. P.: Can assimilation of crowdsourced streamflow observations in hydrological modelling improve flood prediction?, Hydrol. Earth Syst. Sci. Discuss., 12, 11371–11419, doi:10.5194/hessd-12-11371-2015, 2015.

Meier, F., Fenner, D., Grassmann, T., Jänicke, B., Otto, M., and Scherer, D.: Challenges and benefits from crowd-sourced atmo-spheric data for urban climate research using Berlin, Germany, as testbed, in: ICUC9 – 9th International Conference on Urban Climate jointly with 12th Symposium on the Urban Environment, 2015.

Muller, C. L., Chapman, L., Johnston, S., Kidd, C., Illingworth, S., Foody, G., Overeem, A., and Leigh, R. R.: Crowdsourcing for climate and atmospheric sciences: current status and future potential, Int. J. Climatol., 35, 3185–3203, 2015.

Ochoa-Rodriguez, S., Wang, L. P., Gires, A., Pina, R. D., Reinoso-Rondinel, R., Bruni, G., Ichiba, A., Gaitan, S., Cristiano, E., Van Assel, J., Kroll, S., Murlà-Tuyls, D., Tisserand, B., Schertzer, D., Tchiguirinskaia, I., Onof, C., Willems, P., and Ten Veldhuis, J. A. E.: Impact of spatial and temporal resolution of rainfall inputs on urban hydrodynamic modelling outputs: A multi-catchment investigation, J. Hydrol., 531, 389–407, doi:10.1016/j.jhydrol.2015.05.035, 2015.

Overeem, A., Buishand, T. A., and Holleman, I.: Extreme rainfall analysis and estimation of depth-duration-frequency curves using weather radar, Water Resour. Res., 45, doi:10.1029/2009WR007869, 2009a.

Overeem, A., Holleman, I., and Buishand, T. A.: Derivation of a 10-year radar-based climatology of rainfall, J. Appl. Meteorol. Clim., 48, 1448–1463, 2009b.

Overeem, A., Leijnse, H., and Uijlenhoet, R.: Measuring urban rainfall using microwave links from commercial cellular communication networks, Water Resour. Res., 47, doi:10.1029/2010WR010350, 2011.

Overeem, A., Leijnse, H., and Uijlenhoet, R.: Two and a half years of country-wide rainfall maps using radio links from commercial cellular telecommunication networks, Water Resour. Res., 52, 8039–8065, doi:10.1002/2016WR019412, 2016.

Peleg, N., Ben-Asher, M., and Morin, E.: Radar subpixel-scale rainfall variability and uncertainty: lessons learned from observations of a dense rain-gauge network, Hydrol. Earth Syst. Sci., 17, 2195–2208, doi:10.5194/hess-17-2195-2013, 2013.

Schilling, W.: Rainfall data for urban hydrology: what do we need?, Atmos. Res., 27, 5–21, 1991.

Steeneveld, G. J., Koopmans, S., Heusinkveld, B. G., Van Hove, L. W. A., and Holtslag, A. A. M.: Quantifying urban heat island effects and human comfort for cities of variable size and urban morphology in the Netherlands, J. Geophys. Res.-Atmos., 116, doi:10.1029/2011JD015988, 2011.

Tokay, A. and Öztürk, K.: An experimental study of the small-scale variability of rainfall, J. Hydrometeorol., 13, 351–365, 2012.

van de Beek, C. Z., Leijnse, H., Torfs, P. J. J. F., and Uijlenhoet, R.: Climatology of daily rainfall semi-variance in The Netherlands, Hydrol. Earth Syst. Sci., 15, 171–183, doi:10.5194/hess-15-171-2011, 2011.

van de Beek, C. Z., Leijnse, H., Torfs, P. J. J. F., and Uijlenhoet, R.: Seasonal semi-variance of Dutch rainfall at hourly to daily scales, Adv. Water Resour., 45, 76–85, 2012.

Villarini, G., Mandapaka, P. V., Krajewski, W. F., and Moore, R. J.: Rainfall and sampling uncertainties: A rain gauge perspective, J. Geophys. Res.-Atmos., 113, doi:10.1029/2007JD009214, 2008.

Wolters, D. and Brandsma, T.: Estimating the Urban Heat Island in residential areas in the Netherlands using observations by weather amateurs, J. Appl. Meteorol. Clim., 51, 711–721, 2012.

9

Decoupling of dissolved organic matter patterns between stream and riparian groundwater in a headwater forested catchment

Susana Bernal[1,2], Anna Lupon[2,3], Núria Catalán[4], Sara Castelar[1], and Eugènia Martí[1]

[1]Integrative Freshwater Ecology Group, Center for Advanced Studies of Blanes (CEAB-CSIC), Blanes, 17300, Spain
[2]Departament de Biologia Evolutiva, Ecologia i Ciències Ambientals (BEECA), Universitat de Barcelona, Barcelona, 08028, Spain
[3]Department of Forest Ecology and Management, Swedish University of Agricultural Sciences, Umeå, 90183, Sweden
[4]Department of Resources and Ecosystems, ICRA, Catalan Institute for Water Research, Girona, 17003, Spain

Correspondence: Susana Bernal (sbernal@ceab.csic.es)

Abstract. Streams are important sources of carbon to the atmosphere, though knowing whether they merely outgas terrestrially derived carbon dioxide or mineralize terrestrial inputs of dissolved organic matter (DOM) is still a big challenge in ecology. The objective of this study was to investigate the influence of riparian groundwater (GW) and in-stream processes on the temporal pattern of stream DOM concentrations and quality in a forested headwater stream, and whether this influence differed between the leaf litter fall (LLF) period and the remaining part of the year (non-LLF). The spectroscopic indexes (fluorescence index, biological index, humification index, and parallel factor analysis components) indicated that DOM had an eminently protein-like character and was most likely originated from microbial sources and recent biological activity in both stream water and riparian GW. However, paired samples of stream water and riparian GW showed that dissolved organic carbon (DOC) and nitrogen (DON) concentrations as well as the spectroscopic character of DOM differed between the two compartments throughout the year. A simple mass balance approach indicated that in-stream processes along the reach contributed to reducing DOC and DON fluxes by 50 and 30 %, respectively. Further, in-stream DOC and DON uptakes were unrelated to each other, suggesting that these two compounds underwent different biogeochemical pathways. During the LLF period, stream DOC and DOC : DON ratios were higher than during the non-LLF period, and spectroscopic indexes suggested a major influence of terrestrial vegetation on stream DOM. Our study highlights that stream DOM is not merely a reflection of riparian GW entering the stream and that headwater streams have the capacity to internally produce, transform, and consume DOM.

1 Introduction

The transport of dissolved organic matter (DOM) through fluvial networks is of major importance for understanding the links between continental and coastal biogeochemical cycles (Seitzinger and Sanders, 1997; Battin et al., 2008). Stream DOM is a combination of allochthonous (i.e., terrestrially derived) and autochthonous (i.e., in-stream produced) DOM. The former originates mostly from terrestrial systems (i.e., soils, vegetation, and microbes) and it is transported to streams via surface and groundwater flow paths, while the latter derives from in-stream metabolic activity and leachates of litter falling into the stream, especially during the leaf litter fall (LLF) period (Qualls and Haines, 1991, 1992). The bioavailability of DOM can differ substantially between allochthonous and autochthonous sources, and thus, a good assessment of the origin and quality of stream DOM is of great importance to understand the capacity of aquatic ecosystems to store and transform carbon (C) and nitrogen (N) (Cole et al., 2007; Battin et al., 2008; Tranvik et al., 2009). Yet our knowledge of the contribution of allochthonous vs. autochthonous sources to total stream DOM and its variability over time and space is far from complete.

The strong correlation found between dissolved organic carbon (DOC) and nitrogen (DON) in temperate and boreal streams have suggested that the soil organic pool is a major factor controlling the fate and form of stream DOM (Perakis and Hedin, 2002; Hedin et al., 1995; Brookshire et al., 2007; Sponseller et al., 2014). These previous observations are the cornerstone of the passive carbon vehicle hypothesis, which states that soil DOM is stoichiometrically static and behaves almost conservatively when traveling throughout the catchment and stream ecosystems (Brookshire et al., 2007). However, there is an increasing body of studies reporting differences in DOC : DON ratios between allochthonous sources and stream water. For instance, stream DOC : DON ratios can change as a consequence of in-stream heterotrophic DOM production during periods of high ecosystem respiration (Caraco and Cole, 2003; Kaushal and Lewis, 2005; Johnson et al., 2013). Moreover, stream biota can show a strong capacity to process DOM (McDowell, 1985; Bernhardt and McDowell, 2008), with whole-reach DOM uptake rates being even higher than for essential nutrients such as nitrate (Brookshire et al., 2005). The processing of DOM within the stream can lead to a decoupling between stream DOC and DON concentrations because stream DOC is mostly used as an energy source, while DON can alternatively be used as a nutrient source (Kaushal and Lewis, 2005; Lutz et al., 2011; Wymore et al., 2015). Therefore, a significant fraction of stream DOM could be degraded, mineralized, or produced within the stream (either in the stream column or in the hyporheic zone).

Despite the potential role of in-stream biota on processing DOM, its ability to modify DOM concentrations and regulate allochthonous DOM fluxes remains elusive. First, the high variety of molecules used during in situ DOM additions (from monomeric carbohydrates to complex leachate molecules) limits the possibility to compare whole-reach DOM uptake rates among sites and to link manipulative experiments with actual DOM processing under natural conditions (Newbold et al., 2006; Bernhardt and McDowell, 2008). Second, the intrinsic complexity of up-scaling reach scale measurements constrains our understanding of the potential of in-stream processes to modify DOM export at catchment scale (Wollheim et al., 2015). Recent synoptic studies suggest that changes in stream DOC concentrations can be mostly explained by hydrological mixing of different water sources, thus suggesting minimal removal of DOC within streams (Tiwari et al., 2014; Wollheim et al., 2015). Yet these studies are mostly performed during particular periods (usually summer) and in catchments with large wetland and peatland areas that provide large quantities of allochthonous DOM to aquatic ecosystems (Wollheim et al., 2015). Studies with a network perspective are still scarce and usually deal with a high amount of uncertainty because quantity and quality of DOM in groundwater traversing the hyporheic zone and entering the stream is poorly characterized (Tiwari et al., 2014; Casas-Ruíz et al., 2017).

The objective of this study was to investigate the influence of DOM inputs from riparian groundwater (GW) and in-stream processes on the temporal pattern of stream DOC and DON concentrations and quality (DOC : DON stoichiometry and DOM spectroscopic descriptors) in a Mediterranean forested headwater stream. To do so, we assessed the temporal variation of DOM quantity and quality in stream water and riparian GW over 1.5 years. We expected that differences between riparian GW and stream DOM would be small if (i) allochthonous sources dominate the temporal pattern of DOM inputs and (ii) DOM is transported passively along the stream as stated by the carbon vehicle hypothesis (Brookshire et al., 2007). Alternatively, differences between riparian GW and stream water would indicate DOM generation and/or processing of allochthonous DOM within the stream. Specifically, we expected large differences between riparian GW and stream DOM associated with the leaf litter fall period because leachates from fresh material stored in the streambed may increase DOM concentration and fuel heterotrophic stream metabolism.

2   Study site

The study was conducted from October 2010 to December 2011 in the Font del Regàs catchment (14.2 km^2), located in the Montseny Natural Park, northeastern Spain (41°50′ N, 2°30′ E, 300–1200 m a.s.l.). The climate is subhumid Mediterranean, with mild winters and dry summers. Mean annual precipitation (975 mm) and temperature (12.9 °C) during the study period fall within the long-term annual average for this region (Catalan Metereologic Service: http://www.meteo.cat/wpweb/climatologia/serveis-i-dades-climatiques/series-climatiques-historiques/, last access: 15 March 2018).

The catchment is dominated by biotitic granite and it has steep slopes (28 %). Evergreen oak (*Quercus ilex*) and beech (*Fagus sylvatica*) forests cover 54 and 38 % of the catchment area, respectively (Fig. 1). The upper part of the catchment (2 %) is covered by heathlands and grasslands. Population density within the catchment is < 1 person km^{-2}. Hillslope soils (pH ∼ 6) are sandy and have a 3 cm deep organic layer (O-horizon) followed by a 5 to 15 cm deep mineral layer (A-horizon). The riparian zone is relatively flat (slope < 10 %), and it covers 6 % of the catchment area. Riparian soils (pH ∼ 7) are sandy-loam and they have a 5 cm deep O-horizon followed by a 30 cm deep A-horizon. The width of the riparian zone increases from 6 to 32 m from the upper to the lower part of the catchment, whereas the total basal area of riparian trees increases twelvefold (Bernal et al., 2015). *Alnus glutinosa*, *Robinia pseudoacacia*, *Platanus hybrida*, and *Fraxinus excelsior* are the most abundant riparian tree species followed by *Corylus avellana*, *Populus tremula*, *Populus nigra*, and *Sambucus nigra*. During base flow conditions, the riparian GW table is well below the soil sur-

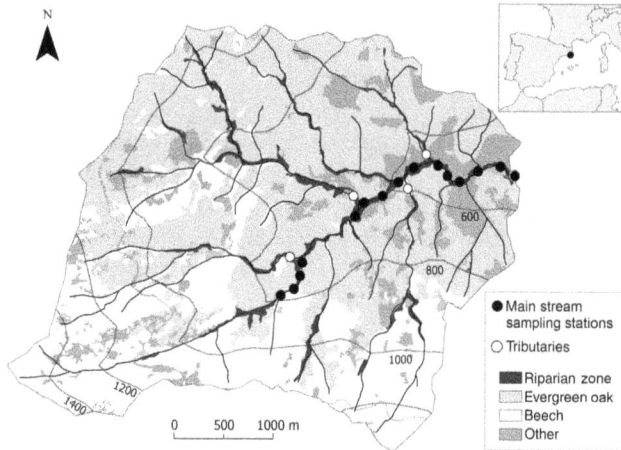

Figure 1. Map of the Font del Regàs catchment within the Montseny Natural Park (northeastern Spain). The vegetation cover and the main stream sampling stations along the 3.7 km reach are indicated. Four permanent tributaries discharged to the main stream from the upstream- to the downstream-most site (white circles). The remaining tributaries were dry during the study period.

face (~ 50 cm), though it can reach the superficial soil organic layers during storm events (Lupon et al., 2016a).

The catchment is drained by a perennial third-order stream. At the headwaters, the streambed is mainly composed of rocks and cobbles (70 %) with a small contribution of sand (~ 10 %). At the valley bottom, sands and gravels represent 44 % of the stream substrate and the presence of rocks is minor (14 %). During base flow conditions, mean stream water velocity is 0.3 m s^{-1}. On average, stream discharge increases along the reach from 20 to 70 L s^{-1}. During the study period, the stream gained water in net terms along the reach, yet it lost water towards the riparian zone in some segments, specifically during summer months. Moreover, mean area-specific stream discharge decreased longitudinally, an indication that hydrological retention was higher at the valley bottom compared to upstream segments. Permanent tributaries comprise about 50 % of the catchment area and contribute 56 % of stream discharge (Bernal et al., 2015).

3   Material and methods

3.1   Field sampling

We selected 15 sampling sites along a 3.7 km reach that were located from 110 to 600 m apart from each other (Fig. 1). At each sampling site, we installed a 1 m long PVC piezometer (3 cm $\varnothing$) in the riparian zone (~ 1.5 m from the stream channel edge). We assumed this water to be representative of the groundwater entering the stream. We collected stream water (from the thalweg) and riparian GW from each sampling site every 2 months from October 2010 to December 2011. Groundwater samples were collected with a 100 mL

syringe connected to a silicone tube. Water samples were collected with pre-acid-washed polyethylene bottles after triple-rinsing them with either stream water or groundwater. Field sampling was conducted during base flow conditions to capture the influence of in-stream processes on DOM dynamics when they are expected to be the highest. Moreover, by avoiding storm flows, we ensured that riparian GW was the main subsurface water source contributing to stream runoff. All field campaigns were performed at least 9 days after storm events, except for October 2011. At each sampling site, we measured stream discharge (Q, in L s^{-1}) by adding 1 L of NaCl-enriched solution to the stream (Gordon et al., 2004). The empirical uncertainty associated with Q was calculated considering pairs of measurements conducted under equal water depth conditions as described in Bernal et al. (2015). On each sampling date, we also collected stream water and measured Q at the four permanent tributaries discharging to Font del Regàs stream, which drained 1.9, 3.2, 1.8, and 1.1 km^2 each (Fig. 1). These data were used for mass balance calculations (see below).

3.2   Laboratory analysis and DOM quality indexes

Water samples were filtered through pre-ashed GF/F filters (Whatman®) and kept cold (< 4 °C) until laboratory analysis (< 24 h after collection). Chloride (Cl$^-$) was used as a conservative hydrological tracer and analyzed by ionic chromatography (Compact IC-761, Metrhom). DOC and total dissolved nitrogen (TDN) concentrations were determined using a Shimadzu TOC-VCS coupled to a TN analyzer. DOC was determined by oxidative combustion infrared analysis and TDN by oxidative combustion chemiluminescence. DON concentration was calculated by subtracting nitrate (NO$_3^-$) and ammonium (NH$_4^+$) concentrations from TDN. Concentrations of NO$_3^-$ and NH$_4^+$ were determined by standard colorimetric methods (details in Bernal et al., 2015).

We used different metrics to assess the quality of DOM and to infer its origin. First, the DOC : DON ratio was used as a general proxy of DOM quality, high values being indicative of plant organic matter sources (Bernal et al., 2005). Then, we assessed DOM properties by optical spectroscopy. Fluorescence excitation–emission spectra were recorded on a Shimadzu RF-5301 PC spectrofluorimeter over an emission range of 270–700 nm (1 nm steps) and an excitation range of 230–430 nm (10 nm steps). Measurements were done at room temperature (20–25 °C) and corrected for instrument baseline offset. A Milli-Q blank was subtracted from each sample to eliminate Raman scattering. Sampling blanks were included to assess for leaching of DOM during the sampling procedure. We followed the procedure in Kothawala et al. (2013) for inner filter correction. Briefly, UV-Vis absorbance spectra (200–800 nm) were obtained in a Shimadzu UV-1700 spectrophotometer, using 1 cm quartz cuvette. Due to fatal circumstances, absorbance spectra could not be recorded for some samples. In these cases, we used the modeled mean absorbance spectra for either riparian GW or surface stream water to apply the inner filter correction. All the corrections were applied using the FDOM correct toolbox for MATLAB (Mathworks, Natick, MA, USA) following Murphy et al. (2010).

We calculated three spectroscopic descriptors: (i) the fluorescence index (FI) which typically ranges from ~ 1.2 to ~ 2 and is linked to the DOM origin with low values being characteristic of terrestrial higher-plant DOM sources and high values of microbial DOM sources (Jaffé et al., 2008), (ii) the biological index (BIX), for which higher values indicate a higher contribution of recently produced DOM (i.e., biological activity or aquatic bacterial origin) (Huguet et al., 2009), and (iii) the humification index (HIX) as a proxy of the humification status of DOM (i.e., higher values indicating higher humification degree) (Ohno, 2002; Fellman et al., 2010).

Parallel factor analysis (PARAFAC) was used to identify the main fluorescence components of DOM (Stedmon et al., 2003). The analysis was performed using the DrEEM toolbox for MATLAB (Mathworks, Inc., Natick, MA) according to Murphy et al. (2013). Scatter peaks and outliers were removed and samples normalized to its total fluorescence prior to fitting the PARAFAC model. The appropriate number of components was determined by visual inspection of both the residual fluorescence and the components' behavior as organic fluorophores. The PARAFAC modeling of EEM spectra from the analyzed samples revealed four independent components (F1–F4; Fig. S1 in the Supplement). Components F2 and F3 corresponded to humic-like materials, while components F1 and F4 corresponded to protein-like fluorescence (Supplement Table S1 and S2). The four-component model was validated by split-half analysis and random initialization with 10 iterations. Finally, the level of coincidence of the obtained model against other PARAFAC models published in the online OpenFluor database (http://www.openfluor.org; June 2017) was assessed, applying a Tucker congruence coefficient of 95 % (Murphy et al., 2014).

3.3 Whole-reach net DOM uptake rates

We investigated the influence of in-stream biogeochemical processes on stream DOM fluxes by applying a mass balance approach for the whole reach. Briefly, we calculated the net flux resulting from in-stream gross uptake and release along the reach (U, in $\mu g \, m^{-2} \, s^{-1}$) by including all hydrological input and output solute fluxes (upstream-most site, tributaries, and riparian GW) in the mass balance. Riparian GW must traverse the hyporheic zone before arriving at the stream water column, and thus, we considered that in-stream net uptake was the result of biogeochemical processes occurring in both the stream water column and the hyporheic zone. For each sampling date, U for either DOC or DON was approximated

with the following:

$$U = (Q_{top} \times C_{top} + \sum_{i=1}^{4} Q_{tr,i} \times C_{tr,i} + \sum_{j=1}^{14} Q_{gw,j} \times C_{gw,j} - Q_{bot} \times C_{bot})/A, \qquad (1)$$

where Q_{top} and Q_{bot} are the discharge at the top and at the bottom of the reach, Q_{tr} is the discharge from tributaries, and Q_{gw} is the net riparian GW inputs (all in $L \, s^{-1}$). Q_{gw} was estimated as the difference in Q between consecutive sampling sites and could be either positive (net gaining) or negative (net losing) (Covino et al., 2010). Top and bottom fluxes were calculated by multiplying Q by stream water solute concentration at the top (C_{top}) and at the bottom (C_{bot}) of the segment, respectively. For each stream segment j, riparian GW fluxes were estimated by multiplying Q_{gw} by solute concentration (C_{gw}) as described in Bernal et al. (2015). Briefly, C_{gw} averaged riparian GW concentration at the top and bottom of the segment for net gaining segments ($Q_{gw} > 0$), while it averaged stream water concentrations at the top and bottom of the segment for net losing segments ($Q_{gw} < 0$). For each tributary i, the input flux to the stream was calculated by multiplying Q_{tr} and solute concentrations (C_{tr}) at the outlet of the tributary. The total active streambed (A) was 8860 m^2 and it was estimated by multiplying the total length of the reach (3.7 km) by the mean wetted width (2.4 m) that varied $< 10 \%$ across the different sampling dates. The values used to calculate U for each sampling date are detailed in Table S3. Finally, we calculated an upper and lower limit of U based on the empirical uncertainty associated with discharge measurements (Q and Q_{gw}) (Bernal et al., 2015).

The mass balance approach used in the present study was similar to that applied for Cl^-, NH_4^+, and NO_3^- for the same reach and period in Bernal et al. (2015). We considered Cl^- as a hydrological reference because this conservative tracer showed $U \sim 0$ for the whole study period (Bernal et al., 2015). For DOC and DON, $U > 0$ indicates that gross uptake prevails over release, $U < 0$ indicates that release prevails over gross uptake, and $U \sim 0$ indicates that gross uptake $\sim$ release. Therefore, we expected $U \neq 0$ if DOM does not behave conservatively and in-stream gross uptake and release processes do not fully counterbalance each other. We assumed that U was indistinguishable from 0 when the range of upper and lower limits contained zero.

To assess the contribution of in-stream net uptake to stream DOM fluxes, we calculated the ratio between $U \times A$ (absolute value) and the total input flux (F_{in}) for each compound (i.e., DOC and DON) and sampling date. F_{in} was the sum of fluxes from upstream ($Q_{top} \times C_{top}$), tributaries ($Q_{tr} \times C_{tr}$), and riparian GW ($Q_{gw} \times C_{gw}$). Riparian GW was included in the calculation only when the main stream was gaining water in net terms (i.e., $Q_{gw} \times C_{gw} > 0$). We interpreted a high $|U \times A|/F_{in}$ ratio as a strong potential of in-stream processes to modify input fluxes (either as a consequence of gross uptake or release). The relative importance of in-stream DOM uptake and release was estimated with

$U > 0/F_{in}$ and $|U < 0|/F_{in}$, respectively. In addition, we calculated the contribution of upstream ($Q_{top} \times C_{top}/F_{in}$) and tributary ($Q_{tr} \times C_{tr}/F_{in}$) inputs to stream DOM fluxes.

3.4 Statistical analysis

The data set was divided in two groups based on the temporal pattern of leaf litter fall because we expected large differences between riparian GW and stream DOM associated with the input of fresh leaf litter to the stream. During the two water years, leaf litter fall began in early October and peaked in early November. In 2010, the litter fall period finished in late November, while it lasted until late December in 2011. There were four sampling dates within the LLF period and six sampling dates during the remaining part of the year (hereafter, non-LLF). Median values for each sampling date were used for analyzing the seasonal pattern of stream DOM concentration and quality (DOC : DON ratio and spectroscopic descriptors). We used a Mann–Whitney test to analyze differences in DOM concentrations and quality between the LLF and non-LLF periods for both stream water and riparian GW (Zar, 2010). Moreover, we used linear regression models to investigate (i) longitudinal patterns of Cl$^-$ and DOM concentrations and (ii) differences in DOM stoichiometry (i.e., the relationship between DOC and DON concentration) between riparian GW and stream water.

We explored the influence of riparian GW on the temporal pattern of stream DOM by analyzing the difference between DOM concentrations in these two water compartments with a Wilcoxon paired rank sum test. Tests were run separately for the LLF and non-LLF periods. Moreover, we compared the temporal variation of longitudinal trends in DOM spectroscopic descriptors between stream water and riparian GW. Longitudinal trends were analyzed by applying linear regression and the standardized regression coefficient (r) was used as a measure of the strength of the longitudinal pattern along the reach. For a particular sampling date, we expected similar longitudinal trends between stream water and riparian GW (and thus similar r) if riparian GW was a major source of DOM to the stream and in-stream processes had a small influence of DOM quality.

Finally, we explored differences in U between LLF and non-LLF periods with a Mann–Whitney test. Moreover, we used Spearman's ρ correlations to test (i) whether U_{DOC} and U_{DON} followed the same temporal pattern and (ii) whether they were behaving conservatively, and thus similar to U_{Cl}.

We chose non-parametric tests for comparing groups of data because the residuals of variables were not always normally distributed (Zar, 2010). All statistical tests were run with JMP v.5.0 statistical software (SAS Institute, Cary, NC).

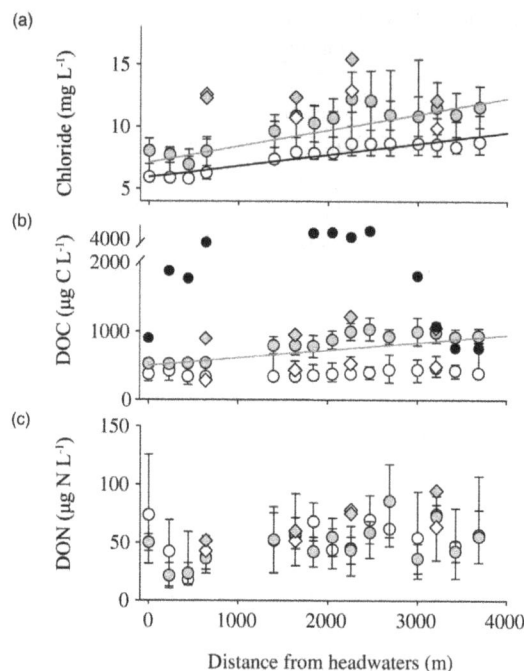

Figure 2. Longitudinal patterns of **(a)** chloride, **(b)** dissolved organic carbon (DOC), and **(c)** dissolved organic nitrogen (DON) concentrations in stream water along the 3.7 km reach. Symbols are median values and whiskers are the interquartile range (25th, 75th percentiles) for the main stream (circles) and tributaries (diamonds). Concentrations are shown separately for the LLF (grey) and non-LLF periods (white). Black circles in **(b)** correspond to the field campaign of November 2010 when DOC concentrations were higher than for the remaining study period. Model regressions are indicated with solid lines only when significant (tributaries not included in the model).

4 Results

4.1 Temporal pattern of chloride and DOM in stream water

During the study period, median Cl$^-$ concentration in the main stream was higher for the LLF (8.6 [7.8, 13.1] [25th, 75th percentiles] mg L^{-1}) than for the non-LLF period (7.8 [7.3, 8.8] mg L^{-1}) (Mann–Whitney test, $Z = 2.82$, d$f = 1$, $p = 0.005$). Stream Cl$^-$ concentrations increased along the reach by 43 and 48 % during the LLF and the non-LLF period, respectively (Fig. 2a). A similar pattern was exhibited by riparian GW (Fig. S2). In the tributaries, median stream Cl$^-$ concentration was 10.2 [8.8, 14.2] mg L^{-1}. For DOC, median concentration in the main stream was higher for the LLF (843 [643, 1243] µg C L^{-1}) than for the non-LLF period (406 [304, 580] µg C L^{-1}) (Mann–Whitney test, $Z = 2.55$, d$f = 1$, $p = 0.008$) (Fig. 3a). Stream DOC concentrations increased along the reach by 58 % during the LLF period (Fig. 2b). In the tributaries, median DOC concentration was 577 [390, 881] µg C L^{-1}. For DON, median concentration in the main stream was 58 [35, 78] μg N L^{-1} and showed no seasonal pattern (Mann–Whitney test, $Z = -0.85$, d$f = 1$, $p > 0.05$) (Fig. 3b). Stream DON concentrations showed no clear longitudinal changes for any of the two study periods (Fig. 2c), though concentrations could vary by 40 % on a single date. No clear longitudinal pattern was found for either DOC or DON in riparian GW (Fig. S2). In the tributaries, median DON concentration was 54 [34, 75] μg N L^{-1}. The median DOC : DON ratio in the main stream was higher during the LLF (DOC : DON = 22 [14, 43]) than during the non-LLF period (DOC : DON = 8 [5, 15]) (Mann–Whitney test, $Z = 1.98$, d$f = 1$, $p = 0.033$) (Fig. 3c).

Median values of FI (> 2) were typical of microbial DOM sources, while low values of HIX (< 2) indicated that the humification of the samples was low (Fig. 3). Regarding the PARAFAC model, the components F1 and F4 (associated with protein-like materials) were responsible for most of the total fluorescence of stream water samples (50 [46, 53] % and 25 [24, 28] %, respectively). The components F2 and F3 (associated with humic-like materials) accounted for 13 [11, 15] % and 11 [9, 13] % of the total fluorescence, respectively (Fig. 4).

There were differences in stream DOM quality between the LLF and non-LLF period, though most of the spectroscopic metrics (BIX, HIX, F1, F2, and F4) were similar between the two periods (in the five cases, Mann–Whitney test, $p > 0.05$). In contrast, values of FI and the humic-like component F3 were higher during the LLF than during the non-LLF period (in the two cases, Mann–Whitney test, $Z < 2.24$, d$f = 1$, $p < 0.05$). The relative contribution of F3 to the total fluorescence was higher during the LLF than during the non-LLF period (Mann–Whitney test, $Z = 3.43$, d$f = 1$, $p < 0.0006$), while the protein-like component F4 showed the opposite pattern (Mann–Whitney test, $Z = -2.23$, d$f = 1$, $p < 0.025$).

4.2 Temporal pattern of chloride and DOM in riparian GW

During the study period, median Cl^{-} concentrations in riparian GW was higher for the LLF (9.8 [7.8, 13.7] mg L^{-1}) than for the non-LLF period (8.7 [7.4, 10.6] mg L^{-1}). DOC in riparian GW showed a similar pattern, with median concentration higher for the LLF (1411 [1133, 2311] μg C L^{-1}) than for the non-LLF period (864 [626, 1414] μg C L^{-1}) (Mann–Whitney test, $Z = 5.49$, d$f = 1$, $p < 0.001$). In contrast, median DON concentrations in riparian GW were lower during the LLF (67 [45, 157] μg N L^{-1}) than during the non-LLF (113 [64, 195] μg N L^{-1}) (Mann–Whitney test, $Z = -1.96$, d$f = 1$, $p = 0.049$). Riparian GW showed higher DOC : DON ratios during the LLF (DOC : DON = 27 [14, 43]) than during the non-LLF period (DOC : DON = 10 [6, 14]) (Mann–Whitney test, $Z = 4.98$, d$f = 1$, $p < 0.001$).

Similar to stream samples, the PARAFAC components related to the protein-like fluorescence (F1 and F4) were re-

sponsible for the major part of the total fluorescence of riparian GW samples (44 [38, 49] % and 26 [23, 29] %, respectively). The fluorescence components associated with humic-like materials, F2 and F3, accounted for 16 [13, 21] % and 12 [9, 17] %, respectively.

Values of FI, BIX, and HIX in riparian GW showed no differences between the LLF and non-LLF period, with medians equaling to 2.49 [2.41, 2.61], 0.67 [0.61, 0.74], and 1.11 [0.85, 1.68], respectively (for the three indexes: Mann–Whitney test, d$f = 1$, $p > 0.05$). Regarding PARAFAC, three out of the four fluorescence components (F1, F3, and F4) showed higher values in riparian GW during the LLF than during the non-LLF period (for the three components: Mann–Whitney test, d$f = 1$, $p < 0.015$). However, the relative contribution of the four components to the total fluorescence did not change between the two periods (for the four components: Mann–Whitney test, d$f = 1$, $p > 0.05$).

4.3 Influence of riparian GW on stream DOM

The paired test comparing stream water and riparian GW samples collected simultaneously along the study reach showed that Cl^{-} concentrations were similar between riparian GW and stream water during the LLF period, but higher in the former than in the latter during the non-LLF period (Table 1). DOC and DON concentrations were higher in riparian GW than in stream water during both the LLF and the non-LLF period (Table 1). However, there were no differences in DOC : DON ratios between riparian GW and stream water in any of the two periods. During the LLF period, concentrations of DOC and DON were uncorrelated to each other, while stream water and riparian GW showed a positive relationship between DOC and DON concentrations during the non-LLF period (Fig. 5).

Spectroscopic descriptors also show differences between the two water bodies, yet those differences were not consistent between the two study periods. During the LLF period, the FI was higher in stream water than in riparian GW, while the opposite trend was observed for indexes associated with both humic-like substances (HIX and F2) and in situ-produced, protein-like compounds (BIX and F4) (Table 1). During the non-LLF period, HIX, F2, F3, and F4 were lower in stream water than in riparian GW, while no differences between the two water bodies were observed for FI, BIX, and F1 (Table 1).

The longitudinal trends in DOM quality differed between stream water and riparian GW. Values of FI in stream water increased along the reach in 8 out of 10 sampling dates, while values of HIX did so in 4 out of 10 cases ($r > 0$ in Fig. 6). Longitudinal trends in stream DOM spectroscopic properties were observed during both the LLF and non-LLF period. In contrast, riparian GW showed no significant longitudinal patterns for either FI, BIX, or HIX in any of the sampling dates. Regarding PARAFAC components, both stream water and riparian GW showed significant changes along the

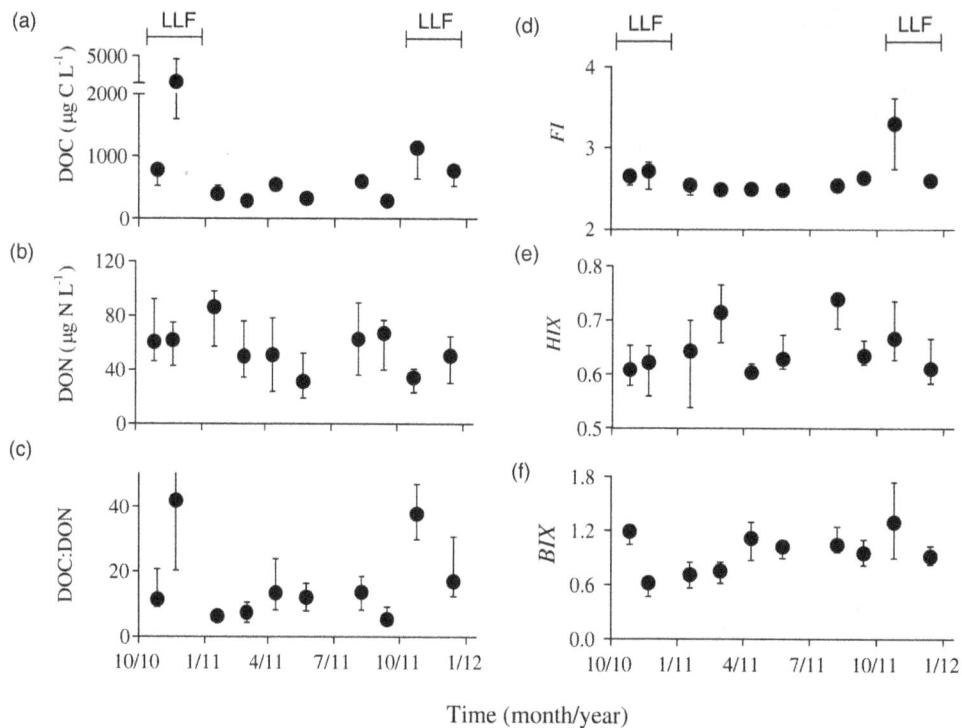

Figure 3. Temporal pattern of (a) dissolved organic carbon (DOC), (b) dissolved organic nitrogen (DON), (c) DOC : DON molar ratio, (d) fluorescence index (FI), (e) humification index (HIX), and (f) biological index (BIX) in stream water. FI, HIX, and BIX were calculated from fluorescence spectroscopy. Symbols are medians and whiskers are 25th and 75th percentiles for samples collected along the main stream. The leaf litter fall (LLF) period is indicated.

Table 1. Characterization of chloride (Cl⁻) and dissolved organic matter (DOM) (both concentrations and quality) in the main stream and in riparian groundwater (riparian GW) for the leaf litter fall (LLF) period and the non leaf litter fall (non-LLF) period at Font del Regàs. Values are medians and interquartile ranges [25th, 75th percentiles] for dissolved organic carbon (DOC) and dissolved organic nitrogen (DON) concentrations, DOC : DON molar ratio, fluorescence index (FI), humification index (HIX), biological index (BIX), and the four PARAFAC components (F1, F2, F3, and F4). The number of cases is shown in parenthesis.

	LLF			non-LLF		
	Stream	Riparian GW	p value*	Stream	Riparian GW	p value
Cl⁻ (mg L⁻¹)	8.6 [7.8, 13.1] (59)	9.8 [7.8, 13.7] (58)	0.2	7.8 [7.3, 8.8] (101)	8.7 [7.4, 10.6] (96)	0.0174
DOC (µgC L⁻¹)	843 [643, 1243] (59)	1411 [1133, 2311] (56)	<0.0001	406 [304, 580] (102)	864 [626, 1414] (93)	<0.0001
DON (µgN L⁻¹)	48 [34, 67] (47)	67 [45, 157] (38)	0.012	63 [36, 87] (97)	113 [64, 195] (82)	<0.0001
DOC : DON	22 [14, 43] (47)	27 [14, 43] (38)	0.8	8 [5, 15] (93)	10 [6, 14] (82)	0.3
Chromophoric indexes						
FI	2.79 [2.56, 2.83] (55)	2.59 [2.44, 2.62] (54)	0.0001	2.54 [2.47, 2.59] (84)	2.53 [2.41, 2.60] (79)	0.211
BIX	0.60 [0.60, 0.67] (55)	0.70 [0.63, 0.75] (54)	0.0072	0.67 [0.61, 0.71] (84)	0.67 [0.60, 0.73] (79)	0.646
HIX	1.03 [0.66, 1.24] (55)	1.51 [0.84, 1.82] (54)	0.0066	0.94 [0.75, 1.09] (84)	1.36 [0.86, 1.63] (79)	<0.0001
PARAFAC components						
F1	1.78 [1.19, 1.87] (55)	1.70 [1.14, 1.90] (54)	0.831	1.24 [0.99, 1.41] (84)	1.31 [1.03, 1.54] (79)	0.373
F2	0.45 [0.32, 0.50] (55)	0.80 [0.40, 0.94] (54)	<0.0001	0.31 [0.27, 0.36] (84)	0.58 [0.36, 0.67] (79)	<0.0001
F3	0.44 [0.28, 0.61] (55)	0.68 [0.28, 0.79] (54)	0.115	0.25 [0.20, 0.29] (84)	0.42 [0.24, 0.47] (79)	<0.0001
F4	0.89 [0.64, 1.02] (55)	1.02 [0.66, 1.16] (54)	0.021	0.65 [0.51, 0.77] (84)	0.83 [0.61, 0.93] (79)	<0.0001

* The p value of the Wilcoxon paired rank sum test is shown in each case.

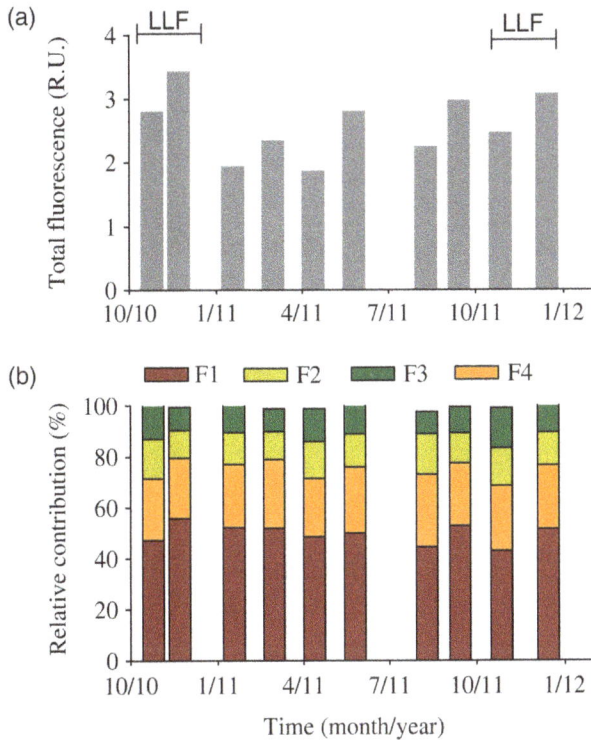

Figure 4. Temporal pattern of (a) total fluorescence of the four PARAFAC components and (b) their relative contribution to total fluorescence in the main stream of the Font del Regàs stream. The fluorescence components F1 and F4 corresponded to protein-like materials, while F2 and F3 corresponded to humic-like materials. Bars are median values for each sampling date. The leaf litter fall (LLF) period is indicated. R.U. are raman units. See more details on the obtained PARAFAC model in Tables S1 and S2 and Fig. S1 (Supplement).

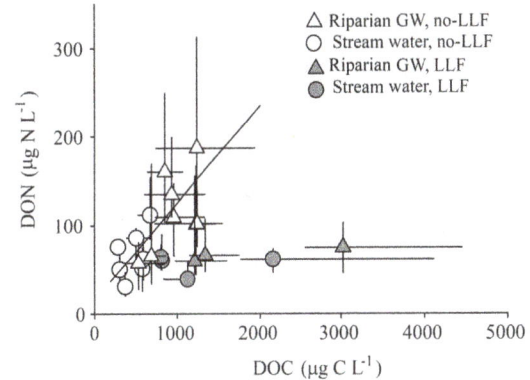

Figure 5. Relationship between dissolved organic carbon (DOC) and dissolved organic nitrogen (DON) concentrations in stream water and riparian groundwater (GW). Symbols are median values and whiskers are 25th and 75th percentiles for each sampling date. The black line shows the DOC vs. DON linear relationship for stream water and riparian GW samples pooled together for the non-LLF period (ANOVA, $F = 16.6$, $df = 13$, $p = 0.0015$). The relationship was not significant for the LLF period.

Table 2. Median and interquartile range [25th, 75th] of the relative contribution of inputs from upstream ($Q_{top} \times C_{top}/F_{in}$), tributaries ($Q_{tr} \times C_{tr}/F_{in}$), net riparian groundwater ($[Q_{gw} \times C_{gw} > 0]/F_{in}$), and in-stream release ($[U \times A < 0]/F_{in}$) to stream solute fluxes at the whole-reach scale. Note that relative contributions from different sources do not add to 100 % because they are medians rather than means.

Relative contribution (%)	Cl$^-$	DOC	DON
Upstream	15 [12, 17]	9 [8, 13]	52 [40, 60]
Riparian groundwater	28 [14, 38]	58 [41, 65]	30 [15, 43]
Tributaries	59 [46, 69]	30 [17, 36]	10 [8, 30]
In-stream release	0 [0, 0.3]	0 [0, 5]	0 [0, 4]

reach in some particular sampling dates. The most consistent pattern was the longitudinal increase in humic-like components (F2 + F3), which was observed in 4 out of 10 sampling dates (Fig. S3).

4.4 Contribution of catchment water sources and in-stream processes to stream DOM fluxes

Riparian GW was the most important source of DOC along the reach (58 % of the total inputs), while upstream sources provided most of the DON to the stream (30 % of the total inputs) (Table 2). The contribution of tributaries to stream DOM fluxes was relatively small compared to stream Cl$^-$ fluxes (Table 2).

Values of $U > 0$ were measured for both DOC and DON, indicating that in-stream processes influenced stream DOM fluxes at Font del Regàs. During the study period, median values of U_{DOC} were 197.7 [58.3, 315] $\mu g\,C\,m^{-2}\,h^{-1}$, whereas values of U_{DON} were 22.3 [4.6, 44.3] $\mu g\,N\,m^{-2}\,h^{-1}$. Differences in the contribution of in-stream processes to stream DOM fluxes between the LLF and the non-LLF pe-

riod were not statistically significant (for both U_{DOC} and U_{DON}, $Z > Z_{0.05}$, $df = 1$, $p > 0.05$). At reach scale, U contributed to modify stream fluxes ($|U \times A|/F_{in}$) by 32 [19, 46] % for DOC and 40.5 [29, 52] % for DON. These values were 10 times higher than for Cl$^-$ (the conservative tracer), for which U_{Cl} represented 3.6 [1.9, 9.4] % of the input fluxes (Fig. 7a). The stream acted as a net sink of DOM ($U > 0$) in 6 and 7 out of 10 sampling dates for DOC and DON, respectively. In these cases, in-stream processes contributed to reducing stream fluxes by 47 [43, 65] % and 37 [28, 40] % for DOC and DON, respectively (Fig. 7b and c, bars).

There was no significant relationship between U for the different compounds considered in this study. No correlation was found between U_{Cl} and either U_{DOC} or U_{DON} (in both cases: $\rho < 0.3$, $p > 0.05$), indicating that both DOC and DON behaved differently than expected from a conservative tracer. Moreover, U_{DOC} and U_{DON} were unrelated to each other (Fig. 8a).

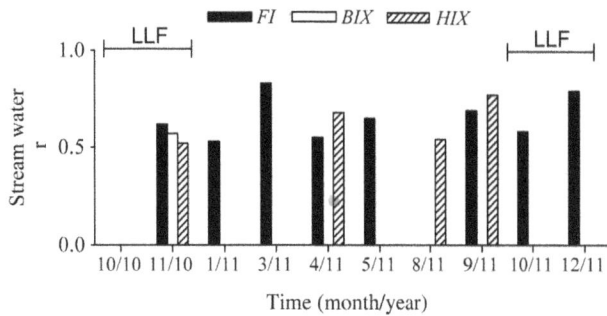

Figure 6. Temporal pattern of the standardized regression coefficient (r) obtained by fitting linear regression models to values of spectroscopic indexes measured along the 4 km study reach. The r is shown for the fluorescence index (FI), biological index (BIX), and humification index (HIX) in stream water. For each sampling date, $r > 0$ indicates that values for a particular spectroscopic index increased significantly in stream water along the study reach. Bars are shown only when the model was significant ($p < 0.05$). The leaf litter fall (LLF) period is indicated. Note that none of the three spectroscopic indexes showed significant longitudinal patterns for riparian groundwater in any of the sampling dates.

5 Discussion

The capacity of streams to mineralize allochthonous DOM, and thus their ability to contribute to the net balance between C storage and emission at global scales, remains elusive, and available results are contradictory. Most of the uncertainties associated with the estimation of biogeochemical processing rates at large scales (reaches > 100 m) rely on the fact that GW inputs are rarely measured (Tiwari et al., 2014; Casas-Ruíz et al., 2017). Our synoptic approach is unique in the sense that it explicitly considers GW inputs, allowing for more reliable C and N budget calculations (Bernal et al., 2015). However, the characterization of the exact DOM chemistry entering from the riparian GW to the stream is a complex issue (e.g., Brookshire et al., 2009). First, the two water bodies (stream and riparian GW) are hydrologically connected throughout the hyporheic zone (Bencala et al., 2011). Thus, hydrological mixing cannot be completely ruled out because stream water can eventually penetrate towards the riparian zone (Bernal et al., 2015). Second, DOM in riparian GW is likely processed while traversing the near-stream and hyporheic zones (Fasching et al., 2015). Hence, by sampling only riparian GW (2 m from the stream channel) and free-flowing water at the thalweg, we could not distinguish whether in-stream processes occurred in the stream water column, the streambed, or the hyporheic zone. Another keen aspect of our study is that we characterized the spectroscopic properties of DOM in both stream water and riparian GW to investigate whether stream DOM reflected allochthonous sources or if in-stream processes modified DOM quality.

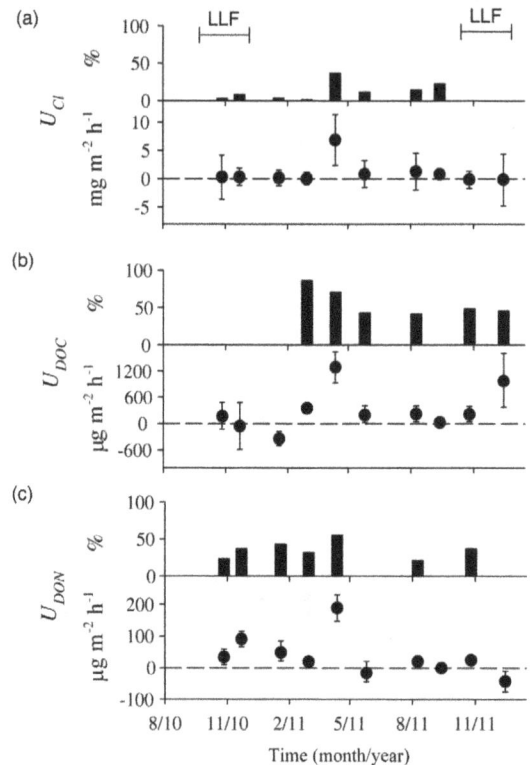

Figure 7. Temporal pattern of in-stream net uptake (U, either in µg or mg m^{-2} h^{-1}) for **(a)** chloride, **(b)** dissolved organic carbon (DOC), and **(c)** dissolved organic nitrogen (DON) at the whole reach scale. Whiskers are the uncertainty associated with the estimation of stream discharge from NaCl slug additions as in Bernal et al. (2015). Values of $U > 0$ indicate that gross uptake prevails over release, while $U < 0$ indicates the opposite. For cases with $U > 0$, the contribution of in-stream net uptake to decrease stream solute fluxes (i.e., $U \times A/F_{in}$, in %) is shown (black bars). The leaf litter fall (LLF) period is indicated.

Our study highlights that DOM in the Font del Regàs stream and riparian GW had an eminently protein-like character, most likely originated from microbial sources and recent biological activity. For instance, the fluorescence of the samples was dominated by F1 and F4 (up to 75 % of the total fluorescence), two PARAFAC components that presented wavelengths typically attributed to tyrosine and tryptophan (Fellman et al., 2010) (Table S1). Moreover, the whole range of BIX values measured in water samples (from 0.4 to 1.63) depicted a strong influence of autochthonous DOM sources (Huguet et al., 2009), while all measured HIX values were < 6, indicating low humification of the samples (Fellman et al., 2010). These values contrast with those reported for stream water samples from boreal and temperate catchments with large peatlands and wetland areas, which usually have high DOC concentrations (> 10 mg C L^{-1}) and highly colored humic materials (e.g., Kothawala et al., 2015). However, similar values of both BIX and HIX to the ones presented here have been reported previously in systems with

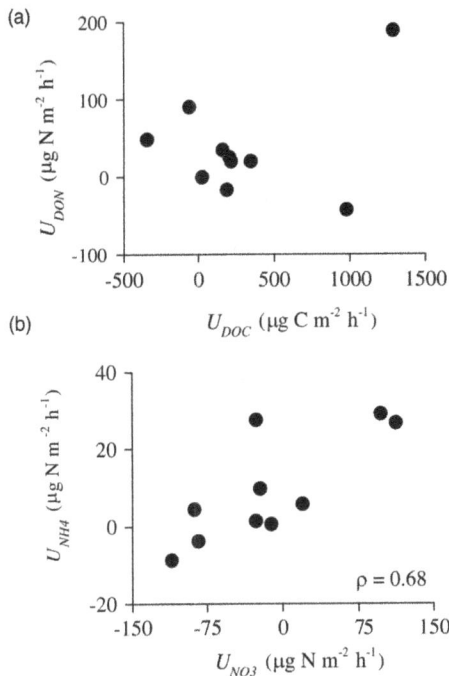

Figure 8. Relationship between in-stream net uptake along the study reach for **(a)** U_{DOC} and U_{DON}, and **(b)** U_{NO_3} and U_{NH_4}. The Spearman coefficient (ρ) is shown only when significant ($p < 0.05$).

low DOC concentrations and not very colored DOM, such as ground caves and spring waters (Birdwell and Engel, 2010; Simon et al., 2010) as well as in soils (Traversa et al., 2014) and some rivers (Huang et al., 2015).

5.1 Empirical evidence of in-stream DOM processing

We found that stream DOM did not exhibit a conservative behavior because the stream showed a large capacity to change DOM fluxes (by 30–40 %) compared to Cl^- fluxes (by 3 %). The predominant protein-like character of stream DOM at Font del Regàs could explain, at least partially, why U_{DOC} and U_{DON} differed from zero during most of the study period. This result indicates that in-stream DOM uptake and release processes were not counterbalancing each other (otherwise U would approach zero). For both DOC and DON, we found that in-stream uptake usually predominated over release (i.e., $U > 0$), suggesting higher DOM consumption than production. Our mass balance calculations indicated that in-stream processes could decrease reach scale fluxes up to 80 and 50 % for DOC and DON, respectively. These findings imply that biogeochemical processes occurring within the stream were able to modify DOC and DON concentrations and fluxes to downstream ecosystems, contrasting with results reported in previous studies (Temnerud et al., 2007; Tiwari et al., 2014; Wollheim et al., 2015). Yet our results are representative of base flow conditions, which represent ca. 60 % of the annual DOC and DON flux in the study catchment (unpublished

data). Moreover, mean water residence time along the reach was relatively low (4 h, unpublished data) because running waters predominated and there were no natural or artificial dams. Further studies including storm flow conditions and/or reaches with small reservoirs would be needed to gain a more complete picture of the role of in-stream processes on DOM dynamics and whether headwater streams shifts from reactors to pipes with changing hydrological conditions (Casas-Ruíz et al., 2017; Raymond et al., 2016).

Noteworthy, median values of in-stream net uptake ($U_{DOC} = 198 \mu g \, C \, m^{-2} \, h^{-1}$ and $U_{DON} = 22.3 \mu g \, N \, m^{-2} \, h^{-1}$) were 10–1000 times lower than rates of in-stream gross uptake and DOM production reported for DOM addition experiments in other headwater streams (Lush and Hynes, 1978; McDowell, 1985; Maranger et al., 2005; Bernhardt and McDowell, 2008; Johnson et al., 2013). These discrepancies could be partially explained by the fact that some of these manipulative experiments used monomeric carbohydrates that are easily bioavailable (Mineau et al., 2016). Moreover, and as previously reported for nutrients, differences between estimates of in-stream gross and net uptake suggest that DOM consumption and production likely occur simultaneously within the stream, and that the former is counterbalanced to some extent by the latter (von Schiller et al., 2015). Supporting this idea, median values of U_{DOC} were > 100 times lower than DOC consumption inferred from measurements of ecosystem respiration calculated from diel cycles of dissolved oxygen concentrations in the same study stream (Lupon et al., 2016b).

The observed differences in the spectroscopic properties of DOM between the stream and riparian GW further support the existence of an autochthonous source of labile DOM in the Font del Regàs stream. For instance, riparian GW presented higher humic-like fluorescence (i.e., higher values of HIX, F2, and F3) than stream water, which is in agreement with a recent study comparing stream and groundwater DOM (Huang et al., 2015). Moreover, the contribution of the protein-like component F1 to the total fluorescence was higher in stream water (50.6 %) than in riparian GW (43.9 %), while the contribution of F2, a ubiquitous humic component related with fulvic acids and re-processed humics, was higher in riparian GW (17.8 %) than in stream water samples (13.1 %). Finally, the lack of longitudinal trends in DOM quality in riparian GW contrasted with the consistent increase in FI observed for stream water along the reach (in 8 out of 10 sampling dates). This finding suggests that stream DOM shifted towards a more microbial origin as it moved downstream, and that this change was more related to in-stream processes than to changes in the spectroscopic character of riparian GW. Altogether, our results highlight that in-stream processes have the potential to change not only the quantity, but also the quality of DOM, which reinforces their potential role as bioreactors rather than C chimneys transforming dissolved inorganic carbon from terrestrial groundwater to CO_2 (Hotchkiss et al., 2015).

5.2   Decoupling between in-stream DOC and DON dynamics

We found that the contribution of tributaries to stream DOM fluxes was relatively small (from 10 to 30 %) compared to stream Cl^- fluxes (> 50 %), suggesting that other sources of DOM predominated within the catchment were more important than tributaries. However, dominant catchment sources differed between DOC and DON: riparian GW was the major contributor of DOC, while most of the DON inputs came from upstream. These differences could be partially explained by changes in vegetation: the upstream sites had no riparian zone and drained beech forests exhibiting low mineralization and nitrification rates, while most of the mid- and downstream sites were flanked by a well-developed riparian forest that holds higher soil N processing rates (Lupon et al., 2016c).

Despite variances in DOM sources, differences in DOC : DON ratios between stream water and riparian GW were small throughout the year. Moreover, water samples showed a positive and moderate relationship between DOC and DON concentrations, especially during the non-LLF period. Similar DOM stoichiometry between terrestrial and aquatic ecosystems has been typically understood as an indication of the recalcitrant and allochthonous nature of organic matter in stream waters (Perakis and Hedin, 2002; Rastetter et al., 2005). Therefore, these results could suggest that allochthonous DOM inputs mostly dominated DOM in stream water. Yet the spectroscopic analysis clearly indicated that the quality of DOM differed between these two compartments, and that stream DOM was likely highly available to biota given the high content of protein-like material, which was higher than in riparian GW entering the stream.

In concordance with the idea that stream DOM was not recalcitrant, we found (i) that U differed from zero for both DOC and DON, and (ii) that U_{DOC} and U_{DON} were unrelated to each other. This finding supports the hypothesis that these two compounds undergo different metabolic and biogeochemical pathways (Kaushal and Lewis, 2005; Lutz et al., 2011): DOC is mostly used as an energy source, while evidence is growing that DON can also be used as a nutrient (Wymore et al., 2015). The dual behavior of DON could partially explain why U_{DON} was unrelated to U_{DOC}, which contrasts with the strong relationship exhibited by instream net uptake rates for the two inorganic forms of N, U_{NO_3} and U_{NH_4}, which are both essential nutrients for biota (Fig. 8b). For DOC, a major fraction of what is taken up (∼ 70 %) follows catabolic pathways (respiration) and is removed to the atmosphere, while the remaining part (∼ 30 %) may be used for microbial growth (del Giorgio and Cole, 1998). Thus, considering that in-stream DOM uptake contributed to reducing allochthonous DOC fluxes by 36–54 %

(25th and 75th percentiles), approximately one-quarter (21–32 %) of the DOC entering or produced within the stream could be released as CO_2 to the atmosphere.

5.3   Influence of leaf litter fall on stream DOM dynamics and spectroscopic properties

Previous studies have reported large increases in stream DOC concentration and ecosystem respiration associated with large inputs of fresh leaf litter in autumn (e.g., Acuña et al., 2004). Thus, we expected large differences in stream DOM concentrations and quality between the LLF and non-LLF period, as well as between riparian GW and stream DOM during the LLF period. Concordantly, the highest stream DOC concentrations and DOC : DON ratios were measured during the LLF period (specially in November 2010). Yet the same pattern was observed for riparian GW, where concentrations of DOC during the LLF period were even higher than in the stream. In this case, higher DOC concentrations could be explained by increases in the groundwater table after autumn rains, which then flow through more superficial organic soil layers (Guarch-Ribot and Butturini, 2016). This idea is supported by the fact that riparian GW showed higher fluorescence during the LLF than the non-LLF period, but no changes in the relative contribution of the four fluorescence components to the total fluorescence. In contrast, the relative contribution of F3 (humic-like component) and F4 (protein-like component) increased and decreased, respectively, in stream water during the LLF period. This result, together with the higher values of FI, bears the idea that leaf litter inputs were a source of humic-like material, but that, at the same time, were fueling microbial activity within the stream. The fact that DOM uptake predominated over release (U_{DOC} and $U_{DON} > 0$) even during some sampling dates within the LLF period supports the hypothesis that fresh particulate organic matter was processed en route and that stream biota was consuming DOM (Battin et al., 2008; Fasching et al., 2014).

6   Conclusions and future direction

Global studies highlight that streams and rivers are important sources of C to the atmosphere (Cole et al., 2007; Raymond et al., 2013). Yet the potential role of streams to mineralize allochthonous DOC and its consequences at the catchment scale is still largely unknown (Hotchkiss et al., 2015). Our study sheds new light on this issue by showing that headwater streams have a strong capacity to internally produce, transform, and consume DOM. The mass balance calculations revealed that in-stream processing substantially modified stream DOC and DON fluxes during base flow conditions. Moreover, we found that DOM concentration and spectroscopic character differed between stream water and riparian GW, which provides evidence that stream DOM is not merely a reflection of riparian DOM entering the stream. On the contrary, our findings suggest that both riparian leaf litter inputs and in-stream DOM cycling are essential controls of DOM dynamics in forested headwater streams. Further work is needed for disentangling the different mechanism underlying DOC and DON processing within the streams as well as for understanding how environmental factors such as nutrient availability and water residence time drive in-stream DOM processing and changes in DOM quality during different hydrological conditions.

Author contributions. SB designed the experiment. SB and AL carried it out. SC, AL, and NC performed all laboratory analysis. SB, AL, and NC analyzed the data set. SB prepared the paper with contributions from AL, NC, SC, and EM.

Competing interests. The authors declare that they have no conflict of interest.

Acknowledgements. We thank Aitana Oltra for assisting with GIS; Francesc Sabater, Sílvia Poblador, Miquel Ribot, Eduardo Martín, and Clara Romero for field assistance; and Montserrat Solé and Alba Guarch for helping with spectrofluorometric analysis. Susana Bernal and Anna Lupon were funded by the Spanish Ministry of Economy and Competitiveness (MINECO) with a Juan de la Cierva contract (JCI-2010-06397) and a FPU grant (AP-2009-3711). Núria Catalán held a Juan de la Cierva (FJCI-2014-23064) and a Beatriu de Pinós (BP-2016-00215) grants. Susana Bernal received additional funds from the Spanish Research Council (CSIC) (JAEDOC027) and the MINECO-funded projects MED_SOUL (CGL2014-59977-C3-2-R) and NICUS (CGL-2014-55234-JIN). The Vichy Catalan Company, the Regàs family and the Catalan Water Agency (ACA) graciously gave us permission for sampling at the Font del Regàs catchment.

Edited by: Matthew Hipsey

References

Acuña, V., Giorgi, A., Munoz, I., Uehlinger, U., and Sabater, S.: Flow extremes and benthic organic matter shape the metabolism of a headwater Mediterranean stream, Freshwater Biol., 49, 960–971, 2004.

Battin, T. J., Kaplan, L. A., Findlay, S., Hopkinson, C. S., Martí, E., Packman, A. I., Newbold, J. D, and Sabater, F.: Biophysical controls on organic carbon fluxes in fluvial networks, Nat. Geosci., 1, 96–100, 2008.

Bencala, K. E., Gooseff, M. N., and Kimball, B. A.: Rethinking hyporheic flow and transient storage to advance understanding of stream-catchment connections, Water Resour. Res., 47, W00H03, https://doi.org/10.1029/2010WR010066, 2011.

Bernal, S., Butturini, A., and Sabater, F.: Seasonal variations of dissolved nitrogen and DOC : DON ratios in an intermittent Mediterranean stream, Biogeochemistry, 75, 351–372, 2005.

Bernal, S., Lupon, A., Ribot, M., Sabater, F., and Martí, E.: Riparian and in-stream controls on nutrient concentrations and fluxes in a headwater forested stream, Biogeosciences, 12, 1941–1954, https://doi.org/10.5194/bg-12-1941-2015, 2015.

Bernhardt, E. S. and McDowell, W. H.: Twenty years apart: comparisons of DOM uptake during leaf leachate releases to Hubbard Brook Valley streams in 1979 versus 2000, J. Geophys. Res., 13, G03032, https://doi.org/10.1029/2007JG000618, 2008.

Birdwell, J. E. and Engel, A. S.: Characterization of dissolved organic matter in cave and spring waters using UV–Vis absorbance and fluorescence spectroscopy, Org. Geochem., 41, 270–280, https://doi.org/10.1016/j.orggeochem.2009.11.002, 2010.

Brookshire, E. N. J., Valett, H. M., Thomas, S. A., and Webster, J. R.: Coupled cycling of dissolved organic nitrogen and carbon in a forest stream, Ecology, 86, 2487–2496, https://doi.org/10.1890/04-1184, 2005.

Brookshire, E. N. J., Valett, H. M., Thomas, S. A., and Webster, J. R.: Atmospheric N deposition increases organic N loss from temperate forests, Ecosystems, 10, 252–262, https://doi.org/10.1007/s10021-007-9019-x, 2007.

Brookshire, J. E., Valett, H. M., and Gerber, S.: Maintenance of terrestrial nutrient loss signatures during in-stream transport, Ecology, 90, 293–299, 2009.

Caraco, N. F. and Cole, J. J.: The importance of organic nitrogen production in aquatic systems: a landscape perspective, in: Aquatic Ecosystems: Interactivity of dissolved organic matter, edited by: Findlay, S. E. G. and Sinsabaugh, R. L., Elsevier Science, San Diego, CA, 263–283, 2003.

Casas-Ruíz, J. P., Catalán, N., Gómez-Gener, L., von Schiller, D., Obrador, B., Kothawala, D. N., López, P., Sabater, S., and Marcé, R.: A tale of pipes and reactors: Controls on the in-stream dynamics of dissolved organic matter in rivers, Limnol. Oceanogr., 62, S85–S94, https://doi.org/10.1002/lno.10471, 2017.

Cole, J. J., Prairie, Y. T., Caraco, N. F., McDowell, W. H., Tranvik, L. J., Striegl, R. G., Duarte, C. M., Kortelainen, P., Downing, J. A., Middelburg, J. J., and Melack, J.: Plumbing the global carbon cycle: integrating inland waters into the terrestrial carbon budget, Ecosystems, 10, 171–184, https://doi.org/10.1007/s10021-006-9013-8, 2007.

Covino, T., McGlynn, B., and Baker, M.: Separating physical and biological nutrient retention and quantifying uptake kinetics from ambient to saturation in successive mountain stream reaches, J. Geophys. Res., 115, G04010, https://doi.org/10.1029/2009JG001263, 2010.

del Giorgio, P. A. and Cole, J. J.: Bacterial growth efficiency in natural aquatic systems, Annu. Rev. Ecol. Evol. S., 29, 503–541, 1998.

Fasching, C., Behounek, B., Singer, G. A., and Battin, T. J.: Microbial degradation of terrigenous dissolved organic matter and potential consequences for carbon cycling in brown-water streams, Scientific Reports, 4, 4981, https://doi.org/10.1038/srep04981, 2014.

Fasching, C., Ulseth, A. J., Schelker, J., Steniczka, G., and Battin, T. J.: Hydrology controls dissolved organic matter export and composition in an Alpine stream and its hyporheic zone, Limnol. Oceanogr., 61, 558–571, https://doi.org/10.1002/lno.10232, 2015.

Fellman, J. B., Hood, E., and Spencer, R. G. M.: Fluorescence spectroscopy opens new windows into dissolved organic matter dynamics in freshwater ecosystems: A review, Limnol. Oceanogr., 55, 2452–2462, 2010.

Gordon, N. D., McMahon, T. A., Finlayson, B. L., Gippel, C. J., and Nathan, R. J.: Stream hydrology: an introduction for ecologists, Wiley, West Sussex, UK, 2004.

Guarch-Ribot, A. and Butturini, A.: Hydrological conditions regulate dissolved organic matter quality in an intermittent headwater stream. From drought to storm analysis, Sci. Total Environ., 571, 1358–1369, 2016.

Hedin, L. O., Armesto, J. J., and Johnson, A. H.: Patterns of nutrient loss from unpolluted, old-growth temperate forests: evaluation of biogeochemical theory, Ecology, 76, 493–509, 1995.

Hotchkiss, E. R., Hall Jr., R. O., Sponseller, R. A., Butman, D., Klaminder, J., Laudon, H., Rosvall, M., and Karlsson, J.: Sources of and processes controlling CO_2 emissions change with the size of streams and rivers, Nat. Geosci., 8, 696–699, https://doi.org/10.1038/NGEO2507, 2015.

Huang, S., Wang, Y., Ma, T., Tong, L., Wang, Y., Liu, C., and Zhao, L.: Linking groundwater dissolved organic matter to sedimentary organic matter from a fluvio-lacustrine aquifer at Jianghan Plain, China by EEM-PARAFAC and hydrochemical analyses, Sci. Total Environ., 529, 131–139, 2015.

Huguet, A., Vacher, L., Relexans, S., Saubusse, S., Froidefond, J. M., and Parlanti, E.: Properties of fluorescent dissolved organic matter in the Gironde Estuary, Org. Geochem., 40, 706–719, https://doi.org/10.1016/j.orggeochem.2009.03.002, 2009.

Jaffé, R., McKnight, D., Maie, N., Cory, R., McDowell, W. H., and Campbell, J. L.: Spatial and temporal variation in DOM composition in ecosystems: the importance of long-term monitoring of optical properties, J. Geophys. Res., 113, G04032, https://doi.org/10.1029/2008JG000683, 2008.

Johnson, L. T., Tank, J. L., Hall Jr., R. O., Mulholland, P. J., Hamilton, S. K., Valett, H. M., Webster, J. R., Bernot, M. J., McDowell, W. H., Peterson, B. J., and Thomas, S. M.: Quantifying the production of dissolved organic nitrogen in headwater streams using ^{15}N tracer additions, Limnol. Oceanogr., 58, 1271–1285, https://doi.org/10.4319/lo.2013.58.4.1271, 2013.

Kaushal S. S. and Lewis, W. M.: Fate and transport of organic nitrogen in minimally disturbed montane streams of Colorado, USA, Biogeochemistry, 74, 303–321, https://doi.org/10.1007/s10533-004-4723-5, 2005.

Kothawala, D. N., Murphy, K. R., Stedmon, C. A., Weyhenmeyer, G. A., and Tranvik, L. J.: Inner filter correction of dissolved organic matter fluorescence, Limnol. Oceanogr.-Meth. 11, 616–630, https://doi.org/10.4319/lom.2013.11.616, 2013.

Kothawala, D. N., Ji, X., Laudon, H., Ågren, A. M., Futter, M. N., Köhler, S. J., and Tranvik, L. J.: The relative influence of land cover, hydrology, and in-stream processing on the composition of dissolved organic matter in boreal streams, J. Geophys. Res.-Biogeo., 120, 1491–1505, https://doi.org/10.1002/2015JG002946, 2015.

Lupon, A., Bernal, S., Poblador, S., Martí, E., and Sabater, F.: The influence of riparian evapotranspiration on stream hydrology and nitrogen retention in a subhumid Mediterranean catchment, Hydrol. Earth Syst. Sci., 20, 3831–3842, https://doi.org/10.5194/hess-20-3831-2016, 2016a.

Lupon, A., Martí E., Sabater, F., and Bernal, S.: Green light:

gross primary production influences seasonal stream N export by controlling fine-scale N dynamics, Ecology, 97, 133–144, https://doi.org/10.1890/14-2296.1, 2016b.

Lupon, A., Sabater, F., Miñarro, A., and Bernal, S: Contribution of pulses of soil nitrogen mineralization and nitrification to soil nitrogen availability in three Mediterranean forests, Eur. J. Soil Sci., 67, 303–313, https://doi.org/10.1111/ejss.12344, 2016c.

Lush, D. L. and Hynes, H. B.: The uptake of dissolved organic matter by small spring stream, Hydrobiologia, 60, 271–275, 1978.

Lutz, B. D., Bernhardt, E. S., Roberts, B. J., and Mulholland, P. J.: Examining the coupling of carbon and nitrogen cycles in Appalachian streams: the role of dissolved organic nitrogen, Ecology, 92, 720–732, 2011.

Maranger, R. J., Pace, M. L., del Giorgio, P. A., Caraco, N. F., and Cole, J. J.: Longitudinal spatial patterns of bacterial production and respiration in a large river-estuary: implications for ecosystem carbon consumption, Ecosystems, 8, 318–330, https://doi.org/10.1007/s10021-003-0071-x, 2005.

McDowell, W. H.: Kinetics and mechanism of dissolved organic carbon retention in a headwater stream, Biogeochemistry, 1, 329–352, https://doi.org/10.1007/BF02187376, 1985.

Mineau, M. M., Wollheim, W. M., Buffam, I., Findlay, S. E. G., Hall Jr., R. O., Hotchkiss, E. R., Koenig, L. E., McDowell, W. H., and Parr, T. B.: Dissolved organic carbon uptake in streams: A review and assessment of reach-scale measurements, J. Geophys. Res.-Biogeo., 121, 2019–2029, https://doi.org/10.1002/2015JG003204, 2016.

Murphy, K. R., Butler, K. D., Spencer, R. G. M., Stedmon, C. A., Boehme, J. R., and Aiken, G. R.: Measurement of Dissolved Organic Matter Fluorescence in Aquatic Environments: An Interlaboratory Comparison, Environ. Sci. Technol., 44, 9405–9412, https://doi.org/10.1021/es102362t, 2010.

Murphy, K. R., Stedmon, C. A., Graeber, D., and Bro, R.: Fluorescence spectroscopy and multi-way techniques. PARAFAC, Anal. Meth., 5, 6557, https://doi.org/10.1039/c3ay41160e, 2013.

Murphy, K. R., Stedmon, C. A., Wenig, P., and Bro, R.: OpenFluor-an online spectral library of auto-fluorescence by organic compounds in the environment, Anal. Meth., 6, 658–661, https://doi.org/10.1039/C3AY41935E, 2014.

Newbold, J. D., Bott, T. L., Kaplan, L. A., Dow, C. L., Jackson, J. K., Aufdenkampe, A. T., Martin, L. A., van Horn, D. J., and Long, A. A.: Uptake of nutrients and organic C in streams in New York City drinking-water-supply watersheds, J. N. Am. Benthol. Soc., 25, 998–1017, 2006.

Ohno, T.: Fluorescence inner-filtering correction for determining the humification index of dissolved organic matter, Environ. Sci. Technol., 36, 742–746, https://doi.org/10.1021/es0155276, 2002.

Perakis, S. S. and Hedin, L. O.: Nitrogen loss from unpolluted South American forests mainly via dissolved organic compounds, Nature, 415, 416–419, 2002.

Qualls, R. G. and Haines, B. L.: Geochemistry of dissolved organic nutrients in water percolating through a forested ecosystem, Soil Sci. Soc. Am. J., 55, 1112–1123, 1991.

Qualls, R. G. and Haines, B. L.: Biodegradability of dissolved organic matter in forest throughfall, soil solution, and stream water, Soil Sci. Soc. Am. J., 56, 578–586, 1992.

Rastetter, E. B., Perakis, S. S., Shaver, G. R., and Agren, G. I.: Terrestrial C sequestration at elevated CO_2 and temperature. The role of dissolved organic nitrogen loss, Ecol. Appl., 15, 71–86, 2005.

Raymond, P. A., Hartmann, J., Lauerwald, R., Sobek, S., McDonald, C., Hoover, M., Butman, D., Strighl, R., Mayorga, E., Humborg, C., Kortelainen, P., Dürr, H., Meybeck, M., Ciais, P., and Guth, P.: Global carbon dioxide emissions from inland waters, Nature, 503, 355–359, https://doi.org/10.1038/nature12760, 2013.

Raymond, P. A., Saiers, J. E., and Sobczak, W. V.: Hydrological and biogeochemical controls on watershed dissolved organic matter transport: pulse-shunt concept, Ecology, 97, 5–16, https://doi.org/10.1890/14-1684.1, 2016.

Seitzinger, S. P. and Sanders, R. W.: Contribution of dissolved organic nitrogen from rivers to estuarine eutrophication, Mar. Ecol.-Prog. Ser., 159, 1–12, 1997.

Simon, K. S., Pipan, T., Ohno, T., and Culver, D. C.: Spatial and temporal patterns in abundance and character of dissolved organic matter in two karst aquifers, Fund. Appl. Limnol., 177, 81–92, 2010.

Sponseller, R. A., Temnerud, J., Bishop, K., and Laudon, H.: Patterns and drivers of riverine nitrogen (N) across alpine, subarctic and boreal Sweden, Biogeochemistry, 120, 105–120, 2014.

Stedmon, C. A., Markanger, S., and Bro, R.: Tracing dissolved organic matter in aquatic environments using a new approach to fluorescence spectroscopy, Mar. Chem., 82, 239–254, 2003.

Temnerud, J., Seibert, J., Jansson, M., and Bishop, K.: Spatial variation in discharge and concentrations of organic carbon in a catchment network of boreal streams in northern Sweden, J. Hydrol., 342, 72–87, https://doi.org/10.1016/j.jhydrol.2007.05.015, 2007.

Tiwari, T., Laudon, H., Beven, K., and Agren, A. M.: Downstream changes in DOC: inferring contributions in the face

of model uncertainties, Water Resour. Res., 50, 514–525, https://doi.org/10.1002/2013WR014275, 2014.

Tranvik, L. J., Downing, J. A., Cotner, J. B., Loiselle, S. A., Strigl, R. G., Ballatore, T. J., Dillon, P., Finlay, K., Fortino, K., Knoll, L. B., Kortelainen, P., Kutser, T., Larsen, S., Laurion, I., Leech, D. M., McCallister, S. L., McKnight, D. M., Melack, J. M., Overholt, E., Porter, J. A., Prairie, Y., Renwick, W. H., Roland, F., Sherman, B., Schindler, D. W., Sobek, S., Tremblay, A., Vanni, M. J., Verschoor, A. M., von Wachenfeldt, E., and Weyhenmeyer, G. A.: Lakes and reservoirs as regulators of carbon cycling and climate, Limnol. Oceanogr., 54, 2298–2314, https://doi.org/10.4319/lo.2009.54.6_part_2.2298, 2009.

Traversa, A., D'Orazio, V., Mezzapesa, G. N., Bonifacio, E., Farrg, K., Senesi, N., and Brunetti, G.: Chemical and spectroscopic characteristics of humic acids and disolved organic matter along two Alfisol profiles, Chemosphere, 111, 184–194, 2014.

von Schiller, D., Bernal, S., Sabater, F., and Martí, E.: A round-trip ticket: the importance of release processes for in-stream nutrient spiraling, Freshwater Sci., 34, 20–30, 2015.

Wollheim, W. M., Stewart, R. J., Aiken, G. R., Butler, K. D., Morse, N. B., and Salisbury, J.: Removal of terrestrial DOC in aquatic ecosystems of a temperate river network, Geophys. Res. Lett., 42, 6671–6679, https://doi.org/10.1002/2015GL064647, 2015.

Wymore, A. S., Rodríguez-Cardona, B., and McDowell, W. H.: Direct response of dissolved organic nitrogen to nitrate availability in headwater streams, Biogeochemistry, 126, 1–10, https://doi.org/10.1007/s10533-015-0153-9, 2015.

Zar, J. H.: Biostatistical analysis, 5th Edn., Prentice-Hall/Pearson, : Essex, England, 2010.

10

Assessing the simple dynamical systems approach in a Mediterranean context: application to the Ardèche catchment (France)

M. Adamovic[1,4]**, I. Braud**[1]**, F. Branger**[1]**, and J. W. Kirchner**[2,3]

[1]Irstea, UR HHLY, Hydrology-Hydraulics Research Unit, Lyon-Villeurbanne, France
[2]Department of Environmental Systems Science, Swiss Federal Institute of Technology, ETH Zurich, Zurich, Switzerland
[3]Swiss Federal Research Institute WSL, Birmensdorf, Switzerland
[4]CNRS, HydroSciences Laboratory, Place Eugene Bataillon, 34095 Montpellier, France

Correspondence to: M. Adamovic (marko.adamovic@hotmail.com)

Abstract. This study explores how catchment heterogeneity and variability can be summarized in simplified models, representing the dominant hydrological processes. It focuses on Mediterranean catchments, characterized by heterogeneous geology, pedology and land use, as well as steep topography and a rainfall regime in which summer droughts contrast with high-rainfall periods in autumn. The Ardèche catchment (Southeast France), typical of this environment, is chosen to explore the following questions: (1) can such a Mediterranean catchment be adequately characterized by a simple dynamical systems approach and what are the limits of the method under such conditions? (2) what information about dominant predictors of hydrological variability can be retrieved from this analysis in such catchments?

In this work we apply the data-driven approach of Kirchner (2009) to estimate discharge sensitivity functions that summarize the behaviour of four sub-catchments of the Ardèche, using low-vegetation periods (November–March) from 9 years of measurements (2000–2008) from operational networks. The relevance of the inferred sensitivity function is assessed through hydrograph simulations, and through estimating precipitation rates from discharge fluctuations. We find that the discharge sensitivity function is downward-curving in double-logarithmic space, thus allowing further simulation of discharge and non-divergence of the model, only during low-vegetation periods. The analysis is complemented by a Monte Carlo sensitivity analysis showing how the parameters summarizing the discharge sensitivity function impact the simulated hydrographs. The resulting dis- charge simulation results are good for granite catchments, which are likely to be characterized by shallow subsurface flow at the interface between soil and bedrock. The simple dynamical system hypothesis works especially well in wet conditions (peaks and recessions are well modelled). On the other hand, poor model performance is associated with summer and dry periods when evapotranspiration is high and low-flow discharge observations are inaccurate. In the Ardèche catchment, inferred precipitation rates agree well in timing and amount with observed gauging stations and SAFRAN climatic data reanalysis during the low-vegetation periods. The model should further be improved to include a more accurate representation of actual evapotranspiration, but provides a satisfying summary of the catchment functioning during wet and winter periods.

1   Introduction

Catchments show a high degree of heterogeneity and variability, both in space and time (McDonnell et al., 2007) raising questions about the degree of complexity that must be used to model their behaviour (Sivapalan, 2003a). Many hydrological models are based on the bottom-up or reductionist approach (Sivapalan, 2003b; Zehe et al., 2006), following the blueprint proposed by Freeze and Harlan (1969). Governing equations such as the Darcy or Richards' equation, which are inherent in many hydrological models, are suitable for point-scale processes (Bloschl and Sivapalan, 1995; Kirchner, 2006). Their use to describe processes at larger scales leads to the calibration of "effective parameters" which are sometimes difficult to link with measurable quantities (Sivapalan, 2003b), although recent methods combining the use of small-scale variability and regionalization techniques were shown to be efficient in preserving spatial patterns of variability (Samaniego et al., 2010). Such "effective" large-scale equations might not, however, describe hydrologic processes realistically, even if they can be calibrated to reproduce observed catchment behaviour (Kirchner, 2006). Klemeš (1983) was one of the first hydrologists proposing the use of alternative modelling concepts. He defines the top-down or downward approach as the "route that starts with trying to find a distinct conceptual node directly at the level of interest (or higher) and then looks for the steps that could have led to it from a lower level". To go in this direction, Sivapalan (2003b) and Kirchner (2006) promote a combination of data analysis and process conceptualization (the top-down approach). This allows understanding of the main drivers of the system functioning (the perceptual model; Beven, 2002) and inferring the system's "emergent properties" (Sivapalan, 2003b) or "functional traits" (McDonnell et al., 2007). Thus, models obtained through this approach are simple, with a limited number of parameters that can be estimated from the available data.

Kirchner (2009) represents a catchment with a simple bucket model in which system parameters are derived directly from measured streamflow fluctuations during recession periods. He based his analysis on storage–discharge relationships with one essential assumption: discharge depends only on the total water stored in the catchment. This approach allows the derivation of a first-order nonlinear differential equation for simulating rainfall–runoff behaviour. Until now, this approach has mostly been applied in small, humid catchments. Kirchner (2009) obtained good results for the Severn (8.70 km^2) and Wye (10.55 km^2) catchments at Plynlimon, in Mid-Wales. Teuling et al. (2010) also applied this approach to the Prealpine Rietholzbach catchment (3.31 km^2) getting good results in wet periods and poor model performance during dry periods. The study of Brauer et al. (2013) showed similar results for the Dutch lowland Hupsel Brook catchment (6.5 km^2) where discharge results were correctly reproduced only in certain periods. Melsen et al. (2014) determined the minimum amount of data required to find robust parameter values for a simple Kirchner (2009) model with two parameters in the Prealpine Rietholzbach catchment (3.31 km^2). They concluded that a two-parameter model is reasonably able to capture high flows but fails to describe the low flows.

Krier et al. (2012) applied the concept of doing hydrology backwards to infer spatially distributed rainfall rates in the Alzette catchment (1092 km^2) in Luxembourg, and found that introducing a soil moisture threshold led to model improvement, especially under the wet conditions. However,

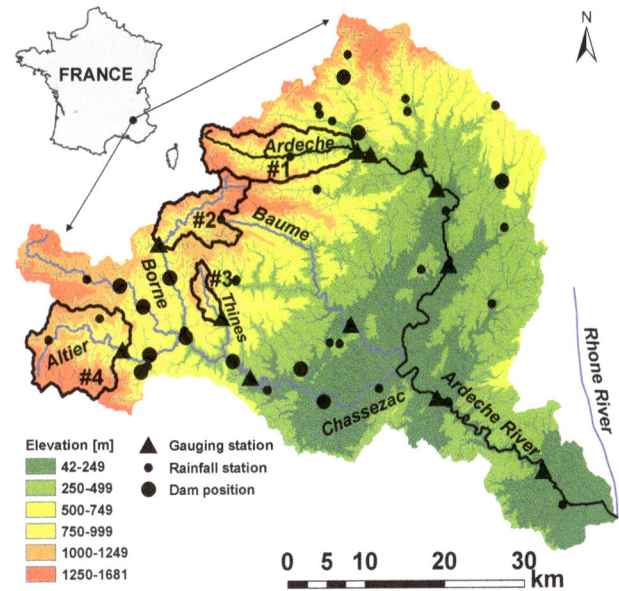

Figure 1. Map of the Ardèche catchment with gauging and rainfall stations, dam locations, and catchments that were examined: #1. Ardèche at Meyras; #2. Borne at Nicolaud Bridge; #3. Thines at Gournier Bridge; #4. Altier at Goulette.

they did not simulate hydrographs. In addition, all those studies used data from well-monitored experimental catchments, and the method has not previously been applied using data from operational networks, which are much more common.

To our knowledge, the simple dynamical system approach (SDSA) proposed by Kirchner (2009) has not been evaluated in a Mediterranean context, where the rainfall regime exhibits strong contrasts between dry conditions in summer and intense rainfall events, often related to stationary Mesoscale Convective Systems (Hernández et al., 1998), during autumns. Wittenberg and Sivapalan (1999), for instance, used recession analyses to estimate groundwater recharge in a Mediterranean type of climate in Australia, but they did not consider the storage–discharge relationship in its implicit differential form, the sensitivity function $g(Q)$.

Mediterranean catchments are also characterized by heterogeneous topography, vegetation and geology. The study of the water cycle in such Mediterranean conditions, as well as a better understanding and modelling of processes triggering flash floods, are central research topics addressed in the HyMeX (Hydrological Cycle in the Mediterranean Experiment; Drobinski et al., 2013) program[1] and in the FloodScale project (Braud et al., 2014)[2] to which this study contributes.

Our study area is the Ardèche catchment (2388 km^2, see location in Fig. 1), which is typical of Mediterranean catchments with highly variable rainfall, steep slopes, and heterogeneous geology and pedology. It is one of the studied

---

[1] www.hymex.org.

[2] http://floodscale.irstea.fr/.

catchments of the Cévennes–Vivarais Hydro-Meteorological
Observatory (OHM-CV, Boudevillain et al., 2011). Previous
studies in this catchment mainly focused on flood forecasting and discharge quantile estimation. Discharge time series from the Ardèche catchment were used to assess the
value of new observations in estimating extreme quantiles,
such as information derived from palaeofloods (Sheffer et
al., 2002), historical floods (Lang et al., 2002; Naulet et al.,
2005) or post-flood survey peak discharge estimates (Gaume
et al., 2009). Flood forecasting studies extended to the whole
Cévennes–Vivarais region are numerous and include work by
Sempere-Torres et al. (1992), Duband et al. (1993), Le Lay
and Saulnier (2007), Saulnier and Le Lay (2009), Tramblay
et al. (2010) and Garambois et al. (2013). Use of distributed
hydrological models for process understanding during flash
floods in the Cévennes–Vivarais region is more recent. Examples of such studies are those of Bonnifait et al. (2009),
Manus et al. (2009), and Braud et al. (2010). Those studies
use a reductionist approach to gain insight into active hydrological processes during floods and highlight a lack of data
or parameter information.

As a complementary approach to the modelling studies
mentioned above, we adopt in this study the data-based approach proposed by Kirchner (2009) to estimate the hydrological water balance of the Ardèche catchment and to
gain insight into the dominant associated processes. Like
in the work of Melsen et al. (2014), we divided our examined period into vegetation period (April–October) and low-
vegetation period (November–March) where evapotranspiration can be considered as a low.

The idea is to use this insight to propose simple models
with very few parameters to learn more about hydrological
functioning at the catchment scale.

In the present paper, we focus on the following questions:
(1) what is the applicability of the simple dynamical systems
approach (SDSA) and what are its limitations in a Mediterranean type catchment like the Ardèche with its particular
conditions (size, climate, geological and pedological heterogeneity), and when data from operational networks are used?
(2) what can we learn about dominant hydrological processes
using this methodology?

To answer those questions, we first estimate the discharge
sensitivity function using the available discharge data. Then
we assess the relevance of the obtained function by testing
how well the simple model based on it can simulate observed discharge, and can retrieve rainfall. The study is complemented by examining the sensitivity of the results to the
parameters of the discharge sensitivity function.

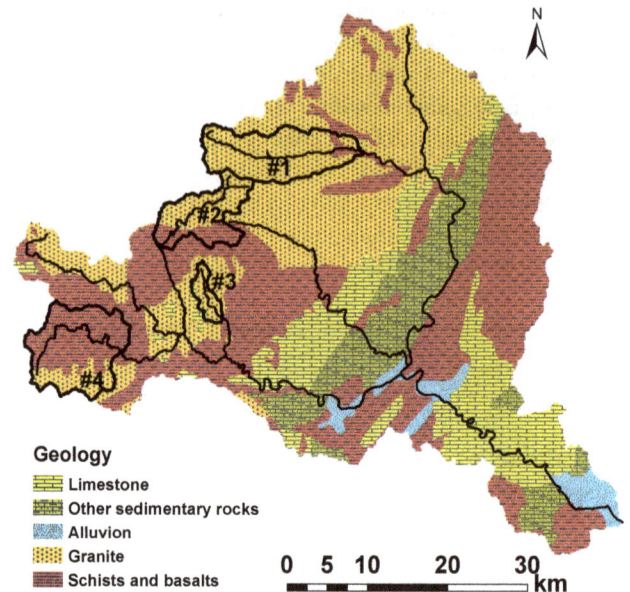

Figure 2. Geological map of the Ardèche catchment (extracted and
processed from geological map of France 1 : 1 000 000 issued by
BRGM (6th edition, 1996).

2 Field site and data

2.1 The Ardèche catchment

The Ardèche catchment is located in southern France
(Fig. 1). The catchment has an area of 2388 km^2, and the
Ardèche River itself has a length of 125 km. There are two
main tributaries in the Ardèche Basin: the Baume and Chassezac rivers, which join the Ardèche River close to one another. Elevation ranges from the mountains of the Massif
Central (highest point: 1681 m) in the northwest, to the con-
fluence with the Rhone River (lowest point: 42 m) in the
southeast.

The main lithologies found in the Ardèche are schist, granite, and limestone (Fig. 2). Upstream, the Ardèche flows from
west to east in a deep granite valley, then flows through basalt
formations and schist in a north–south direction. Downstream, it flows through bedded and massive limestone before flowing into the Rhone River (see for example the description provided by Naulet et al., 2005).

Among the land use types found in the Ardèche, forest dominates throughout the basin (Corine Land Cover
database[3]). Forest is represented by a mix of coniferous
(27 %), broadleaf (13 %) and Mediterranean trees (17 %).
Shrubs and bushes are also well represented in the catchment,
occupying a significant portion of the area (17 %). We also
distinguish significant areas of bare soil in the central and
southern part of the Ardèche, as well as a few small urban
areas and areas of early and late crops.

[3]http://sd1878-2.sivit.org/

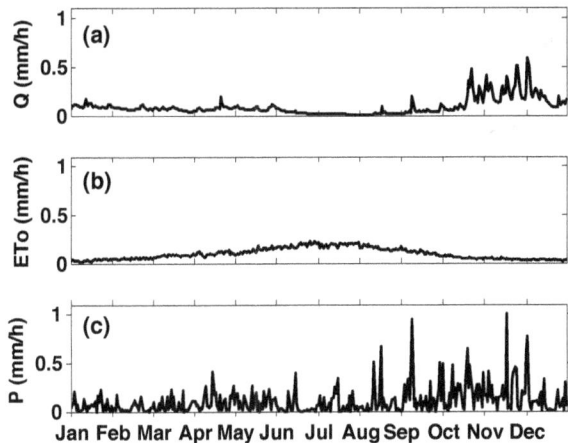

Figure 3. Average hourly discharge (**a**), reference ET_0 (**b**) and rainfall (**c**) in $(mm\,h^{-1})$ at the Ardèche outlet for all months between 2000 and 2008. (**b**) and (**c**) are calculated from the SAFRAN reanalysis. In red: vegetation period; in blue: low-vegetation period.

In the Ardèche Basin, there is a strong influence of the Mediterranean climate with seasonally heavy rainfall events during autumn. Historical data show that these events usually lead to flash floods: Lang et al. (2002) mention seven rainfall events locally exceeding 400 mm during the 1961–1996 period. They also comment on the relatively quick flow response (a couple of hours) to precipitation due to the steepness of the upstream part of the catchment and presence of granitic and basaltic rocks.

Figure 3 shows the average hourly regime of the main terms of the water balance equation for all months between 2000 and 2008, differently coloured with respect to vegetation (red) and low-vegetation periods (blue). Under the main terms of the water balance we consider discharge, evapotranspiration and precipitation. As we consider interannual values, change in water storage is assumed to be zero. This hypothesis is consistent with the lack of a regional aquifer in the Ardèche catchment. The hydrological year consists mainly of two periods. There is a rainy season (September–February) with maximum precipitation intensity in autumn, characterized by rainfall amounts greatly exceeding reference evapotranspiration ET_0 (calculated based on the SAFRAN reanalysis of Quintana-Seguí et al., 2008: see next section), and by high discharge. On the other hand, during the dry season (March–August), on average ET_0 is much larger than precipitation and runoff is low. Evapotranspiration is influenced by the seasonal cycles of temperature, radiation and vegetation, the latter being particularly marked in the Ardèche catchment, which is mostly covered by forests (around 60 % of the total catchment area, with 27 % of the forest being coniferous and thus remaining green even in winter).

2.2 Available data and first data consistency analysis

2.2.1 Observations used in the study

In the Ardèche catchment, measurements of the hydrological state variables were mainly started in the 1960s for the purpose of flood forecasting. In our study, we use hourly data of precipitation (P), reference evapotranspiration (ET_0) and discharge (Q) from the period 1 January 2000 until 31 December 2008. These data come from operational networks, and not from research catchments as in previous applications of the SDSA, which renders the study challenging and interesting, as operational networks account for a large fraction of the available discharge data in many regions.

The analysis is mostly constrained by the availability of discharge data, which were obtained from the national Banque Hydro website (www.hydro.eaufrance.fr) and Electricité de France (france.edf.com/). Unfortunately, numerous dams and hydro-power stations are located in the upper parts of the Ardèche and Chassezac catchments (Fig. 1). These dams are used to regulate the water level throughout the year, in particular to ensure a sufficient discharge in the river for recreational use in the summer period. Data to reconstruct natural discharge at the hourly time step were not available. Thus we had to discard several gauging stations located downstream of the dams in order to apply the simple dynamical system approach to data where the water balance can be closed.

As the stations were not designed and managed for low-flow measurements, the low-flow time series were investigated by contacting the operational services in charge of the stations. Consequently, two stations had to be removed from further analysis due to unreliable measurements and agriculture water withdrawals in summer periods. Ultimately, four sub-catchments could be examined: the Ardèche at Meyras (#1), the Borne at Nicolaud Bridge (#2), the Thines at Gournier Bridge (#3), and the Altier at Goulette (#4); see locations in Fig. 1. These four sub-catchments are characterized by steep slopes (> 15 %), average altitude of around 1000 m and igneous and metamorphic bedrock. We have also computed Strahler stream order and channel length using TauDEM tools (Tarboton et al., 2009) in order to classify and measure the size of the river network. The analysis was conducted using the 25 m resolution IGN DTM and the D8 flow direction algorithm, so the resulting network statistics may only loosely resemble those that would be obtained from more accurate procedures such as field mapping. Main physiographic catchment characteristics are summarized in Table 1.

The discharge data were available at varying time intervals, and were aggregated to hourly sums. Two types of precipitation data have been examined and are used throughout the analysis. Local rain gauges at the hourly time step provided by the OHM-CV database (Boudevillain et al., 2011) are used as the primary source of rainfall data for the catch-

Table 1. Physiographic characteristics of the four examined Ardèche sub-catchments. Strahler stream order, channel length and drainage density are calculated from the 25 m IGN DTM using TauDEM tools (Tarboton et al., 2009).

Catchment ID River and catchment name	#1 Ardèche at Meyras	#2 Borne at Nicolaud Bridge	#3 Thines at Gournier Bridge	#4 Altier at Goulette
River name	Ardèche	Borne	Thines	Altier
Drainage area (km^2), A	98.43	62.6	16.73	103.42
Average altitude (m)	898.54	1113	892.75	1149.13
Average slope (%)	23.43	20.13	16.72	17.13
Forest cover (%)	68	68	51	42
Strahler stream order	4	3	3	5
Channel length (km), L	94.31	59.26	13.51	97.38
Drainage density ($km\,km^{-2}$), $D = L/A$	0.96	0.95	0.81	0.94

ment Ardèche at Meyras (#1). For the catchments Borne at Nicolaud Bridge (#2), Thines at Gournier Bridge (#3) and Altier at Goulette (#4) we use the SAFRAN reanalysis of Météo-France, based on 8 by 8 km^2 grids (Quintana-Seguí et al., 2008; Vidal et al., 2010) since either the local rain gauge shows lack of data and time gaps, or there is no rain gauging station in the catchment (e.g. Thines at Gournier Bridge (#3). These precipitation data are calculated as catchment averages at hourly time steps. To compute the reference evapotranspiration ET_0, we also used the climate variables of the SAFRAN reanalysis of Météo-France at an hourly time step. ET_0 is calculated using the Penman–Monteith formula according to FAO recommendations (Allen et al., 1998). In order to account for vegetation type, we compute potential evapotranspiration (PET) as reference evapotranspiration ET_0 modulated by a crop coefficient depending on the nature of vegetation for each catchment (Eq. 1).

$$PET = K_C \cdot ET_0 \qquad (1)$$

We also took into account the seasonal variability of vegetation through the definition of three crop coefficient stages: initial (1 January–1 April), mid-season (15 April–15 October) and late season (1 November–31 December). Periods between initial and mid-season as well as between mid-season and late season are interpolated linearly. The values of crop coefficients for the Ardèche catchments were obtained through the FAO database (Allen et al., 1998). For each catchment we determined the cover estimates for each vegetation type (Broad-leaf forest, Mediterranean forest, Coniferous forest, Early crops, Late crops, Shrubs and bushes and Bare soil) and we calculated a weighted average crop coefficient per sub-catchment for each stage (see Table 2). Reference evapotranspiration ET_0 and ET_0 modulated by the crop coefficient ($K_C ET_0$) over the examined period (2000–2008) are given in Table 3.

2.2.2 Data consistency

To further assess data quality, we evaluated the consistency of the local rainfall station with SAFRAN data for the Ardèche at Meyras (#1) catchment at the hourly time step. The resulting coefficient of determination was 0.99. For the rest of the sub-catchments, we first assumed that SAFRAN rainfall is representative of the catchment average. However, by looking at the mean annual water fluxes (Table 3) and estimated runoff coefficients, we infer that the mass balances for catchments #2, #3 and # 4 are implausible.

For these reasons, two actual evapotranspiration (AET) estimates and runoff coefficients are provided to gain useful insight about data uncertainty. In Table 3 the first evapotranspiration estimate comes from the water balance $AET_{WB} = P - Q$, where P is the average annual precipitation and Q the annual runoff, assuming that change in water storage is null. In Table 3, the first runoff coefficient (C) is calculated as the ratio between Q and P. We also note that AET_{WB} shows either high underestimation (#2 and #4) or overestimation (#3) in comparison with the $K_C ET_0$ data, which once again points out the water balance closure issue.

In Table 4 the second AET estimate corresponds to Turc (1951) annual actual evapotranspiration, which is calculated using the following formula:

$$AET_{Turc} = \frac{P}{\sqrt{0.9 + \frac{P^2}{L^2}}}, \qquad (2)$$

where P is annual precipitation in $mm\,yr^{-1}$ and $L = 300 + 25\,T + 0.05\,T^3$ (T is the average annual temperature in °C). Here, the second runoff coefficient (C_{Turc}) is calculated using the following equation:

$$C_{Turc} = \frac{P - AET_{Turc}}{P}, \qquad (3)$$

where C_{Turc} is the runoff coefficient, P is precipitation ($mm\,yr^{-1}$) and AET_{Turc} is the actual Turc evapotranspiration

Table 2. Weighted average crop coefficient for each examined catchment per growing stage.

Catchment name	K_c_initial (Jan–Apr)	K_c_mid_season (May–Oct)	K_c_late_season (Nov–Dec)
The Ardèche at Meyras (#1)	0.74	0.94	0.79
Borne at Nicolaud Bridge (#2)	0.73	0.96	0.80
Thines at Gournier Bridge (#3)	0.68	0.94	0.75
Altier at Goulette (#4)	0.62	0.97	0.75

Table 3. Hydro-climatic characteristics of the four examined Ardèche sub-catchments (2000–2008).

Catchment ID	#1	#2	#3	#4
Catchment name	Ardèche at Meyras	Borne at Nicolaud Bridge	Thines at Gournier Bridge	Altier at Goulette
Precipitation (mm yr^{-1}), P	1621	1633	1892	1176
Streamflow (mm yr^{-1}), Q	1057	1579	970	932
Runoff coefficient, C	0.65	0.97	0.51	0.79
Actual evapotranspiration (mm yr^{-1}), $AET_{wb} = P - Q$	564	54	922	244
ET_0 SAFRAN (mm yr^{-1})	809	792	860	775
K_c ET_0 (mm yr^{-1})	731	729	762	699

(mm yr^{-1}). We use AET_{Turc} in this formula along with precipitation in order to estimate annual runoff coefficients in the examined catchments.

The values of the water balance components differ from catchment to catchment as illustrated in Table 3. By comparing Tables 3 and 4, we note that the mass balance AET_{WB} and AET_{Turc} estimates are only consistent for the Ardèche at Meyras (#1) catchment; at the other three sites they differ greatly, leading to inconsistent runoff coefficients for the same catchment. This suggests that either the rainfall or ET_0 (or possibly both) are not representative at the other catchments. Regarding rainfall, the gridded SAFRAN product is known to underestimate precipitation in mountainous areas and to underestimate the occurrence of strong precipitation, which could help to explain the water balance closure problems (see Sect. 5.1 for more details).

Discharge data uncertainty has been addressed in many works and sometimes it can be quite large, especially in catchments where high flows are seldom gauged due to safety reasons (Le Coz et al., 2010) or where low flows may be difficult to measure accurately. Nevertheless, here we decided to go ahead with the available operational discharge data, to assess whether the SDSA can provide useful information about catchment hydrological functioning in a Mediterranean context, even in the presence of some uncertainty in the discharge data.

However, in order to apply the SDSA with data where water balance closure is more representative, we rescaled precipitation and $K_c ET_0$ values for catchments (#2, #3 and #4). Our rescaling scheme (see next section for more details) assumes that the discharge data were accurate enough for the

application of the SDSA, which relies mainly on discharge data.

2.3 Rescaling of water balance fluxes

The first step in the rescaling analysis was to obtain a robust estimate of actual evapotranspiration.

We used the following equation of Fu (1981) to draw Budyko (1974) type curves for the Ardèche catchments:

$$\frac{AET}{P} = 1 + \frac{ET_0}{P} - \left[1 + \left(\frac{ET_0}{P}\right)^w\right]^{\frac{1}{w}}, \qquad (4)$$

where AET / P is the evapotranspiration ratio, ET_0/P is the dryness index and w is a catchment parameter.

The parameter w was empirically derived by Fu (1981) and it can have values from $1 \sim$ to ∞. Zhang et al. (2004) defined parameter w as a coefficient representing "the integrated effects of catchment characteristics such as vegetation cover, soil properties and catchment topography on the water balance".

In our study, we drew Fu curves with parameter w ranging between 1.5 and 5 to gain insight about evapotranspiration ratios in the Ardèche. The next step was to compare those curves with mean actual annual evapotranspiration ratios obtained using the Turc (1951), Schreiber (1904), Pike (1964) and Budyko formulae (see Table 5). We note from Fig. 4 that almost all calculated AET/P ratios lie in a w range between 1.7 and 3. On the other hand, the AET estimates derived using $AET_{WB} = P - Q$ (cyan colour in Fig. 4) for catchments #2, #3 and #4 were found to lie outside the range of values given by the various formulae, highlighting the water balance problem. Finally, to assess and adjust our data sets (P

Table 4. Scaling hydro-climatic characteristics of the four examined Ardèche sub-catchments (2000–2008).

| Catchment ID | #1 | #2 | #3 | #4 |
Catchment name	Ardèche at Meyras	Borne at Nicolaud Bridge	Thines at Gournier Bridge	Altier at Goulette
Turc actual evapotranspiration (mm yr^{-1}), AET$_{\text{Turc}}$	609	505	571	475
Runoff coefficient, C_{Turc}	0.62	0.69	0.70	0.60
Temperature (°C), T	11.2	8.0	9.9	7.7
P_{Turc} (mm yr^{-1})	–	2084	1541	1407
Scaling P coefficient, α_P	–	1.27	0.81	1.2
Scaling AET coefficient, α_{AET}	–	0.69	0.75	0.68
New runoff coefficient, C_n	0.65	0.76	0.63	0.66

Table 5. Description of different empirical formulas for estimating mean annual actual evapotranspiration: AET is actual evapotranspiration (mm yr^{-1}), P is precipitation (mm yr^{-1}), ET$_0$ is potential evapotranspiration (mm yr^{-1}), and T is mean air temperature (°C).

Equation	Reference
$\text{AET} = P[1 - \exp\left(-\frac{\text{ET}_0}{P}\right)]$	Schreiber (1904)
$\text{AET} = \dfrac{P}{\sqrt{0.9 + \left(\frac{P}{L}\right)^2}}$ where $L = 300 + 25T + 0.05T^3$	Turc (1951)
$\text{AET} = P / \left[1 + \left(\frac{P}{\text{ET}_0}\right)^2\right]^{0.5}$	Pike (1964)
$\text{AET} = \left[P\left(1 - \exp\left(-\frac{\text{ET}_0}{P}\right)\right)\text{ET}_0\tanh\left(\frac{P}{\text{ET}_0}\right)\right]^{0.5}$	Budyko (1974)

and AET), we chose Turc-inferred evapotranspiration as representative for future analysis. In the 1951 paper, Turc reports an evaluation of his formula by comparing measured interannual discharge to values estimated through $P - \text{AET}$ where AET is estimated by Eq. (2) of the paper with generally good performance. The considered data set covered countries all over the world. Thus, relying only on the P and T and not on ET$_0$, we could avoid the use of evapotranspiration and reduce uncertainty in estimating AET. In addition, the Turc equation is widely used in France to estimate AET, and thus our results can be compared to other studies.

We then make the following assumptions. We assume that the long-term average Q is valid. We also assume that the "relative" day-to-day variations of K_c ET$_0$ and P are valid, but that the mean P does not reflect the whole-catchment P, and the mean K_cET$_0$ does not reflect the mean AET. Therefore the means need to be rescaled to achieve a consistent set of measurements. As mentioned before, we assume that the Turc (1951) formula correctly describes the relationship between average AET and average P. Then we iteratively solve the Turc formula to find long-term average AET$_{\text{Turc}}$ and P_{Turc} that are consistent with one another, and consistent with the average Q.

The hourly precipitation values are then rescaled by multiplying them by the ratio found in the previous step between the average P_{Turc} and the average measured P. Secondly,

the ET$_0$ values are also rescaled by multiplying the hourly K_cET$_0$ by the ratio found between the average AET$_{\text{Turc}}$ and the initial K_cET$_0$ estimate. The improved AET estimate is $\text{AET} = \alpha_{\text{AET}} \cdot K_c\text{ET}_0$ where α_{AET} is the scaling AET factor provided in Table 4. While this scaling factor is assumed to be constant throughout the year, hourly variation (hourly ET$_0$ signal) and seasonal variations (seasonal K_c) of AET are considered. Assuming one mean annual value of α_{AET} is coarse, as strong seasonal variations in AET signal are expected due to the seasonal variations of ET$_0$ and vegetation activity, but water balances (and thus α_{AET} estimates) would be more uncertain over shorter periods. Table 4 shows the results of data rescaling for catchments #2, #3 and #4 that have unrealistic mass balances. It gives the values of the computed rescaled AET$_{\text{Turc}}$ and the corresponding computed mean annual precipitation (P_{Turc}). In addition, scaling parameter values α_{AET} and α_P are given for each considered catchment. We consider the rescaled runoff coefficients to be more realistic, as they are closer to those of catchment #1, where the water balance is consistent with Turc (1951) AET.

The new precipitation and new AET values for catchments #2, #3 and #4 are then used in further analysis, whereas original data were conserved only for catchment #1.

3 Methodology

In this part, we first present the estimation of the discharge sensitivity function, $g(Q)$, which is used to characterize the catchment hydrological response. Then we assess whether the estimated $g(Q)$ is really representative of the catchment behaviour using two additional calculations. First, a simple bucket deterministic model is built for the various examined sub-catchments and simulated discharge is compared to observations. Second, rainfall catchment amounts are retrieved from discharge fluctuations ("doing hydrology backwards") and compared to independent observations. Afterwards, we present a sensitivity analysis showing the impact of the parameters of the $g(Q)$ function on the results. Finally, simple dynamical systems approach is used with non-rescaled precipitation and evapotranspiration data to show how data inconsistency problems may affect discharge simulations.

3.1 Estimation of the sensitivity function $g(Q)$

Kirchner (2009) proposed a method for determining nonlinear reservoir parameters for a simple bucket model with the assumption that discharge Q depends uniquely on total water storage S in the catchment. The analysis starts, as many parametric rainfall–runoff models do, with the water balance equation, in which the total catchment storage variation is estimated using

$$\frac{dS}{dt} = P - \text{AET} - Q, \tag{5}$$

where S is water storage volume (L) and P, AET and Q are rates of precipitation, actual evapotranspiration and discharge, respectively ($L\,T^{-1}$). Q, P, AET and S are considered as functions of time and considered to be averaged over the whole catchment (Kirchner, 2009).

It is known that precipitation measurements are spatially variable. Rain gauges reflect precipitation on areas much smaller than the catchment itself. The same comment is valid for evapotranspiration estimates, which are typically representative of much smaller areas than the catchment.

In Eq. (5), only discharge can be considered as a state variable that characterizes the entire catchment. This observation led Kirchner (2009) to make the fundamental assumption that discharge is uniquely dependent on total water storage S in the catchment, and that therefore

$$Q = f(S) \text{ or } S = f^{-1}(Q). \tag{6}$$

Differentiating Eq. (6) with respect to time, one obtains:

$$\frac{dQ}{dt} = \frac{dQ}{dS}\frac{dS}{dt} = \frac{dQ}{dS}(P - \text{AET} - Q), \tag{7}$$

and dQ/dS can also be expressed as a function of Q, following Kirchner (2009), as

$$\frac{dQ}{dS} = f'(S) = f'(f^{-1}(Q)) = g(Q), \tag{8}$$

where $g(Q)$ is the "sensitivity function" as defined in Kirchner (2009). It describes the sensitivity of discharge to changes in storage, as a function of discharge itself. This is useful because discharge is directly measurable whereas whole-catchment storage is not.

Combining Eqs. (7) and (8) we can express $g(Q)$ as Kirchner (2009):

$$g(Q) = \frac{dQ}{dS} = \frac{dQ/dt}{dS/dt} = \frac{dQ/dt}{P - \text{AET} - Q}, \tag{9}$$

where the sensitivity function can be described using precipitation (P), actual evapotranspiration (AET), discharge (Q) and rate of change of discharge (dQ/dt).

Following the approach of Kirchner (2009), we consider periods when precipitation and actual evapotranspiration are relatively small compared to discharge, obtaining the following equation, which shows that under these conditions the discharge sensitivity function can be estimated from discharge data alone:

$$g(Q) = \frac{dQ}{dS} \approx -\frac{dQ/dt}{Q}\Big|_{P \ll Q, \text{AET} \ll Q} \cdot \tag{10}$$

We select hourly records for nighttime (defined as the period between sunset and sunrise) during which the total rainfall is less than 0.1 mm within the preceding 6 h and following 2 h (Krier et al., 2012). We also tested larger time windows (10 and 12 h instead of 8 h) which did not improve $g(Q)$ estimation.

The sensitivity function $g(Q)$ is estimated using discharge records from low-vegetation periods (from November to March) from 2000 until 2008, when vegetation and ET_0 could be considered to have a smaller impact on stream discharge. Melsen et al. (2014) also pointed out the importance of selecting the low-vegetation periods for estimating the $g(Q)$ function due to the high evapotranspiration conditions in the rest of the year. They also suggest that one winter season could be enough to get a robust estimate of $g(Q)$. Given the larger catchment size and larger climate variability in our catchment, we use the whole low-vegetation period from 2000–2008 for this estimation. Later, the resulting $g(Q)$ function was nevertheless used for precipitation retrieval and discharge simulation during both low-vegetation and vegetation periods (April–October).

We avoid the vegetation period for the estimation of the $g(Q)$ function since, as Fig. 3 shows, during this period ET_0 is much larger than discharge, and the Ardèche catchments clearly respond to ET_0 forcing during the entire 24 h period. In addition, in the Ardèche Basin, the diurnal amplitude (computed as half the difference between the daily maximum and minimum flow) often exceeds 20 % of the daily average flow.

These rainless nighttime hours are further used to determine the sensitivity function $g(Q)$ by constructing "recession plots" (Brutsaert and Nieber, 1977) of the flow recession rate ($-dQ/dt$) as a function of discharge. Following Brutsaert and Nieber (1977) and Kirchner (2009), the flow recession rate is estimated as the difference between two successive hours as:

$$-\frac{dQ}{dt} = \frac{(Q_{t-\Delta t} - Q_t)}{\Delta t}. \tag{11}$$

Then, the discharge is averaged over those 2 hours as $(Q_{t-\Delta t} + Q_t)/2$. Binning is then done by grouping the individual hourly data into ranges of Q and then calculating the mean and standard error for $-dQ/dt$ and Q for each bin. Following Kirchner (2009), values of $-dQ/dt \leq 0$ are also included in the binning analysis to avoid the introduction of bias. The bin size was initially set at 1 % of the logarithmic range in Q but was locally increased if necessary to bring the standard error of $-dQ/dt$ down to 50% of the mean $-dQ/dt$ (Kirchner, 2009).

A quadratic function (Kirchner, 2009) is then fitted to the binned means leading to the following empirical equation in log space:

$$\ln(g(Q)) = \ln\left(-\frac{dQ/dt}{Q}\bigg|_{P \ll Q, AET \ll Q}\right)$$
$$\approx C_1 + C_2 \ln(Q) + C_3(\ln(Q))^2. \quad (12)$$

As noted by Kirchner (2009), the C_2 parameter in Eq. (12) is one less than the linear term in a regression fit to the binned $\ln(-dQ/dt)$ versus $\ln(Q)$ plot.

3.2 Discharge simulation

Discharge sensitivity functions can be used to simulate discharge (Kirchner, 2009) by combining Eqs. (9) and (10), resulting in the following expression, where the quadratic function of Eq. (12) is used to describe $g(Q)$:

$$\frac{dQ}{dt} = \frac{dQ}{dS}\frac{dS}{dt} = g(Q)(P - AET - Q). \quad (13)$$

In solving Eq. (13), attention is paid to two details: time lags and numerical instabilities (Kirchner, 2009). A time lag is introduced to account for flow routing delays between changes in catchment storage and changes in discharge at the outlet. Changes in subsurface storage could also lag behind rainfall inputs due to the delays necessary for rainfall to infiltrate and change discharge at the outlet. However, these time lags do not affect the estimation $g(Q)$ since Q and dQ/dt are measured simultaneously at the catchment outlet.

Equation (13) indicates that dQ/dt depends on the balance between precipitation, actual evapotranspiration and discharge. However, variations in $P - AET - Q$ are mainly forced by variations in precipitation. For instance, in the Ardèche at Meyras (#1) catchment, the variance of hourly precipitation is over 15 times larger than the variance of hourly discharge and around 80 times larger than the variance of hourly evapotranspiration. In discharge simulations, lag time is not of such importance since discharge is highly auto-correlated. However, in precipitation retrieval, lag time is taken into account to enhance model performance (see Sect. 3.3 for more details) because precipitation varies more on short timescales.

In order to minimize numerical instabilities, Eq. (13) is solved using its log transform (Kirchner, 2009):

$$\frac{d(\ln(Q))}{dt} = \frac{1}{Q}\frac{dQ}{dt} = \frac{g(Q)}{Q}(P - AET - Q)$$
$$= g(Q)\left(\frac{P - AET}{Q} - 1\right). \quad (14)$$

Equation (14) is then computed using fourth-order Runge–Kutta integration, iterating on an hourly time step. A single value of measured discharge is used to initialize the simulation. In addition, Kirchner (2009) also remarked that solution can be unstable unless the parameter C_3 of Eq. (12) is less than 0.

To estimate the AET term in Eq. (14), Kirchner (2009) originally used Penman–Monteith reference evapotranspiration and a rescaling effective parameter (k_e) that was calibrated for the entire study period. Other authors have used slight variants of this approach: Teuling et al. (2010) used the Priestley–Taylor equation to estimate catchment-scale evapotranspiration, defining the evaporation efficiency as a fitting parameter; Brauer et al. (2013) used a parameter f that takes into account the difference between potential and actual evapotranspiration on a monthly basis.

In our study, we assumed that actual evapotranspiration is equal to potential evapotranspiration (PET) throughout the year, being defined as reference evapotranspiration ET_0 modulated by a crop coefficient depending on the nature of vegetation for each catchment (Eq. 1). The strong hypothesis that $AET = PET$ is likely to be more relevant in winter, when there is sufficient water content in the air and soils, than in summer. For example, Boronina et al. (2005) found that in Cyprus, actual evapotranspiration was close to potential rate during the November–March period since there was always water present in the air and soils. Nonetheless we use this assumption, even in summer, as a first rough approximation in order to assess the feasibility of such a simple modelling concept. For the application to the Ardèche catchment, as mentioned in Sect. 2.2.1, we assumed that AET was given by Eq. (1), computed on an hourly time step. According to the catchment, either the original $K_C ET_0$ (catchment #1) or rescaled $\alpha_{AET} K_C ET_0$ (catchments #2, #3, #4) were used.

To show how data inconsistency problems may affect the performance of discharge simulation, we also ran the model with non-rescaled values of precipitation and evapotranspiration. The resulting model performance is reported in Sect. 4.5.

3.3 Rainfall retrieval based on $g(Q)$

Until recently, it was considered infeasible to infer precipitation from streamflow fluctuations. Spatial variability of precipitation is high and conventional rain gauges can only measure precipitation over an area that is many orders of magnitude smaller than a catchment itself. We assess the relevance of the inferred storage–discharge relationship for the examined catchments in the Ardèche using the rainfall retrieval scheme ("doing hydrology backward") as proposed by Kirchner (2009) and further tested by Krier et al. (2012).

Assuming that the assumptions of the SDSA are valid, we can infer temporal patterns of precipitation rates from streamflow fluctuations using the following inversion of Eq. (13), as outlined by Kirchner (2009):

$$P - AET = \frac{dS}{dt} + Q = \frac{dQ/dt}{dQ/S} + Q = \frac{dQ/dt}{g(Q)} + Q. \quad (15)$$

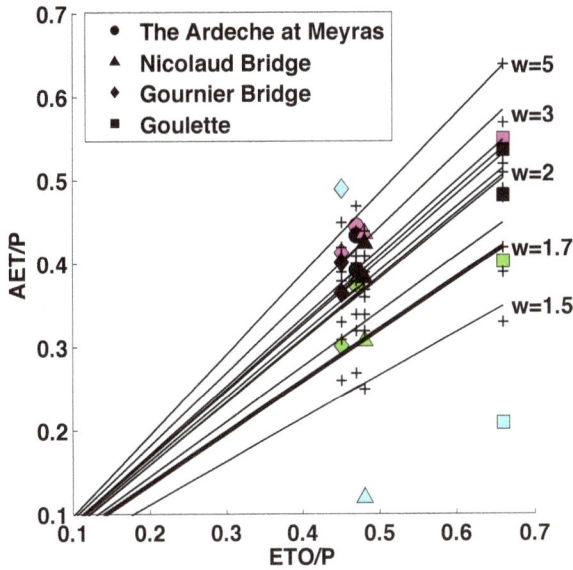

Figure 4. Mean annual evapotranspiration ratio AET / P as a function of the dryness index ET_0 / P for different values of parameter w, using the Fu (1981) curve and different formulas (Turc, Schreiber, Pike, Budyko; see Table 5). Colours correspond to different formulas (cyan = original data; green = Turc, blue = Schreiber, pink = Pike, red = Budyko) and shapes represent different examined catchments.

To apply this concept, one must take account of the travel time lag between changes in discharge from the hillslope and changes in streamflow at the outlet. A time lag l is used for this purpose leading to the following equation (Kirchner, 2009):

$$P - AET \approx \frac{(Q_{t+l+1} - Q_{t+l-1})/2}{[g(Q_{t+l+1}) + g(Q_{t+l-1})]/2} + (Q_{t+l+1} - Q_{t+l-1})/2, \tag{16}$$

where l is the travel time lag.

The time lag is optimized for each sub-catchment by calculating the correlation coefficient between estimated and measured rainfall using the lag times of 1, 2, 3, 4, 5, 6, 12, 24 and 48 h. The lag time that shows the best correlation is used. This approach is similar to the one used by Krier et al. (2012).

To make this concept of "doing hydrology backward" feasible, we identify periods when the contribution of evapotranspiration in the water balance equation can be neglected. This includes rainy periods when relative humidity should be relatively high, resulting in low evapotranspiration fluxes and thus P-AET ≈ P. Based on this assumption, precipitation rates can be directly deduced from the streamflow fluc-

tuations using the following formula (Kirchner, 2009):

$$P \approx MAX\left(0, \frac{(Q_{t+l+1} - Q_{t+l-1})/2}{[g(Q_{t+l+1}) + g(Q_{t+l-1})]/2} + (Q_{t+l+1} - Q_{t+l-1})/2 \right), \tag{17}$$

where P is the precipitation rate retrieved from discharge fluctuations with time lag l.

To measure the agreement between the reference values and the retrieved values we use the coefficient of determination R^2 (see Sect. 3.4 for more details). The reference precipitation is defined as a combination of local rain gauging and SAFRAN estimates depending on the sub-catchment being examined (see Sect. 2.2).

3.4 Comparison between observed and simulated/retrieved values

To assess model efficiency, we use Nash–Sutcliffe efficiency and percent bias as model evaluation criteria for discharge simulations, and coefficient of determination for rainfall retrieval. Nash–Sutcliffe efficiency, NSE (Nash and Sutcliffe, 1970) is used as a dimensionless model evaluation statistic indicating how well the simulated discharges fit the observations. We compute the NSE to emphasize the high flows as shown in the following equation:

$$NSE = 1 - \left(\frac{\sum_{i=1}^{n} \left(Y_i^{obs} - Y_i^{sim} \right)^2}{\sum_{i=1}^{n} \left(Y_i^{obs} - Y^{mean} \right)^2} \right), \tag{18}$$

where Y_i^{obs} is the ith observation of discharge data, Y_i^{sim} is the simulated discharge value for ith time step, Y^{mean} is the mean of all observed data and n represents the number of observations.

NSE values range between $-\infty$ and 1.0, with 1 representing the optimal value (see Moriasi et al., 2007, for a recent review of performance criteria). We also computed NSE on the logarithm of the discharge to give less weight to the peaks.

In addition, percent bias (PBIAS) was also calculated as a part of the model evaluation statistics. It measures total volume difference between two time series, as Eq. (19) indicates:

$$PBIAS = \frac{\sum_{i=1}^{n} (Y_i^{obs} - Y_i^{sim}) \cdot 100}{\sum_{i=1}^{n} (Y_i^{obs})}, \tag{19}$$

where Y_i^{obs} is the ith observation of discharge data, Y_i^{sim} is the simulated discharge value for the ith time step, n represents the number of observations and 100 converts the result to percent.

The optimal value of PBIAS is 0.0 where positive values indicate model overestimation bias, and negative values indicate model underestimation bias (e.g. Gupta et al., 1999).

In rainfall retrieval, model performance is assessed by using the coefficient of determination (R^2) to quantify the linear correlation between observed and inferred precipitation. R^2 ranges from 0 to 1, where higher values indicate smaller error variance (e.g. Moriasi et al., 2007). Although the inversion formula yields individual hourly values (Eq. 14), we use daily averages to compute R^2. This is done to reduce the effects of small discrepancies in timing that become less consequential when R^2 is calculated on a daily time step (Kirchner, 2009).

3.5   Sensitivity analysis

In this part, we performed a Monte Carlo analysis to sample the parameter space defined by the three parameters C_1, C_2 and C_3 and investigate further whether the values derived from streamflow fluctuations are representative, and how these parameters impact streamflow simulations. This Monte Carlo sensitivity study was conducted for the Ardèche at Meyras (#1) catchment.

A representative set of 10 000 (C_1, C_2, C_3) triplets was sampled randomly from the a priori defined parameter ranges (see Sect. 4.4 for more details) using Monte Carlo methods. Then the discharge was simulated using the model presented in Sect. 3.2 and Eq. (14). We used the NSE (ln for low flow and linear for high flows) to measure the similarity between the simulated and observed discharge. Then we verified that the parameter set derived from data is in the range of the sets leading to the best agreement between model and observations.

The number of simulations (10 000) was assumed to be adequate in view of the relative simplicity of the parametric model, and because the best-fit NSE did not change significantly beyond 10 000 simulations. For comparison, Zhang et al. (2008) and Tekleab et al. (2011) used 20 000 simulations for a four-parameter dynamic water balance model, and Uhlenbrook et al. (1999) used more than 400 000 model runs for the much more complex HBV model with 12 parameters.

4   Results

The results section is divided into five parts. In the first part, results concerning estimation of $g(Q)$ function and its sensitivity analysis are given. Then we present the assessment of the relevance of this estimated $g(Q)$ function by examining the accuracy of the simulated discharge (Sect. 4.2) and retrieved precipitation (Sect. 4.3). In Sect. 4.4, the impact of parameter variations on the simulated hydrographs and results of the Monte Carlo simulations are shown. Finally, the results with non-scaled original data are presented in Sect. 4.5.

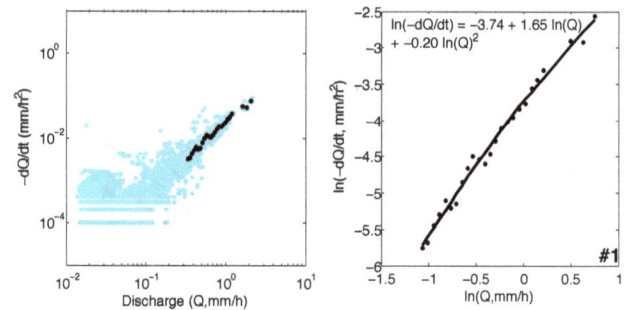

Figure 5. Recession plots for the Ardèche at Meyras (#1) catchment for all low-vegetation periods between 2000 and 2008: left, flow recession rates ($-\mathrm{d}Q/\mathrm{d}t$) as a function of flow (Q) for individual rainless night hours (blue dots) and their binned averages (black dots); right, quadratic curve fitting with binned means.

4.1 $g(Q)$ estimation

Figure 5 shows an example of a recession plot for the Ardèche at Meyras (#1) catchment for the all low-vegetation periods between 2000 and 2008. We observe that the recession plot exhibits large scatter at low discharge. This result is consistent with the findings of Kirchner (2009) and Teuling et al. (2010). They argue that this is possibly due to measurement errors or differences between the modelling concept and reality.

Table 6 provides values of the recession plot parameters for all four catchments during low-vegetation periods between 2000 and 2008. It shows one parameter set for each catchment. We observe that our choice of the low-vegetation period for estimation of $g(Q)$ gives consistent results amongst different catchments, with similar values of parameters C_1 and C_2. We also observe that the C_3 parameter, which controls the downward/upward curving of the $g(Q)$ function, is always negative, ranging from -0.02 up to -0.2. This is important because Kirchner (2009) obtained realistic simulated discharge only when recession plots are downward-curving on a log–log scale (meaning the C_3 parameter is negative). Eventually, these parameter sets allowed stable discharge simulation as can be seen in Sect. 4.2.

We have also tested $g(Q)$ estimation for all vegetation periods between 2000 and 2008; during these periods, the C_3 parameter tended to be positive. In this case, when the $g(Q)$ function is extrapolated to very low discharges, very high values of $g(Q)$ are obtained, and thus, numerical instabilities appear that lead to model non-functionality. This is also probably due to the distortion of the discharge time series by evapotranspiration as explained in Sect. 3.1. Melsen et al. (2014) concluded that a two-parameter 'bucket' model is reasonably able to capture high flows but not low flows. In our analysis we used the three-parameter model where the third parameter C_3 is essentially related to the low flows (see Sect. 4.4.1) in order to capture the catchment behaviour in that flow regime.

Table 6. Parameter values for the examined catchments for all low-vegetation periods (2000–2008).

Catchment name (ID)	C_1	C_2	C_3
The Ardèche at Meyras (#1)	−3.74	0.65	−0.2
Borne at Nicolaud Bridge (#2)	−4.08	0.74	−0.15
Thines at Gournier Bridge (#3)	−3.71	0.72	−0.13
Altier at Goulette (#4)	−3.80	0.82	−0.02

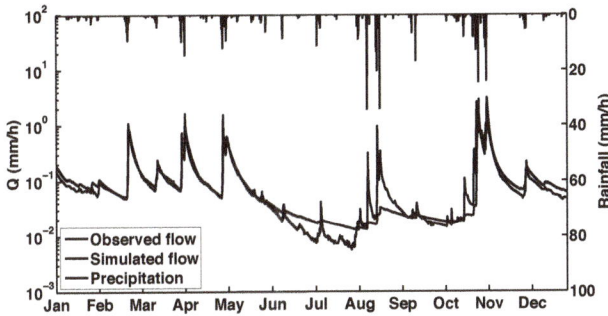

Figure 6. Series of simulated hourly hydrographs (red) for the Ardèche at Meyras (#1) catchment for the year 2004, compared with observed discharge (blue).

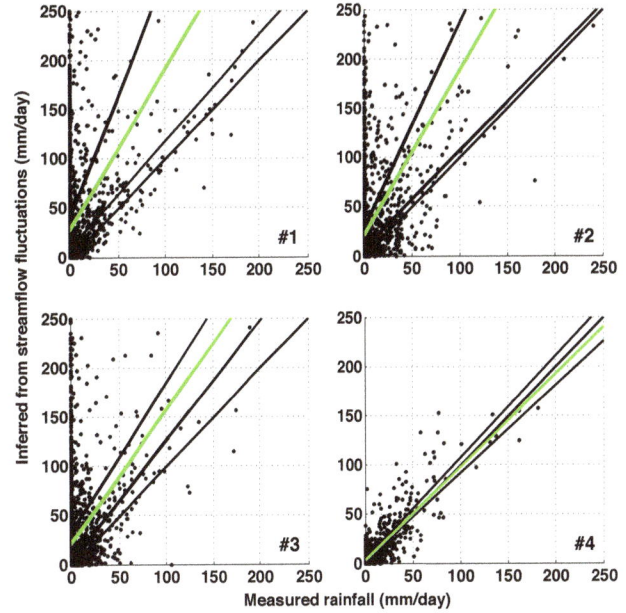

Figure 7. Inferred versus measured daily precipitation for the study catchments: #1. Ardèche at Meyras; #2. Borne at Nicolaud Bridge; #3. Thines at Gournier Bridge; #4. Altier at Goulette. Blue dots correspond to the inferred daily totals from low-vegetation periods; red points correspond to the inferred daily totals from vegetation periods; blue line is correlation for low-vegetation periods, red line for vegetation periods and green line for total examined periods.

4.2 Discharge simulations

Continuous discharge simulations were performed for 2000–2008. Figure 6 presents a simulation extract (year 2004) for the Ardèche at Meyras (#1) catchment. Table 7 presents a model performance summary (NSE, NSE of the logarithm of discharge, and PBIAS) for each catchment and each year.

Looking at Fig. 6, we can see that discharge simulations reproduce the observed hydrograph behaviour better in winter and low-vegetation periods. The low-flow (summer) periods are less well reproduced, even if the overall performance of the simulation is good. The influence of evapotranspiration in summer periods can be one of the explaining factors for that. It should be noted that high evapotranspiration influence is visible only when discharge is evaluated in log space. In linear space, evapotranspiration has a negligible influence on (already quite small) discharge, and the model runs well under dry conditions.

We note in Table 7 that the Ardèche at Meyras (#1) catchment shows satisfactory performance with NSE = 0.68, NSE log = 0.74 and PBIAS of 7.9 % for the 9-year simulation period. Unsatisfactory performance is observed for 2005 (NSE = −0.15, NSE log = 0.07 and PBIAS of 62.2 %). Year 2005 in general can be characterized as a dry year with annual precipitation of 775 mm and annual reference evapotranspiration of 947 mm for this catchment. A mean annual precipitation of 1621 mm and mean evapotranspiration of 809 mm across the examined period (2000–2008) clearly confirms that year 2005 can be considered unusually dry. Table 7 also shows that the year-to-year variations in NSE are

very large with some very good results in some years and poor results in other years. This could be a major challenge if the model were to be used for operational purposes.

Furthermore, Gupta et al. (1999) show that PBIAS values for streamflow tend to vary more than other performance criteria between dry and wet years. This could be another possible explanation of the overall poor model performance in 2005 for the Ardèche at Meyras catchment. The Borne at Nicolaud Bridge (#2) and Thines at Gournier Bridge (#3) catchments show good overall performance for the 9-year period with NSE = 0.67 and NSE log = 0.61 and NSE = 0.55 and NSE log = 0.78 respectively. These catchments have stronger variations in PBIAS, however. The last catchment, Altier at Goulette (#4) shows satisfactory model performance with NSE = 0.74 and NSE log = 0.18. It is not known whether the low NSE log value reflects poor model performance or unreliable low-flow discharge data.

4.3 Precipitation retrieval

Following SDSA we retrieved precipitation from discharge fluctuations. We used the same $g(Q)$ derived from the low-vegetation periods (2000–2008) to infer precipitation rates in both vegetation and low-vegetation periods.

The coefficient of determination, mean bias, and slope of the relationship between inferred and measured rainfall for

Table 7. Summary statistics of computed NSE, NSE log and PBIAS for each examined catchment in the Ardèche Basin.

Year	The Ardèche at Meyras (#1)			Borne at Nicolaud Bridge (#2)			Thines at Gournier Bridge (#3)			Altier at Goulette (#4)		
	NSE linear	NSE log	PBIAS (%)	NSE linear	NSE log	PBIAS (%)	NSE linear	NSE log	PBIAS (%)	NSE linear	NSE log	PBIAS (%)
2000	0.60	0.85	−20.7	0.76	0.83	5.02	0.49	0.86	−18.14	0.53	0.70	−1.58
2001	0.61	0.85	5.7	0.59	0.74	33.56	0.27	0.85	−1.43	0.67	0.62	1.86
2002	0.82	0.82	−1.2	0.63	0.53	−12.77	0.68	0.83	−15.05	0.65	0.44	−17.88
2003	0.76	0.72	13.	0.73	0.63	5.78	0.79	0.82	14.27	0.89	−0.19	12.43
2004	0.69	0.86	5.1	−0.07	0.37	−35.28	−0.26	0.78	−18.38	0.42	0.05	−11.09
2005	−0.15	0.07	62.2	0.66	0.64	18.16	0.21	0.53	48.22	0.70	−0.86	0.04
2006	0.51	0.71	19.6	0.68	0.58	0.58	0.36	0.72	17.67	0.18	−0.61	6.90
2007	0.11	0.67	21.8	0.51	0.28	−23.67	0.30	0.71	24.47	−1.22	0.34	−14.48
2008	0.76	0.85	8.2	0.75	0.43	−9.35	0.69	0.79	−6.89	0.83	0.62	6.04
2000–2008	0.68	0.74	7.9	0.67	0.61	0.75	0.55	0.78	0.98	0.74	0.18	−0.29

Table 8. Model performance of inferred versus measured daily rainfall in four sub-catchments for all low-vegetation periods 2000–2008.

Gauging station	R^2	Mean bias (mm day^{-1})	Slope	Time lag (h)
Ardèche at Meyras (#1)	0.41	7.9	1.1	2 (optimized)
Ardèche at Meyras (#1)	0.41	7.9	1.1	1
Borne at Nicolaud Bridge (#2)	0.56	7.4	1.01	2 (optimized)
Thines at Gournier Bridge (#3)	0.61	4.7	1.22	2 (optimized)
Altier at Goulette (#4)	0.71	2	1.09	2
Altier at Goulette (#4)	0.72	2	1.09	1 (optimized)

examined catchments and low-vegetation periods, as well as information about lag time, can be found in Table 8. Other lag times (>2 h) showed poor model performance and are not discussed further in the paper.

Figure 7 shows daily precipitation retrieval for the four studied sub-catchments of the Ardèche during low-vegetation periods, vegetation periods and for the entire study period 2000–2008 using the same $g(Q)$ function estimated from low-vegetation periods (Table 6).

Good correlation between retrieved precipitation and observed precipitation can be observed for low-vegetation periods where the slope of the regression line shows a modest degree of overestimation. Figure 7 illustrates that the inferred precipitation daily totals from low-vegetation periods (blue line) agree quite well with the precipitation measurements in the Altier at Goulette (#4) catchment, yielding R^2 of 0.72. In the other catchments, the inferred precipitation daily totals are well correlated with the either local precipitation measurements or SAFRAN data, showing however sometimes a strong tendency toward overestimation (e.g. the Ardèche at Meyras, #1). Figure 7 also shows strong precipitation overestimation for three examined catchments #1, #2 and #3 in summer periods (red line) and consequently for total examined period, too (green line).

The optimized time lags are generally very small (less than 2 h), which confirms the very short response time in the Ardèche catchment. In order to see whether the retrieved daily rainfalls were sensitive to the lag time, we compared the results obtained with different lag times for two catchments: the Ardèche at Meyras (#1) and Altier at Goulette

(#4). The Ardèche at Meyras (#1, 98 km^2) has an optimized lag time of 2 h. We tested the retrieval behaviour with lag times of 1 and 2 h and we observe almost no change in the performance (Table 8): we obtain the same coefficient of determination of 0.41 and a bias of 7.9 mm day^{-1} at a lag time of 1 and 2 h. Similar results are obtained for the Altier at Goulette (#4) catchment, where we observed a slightly better precipitation modelling performance with lag time of 1 h ($R^2 = 0.72$) rather than with a lag time of 2 h ($R^2 = 0.71$).

4.4 Sensitivity analysis

4.4.1 Impact of parameter variations on the simulated hydrographs

As a first approach, a manual sensitivity analysis was done by successively varying the values of each parameter and plotting the corresponding simulated hydrographs (grey areas in Figs. 8 and 9). The results for the Ardèche at Meyras (#1) catchment (year 2004) are presented; see Figs. 8 and 9 for the C_3 and C_1 parameters, respectively. The results for the parameter C_2 are not presented here since this parameter only varies slightly when estimated from low-vegetation periods in each year (see Sect. 4.1) and the results are graphically quite similar to those for the parameter C_1 (but peaks are less affected). The NSE values of log discharge are also calculated (Table 9).

We can see that C_3 seems to be influential during the low-flow summer period and also during recessions of events following low-flow periods (Fig. 8). However, it does not play

Table 9. NSE values of log discharge for the Ardèche at Meyras (#1) catchment, illustrating sensitivity to changes in the C_1 and C_3 parameters. In bold: values obtained from data.

C_1 parameter	NSE on log of discharge	C_3 parameter	NSE on log of discharge
−4	0.81	−0.3	0.68
−3.8	0.85	−0.25	0.79
−3.74	0.86	−0.21	0.85
−3.7	0.86	**−0.2**	0.86
−3.6	0.86	−0.19	0.86
−3.5	0.86	−0.17	0.86
−3.4	0.85	−0.16	0.85
−3.3	0.83	−0.15	0.83
−3.2	0.81	−0.1	0.45
−3	0.71	−0.09	0.26

a significant role in the peaks and in well-established high-flow conditions. In contrast, the C_1 parameter has an important influence on the whole hydrograph (Fig. 9), including the peaks. Low values of C_3 tend to flatten the model response, causing overestimated low-flow values and underestimated peaks.

From Table 9 we can also observe that the model efficiency for the parameter values that were obtained from the recession plots is close to optimal (at least for this year at this site), and cannot be substantially improved by manual parameter adjustments.

4.4.2 Exploration of parameter range using Monte Carlo simulations

In order to complement the manual sensitivity analysis presented above, to explore the range of these parameters and to assess whether the parameters of the $g(Q)$ function derived from data analysis are representative, we performed Monte Carlo simulations using the model described by Eq. (14) with randomly sampled values of the three parameters C_1, C_2 and C_3. The parameters were sampled randomly from the a priori defined parameter range given in Table 10. For each simulation, the NSE and NSE log (on the log of discharge) were calculated to assess the performance of the parameter set. The results are presented using dotty plots for the Ardèche at Meyras (#1) catchment in Fig. 10. Table 10 also indicates the range of "behavioural" values for each parameter as derived from the dotty plots, defined as the range where NSE is higher than 0.7, along with the values derived from the recession plots.

The results show that when the parameters are calibrated to discharge simulations, their ranges are quite large. The maximum model performance appears to be around 0.8 for all three parameters and both indicators. Low-flow performance (NSE log) is not very sensitive to the variations of the parameters. Giving peak flow more weight (NSE) allows the identification of clear optima and a narrower range for the C_1

parameter. Concerning the C_2 parameter, although the initial guess of the parameter range was quite narrow (see Table 10), the final "optimized" range is almost the same, with no clear optimum. For the C_3 parameter, the final "optimized" range is found to be half of the initial one. These two parameters appear thus to be not very sensitive, although the sign of the C_3 parameter was already identified as a key element of successful discharge simulations. Finally, the parameter values obtained from recession plots are in the optimized parameter range, thus suggesting that the analysis of discharge recessions is sufficiently informative and that there is no need of additional model calibration for discharge simulation. Beven and Binley (1992) have argued that having too many parameters increases the degrees of freedom beyond what data can properly deal with; this results in having different sets of parameters that give similar results (the equifinality problem). Figure 10 shows that although conventional parameter calibration leads to substantial equifinality (particularly for the C_2 parameter), the parameter values obtained from the recession plots fit well within the "behavioural" parameter range from the Monte Carlo analysis. Our analysis shows that the recession plots yield parameter estimates that are consistent with (and arguably better constrained than) parameter values obtained from conventional model calibration methods.

4.5 Modelling performance with non-scaled original data

In Sect. 2.3 we introduced a rescaling technique to obtain more representative water balances for catchments #2, #3 and #4. Here, we show the consequences of foregoing this rescaling for those three catchments that showed unrealistic mass balances (Table 3). Figure 11 shows observed discharge and simulated hourly hydrographs for the Altier at Goulette (#4) catchment for the year 2000, obtained with non-scaled data, rescaling of precipitation alone, and rescaling of both precipitation and evapotranspiration.

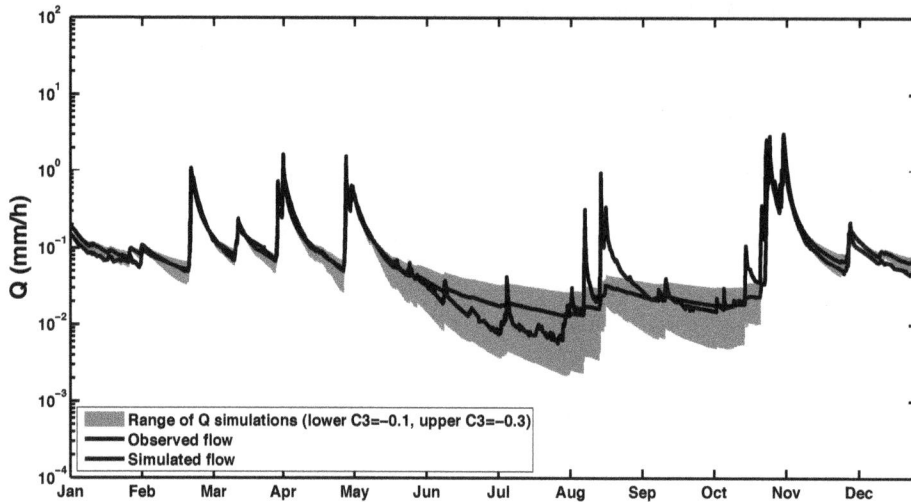

Figure 8. Observed versus simulated hydrograph ($C_3 = -0.2$) for the Ardèche at Meyras (#1) catchment (year 2004), with C_3 parameter variations (C_1 and C_2 values are kept constant at (-3.74) and 0.65, respectively). The grey area shows the range of discharge simulations.

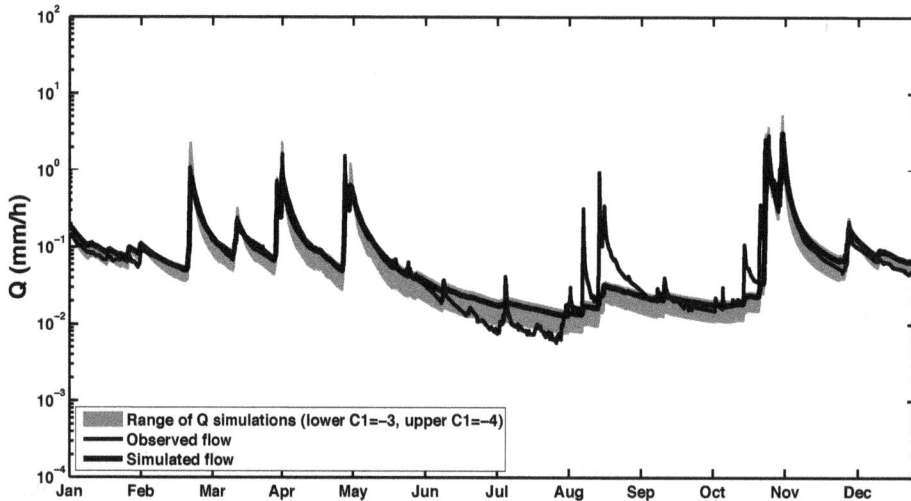

Figure 9. Observed versus simulated hydrograph ($C_1 = -3.74$) for the Ardèche at Meyras (#1) catchment (year 2004) with C_1 parameter variations (C_2 and C_3 values are kept constant at 0.65 and -0.2, respectively). The grey area shows the range of discharge simulations.

The lack of water balance closure may contribute substantially to poor model performance, as can be seen from Fig. 11. The simple dynamical systems approach, like many modelling approaches, is based on conservation of mass; it is therefore unsurprising that it may perform poorly when tested against data sets that violate mass conservation. We observe that when the original non-scaled data are used, discharge is generally underestimated. By introducing the rescaled precipitation, flow peaks can be better reproduced, but model performance is still poor during the vegetation period. If both the rescaled evapotranspiration and rescaled precipitation are used, significantly better results are obtained in both vegetation and low-vegetation periods.

As a complement to assessing modelling performance with non-scaled data, we re-ran the SDSA model for these catchments to see how this affects the hydrograph simulation and performance indicators. Table 11 compares model performance with the original operational data and the rescaled data, using NSE, NSE on log of discharge and PBIAS as performance metrics. We observe that model performance is markedly improved by using the rescaled precipitation as forcing (runoff coefficients are more representative as shown in Table 4). In addition, model performance is improved by also introducing rescaled evapotranspiration (better NSE and lower PBIAS values are obtained).

Table 10. Comparison of the chosen parameter range and parameters obtained from low-vegetation periods for the Ardèche at Meyras (#1) catchment.

Parameters	C_1	C_2	C_3
Parameter range	$[-1]-[-6]$	$[0.1-1]$	$[-0.001]-[-0.5]$
The range of "behavioural" values	$[-3.5]-[-4.5]$	$[0.1-0.9]$	$[-0.001]-[-0.25]$
Reference (from recession plots)	-3.74	0.65	-0.2

Table 11. Model performance for three examined catchments over the whole examined period (2000–2008), comparing the original operational data and rescaled precipitation and evapotranspiration data.

Catchment	Performance	Operational	Rescaled P	Rescaled P and AET
Borne at Nicolaud Bridge (#2)	NSE	0.45	0.65	0.67
	NSE log	0.58	0.70	0.61
	PBIAS	42	14.2	0.75
Thines at Gournier Bridge (#3)	NSE	0.36	0.50	0.55
	NSE log	0.79	0.62	0.78
	PBIAS	-13.8	22	0.98
Altier at Goulette (#4)	NSE	0.54	0.79	0.74
	NSE log	-4.90	-2.99	0.18
	PBIAS	49	23.65	-0.29

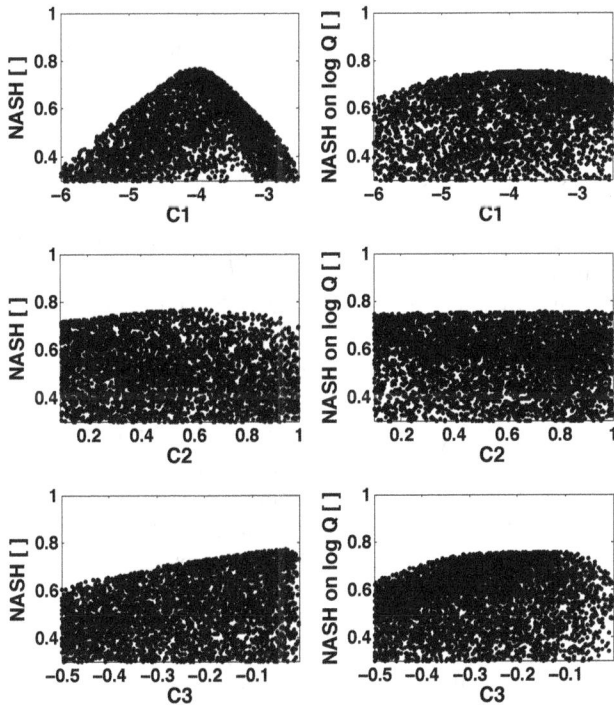

Figure 10. Dotty plots for the Ardèche at Meyras (#1) catchment (left: plots with NSE efficiencies; right: plots with NSE efficiencies calculated on log Q).

5 Discussion

In this study, the SDSA method was applied to four sub-catchments in the Ardèche catchment (France), representative of Western Mediterranean catchments. We first discuss the advantages and limits of the method for this type of catchment. Then we discuss how the application of this approach was useful in deriving information about the catchment functioning and possible dominant processes.

5.1 About the applicability of the SDSA to Mediterranean type catchments

The application of this method to the Ardèche catchment was at first quite challenging. In particular, the basins are larger and less humid than those of the original case studies; in addition, data availability is more limited and data quality is distinctly lower.

5.1.1 Drainage area

The drainage area does not seem to be a limiting factor at the scale of our catchments. The catchments where this theory has been applied so far in order to reproduce the hydrograph were typically smaller than $\sim 10\,\text{km}^2$. In our study, the sizes of the studied catchments varied from $16\,\text{km}^2$ to $103\,\text{km}^2$ and SDSA performance was not correlated to the size of the catchment. Krier et al. (2012) report that when this approach is used for "doing hydrology backward" to retrieve rainfall amounts, the model performance in larger basins is as good as or sometimes even better than in smaller catch-

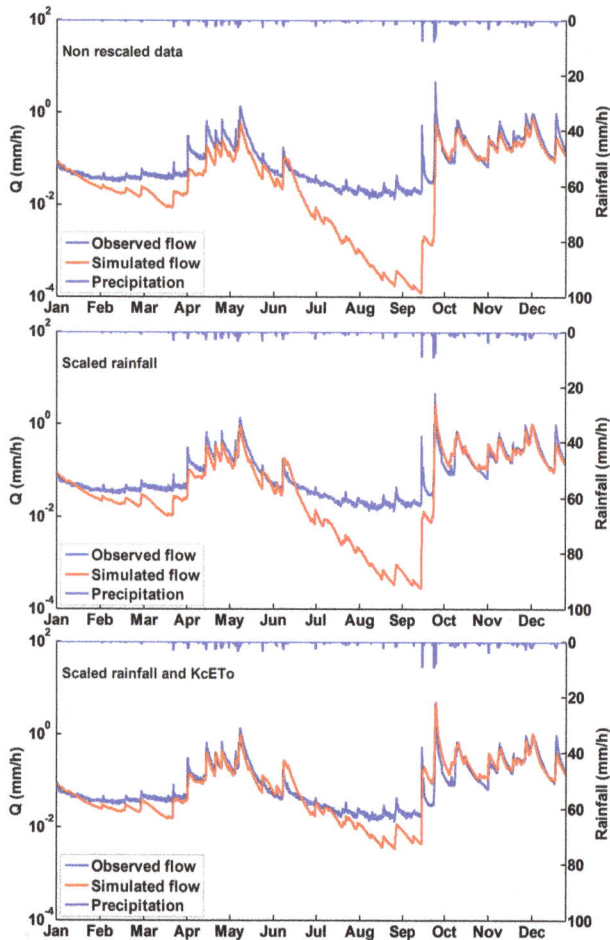

Figure 11. Series of simulated hourly hydrographs (red) for Altier at Goulette (#4) catchment for the year 2000 and its comparison with observed discharge (blue), using original non-scaled data (top), with rescaled P only (middle), and rescaled P and K_c ET_0 (bottom).

ments. Kirchner (2009) also addressed this issue arguing that the approach was unlikely to work for catchments that are too big (e.g. more than $1000\,km^2$). This is due to the lag times required for changes in discharge to reach the outlet; in such large catchments these lag times would be so long and variable that the model would be likely to fail. In addition, the theory presented here could not be expected to work in the catchments that are bigger than the scale of individual storms (Kirchner, 2009). Suggestions for how to deal with large river basins are given in Sect. 6.

5.1.2 Data quality

Our study demonstrates that data quality is particularly important for the application of this method. Concerning discharge data, the method is based on the discharge-sensitivity function $g(Q)$, and discharge measurement errors consequently will lead to biases in the appraisal of the catchment

functioning. In catchments with artificial reservoirs/dams the assumption of a unique storage–discharge relation will not hold from the SDSA point of view. Thus this will limit the applicability of the SDSA method (and many other catchment models) in practice.

In the present study, we used discharge data from operational networks. We have shown in Sect. 2.2 that there are known issues with the quality of these data for our purposes. Nevertheless, when data consistency is sufficient (e.g. Ardèche at Meyras (#1) station), a robust estimation of the $g(Q)$ function from low-vegetation periods can be obtained, leading to accurate simulation of the discharge. In addition, we would like to emphasize that the discharge sensitivity function $g(Q)$ only depends on discharge. Its estimation is therefore not dependent on the rescaling performed on rainfall and ET_0 data. There is only a minor impact of this rescaling on the selection of points retained for the recession plots; rainfall thresholds are used in this selection, but the results are only marginally impacted by rescaling. On the other hand, rescaling is of paramount importance in the evaluation of the relevance of the estimated $g(Q)$ function using discharge simulation, as shown in Sect. 4.5., because accurate discharge simulations require that mass is conserved.

The quality of the rainfall data was questioned early in our work, and rescaling of precipitation was needed to obtain realistic results. As mentioned in Sect. 2.2, the gridded SAFRAN product underestimates precipitation especially in mountainous areas and underestimates the occurrence of strong precipitation ($P > 20\,mm\,day^{-1}$ (Quintana-Seguí et al., 2008; Vidal et al., 2010). As the SAFRAN reanalysis is performed on so-called "symposium zones", assumed to be homogeneous in terms of climate characteristics, overestimation of rainfall is also possible if those zones are inaccurately delineated, as is probably the case for the Thines at Gournier bridge (#3) catchment. Some authors tried to overcome the rainfall underestimation problem in mountainous areas by interpolating the SAFRAN data across altitude bands (Etchevers et al., 2001; Lafaysse et al., 2011; Thierion et al., 2012), but these data were not available for the present study. In addition, SAFRAN re-analyses are based on existing rain gauges. In mountainous areas, the few rain gauges that do exist are generally located in lower, flatter terrain, and may not capture the increase of rainfall with altitude that has been identified in this region (Molinié et al., 2011). It would be interesting to assess the performance of the "hydrology backwards" rainfall inversion using more accurate rainfall estimates as reference. As reference daily rainfall, we propose to use the SPAZM reanalysis (Gottardi, 2009), which improves rainfall estimation in mountainous area, when it becomes available to us.

Assuming that the discharge data are reliable, it was shown that when input rainfall and ET_0 consistent with the water balance closure are used, the discharge simulated using the $g(Q)$ function is much more accurate than with the original input data. Coussot (2015) generalized the study presented in this paper to about 20 catchments of the Cévennes region and found the same kind of water balance closure problems as in our study. Once rescaling of rainfall and ET_0 was performed, he obtained similar results as those presented in this paper.

One assumption behind the rescaling approach proposed in Sect. 2.2.2 is that discharge data are reliable enough to provide an accurate estimate of annual runoff. This is of course questionable, because stage–discharge relationships are known to be highly extrapolated in this region due to the difficulty of gauging high discharges (e.g. Le Coz et al., 2010). As also mentioned in Sect. 2.2.1, low discharges are also highly uncertain, because these stations were often designed for flood warning purposes. Work is currently in progress in order to quantify the runoff data accuracy. This work is based on the BaRatin method (Le Coz et al., 2014) which provides an uncertainty range on the estimated discharge. The uncertainty can be propagated to the whole discharge time series (Branger et al., 2015) and the next step will be the propagation to the hydrological water balance and the quantification of uncertainty for the annual and monthly values. This work will help quantify which of the data (rainfall, discharge or both) need to be improved.

In addition, the operational discharge measurement network has recently been complemented by research instrumentation covering nested scales (see Braud et al., 2014 for details). In particular, small catchments ranging from 0.5 to 100 km^2 have been monitored continuously since 2010. The data set was not long enough to be used in the present study, but these new data are expected to be of higher accuracy than the operational data used in this study, so that they can provide additional insight into the hydrological response of the catchments.

Regarding discharge uncertainty, if data have to be rescaled, an approach like the one proposed by Yan et al. (2012) should be preferred, as it allows a consolidation of the water balance at the scale of the whole Ardèche catchment, taking into account data uncertainties on all the components, and constraining the results with the water balance equation along the river network.

The simulation results show that additional effort must be put into quantifying data uncertainty in both discharge and rainfall. The derivation of more accurate rainfall fields combining various data sources (such as radar data and in situ gauges (see, for instance, Delrieu et al., 2014) should also be encouraged. It could also be interesting to use actual evapotranspiration estimates derived from remote sensing techniques adapted to complex topography (e.g. Gao et al., 2011; Seiler and Moene, 2010) to obtain independent estimates of AET and better constrain this component in hydrological modelling.

5.1.3 Adequacy of SDSA in our catchment

The sampling strategy of deriving the $g(Q)$ functions from low-vegetation periods appeared to be adequate in our case.

We estimated $g(Q)$ by using the streamflow data from low-vegetation periods of the 9-year time series (2000–2008) and then used the resulting parametrization to reproduce the hydrographs (continuous simulations) for the rest of the 9-year interval. This procedure can be understood as a "differential split-sample test" (Klemes, 1986) where the 9-year-long period encompasses different seasonal precipitation variations including wet and dry periods. The results show that the information retrieved from only a fraction of the discharge time series is relevant also for periods with very different characteristics.

Independently from the data quality issues, we also showed that the SDSA model performs better during the wet, winter periods than the dry, summer periods and dry years (see Sect. 4.2). We interpret these results as an indication that the current model is not fully adapted to the high evapotranspiration conditions of our Mediterranean catchments. We must also point out that, when assessing the relevance of the estimated $g(Q)$ function using continuous discharge simulations (Sect. 3.2), it is necessary to provide an estimate of AET. In this first approach, we used the hypothesis that AET = PET where PET is the rescaled $K_C \cdot ET_0$. This assumption is crude because an average annual rescaling factor is used, whereas a monthly value would certainly be more relevant. The method of Thornthwaite and Mather (1955) cited by Gudulas et al. (2013) which provides monthly estimates of AET could be a way to improve our simulations in future works, and an example of application to the Ardèche at Meyras (#1) catchment is provided in Adamovic et al. (2015). Nevertheless, we show in Sect. 4.3 that rainfall retrieval during the vegetation period is poor, confirming the lower performance of the SDSA in this period. The method is therefore less reliable when discharge is low, especially in summer. This is one limitation of the SDSA for dry catchments.

In addition, the recent study of Brauer et al. (2013) showed that the two-parameter model they used cannot deal with complexity of hydrological processes in their catchment (only 39 % of the hydrographs had NSE over 0.5). In the Ardèche catchments, the three-parameter model succeeds in capturing the catchment behaviour, with quite good response of discharge to rainfall in low-vegetation periods (peaks and recession were nicely reproduced).

5.1.4 Interest of the SDSA as compared to other hydrological modelling approaches

Recession analysis has been used to build hydrological models for many years (e.g. Brutsaert and Nieber, 1977). What is new in the SDSA is not the reservoir itself, but the manner to derive its structure and parameters from the data analysis: in particular, here the functional form of the storage–discharge relationship is not specified a priori, but determined directly from data without calibration (Kirchner, 2009). This is the very definition of the top-down or data-driven modelling approach, that is acknowledged to be a major paradigm shift in modelling by the hydrological community (and which was a major emphasis of the PUB decade; see, for instance, Sivapalan, 2003b and Hrachowitz et al., 2013). We argue that testing this kind of approach on new data sets, for various climatic conditions, contributes to the advance of hydrological science in itself. We have also compared the model results with other models that are based on similar data-driven methodology (e.g. Brauer et al., 2013 and Melsen et al., 2014) and obtained similar results.

The major limitation of the SDSA is of course the availability of good quality discharge data with a short time step, in catchments representative of the spatial variability of hydro-climatic conditions. Discharge must also be representative of natural conditions, which could also limit its applicability in catchments impacted by human activity.

5.2 Catchment functioning hypotheses derived from the analysis

The most important output from our application of the simple dynamical systems approach is the validation of underlying hypotheses and information about the dominant processes that can be derived from the model parametrization.

5.2.1 General considerations

The SDSA model is based on an underlying hypothesis that regards a catchment as a single nonlinear bucket model. In our study we note the good performance of the model in each sub-catchment which suggests that SDSA, although it was developed for humid regions, remains valid for these Mediterranean sub-catchments as well. We can thus interpret that these sub-catchments do follow the model's functioning hypotheses, especially in winter and low-vegetation periods. These results are consistent with the findings of Brauer et al. (2013) for the Hupsel Brook catchment, Kirchner (2009) for Plynlimon and Teuling et al. (2010) for the Rietholzbach catchment. In contrast, during the vegetation period the model seems to be less adapted to our Mediterranean setting. The catchments seem to behave differently when they are dry. This is probably due to the strong influence of evapotranspiration. In our hydroclimatic context (see details in next section), and taking into account that no regional groundwater exists in the Ardèche catchment, discharge provided by the SDSA can be associated with subsurface flow (generally assumed to occur via lateral flow along perched water tables in shallow soils), which is less active in summer and when evapotranspiration is high. It could be necessary to consider another storage, probably more superficial than the "SDSA" storage, which could be used to supply evapotranspiration over shorter timescales, and which may be largely decoupled from subsurface lateral flow that sustains base flows.

5.2.2 Links with physiographic characteristics of the catchments

The model works better in the Ardèche at Meyras (#1) and Thines at Gournier Bridge (#3) catchments, which both are granitic (see Fig. 2). The hypothesis of shallow subsurface flow caused by saturation of at interface between soil and bedrock makes particular sense in this geology (e.g. Cosandey and Didon-Lescot, 1989; Tramblay et al., 2010).

In the forested granitic catchments of this region, infiltration capacity is generally very high and runoff occurs due to soil saturation (e.g. Tramblay et al., 2010). However, this saturation mostly occurs at the interface between the very thin soil and the large altered bedrock, where contrasts of hydraulic conductivity can be encountered, leading to quick lateral subsurface flow. Experiments are currently being conducted on infiltration plots to quantify the velocity of this lateral flow (see Braud et al., 2014 for their description). Therefore the main mechanism we are speaking about is quick lateral subsurface flow which transits through the reservoir considered in the Simple Dynamical Systems Approach. On agricultural areas, in the intermediate part of the Ardèche catchment, infiltration excess surface runoff is likely to occur (and has been observed in the field). Its contribution is also under investigation using detailed experiments (see Braud et al., 2014).

In addition, unaltered bedrock tends to be impermeable, but flow pathways are created in the many fractures, joints and fissures of the altered horizons. During extended rainfall those flow pathways might become connected, generating rapid subsurface flow (Krier et al., 2012). Moreover, the parameter values of the granite catchments are quite similar (see Table 6).

To quantify the relative influence of several predictors of the catchment response (and values of C_1, C_2, C_3 parameters), Adamovic (2014) used factor analysis of mixed data (FAMD). By using this statistical technique along with HCPC (hierarchical clustering on principal components) analysis, geology was found to be the only dominant predictor of runoff variability. The role of geology is more thoroughly demonstrated in Adamovic (2014) for the catchments studied in this paper and in Coussot (2015) for a larger set of catchments from the Cévennes region, but a review of this work is beyond the scope of the present paper.

This is also consistent with the contemporary literature, as geology has been invoked in numerous recent studies as a controlling factor of flood response (Gaál et al., 2012; Garambois et al., 2013; Krier et al., 2012; Vannier et al., 2014). As also discussed by Kirchner (2009), the theory is challenged by catchments with heterogeneous geology and thus with many disconnected subsurface storage reservoirs. This might explain the good modelling performance in granite catchments (see also Vannier (2013) for similar conclusions using a reductionist modelling approach).

6 Conclusion and perspectives

Our study describes in detail the application of SDSA methodology to four catchments of the Ardèche Basin ranging from 16 to 103 km², typical of the Western Mediterranean environment.

To have more representative water balance fluxes, we rescaled precipitation and evapotranspiration for three subcatchments (#2, #3 and #4). In our work we used average annual scaling coefficients for the whole time series (for both precipitation and evapotranspiration). In the future, varying the scaling coefficients according to different seasons could possibly lead to a better approximation of hourly precipitation and evapotranspiration fluxes.

We calculated the discharge sensitivity functions from low-vegetation periods and performed continuous discharge simulations with an hourly time step for the period 2000–2008. We also inferred precipitation and performed sensitivity analyses of the three parameters of the discharge sensitivity function.

Our results show that good results for discharge simulation can be obtained, especially under winter humid conditions and for catchments characterized by predominantly granitic lithology. Under dry conditions, poor model performance is mainly related to the disturbed water balance terms, high influence of AET and imprecise discharge measurements. Improving AET estimation is recommended for better model performance in summer periods when evapotranspiration is high and when the unsaturated zone has a significant role in attenuating the precipitation input. Working on the quantification of data accuracy and error reduction is also recommended in order to get more robust and reliable results.

As a perspective to this study, dominant predictors of runoff variability other than geology (such as land use, soil properties, drainage density, topographic steepness etc.) still need to be explored and linked to catchment hydrological behaviour. Relating the obtained parameters of the discharge sensitivity function to the catchment characteristics using different statistical classification techniques (e.g. principal component analysis (PCA) and factor analysis of mixed data (FAMD) or self-organized maps) could allow us to apply the method also to ungauged basins, thus contributing to the PUB initiative (Hrachowitz et al., 2013). Another step would be then to create a distributed "Kirchner type" hydrological model where a parameter set would be attributed to "regions" discretized on the basis of their physiographic characteristics. This would allow us to determine the rainfall–runoff behaviour in large scale river basins by taking into account the precipitation spatial distribution and flood flow routing through the channel network. We would then be able to broaden our understanding of nonlinear catchment response and travel time lags as suggested by Kirchner (2009).

Acknowledgements. The study is conducted within the FloodScale project, funded by the French National Research Agency (ANR) under contract no. ANR 2011 BS56 027, which contributes to the HyMeX program. The HyMeX database teams (ESPRI/IPSL and SEDOO/Observatoire Midi-Pyrénées) helped in accessing the data. The authors acknowledge Brice Boudevillain for providing the OHM-CV rainfall data, Météo-France for their rainfall and SAFRAN climate data. EdF-DTG provided discharge data from three of the gauges used in this study. We thank the Region Rhône-Alpes for its funding of the PhD grant of the first author. We thank R. Krier for providing us the codes used to perform the recession analysis and E. Leblois for constructive comments on the paper.

Edited by: M. Mikos

References

Adamovic, M.: Development of a data-driven distributed hydrological model for regional catchments prone to Mediterranean flash floods. Application to the Ardèche catchment (France), PhD thesis, University of Grenoble, France, 2014.

Adamovic, M. et al.: Interactive comment on "Does the simple dynamical systems approach provide useful information about catchment hydrological functioning in a Mediterranean context? Application to the Ardèche catchment (France)" by M. Adamovic et al., Hydrol. Earth Syst. Sci. Discuss., 11, C6170–C6171, 2015.

Allen, R., Pereira, L., Raes, D., and Smith, M.: Crop evapotranspiration - Guidelines for computing crop water requirements – FAO Irrigation and drainage paper 56, citeulike-article-id:10458368, 1998.

Beven, K. and Binley, A.: The future of distributed models: Model calibration and uncertainty prediction, Hydrol. Proc., 6, 279–298, doi:10.1002/hyp.3360060305, 1992.

Beven, K.: Towards a coherent philosophy for modelling the environment, Proc. Roy. Soc. Lnd. A, 458, 2465–2484, doi:10.1098/rspa.2002.0986, 2002.

Bloschl, G. and Sivapalan, M.: Scale issues in hydrological modelling: a review, Hydrol. Proc., 9, 251–290, 1995.

Bonnifait, L., Delrieu, G., Lay, M. L., Boudevillain, B., Masson, A., Belleudy, P., Gaume, E., and Saulnier, G.-M.: Distributed hydrologic and hydraulic modelling with radar rainfall input: Reconstruction of the 8–9 September 2002 catastrophic flood event in the Gard region, France, Adv. Water Res., 32, 1077–1089, doi:10.1016/j.advwatres.2009.03.007, 2009.

Boronina, A., Golubev, S., and Balderer, W.: Estimation of actual evapotranspiration from an alluvial aquifer of the Kouris catchment (Cyprus) using continuous streamflow records, Hydrol. Proc., 19, 4055–4068, 2005.

Boudevillain, B., Delrieu, G., Galabertier, B., Bonnifait, L., Bouilloud, L., Kirstetter, P.-E., and Mosini, M.-L.: The Cévennes-Vivarais Mediterranean Hydrometeorological Observatory database, Water Resour. Res., 47, W07701, doi:10.1029/2010wr010353, 2011.

Branger, F., Dramais, G., Horner, I., Le Boursicaud, R., Le Coz, J., and Renard, B.: Improving the quantification of flash flood hydrographs and reducing their uncertainty using noncontact streamgauging methods, EGU General Assembly 2015, Vienna, 12–17 April 2015, Geophys. Res. Abstr., Vol. 17, EGU2015-5768, 2015.

Braud, I., Roux, H., Anquetin, S., Maubourguet, M.-M., Manus, C., Viallet, P., and Dartus, D.: The use of distributed hydrological models for the Gard 2002 flash flood event: Analysis of associated hydrological processes, J. Hydrol., 394, 162–181, doi:10.1016/j.jhydrol.2010.03.033, 2010.

Braud, I., Ayral, P. A., Bouvier, C., Branger, F., Delrieu, G., Le Coz, J., Nord, G., Vandervaere, J. P., Anquetin, S., Adamovic, M., Andrieu, J., Batiot, C., Boudevillain, B., Brunet, P., Carreau, J., Confoland, A., Didon-Lescot, J. F., Domergue, J. M., Douvinet, J., Dramais, G., Freydier, R., Gérard, S., Huza, J., Leblois, E., Le Bourgeois, O., Le Boursicaud, R., Marchand, P., Martin, P., Nottale, L., Patris, N., Renard, B., Seidel, J. L., Taupin, J. D., Vannier, O., Vincendon, B., and Wijbrans, A.: Multi-scale hydrometeorological observation and modelling for flash-flood understanding, Hydrol. Earth Syst. Sci., 11, 1871–1945, doi:10.5194/hessd-11-1871-2014, 2014.

Brauer, C. C., Teuling, A. J., Torfs, P. J. J. F., and Uijlenhoet, R.: Investigating storage-discharge relations in a lowland catchment using hydrograph fitting, recession analysis, and soil moisture data, Water Resour. Res., 49, 4257–4264, doi:10.1002/wrcr.20320, 2013.

Brutsaert, W. and Nieber, J. L.: Regionalized drought flow hydrographs from a mature glaciated plateau, Water Resour. Res., 13, 637–643, doi:10.1029/WR013i003p00637, 1977.

Budyko, M. I.: Climate and life, English Edition, edited by: Miller, D. H., Academic Press, New York, 508 pp., 1974.

Cosandey, C. and Didon-Lescot, J. F.: Etude des crues cevenoles: conditions d'apparition dans un petit bassin forestier sur le versant sud du Mont Lozere, France, International Association of Hydrological Sciences, Wallingford, ROYAUME-UNI, 13 pp., 1989.

Coussot, C.: Assessing and modelling hydrological behaviours of Mediterranean catchments using discharge recession analysis. Master Thesis, HydroHazards, University of Grenoble, France, 54 pp., 2015.

Delrieu, G., Wijbrans, A., Boudevillain, B., Faure, D., Bonnifait, L., and Kirstetter, P.-E.: Geostatistical radar–raingauge merging: A novel method for the quantification of rain estimation accuracy, Adv. Water Resour., 71, 110–124, doi:10.1016/j.advwatres.2014.06.005, 2014.

Drobinski, P., Ducrocq, V., Alpert, P., Anagnostou, E., Béranger, K., Borga, M., Braud, I., Chanzy, A., Davolio, S., Delrieu, G., Estournel, C., Boubrahmi, N. F., Font, J., Grubisic, V., Gualdi, S., Homar, V., Ivancan-Picek, B., Kottmeier, C., Kotroni, V., Lagouvardos, K., Lionello, P., Llasat, M. C., Ludwig, W., Lutoff, C., Mariotti, A., Richard, E., Romero, R., Rotunno, R., Roussot, O., Ruin, I., Somot, S., Taupier-Letage, I., Tintore, J., Uijlenhoet, R., and Wernli, H.: HyMeX, a 10-year multidisciplinary program on the Mediterranean water cycle, B. Am. Meteor. Soc., 95, 1063–1082, doi:10.1175/BAMS-D-12-00242.1, 2013.

Duband, D., Obled, C., and Rodriguez, J. Y.: Unit hydrograph revisited: an alternate iterative approach to UH and effective precipitation identification, J. Hydrol., 150, 115–149, doi:10.1016/0022-1694(93)90158-6, 1993.

Etchevers, P., Durand, Y., Habets, F., Martin, E., and Noilhan, J.: Impact of spatial resolution on the hydrological simulation of the Durance high-Alpine catchment, France, Ann. Glaciol., 32, 87–92, 2001.

Freeze, R. A. and Harlan, R. L.: Blueprint for a physically-based, digitally-simulated hydrologic response model, J. Hydrol., 9, 237–258, doi:10.1016/0022-1694(69)90020-1, 1969.

Fu, B. P.: On the calculation of the evaporation from land surface (in Chinese), Sci. Atmos. Sin., 5, 23–31, 1981.

Gaál, L., Szolgay, J., Kohnová, S., Parajka, J., Merz, R., Viglione, A., and Blöschl, G.: Flood timescales: Understanding the interplay of climate and catchment processes through comparative hydrology, Water Resour. Res., 48, W04511, doi:10.1029/2011WR011509, 2012.

Gao, Z. Q., Liu, C. S., Gao, W., and Chang, N. B.: A coupled remote sensing and the Surface Energy Balance with Topography Algorithm (SEBTA) to estimate actual evapotranspiration over heterogeneous terrain, Hydrol. Earth Syst. Sci., 15, 119–139, doi:10.5194/hess-15-119-2011, 2011.

Garambois, P. A., Roux, H., Larnier, K., Castaings, W., and Dartus, D.: Characterization of process-oriented hydrologic model behavior with temporal sensitivity analysis for flash floods in Mediterranean catchments, Hydrol. Earth Syst. Sci., 17, 2305–2322, doi:10.5194/hess-17-2305-2013, 2013.

Gaume, E., Bain, V., Bernardara, P., Newinger, O., Barbuc, M., Bateman, A., Blaškovičová, L., Blöschl, G., Borga, M., Dumitrescu, A., Daliakopoulos, I., Garcia, J., Irimescu, A., Kohnova, S., Koutroulis, A., Marchi, L., Matreata, S., Medina, V., Preciso, E., Sempere-Torres, D., Stancalie, G., Szolgay, J., Tsanis, I., Velasco, D., and Viglione, A.: A compilation of data on European flash floods, J. Hydrol., 367, 70–78, doi:10.1016/j.jhydrol.2008.12.028, 2009.

Gottardi, F.: Estimation statistique et reanalyse des precipitations en montagne – Utilisation d'ébauches par types de temps et assimilation de donnees d'enneigement : Application aux grands massifs montagneux francais, Hydrology, Institut Polytechnique de Grenoble-INPG, French, https://tel.archives-ouvertes.fr/, 261 pp., 2009.

Gudulas, K., Voudouris, K., Soulios, G., and Dimopoulos, G.: Comparison of different methods to estimate actual evapotranspiration and hydrologic balance, Desalination Water Treat., 51, 2945–2954, doi:10.1080/19443994.2012.748443, 2013.

Gupta, H. V., Sorooshian, S., and Yapo, P. O.: Status of automatic calibration for hydrologic models: Comparison with multilevel expert calibration, J. Hydrol. Eng., 4, 135–143, 1999.

Hernández, E., Cana, L., Díaz, J., García, R., and Gimeno, L.: Mesoscale convective complexes over the western Mediterranean area during 1990–1994, Meteorl. Atmos. Phys., 68, 1–12, 1998.

Hrachowitz, M., Savenije, H. H. G., Blöschl, G., McDonnell, J. J., Sivapalan, M., Pomeroy, J. W., Arheimer, B., Blume, T., Clark, M. P., Ehret, U., Fenicia, F., Freer, J. E., Gelfan, A., Gupta, H. V., Hughes, D. A., Hut, R. W., Montanari, A., Pande, S., Tetzlaff, D., Troch, P. A., Uhlenbrook, S., Wagener, T., Winsemius, H. C., Woods, R. A., Zehe, E., and Cudennec, C.: A decade of Predictions in Ungauged Basins (PUB) – a review, Hydrol. Sci. J., 58, 1198–1255, doi:10.1080/02626667.2013.803183, 2013.

Kirchner, J. W.: Getting the right answers for the right reasons: Linking measurements, analyses, and models to advance the science of hydrology, Water Resour. Res., 42, W03S04, doi:10.1029/2005wr004362, 2006.

Kirchner, J. W.: Catchments as simple dynamical systems: Catchment characterization, rainfall-runoff modeling, and doing hydrology backward, Water Resour. Res., 45, W02429, doi:10.1029/2008WR006912, 2009.

Klemes, V.: Operational testing of hydrological simulation models, Hydrol. Sci. J., 31, 13–24, 1986.

Klemeš, V.: Conceptualization and scale in hydrology, J. Hydrol., 65, 1–23, doi:10.1016/0022-1694(83)90208-1, 1983.

Krier, R., Matgen, P., Goergen, K., Pfister, L., Hoffmann, L., Kirchner, J. W., Uhlenbrook, S., and Savenije, H. H. G.: Inferring catchment precipitation by doing hydrology backward: A test in 24 small and mesoscale catchments in Luxembourg, Water Resour. Res., 48, W10525, doi:10.1029/2011WR010657, 2012.

Lafaysse, M., Hingray, B., Etchevers, P., Martin, E., and Obled, C.: Influence of spatial discretization, underground water storage and glacier melt on a physically-based hydrological model of the Upper Durance River basin, J. Hydrol., 403, 116–129, 2011.

Lang, M., Moussay, D., Recking, A., and Naulet, R.: Hydraulic modelling of historical floods: a case study on the Ardeche river at Vallon-Pont-d'Arc, 183–189, 2002.

Le Coz, J., Hauet, A., Pierrefeu, G., Dramais, G., and Camenen, B.: Performance of image-based velocimetry (LSPIV) applied to flash-flood discharge measurements in Mediterranean rivers, J. Hydrol., 394, 42–52, doi:10.1016/j.jhydrol.2010.05.049, 2010.

Le Coz, J., Renard, B., Bonnifait, L., Branger, F., and Le Boursicaud, R.: Combining hydraulic knowledge and uncertain gaugings in the estimation of hydrometric rating curves: A Bayesian approach, J. Hydrol., 509, 573–587, doi:10.1016/j.jhydrol.2013.11.016, 2014.

Le Lay, M. and Saulnier, G. M.: Exploring the signature of climate and landscape spatial variabilities in flash flood events: Case of the 8–9 September 2002 Cévennes-Vivarais catastrophic event, Geophys. Res. Lett., 34, L13401, doi:10.1029/2007GL029746, 2007.

Manus, C., Anquetin, S., Braud, I., Vandervaere, J. P., Creutin, J. D., Viallet, P., and Gaume, E.: A modeling approach to assess the hydrological response of small mediterranean catchments to the variability of soil characteristics in a context of extreme events, Hydrol. Earth Syst. Sci., 13, 79–97, doi:10.5194/hess-13-79-2009, 2009.

McDonnell, J. J., Sivapalan, M., Vaché, K., Dunn, S., Grant, G., Haggerty, R., Hinz, C., Hooper, R., Kirchner, J., Roderick, M. L., Selker, J., and Weiler, M.: Moving beyond heterogeneity and process complexity: A new vision for watershed hydrology, Water Resour. Res., 43, W07301, doi:10.1029/2006WR005467, 2007.

Melsen, L. A., Teuling, A. J., van Berkum, S. W., Torfs, P. J. J. F., and Uijlenhoet, R.: Catchments as simple dynamical systems: A case study on methods and data requirements for parameter identification, Water Resour. Res., 50, 5577–5596, doi:10.1002/2013WR014720, 2014.

Molinié, G., Ceresetti, D., Anquetin, S., Creutin, J. D., and Boudevillain, B.: Rainfall Regime of a Mountainous Mediterranean Region: Statistical Analysis at Short Time Steps, J. Appl. Meteorol. Climatol., 51, 429–448, doi:10.1175/2011JAMC2691.1, 2011.

Moriasi, D. N., Arnold, J. G., Van Liew, M. W., Bingner, R. L., Harmel, R. D., and Veith, T. L.: Model evaluation guidelines for systematic quantification of accuracy in watershed simulations, Trans. ASABE, 50, 885–900, 2007.

Nash, J. E. and Sutcliffe, J. V.: River flow forecasting through conceptual models part I – A discussion of principles, J. Hydrol., 10, 282–290, 1970.

Naulet, R., Lang, M., Ouarda, T. B. M. J., Coeur, D., Bobée, B., Recking, A., and Moussay, D.: Flood frequency analysis on the Ardèche river using French documentary sources from the last two centuries, J. Hydrol., 313, 58–78, 2005.

Pike, J. G.: The estimation of annual run-off from meteorological data in a tropical climate, J. Hydrol., 2, 116–123, doi:10.1016/0022-1694(64)90022-8, 1964.

Quintana-Seguí, P., Le Moigne, P., Durand, Y., Martin, E., Habets, F., Baillon, M., Canellas, C., Franchisteguy, L., and Morel, S.: Analysis of near-surface atmospheric variables: Validation of the SAFRAN analysis over France, J. Appl. Meteorol. Climatol., 47, 92–107, 2008.

Samaniego, L., Kumar, R., and Attinger, S.: Multiscale parameter regionalization of a grid-based hydrologic model at the mesoscale, Water Resour. Res., 46, W05523, doi:10.1029/2008WR007327, 2010.

Saulnier, G. M. and Le Lay, M.: Sensitivity of flash-flood simulations on the volume, the intensity, and the localization of rainfall in the Cévennes-Vivarais region (France), Water Resour. Res., 45, W10425, doi:10.1029/2008WR006906, 2009.

Schreiber, P.: Über die Beziehungen zwischen dem Niederschlag und der Wasserführung der Flüsse in Mitteleuropa, Zeitschr. Meteorol., 21, 441–452, 1904.

Seiler, C. and Moene, A. F.: Estimating Actual Evapotranspiration from Satellite and Meteorological Data in Central Bolivia, Earth Interact., 15, 1–24, doi:10.1175/2010EI332.1, 2010.

Sempere-Torres, D., Rodriguez-Hernandez, J. Y., and Obled, C.: Using the DPFT approach to improve flash flood forecasting models, Nat. Hazards, 5, 17–41, 1992.

Sheffer, N. A., Enzel, Y., and Benito, G.: Paleofloods in southern France-the Ardeche River, PHEFRA workshop, Barcelona, 25–31, 2002.

Sivapalan, M.: Process complexity at hillslope scale, process simplicity at the watershed scale: is there a connection?, Hydrological Processes, 17, 1037–1041, doi:10.1002/hyp.5109, 2003a.

Sivapalan, M.: Prediction in ungauged basins: a grand challenge for theoretical hydrology, Hydrol. Proc., 17, 3163–3170, doi:10.1002/hyp.5155, 2003b.

Sivapalan, M.: Pattern, Process and Function: Elements of a Unified Theory of Hydrology at the Catchment Scale, in: Encyclopedia of Hydrological Sciences, John Wiley & Sons, Ltd, Chichester, UK, 193–219, doi:10.1002/0470848944.hsa012, 2006.

Tarboton, D. G., Schreuders, K. A. T., Watson, D. W., and Baker, M. E.: Generalized terrain-based flow analysis of digital elevation models, 18th World IMACS Congress and MODSIM09 International Congress on Modelling and Simulation, 2000–2006, 2009.

Tekleab, S., Uhlenbrook, S., Mohamed, Y., Savenije, H. H. G., Temesgen, M., and Wenninger, J.: Water balance modeling of Upper Blue Nile catchments using a top-down approach, Hydrol. Earth Syst. Sci., 15, 2179–2193, doi:10.5194/hess-15-2179-2011, 2011.

Teuling, A. J., Lehner, I., Kirchner, J. W., and Seneviratne, S. I.: Catchments as simple dynamical systems: Experience from a Swiss prealpine catchment, Water Resour. Res., 46, W10502, doi:10.1029/2009WR008777, 2010.

Thierion, C., Longuevergne, L., Habets, F., Ledoux, E., Ackerer, P., Majdalani, S., Leblois, E., Lecluse, S., Martin, E., Queguiner, S., and Viennot, P.: Assessing the water balance of the Upper Rhine Graben hydrosystem, J. Hydrol., 424–425, 68–83, doi:10.1016/j.jhydrol.2011.12.028, 2012.

Thornthwaite, C. and Mather, J.: The water balance, Climatology, VIII, New Jersey, NY, 1–37, 1955.

Tramblay Y., Bouvier C., Crespy A., and Marchandise A.: Improvement of flash flood modelling using spatial patterns of rainfall: a case study in south of France. Global Change: Facing Risks and Threats to Water Resources Proceedings of the Sixth World FRIEND Conference, Fez, Morocco, October 2010, IAHS Publ. 340, 172–178, http://y.tramblay.free.fr/doc/Tramblay-redbook.340.pdf, 2010.

Turc, L.: Nouvelles formules pour le bilan d'eau en fonction des valeurs moyennes annuelles de précipitations et de la température, Comptes Rendus de l'Académie des Sciences, Paris, 233, 633–635, 1951.

Uhlenbrook, S., Seibert, J. A. N., Leibundgut, C., and Rodhe, A.: Prediction uncertainty of conceptual rainfall-runoff models caused by problems in identifying model parameters and structure, Hydrol. Sci. J., 44, 779–797, doi:10.1080/02626669909492273, 1999.

Vannier, O.: Apport de la modélisation hydrologique régionale à la compréhension des processus de crue en zone méditerranéenne, Thèse de l'Ecole doctorale Terre, Univers, Environnement, University of Grenoble, Grenoble, France, 22 November 2013, 274 pp., 2013.

Vannier, O., Braud, I., and Anquetin, S.: Regional estimation of catchment-scale soil properties by means of streamflow recession analysis for use in distributed hydrological models, Hydrol. Proc., 28, 6276–6291, doi:10.1002/hyp.10101, 2014.

Vidal, J. P., Martin, E., Franchistéguy, L., Baillon, M., and Soubeyroux, J. M.: A 50-year high-resolution atmospheric reanalysis over France with the Safran system, Int. J. Climatol., 30, 1627–1644, 2010.

Yan, Z., Gottschalk, L., Leblois, E., and Xia, J.: Joint mapping of water balance components in a large Chinese basin, J. Hydrol., 450–451, 59–69, doi:10.1016/j.jhydrol.2012.05.030, 2012.

Wittenberg, H. and Sivapalan, M.: Watershed groundwater balance estimation using streamflow recession analysis and baseflow separation, J. Hydrol., 219, 20–33, doi:10.1016/s0022-1694(99)00040-2, 1999.

Zehe, E., Lee, H., and Sivapalan, M.: Dynamical process upscaling for deriving catchment scale state variables and constitutive relations for meso-scale process models, Hydrol. Earth Syst. Sci., 10, 981–996, doi:10.5194/hess-10-981-2006, 2006.

Zhang, L., Hickel, K., Dawes, W. R., Chiew, F. H. S., Western, A. W., and Briggs, P. R.: A rational function approach for estimating mean annual evapotranspiration, Water Resour. Res., 40, W02502, doi:10.1029/2003WR002710, 2004.

Zhang, L., Potter, N., Hickel, K., Zhang, Y., and Shao, Q.: Water balance modeling over variable time scales based on the Budyko framework – Model development and testing, J. Hydrol., 360, 117–131, 2008.

11

Large-scale analysis of changing frequencies of rain-on-snow events with flood-generation potential

D. Freudiger, I. Kohn, K. Stahl, and M. Weiler

Chair of Hydrology, University of Freiburg, Freiburg, Germany

Correspondence to: D. Freudiger (daphne.freudiger@hydrology.uni-freiburg.de)

Abstract. In January 2011 a rain-on-snow (RoS) event caused floods in the major river basins in central Europe, i.e. the Rhine, Danube, Weser, Elbe, Oder, and Ems. This event prompted the questions of how to define a RoS event and whether those events have become more frequent. Based on the flood of January 2011 and on other known events of the past, threshold values for potentially flood-generating RoS events were determined. Consequently events with rainfall of at least 3 mm on a snowpack of at least 10 mm snow water equivalent (SWE) and for which the sum of rainfall and snowmelt contains a minimum of 20 % snowmelt were analysed. RoS events were estimated for the time period 1950–2011 and for the entire study area based on a temperature index snow model driven with a European-scale gridded data set of daily climate (E-OBS data). Frequencies and magnitudes of the modelled events differ depending on the elevation range. When distinguishing alpine, upland, and lowland basins, we found that upland basins are most influenced by RoS events. Overall, the frequency of rainfall increased during winter, while the frequency of snowfall decreased during spring. A decrease in the frequency of RoS events from April to May has been observed in all upland basins since 1990. In contrast, the results suggest an increasing trend in the magnitude and frequency of RoS days in January and February for most of the lowland and upland basins. These results suggest that the flood hazard from RoS events in the early winter season has increased in the medium-elevation mountain ranges of central Europe, especially in the Rhine, Weser, and Elbe river basins.

1   Introduction

Rain-on-snow (RoS) events are relevant for water resources management, particularly for flood forecasting and flood risk management (McCabe et al., 2007). RoS events have the potential to cause large flood events during the winter season. They represent one of five flood process types defined by Merz and Blöschl (2003) that occur in temperate-climate mountain river systems and are strongly elevation dependent. These events are complex as they do not only depend on the rain intensity and amount, but also on the prevailing freezing level, the snow water equivalent (SWE), the snow energy content, the timing of release, and the areal extent of the snowpack (Kattelmann, 1997; McCabe et al., 2007). Snowpacks are water reservoirs of large regional extent and storage capacity, which can produce rapid melt in combination with warm air temperatures and high humidity (e.g. Singh et al., 1997; Marks et al., 1998). Consequently, cumulative rainfall and snowmelt can increase the magnitude of runoff and can thus generate much greater potential for flooding than a usual snowmelt event (Kattelmann, 1985; Marks et al., 1998). Besides their large damage potential, such events are also very difficult to forecast as shown by Rössler et al. (2014) for a RoS-driven flood event in October 2011 in the Bernese Alps in Switzerland. Scientific interest has therefore increased in the last decades, and a number of different methods of analysis have been developed to better understand and quantify the physical processes by studying individual events in different locations (e.g. Blöschl et al., 1990; Singh et al., 1997; Floyd and Weiler, 2008; Garvelmann et al., 2013).

Many studies observed an increase in the occurrence of rainfall in the wintertime and a trend to earlier snowmelt due to an increase of air temperatures in Europe (e.g. Birsan et al., 2005; Renard et al., 2008). Furthermore, Köplin et al. (2014) predicted a shift from snowmelt-dominated runoff to a more variable snow- and rain-fed regime in the future in Switzerland. These meteorological changes are very likely to influence the occurrence and magnitude of RoS events, and Köplin et al. (2014) predict a diversification of flood types in the wintertime as well as an increase of RoS flood events in the future in Switzerland.

Although Merz and Blöschl (2003) observed that 20 % of the flood events in Austria were RoS-driven during the period 1971–1997 and hence showed the importance of such events in central Europe, only few studies have specifically analysed the changes of the frequency of RoS events over time and especially over large areas. Ye et al. (2008) observed an increase in RoS days in northern Eurasia, which they were able to correlate with the observed increase in air temperature and rainfall in the wintertime. Sui and Koehler (2001) attributed an increase in peak flows in the northern Danube tributaries in Germany to an increase in RoS events, based on the combination of decreasing SWE and increasing maximum daily winter precipitation sums they found at a number of climate stations in the area. McCabe et al. (2007) found disparate trends in the western USA with generally positive temporal trends of RoS events frequencies for high-elevation sites and negative trends for low-elevation sites. In these areas, the increase of temperature appears to affect the occurrence of snow, contributing therefore to a lower frequency of RoS events (McCabe et al., 2007). Similarly, Surfleet and Tullos (2013) predicted with a model experiment that an increase in air temperature due to climate change would lead to a decrease of high peak flow due to RoS events for low- and middle-elevation zones, while at high-elevation bands these kinds of events would increase. All these studies show the correlation between the frequency of RoS events, the changes in air temperature, and the importance of the elevation range. They therefore stress the need for a more accurate trend analysis of those events in the context of climate change in central Europe, where discharges mainly depend on alpine and mid-elevation tributaries.

Previous studies differ on the definition of RoS events. McCabe et al. (2007) and Surfleet and Tullos (2013) defined an event as RoS-driven if simultaneously rainfall occurs, maximum daily temperature is greater than $0\,°C$, and a decrease in snowpack can be observed; while for Ye et al. (2008), a RoS event takes place only when at least one of four daily precipitation measurements is liquid and the ground is covered by $\geq 1\,cm$ of snow. Sui and Koehler (2001) found that most RoS events in southern Germany occurred when snowmelt was larger than the rainfall depth. These definitions allow identifying all possible RoS events but are insufficient if one focusses on the events that can effectively cause flood events.

Due to the great hydrologic impact that RoS events can have, there is a real need for assessing the changes in frequencies of those RoS events that may generate large floods. A good example of such an event is the flood in January 2011 in central Europe. During a strong negative phase of the

North Atlantic Oscillation, temperature anomalies in December 2010 reached $-4\,°C$ in central and northern Europe (Lefebvre and Becker, 2011), and record snowpacks were observed nearly all over Germany for this time of the year (e.g. Böhm et al., 2011; LHW, 2011; Besler, 2011). January 2011 then brought thawing temperatures in combination with rainfall events, and from 6 to 16 January very high flows were observed at nearly all German gauging stations (e.g. Böhm et al., 2011; Bastian et al., 2011; Karuse, 2011; LHW, 2011; Fell, 2011; Besler, 2011). Kohn et al. (2014) identified the simultaneous occurrence of rainfall and snowmelt as the driving factor for those flood events, which led, beside other impacts, to a restriction of navigation on the river Rhine and large inundations in the lower Elbe river basin. The aims of this study are therefore (i) to derive criteria for RoS-driven events that have the potential to cause floods, using the case study of January 2011 in Germany, and (ii) to analyse the changes in frequencies and magnitudes of these types of events during the time period 1950–2011 in six major central European river basins, i.e. Rhine, Danube, Elbe, Weser, Oder, and Ems.

2   Materials and methods

2.1   Study area

The study area embodies the six major river basins of the German fluvial network. Since only German streamflow records were used, the basins of the rivers Rhine, Danube, Elbe, Weser, Oder, and Ems are considered only upstream of the most downstream station in German territory (Fig. 1). According to the Hydrological Atlas of Germany (HAD, Bundesanstalt für Gewässerkunde), the basins were divided into alpine, upland, or lowland sub-basins. This classification is motivated by the elevation of the main tributaries, but is not strictly guided by elevation as the main rivers drop quickly to lower elevations.

Only the basins of the rivers Rhine and Danube have an "alpine" section in this classification. The alpine portion of the Rhine encompasses the basin ca. above Basel, and besides the entire basin area in Austria and Switzerland with high mountains up to 4000 m a.s.l. it also includes the southern Black Forest with elevations of only up to 1500 m a.s.l. The alpine section of the Danube basin consists also of a small part in the Black Forest near the source, but then includes mainly the southern tributaries of the Danube from the Alps. They comprise the river Inn, which originates from the Swiss and Austrian Alps with elevations over 3000 m a.s.l., as well as a number of tributaries from the northern range of the Alps along the German–Austrian border with elevations below 3000 m a.s.l.

All river basins contain upland areas. These stretch from near the southern border of Germany to the southern boundary of the northern German lowlands, as well as in the

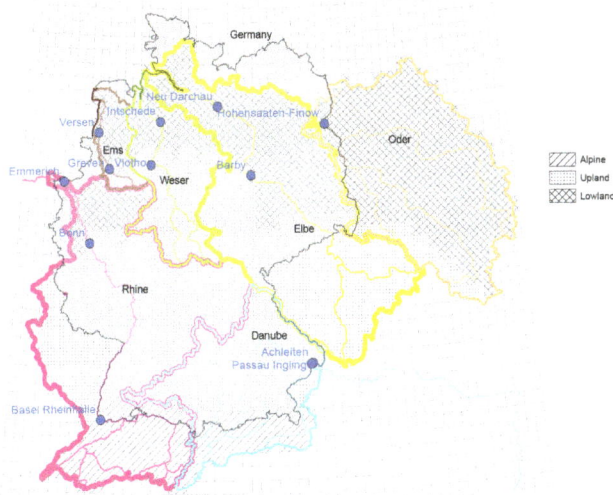

Figure 1. Study area: delimitation of the basin boundaries of the Rhine, Danube, Elbe, Weser, Oder, and Ems basins and subdivision into alpine, upland, and lowland. The background raster corresponds to the E-OBS data set. Gauging stations at the outlet of each sub-basin are represented with blue dots.

northern part of the Czech Republic (Elbe basin). The landscape can be described as upland with elevation ranges from 200–300 m a.s.l. to up to Feldberg, 1493 m a.s.l., the highest mountain in Germany outside the Alps.

With the exception of the Danube, which is only considered to the German border, all river basins contain a lowland section. These areas in northern Germany and western Poland (Oder basin) are mainly constituted of lowland areas with altitudes ranging from 0 to 200 m a.s.l.

2.2 Meteorological and hydrometric data

Daily mean temperature and precipitation sums were obtained from the European Climate Assessment and Dataset project (ECA&D, http://www.ecad.eu) and the EU-FP6 project ENSEMBLES (http://ensembles-eu.metoffice.com). The so-called E-OBS data set (version 6.0) was interpolated from climate stations all across Europe into a $0.25° \times 0.25°$ regular latitude–longitude grid (Fig. 1; Haylock et al., 2008). The time series are available from 1 January 1950 to 31 December 2011 and cover the study area 46.00–55.25° N, 5.25–19.75° E.

Daily mean discharge data from more than 300 gauging stations (Fig. 2) in Germany were provided by German public authorities. Details on the assembled data set can be found in Kohn et al. (2014). The time series are of different lengths, but most of them cover the period 1950–2011. Since authorities usually correct peak discharge values with hydraulic modelling or revise rating curves during the years after a flood event, some of the most recent data used in this study were raw (as yet uncorrected) data. However, later correc-

tions to the peaks are not expected to change the relative ranking and hence the results of the trend analysis, and the discharge data were found suitable for this study. To assess the accuracy of the data, all data included in this analysis passed a visual quality control.

2.3 Estimation of snowpack and snowmelt

Snow accumulation and melt were estimated based on daily E-OBS mean temperature and precipitation sum data for the entire study area and are given in mm SWE. Precipitation is assumed to be solid if air temperature $T_a < 1 °C$ and liquid if $T_a \geq 1 °C$. Snowmelt M (mm) is estimated using a temperature index model, which assumes a relationship between ablation and air temperature (Eq. 1; e.g. Finsterwalder and Schunk, 1887; Collins, 1934; Corps of Engineers, 1956).

$$M = M_f \cdot (T_a - T_b) \tag{1}$$

Martinec and Rango (1986) calculated degree-day factors M_f for open areas depending on the snow density of the snowpack. They suggested M_f values from 3.5 to $6 \, \text{mm} \, °C^{-1} \, \text{day}^{-1}$ and even smaller for fresh snow. They also observed that M_f increases over the melt period. Hock (2003) listed M_f values for snow in high-elevation areas between 2.5 and $6 \, \text{mm} \, °C^{-1} \, \text{day}^{-1}$. For sake of simplicity of the large-scale analysis, a constant conservative value of $M_f = 3 \, \text{mm} \, °C^{-1} \, \text{day}^{-1}$ was chosen for the entire study area and melt period. This value was found to represent the area well, since snow melts very fast in upland and lowland regions in Germany and the snowpack consists therefore mainly of fresh snow. The base temperature T_b represents the threshold temperature for melting snow. Most studies set T_b to 1 °C, since energy is needed to bring the snow to 0 °C to start melting (e.g. Hock, 2003). T_b was therefore set to 1 °C. The sensitivity of the subsequent trend calculation to the choice of T_b and M_f was tested ranging from 0 to 2 °C for T_b and from 2 to $5 \, \text{mm} \, °C^{-1} \, \text{day}^{-1}$ for M_f. Both parameters were found to have an impact on the calculated snow depth but to be rather insensitive for the trends calculation, since the trend analysis considers relative changes to the mean (Kohn et al., 2014).

For every grid cell, daily SWE of the snowpack was calculated for day i as the sum of the SWE of the day before and the snowmelt M (mm) or snowfall S (mm SWE) of the actual day (Eq. 2) and is given in mm SWE.

$$\text{SWE}_i = \begin{cases} \text{SWE}_{i-1} + S_i, & \text{if} \quad T_a < T_b \\ \text{SWE}_{i-1} - M_i, & \text{if} \quad T_a \geq T_b \end{cases} \tag{2}$$

Snowpack was estimated for a period from 2 to 1 August of the following year. Since August is the month with the least likely snow accumulation, SWE was re-set to zero on 2 August of each year. Therefore the model only accounts for annual snow and no multi-year snow is taken into account in the high alpine areas. The results show the winter period

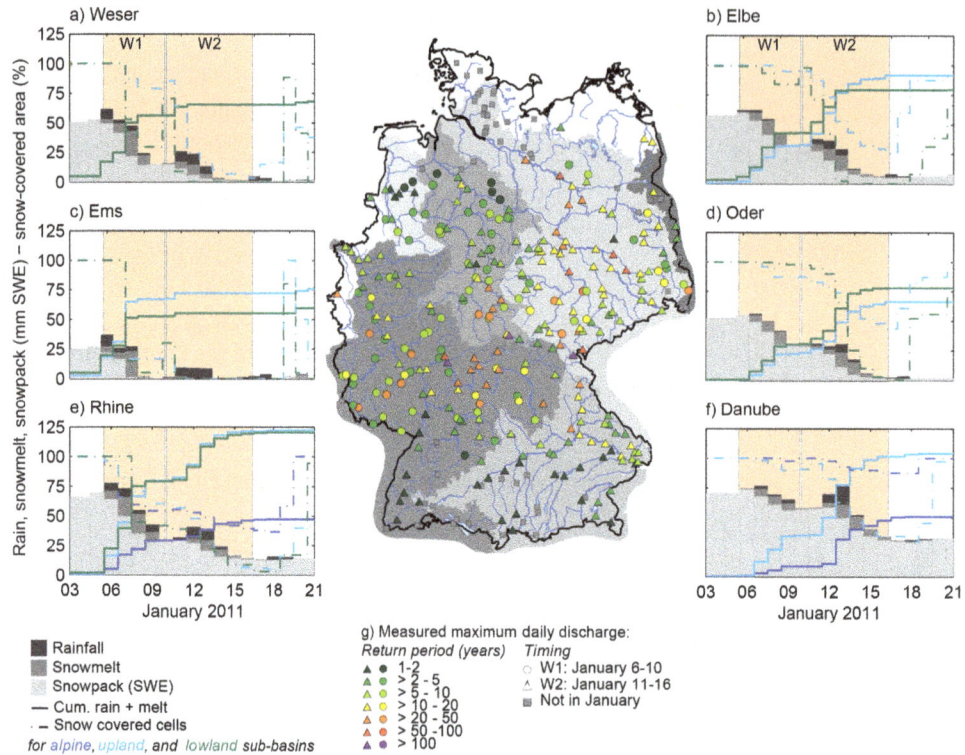

Figure 2. Large-scale analysis of the RoS event of January 2011. (**a–f**) Mean basin-wide daily snowpack, rainfall, and snowmelt (mm SWE) from 1 to 31 January, as well as the percentage of snow-covered cells. Map: return period and occurrence period of the maximum daily discharge in the calender year 2011 at all gauging stations (modified from Kohn et al., 2014).

from 1 November to 31 May taking into account potential snow season extents in the entire study area. As snow measurements in Germany are available in few locations only, the calculated SWE was compared by (Kohn et al., 2014) to the products of the snow model SNOW4 (Germany's National Weather Service, DWD), which is based on the interpolation of ground-based snow measurements, for the winter 2010–2011. A frequency analysis was performed on the occurrence and amount of snow per area every day, and both model outputs were found to be very similar.

2.4 Definition of RoS events

The aim of this study is to identify those RoS events in time series of rainfall occurrence and snowpack existence that have the potential to cause floods. Thus selection criteria need to be defined. The general variables for a RoS-driven runoff generation event are as follows:

1. Rainfall R: the amount of R must be substantial, otherwise the event may be only snowmelt-driven.

2. Snowpack SWE: SWE needs to be large enough to be able to substantially contribute to runoff.

3. Snowmelt M: the amount of M must be large enough compared to R; otherwise the event may be rather rain-driven.

The time and magnitude of a flood driven by a RoS event also depend on the response time of a basin. In this study we distinguish between a RoS day and a RoS event. A RoS day is defined as a day when all hydrometeorologic conditions (R, SWE, and M) for a RoS event are met. A RoS event always starts with a RoS day and may contain RoS days and non-RoS days. A RoS event lasts from the initial RoS day to the day when the maximum discharge is observed after the first or after additional RoS days and within an assumed maximum response time following a RoS day. A RoS event with several RoS days is then considered as one event, even if it consists of multiple flood waves. We then defined the equivalent precipitation depth P_{eq} of an event as corresponding to the sum of daily rainfall and snowmelt during the RoS event.

The 2011 RoS event in Germany is a good example of the flood-generation potential of such events. Based on the reanalysis of this 2011 RoS event in Germany (Kohn et al., 2014), selection criteria in the form of threshold values for the variables described above were defined. For the reanalysis, the return periods of the annual maximum daily discharge at all available gauging stations were calculated with

Table 1. Historical RoS events (sources: Wetterchronik, 2001; Kohn et al., 2014).

Date	Province	Basin	Event description
18 Mar 1970	Lower Saxony	Elbe	Snowmelt and rainfall led to flooding all over the river, especially in Uelzen and Lüneburg.
27–28 Feb 1987	Bavaria	Danube	Large flooding due to snowmelt and incessant rainfalls in Schambachtal.
20 Jan 1997	Rhineland-Palatinate	Rhine	Small flooding after a snow-rich January and a very wet February.
6–10 Jan 2011	Central Europe	Western basins: Rhine, Weser, Ems	First wave of large-scale flooding due to snow-rich December 2010 followed by thawing temperatures in January 2011.
11–16 Jan 2011	Central Europe	All basins	Second wave of large-scale flooding due to snow-rich December 2010 followed by thawing temperatures in January 2011.

the generalised extreme value distribution and the parameters were estimated with the maximum likelihood method (Venables and Ripley, 2002). This allowed proving the importance of the associated flood peaks. To clearly identify the flood event as RoS-driven, the measured peak discharge during the event was then compared to rainfall, modelled snow depth, and modelled snowmelt for each sub-basin, which allowed defined threshold values for R, SWE, and M. Additionally, documented historical events, for which RoS processes were identified in the literature as the main cause for flooding within the study area, are listed in Table 1. These kinds of events are overall not well documented and the information comes mostly from diverse textual information sources, but it gives us the location and the day of occurrence of past RoS floods. This information was used to "validate" the selection criteria for RoS events that have been set on the flood event of January 2011, since it allows checking if the criteria are representative for other past events.

To compare P_{eq} with the observed river discharge of RoS events that caused floods at the sub-basin scale, 12 gauging stations were selected (Fig. 1). In the case of nested basins, the difference in discharge between the lower and upper station was considered. No discharge data were available for the Oder basin outside Germany; therefore only one station at the outlet of the lowland sub-basin was considered. Finally, the probability distribution of the daily discharge values was classified into quantiles for each month and sub-basin. This expression gives an overview of the seasonal anomaly of RoS-driven peak discharge values. The 0.5 quantile represents the median, and the closer to 1 the quantile is, the higher is the discharge and the higher is its flood-generating potential.

2.5 Trend analysis

For all alpine, upland, and lowland sub-basins, the P_{eq}, the corresponding observed peak discharge, as well as the other variables influencing the occurrence of RoS events – namely R, S, SWE, and M – were analysed for temporal trends. These trends were calculated as the slope of a linear regression with time and are expressed in percent to the mean value of the time series. This allows comparing the importance of the changes in time in one basin with the changes in other basins. The statistical significance of the trends was tested at a 5 % significance level using the non-parametric Mann–Kendall test (Mann, 1945). Trends were calculated and compared for the time periods 1950–2011 and 1990–2011, hereafter referred to as long-term and short-term trends.

3 Results

3.1 Reanalysis of the large-scale RoS event in January 2011 in central Europe

The return periods of January 2011 peak flows illustrate the severity of large-scale flooding across Germany (Fig. 2g). Maximum annual daily discharge was observed from 6 to 16 January at all gauging stations except in the northern part of the Elbe lowland sub-basin and in the Danube alpine sub-basin. With few exceptions discharge peaks reached at least a 1–2-year flood level, with large areas being affected by 20–50-year floods along the main rivers and a few exception with 100-year floods in headwaters. Since two distinct discharge peaks were observed at nearly all gauging stations, the flood can be described as a two-wave flood event that spread from west to east. Figure 2a–f further shows the evolution of the mean basin-wide daily R, SWE, and the percentage of snow-covered grid cells in the Rhine, Danube, Elbe, Weser, Ems, and Oder river basins during the flood event. On 5 January 2011, snow covered 100 % of the grid cells of the study area and the mean SWE varied from 25 mm (Ems) to 70 mm (Danube). The discharge peaks correspond very well to two phases of rainfall combined with snowmelt, and the event was therefore identified as a RoS-driven flood.

During the first flood wave (W1, from 6 to 10 January), mean basin R of 2–10 mm day^{-1} fell on the western basins (Fig. 2a, c, d), while in the eastern basins R never exceeded 2 mm day^{-1} (Fig. 2b, d, f). W1 generated in total P_{eq} between 51 and 71 mm in the western basins Rhine, Weser, and Ems, and between 22 and 48 mm in the eastern basins Danube, Elbe, and Oder. In both regions the snowmelt content in P_{eq} exceeded 25 %. Rain and snowmelt played an equally important role in the western basins, whereas snowmelt played the most important role in the eastern basins.

During the second wave (W2, from 11 to 16 January), mean basin R reached 0–10 mm day^{-1} in the western basins and 2–17 mm day^{-1} in the eastern basins. W2 generated in total P_{eq} between 27 and 48 mm in the western basins, and between 49 and 26 mm in the eastern basins. On 10 January, most of the snow had already melted in the northwestern part of Germany and W2 was therefore rather rain-driven in this area, especially in the Ems river basin, where the second flood wave was only rain-driven since all snow was already melted at this time. However, the snowpack was still substantial at the time in central Germany. After W2, nearly all snow was melted in the Ems, Weser, and Oder basins. In the Rhine and Danube river basins, most of the remaining snow was located in the alpine region.

In Fig. 2a–f the cumulative P_{eq}s are also compared for all sub-basins according to their elevation classification. The alpine sub-basins of the Rhine and Danube produced on average 4 mm day^{-1} equivalent precipitation depth during W1 and W2, which was only half of the other sub-basins. However, on 13 January the cumulative equivalent precipitation depth increased by 21 mm in the Danube alpine sub-basin. The upland sub-basins reacted very strongly to the RoS events, with P_{eq} of almost 25 mm day^{-1} during W1 in all basins and especially in the west. The Weser upland areas were once again strongly affected during W2. After W2, nearly all the grid cells were free of snow. During W1 P_{eq} was mostly caused by snowmelt at all elevation ranges in the east and showed very similar reaction for all lowland and upland sub-basins. W2 resulted in a very fast increase of the cumulative P_{eq} within few days in the east, especially for the Elbe and Danube upland sub-basins (Fig. 2b, f). The Weser and Ems lowland sub-basins, in the western half of Germany, were also strongly impacted during W1, since the amount of snow was substantial and unusual, and it had nearly completely melted by the end of W1. W2 therefore had only little impact on these elevation ranges in these sub-basins (Fig. 2a, c). The Rhine basin usually has a lot of snow during the wintertime, due to its larger upland elevation ranges and the influence of its alpine part. For this reason, there were not many differences in the snowmelt processes of the Rhine lowland and upland sub-basins, and both elevation ranges were strongly impacted by both RoS flood waves.

3.2 Criteria for potentially flood-generating RoS events

Selection criteria for RoS events that have the potential to cause floods were chosen based on the reanalysis of the RoS-driven flood event in January 2011 in central Europe (Sect. 3.1). At the beginning of each flood wave, W1 and W2, the average basin SWE had reached at least 10 mm in all river basins but the Ems, where all snow was already melted after W1. The average basin rainfall depth was at least 2 mm, and the average basin snowmelt was at least 25 % of P_{eq}. We therefore chose conservative threshold values of 3 mm for rainfall, 10 mm SWE for snowpack, and 20 % for the snowmelt amount in P_{eq} to define a RoS day. During the 2011 event, the longest flood wave lasted 6 days (W2), and therefore the maximum response time of all basins was set to 6 days. Thus, the duration of a RoS flood event is limited to 6 days after the last RoS day.

Figure 3 shows P_{eq} for all RoS events ($R \geq 0$ mm, SWE > 0 mm, $M > 0$ %), potentially flood-generating RoS events according to the criteria above ($R \geq 3$ mm, SWE > 10 mm, $M > 20$ %), and documented historical RoS-driven flood events against the corresponding measured discharge for all river sub-basins. Most of the sample "all events" have only little impact on the discharge and will not cause floods or are more rain-driven than rain-on-snow-driven and are therefore not of interest for this analysis. The selection of RoS events according to the above threshold values reveals that most of the RoS events in the alpine sub-basins have the potential to generate floods, while in the lowland sub-basins only few events fall into this category. The criteria $R \geq 3$ mm, SWE > 10 mm, and $M > 20$ % were able to select all documented historical RoS-driven flood events (Table 1) except for those of W1 in the lowland sub-basins of Oder and Elbe in January 2011. Less than 3 mm of rain fell in total, while the high temperatures generated more than 25 mm of snowmelt in these sub-basins. Thus, W1 was rather snowmelt-driven than RoS-driven in the lowland sub-basins, but it was RoS-driven at the scale of the whole Oder and Elbe basins.

Figure 3 also gives the correlation coefficients between P_{eq} and the corresponding measured peak discharges for each sub-basin. The correlation shows how strongly runoff generation is RoS-driven. The higher the correlation, the more the discharge is influenced by RoS events. The alpine and upland sub-basins of the Rhine, Weser, and Danube showed the highest positive correlation, with coefficients between 0.68 and 0.75. In the other upland sub-basins Elbe and Ems, correlation coefficients of 0.48 and 0.51 were found, and in all lowland sub-basins none of the correlation coefficients was higher than 0.48. A few large events in Fig. 3h–l suggest that a RoS event can emphasise a flood event if discharge is already high at the time of occurrence.

The analysis of the discharge quantiles that correspond to the selected potentially flood-generating RoS events further supports these differences. Figure 4 shows the percentage of RoS event in each month's quantile range. In the alpine sub-basins of the Rhine and Danube, all selected RoS events correspond to high discharges (quantiles of 0.7–1) for every month except May. The percentage of events in the highest discharge class (quantiles of 0.9–1) is large for all winter months, and progressively increasing from March to May, which means that the RoS events led to very high daily discharges for the given months in these areas. In the upland

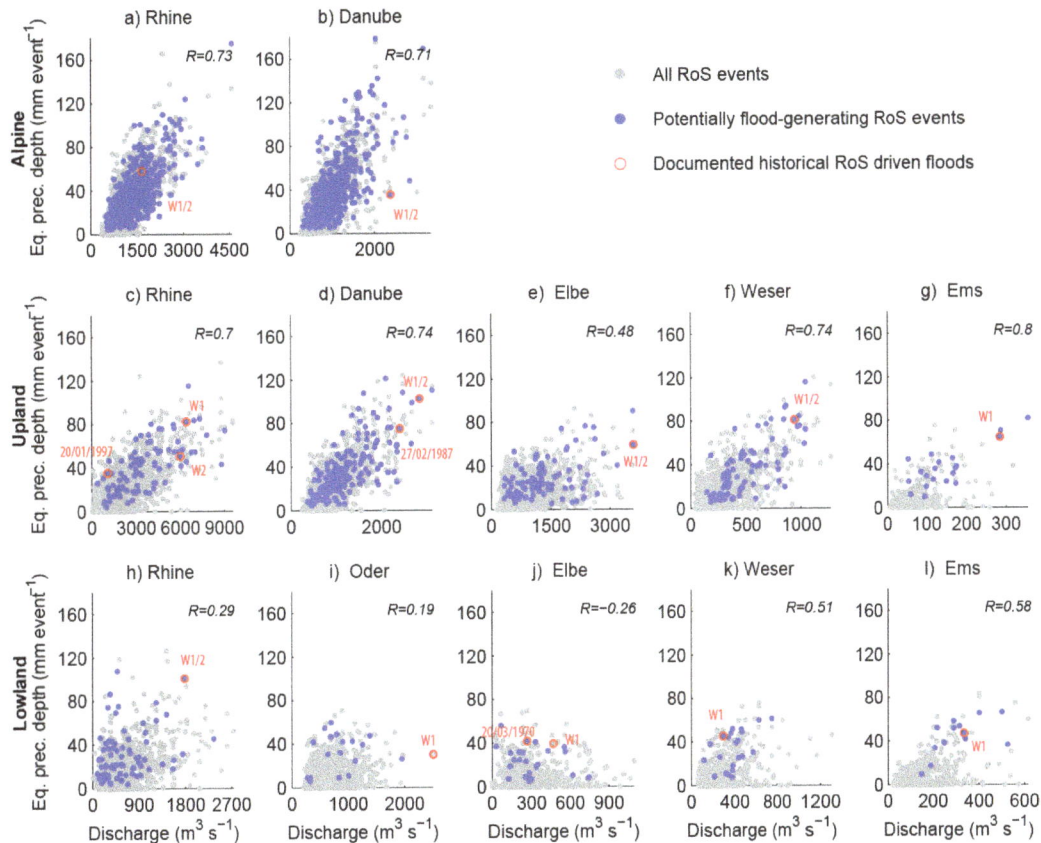

Figure 3. Total equivalent precipitation depth and corresponding peak discharge for all possible RoS events (SP > 0 mm SWE, M > 0 %, $P_L \geq 0$ mm), for all potentially flood-generating events (SP > 10 mm SWE, M > 20 %, $P_L \geq 3$ mm), and for documented historical RoS-driven floods. The correlation coefficient R is given for the potentially flood-generating RoS events.

sub-basins, RoS events occurred from December to April, with most events in January–March again in the highest discharge class (quantiles of 0.9–1). Only few events correspond to discharge quantiles < 0.7, confirming the strong flood-generating potential of the selected RoS events in this elevation range. In the lowland sub-basins, correspondence between RoS events and discharge quantiles is more heterogeneous. RoS events occurred only from December to March and had corresponding peak discharges of all quantiles, which supports the observations from Fig. 3 that RoS events do not necessarily cause the highest floods in these regions, the discharge being mostly rain-driven. However, in the Weser and Ems lowland sub-basins, RoS events were very infrequent, but the few events that occurred between December and March led mostly to relatively high discharge peaks (quantiles of 0.8–1).

3.3 Trends in magnitude and frequency of RoS events

Figure 5 shows the annual sum of P_{eq} of all selected RoS events according to the thresholds described in Sect. 3.2 over the entire winter season, the early winter season (November–February), and the late winter season (March–May) from winter 1950–1951 to 2010–2011. The trends were calculated only from years with RoS events and therefore represent the change in the magnitude of RoS events. In the alpine sub-basins of the Rhine and Danube, RoS events of the whole winter season generated P_{eq} between 100 and 600 mm year^{-1}. This corresponds on average to 45, 22, and 72 % of the total winter, early winter, and late winter precipitation (rain and snow) respectively. In both basins P_{eq} was greater in late winter than in early winter. No clear trends were identified in the magnitude for P_{eq} of the early winter season, but the late winter season showed decreasing magnitudes over time, also leading to decreasing trends for the entire winter. Table 2 shows the frequency of RoS events for time slices of 10 years for the early and late winter seasons as the fraction of the total number of the RoS events that occurred within the period 1950–2011. On average, one-third of the RoS events occurred in the early winter as opposed to two-thirds in the late winter in the alpine basins. In the Danube river sub-basin, early winter events became more frequent in the period 1990–2011 than in 1950–1990. In the Rhine sub-basins, no changes in RoS frequency were observed between the two time periods.

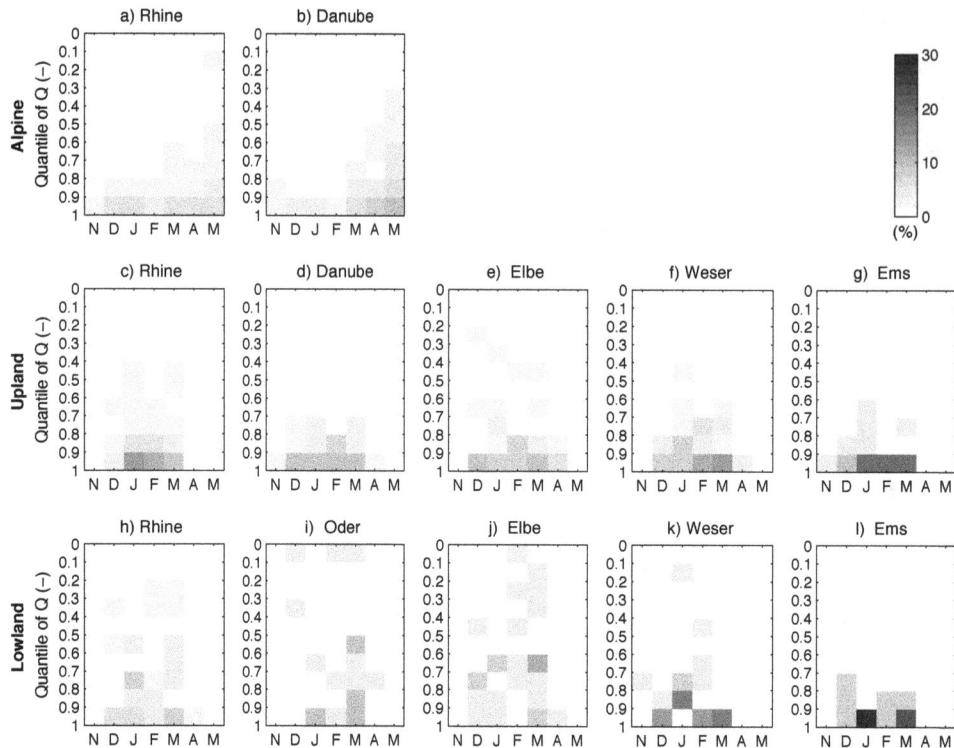

Figure 4. Percentage of potentially flood-generating RoS events from 1950 to 2011 by month of occurrence (November–May) and corresponding peak discharge quantile.

In the upland river sub-basins, RoS events generated maximum annual sum of P_{eq} from 90 mm year^{-1} in the Ems basin to up to 400 mm year^{-1} in the Danube basin (Fig. 5c–g), corresponding to an average of 21, 28, and 35 % of the entire winter, early winter, and late winter precipitation respectively. The Danube sub-basin showed decreasing trends in the magnitude of RoS events for the early and late winter seasons. In the Rhine and Weser sub-basins, the magnitude of the RoS events increased in the late winter season and decreased or remained constant in the early winter season. The strong increasing trend in late winter in the Rhine upland basin is influenced by a large RoS event that occurred in the 1980s. In the Elbe sub-basin, both early and late winter seasons showed increasing trends in the magnitude of the RoS events. In the Ems upland sub-basin, RoS events were rare and occurred mostly only during the early winter season. A decreasing trend in the magnitude of those events was observed (Fig. 5g). In all upland sub-basins, RoS events occurred more often in the early winter season (on average 70 % of all RoS events) than in the late winter season (30 %, Table 2). While the frequency of RoS events in the early winter season remained constant between the periods 1950–1990 and 1990–2011, the late winter events in all upland sub-basins became less frequent in the second time period.

In the lowland basins, RoS events were rare and generated maximum equivalent precipitation depths from

70 mm year^{-1} in the Oder basin to up to 250 mm year^{-1} in the Rhine basin (Fig. 5h–l), corresponding to an average of 13, 18, and 29 % of the total winter, early winter, and late winter precipitation respectively. Since the occurrence of RoS events is infrequent, they depend on very specific meteorological conditions and can occur either in the early or late winter seasons. The Rhine lowland showed the largest P_{eq} of RoS events in the winter season (Fig. 5h), due to the runoff contribution from a small part of the basin located in the medium-elevation mountain ranges. For all lowland sub-basins except the Oder, the magnitude of the events decreased in the late winter season. In the early winter season, the magnitude increased in the Rhine, Weser, and Ems sub-basins and decreased in the Oder and Elbe sub-basins (Fig. 5h–l). Comparing the period 1950–1990 to 1990–2011 in the lowland sub-basins, RoS events became less frequent in both the early and late winter seasons (Table 2).

3.4 Trends in RoS compounds and discharge

In Fig. 6 long-term trends of the rainfall and snowfall sums, of the average SWE, of the total equivalent precipitation depths of all possible RoS days ($R \geq 0$ mm, SWE > 0 mm, $M > 0$ %), of the selected potentially flood-generating RoS days ($R \geq 3$ mm, SWE > 10 mm, $M > 20$ %), and of the corresponding peak discharge are shown in percent change per year relative to the mean of the time period. The trends were

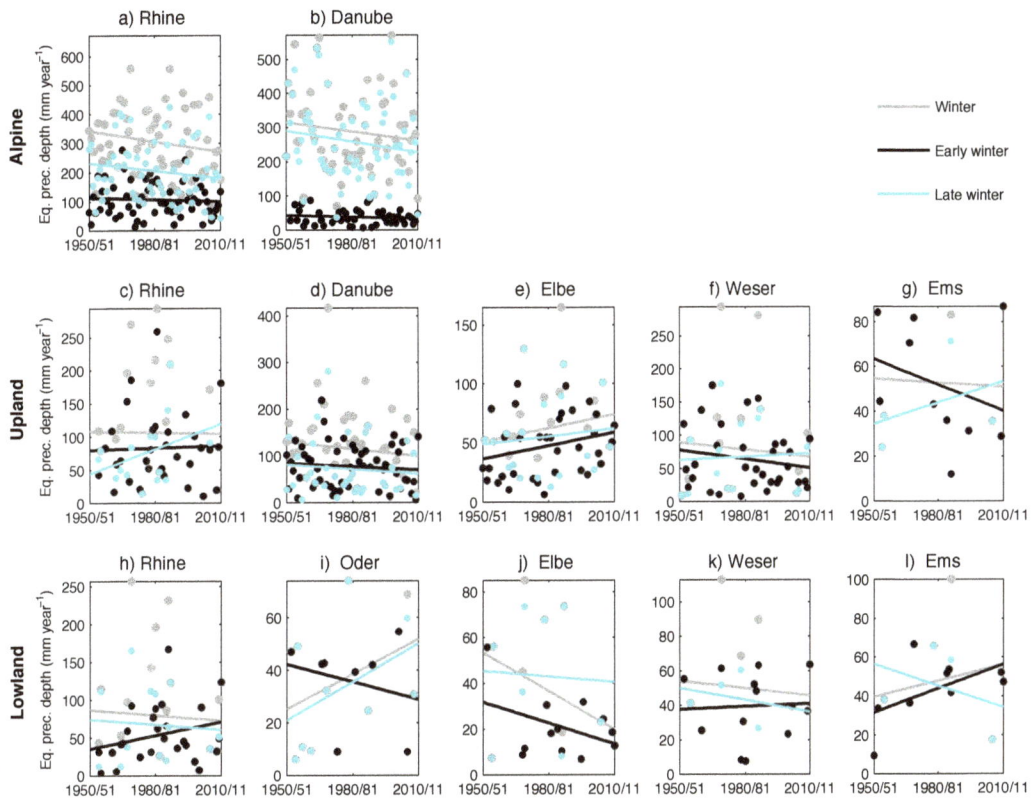

Figure 5. Total equivalent precipitation depth for all selected RoS events in the winter (November–May), in the early winter (November–February), and in the late winter (March–May) for the period 1950–2011. Only the years with RoS events are represented.

Table 2. Percentage of RoS events (1950–2011) in each sub-basin that occurred during time slices of 10 years during the early and late winter seasons. The italicised columns highlight the period 1990–2011, thus corresponding to the short-term trend analysis performed for RoS magnitudes.

Period	Early winter						Late winter					
	50/51 – 59/60	60/61 – 69/70	70/71 – 79/80	80/81 – 89/90	*90/91 – 99/00*	*00/01 – 10/11*	50/51 – 59/60	60/61 – 69/70	70/71 – 79/80	80/81 – 89/90	*90/91 – 99/00*	*00/01 – 10/11*
ALP Rh	6.9	7.9	6.2	5.8	*7.9*	*6.6*	8.6	12.4	9.8	9.8	*9.0*	*9.2*
ALP Do	4.3	1.9	4.1	4.3	*5.5*	*5.5*	11.9	13.6	10.4	12.2	*13.0*	*13.2*
UPL Rh	2.1	21.3	11.7	17.0	*8.5*	*14.9*	2.1	6.4	3.2	10.6	*0.0*	*2.1*
UPL El	11.0	15.4	7.7	14.3	*8.8*	*8.8*	3.3	13.2	3.3	7.7	*1.1*	*5.5*
UPL We	11.5	19.2	2.6	14.1	*10.3*	*12.8*	3.8	10.3	5.1	6.4	*0.0*	*3.8*
UPL Do	12.2	15.6	11.7	8.9	*11.1*	*10.6*	4.4	8.3	4.4	6.1	*2.2*	*4.4*
UPL Em	20.0	15.0	10.0	10.0	*5.0*	*15.0*	10.0	0.0	0.0	10.0	*0.0*	*5.0*
LOW Rh	2.7	13.3	9.3	22.7	*8.0*	*10.7*	4.0	10.7	2.7	12.0	*0.0*	*4.0*
LOW Od	8.0	12.0	4.0	12.0	*8.0*	*12.0*	8.0	12.0	8.0	12.0	*0.0*	*4.0*
LOW El	13.6	9.1	9.1	22.7	*0.0*	*27.3*	4.5	4.5	4.5	4.5	*0.0*	*0.0*
LOW We	15.4	15.4	0.0	23.1	*0.0*	*15.4*	7.7	0.0	7.7	7.7	*0.0*	*7.7*
LOW Em	5.6	11.1	5.6	11.1	*0.0*	*11.1*	16.7	11.1	11.1	5.6	*0.0*	*11.1*

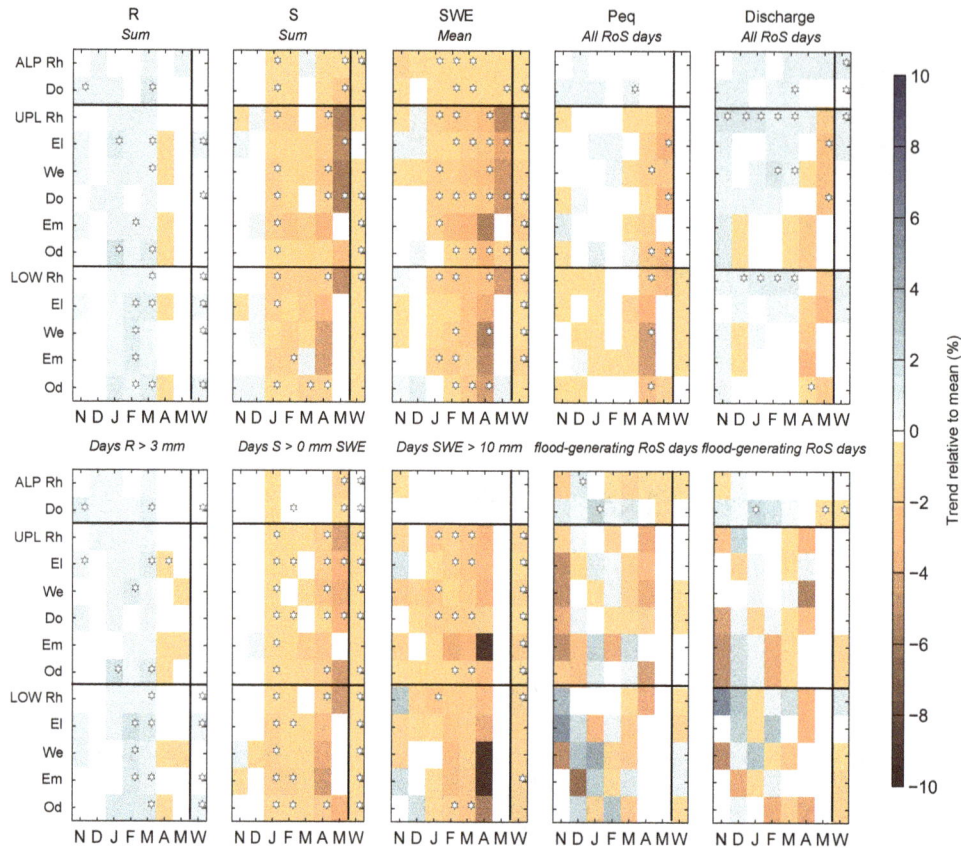

Figure 6. Comparison of the long-term trends (1950–2011) of the total equivalent precipitation depth, the snow water equivalent, the snowfall, and the rainfall for the individual months November–May (N–M) and the winter season (W) for the alpine (ALP), upland (UPL) and lowland (LOW) sub-basins of the Rhine (Rh), Danube (Do), Elbe (El), Weser (We), Oder (Od), and Ems (Em). Trends are given as the yearly change relative to the mean of the period 1950–2011. Statistically significant trends at $p < 5\%$ are shown with a red star.

calculated for the individual months from November to May and for the sum over the entire winter season for all basins. In contrast to Fig. 5, the trends were calculated including years without RoS events and thus also account for changes in the frequency of occurrence. For R around 20% of the trends were statistically significant, for S 30%, and for mean SWE 42%. For the sample of all events, around 10% of the trends in P_{eq} and 17% of the trends in discharge were statistically significant at $p < 0.05$, while for the selected potentially flood-generating events only around 2% of the trends were statistically significant. Overall, the detected long-term trends range between −4 and +4% for all variables. R increased in November, December, and April in the alpine sub-basins and from January to March in the upland and lowland sub-basins. In contrast, S showed overall decreasing trends. Similar to S, SWE decreased for all elevation ranges and all winter months, with especially large negative trends in April in some upland and all lowland basins. Similarly, the number of days with SWE > 10 mm SWE decreased between January and April in the upland and lowland sub-basins, especially in April in the Ems and Weser sub-basins, indicating a shortening of the winter duration in these regions. No clear

trends in the number of days with SWE > 10 mm SWE were identified in the alpine regions. In the alpine sub-basins, the trends in P_{eq} were overall positive during the early winter season and negative in April and May. In the upland sub-basins the trends P_{eq} were positive in January and February and negative from March to May. In the lowland sub-basins these trends were negative for all winter months. The trends in P_{eq} for the selected RoS days were very similar to those for all RoS days in the alpine and upland sub-basins, but the lowland trends differed with positive trends from November to January. The trends in corresponding peak discharges were very similar to the trends in P_{eq}, with slightly more positive values for the entire winter season for all basins and elevation ranges except for the Ems and Oder sub-basins, where trends were negative.

Compared to the long-term trends in Fig. 6, the short-term trends (Fig. 7) overall showed stronger negative and positive trends for all variables ranging from less than −10% to more than +10%. A maximum of 20% of all trends were statistically significant. Opposite to the long-term trends, the short-term trends in R and in the occurrence of days with a rainfall sum of at least 3 mm were positive in February

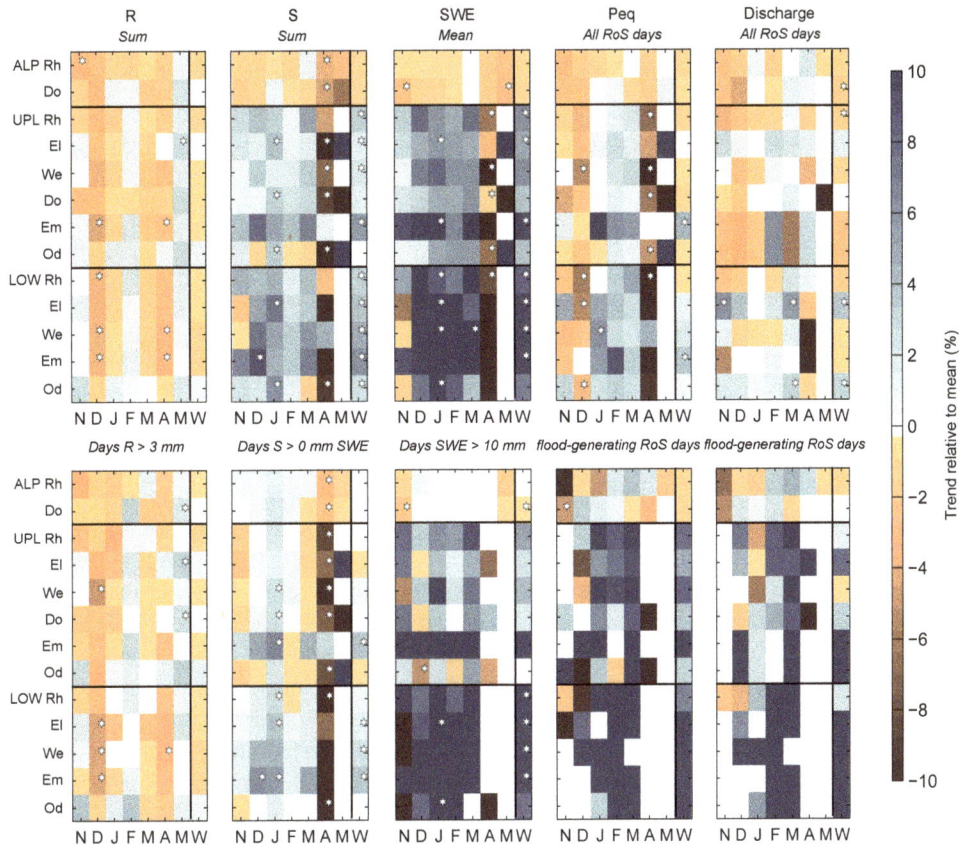

Figure 7. Comparison of the short-term trends (1990–2011) of the total equivalent precipitation depth, the snow water equivalent, the snowfall, and the rainfall for the individual months November–May (N–M) and the winter season (W) for the alpine (ALP), upland (UPL) and lowland (LOW) sub-basins of the Rhine (Rh), Danube (Do), Elbe (El), Weser (We), Oder (Od), and Ems (Em). Trends are given as the yearly change relative to the mean of the period 1990–2011. Statistically significant trends at $p < 5\%$ are shown with a red star.

and May and negative in the rest of the winter for all basins. Also opposite to the long-term trends, short-term trends in S were positive from November to March and negative in April in all upland and lowland sub-basins. The alpine sub-basins showed, in contrast, negative trends in S for the long-term and short-term periods. The same differences are reflected in SWE, with high positive short-term trends from December to March, only negative trends in May in the upland and lowland sub-basins, and negative trends for all winter months in the alpine sub-basins. Trends in the P_{eq} of all RoS days were negative in the alpine sub-basins for all winter months, but the trends in potentially flood-generating RoS days were mostly positive. In the upland and lowland sub-basins, trends were negative in November, December, and April and positive from January to March, and a particularly strong increase in the P_{eq} of the selected RoS days was observed from January to March. This increase was also reflected in the trends of the corresponding peak discharge in all upland and lowland sub-basins. In January and February, the trends in discharge had a similar direction to the trends in P_{eq} but were smaller.

The sensitivity analysis of the base temperature showed that increasing or decreasing T_b by $\pm 1\,°C$ had little impact on the direction of the trends. On average, 95% of the calculated trends showed the same direction. The P_{eq} of the selected potentially flood-generating RoS days was the most sensitive with still around 80% of the trends having the same direction. Rainfall trends were the least sensitive, with 100% showing the same direction. The calculated trends were also insensitive to the degree-day factor M_f when testing values ranging from 2 to 5 mm °C^{-1} day^{-1}. This low sensitivity can be explained by the fact that the trend analysis considers the relative changes to the mean.

4 Discussion

The RoS-driven flood event that spread over Germany and central Europe in January 2011 is a good example of the large-scale impact that RoS events can have. From 6 to 16 January nearly all gauging stations in Germany observed the maximum water level in 2011. On the one hand, the large-scale impact of this RoS event was due to the extremely widespread snow cover over the study area in the beginning of January, and its remarkable depth with extreme values at several locations in Germany (e.g. Böhm et al., 2011; Bastian et al., 2011; Karuse, 2011; LHW, 2011; Fell, 2011; Besler, 2011). The snowpack therefore represented a very large water reservoir available for runoff. On the other hand, the climatic conditions with thawing temperatures and rainfall in January provided the required energy to melt the snowpack. Runoff was therefore generated during a very short time, with a maximum of 6 days for each flood wave, and simultaneously reached the upper and lower parts of the Rhine, Danube, Weser, Elbe, Oder, and Ems basins, thus causing floods in most areas. Even if most of these discharges correspond to return periods of less than 10 years and thus are not statistically extreme (Kohn et al., 2014), this RoS event emphasises the large-scale impact of such events and their potential of shifting the annual peak flow from the late winter season to early winter season. This event therefore represents a good reference for the characterisation of RoS events with flood-generation potential.

The runoff generation from RoS events is influenced by many antecedent conditions and physical processes such as the thermal, mass, and wetness conditions of the snowpack, or the snow metamorphism, the water movement through the wet snow, the interaction of melt water with underlying soil, or the overland flow at the snow base (Singh et al., 1997; Marks et al., 1998). A physically based model would be needed to continuously simulate the development of the snowpack as well as evapotranspiration, sublimation, and infiltration, and thus to estimate the actual runoff generation. In a large-scale analysis however, parameterisation for all processes would be difficult. The almost continental scale of the study requires considering data availability and conceptualisation of processes dominant at that scale. In the case of this study, these are the hydrometeorological magnitude of rain-on-snow events and the temporal scale relevant to flood generation at the large scale. The aim of this study is therefore not to improve the description or understanding of runoff generation at the hillslope or small catchment scale during a RoS event, but to analyse RoS events' long-term evolution at the large scale. Therefore, no runoff is calculated, but the equivalent precipitation depth is related to the RoS event by its comparison to measured peak discharge during the event. However, even at the small catchment scale, Rössler et al. (2014) concluded that the hydrometeorological conditions are the main factors quantifying RoS-driven flood events.

Despite its apparent simplicity compared to energy balance methods, Ohmura (2001) showed that the temperature-based melt index model was sufficiently accurate for most practical purposes, and was justified on physical grounds since the air temperature is the main heat source for the atmospheric longwave radiation. This model has the advantage that it needs only daily precipitation and temperature data, often the only available data, and has already proved to be accurate enough for large-scale modelling if the inten-

tion is to identify basin-scale processes (Merz and Blöschl, 2003). The conceptual temperature index model employed in this study allowed estimating the potential snowpack and snowmelt. Even when the degree-day factor was generalised for the entire study area, the method led to a good estimation of the equivalent precipitation depth, was accurate enough to recognise known historical events, and was supported by the comparison to measured peak discharges during the RoS event. This validation of RoS events and the selection and analysis of potential flood-generating RoS events add to previous studies, which have mostly looked only at RoS days.

Using the January 2011 RoS-driven flood events as a reference, it was therefore possible to identify the magnitude of rainfall, the snowpack, and the percentage of snowmelt in the equivalent precipitation depth as the main characteristics of a RoS event and as the major characteristics for runoff generation at the large scale. The resulting threshold values of 3 mm for rainfall, 10 mm SWE for snowpack, and 20 % for snowmelt content in P_{eq} were able to detect all documented historical RoS-driven flood events in the time series and to specifically select only potentially flood-generating RoS events. These thresholds have therefore proved to be good indicators for RoS-driven flood events. The snowpack threshold (10 mm SWE) is not only representative for the event of January 2011, but also corresponds to the definition of the beginning of the winter by Beniston (2012) and Bavay et al. (2013). The advantage of the approach of identifying RoS events with threshold values is that it can easily be applied to other basins where discharge measurements are available, and it represents a useful tool for analysing changes in frequencies and magnitudes of those events.

The results showed an elevation dependence of RoS events in all basins, confirming the observations of different previous studies (e.g. Merz and Blöschl, 2003; Pradhanang et al., 2013). RoS events generally have a high impact on discharge peaks in alpine and upland basins. These events are most likely to lead to high discharges (quantiles of 0.7–1 in the alpine sub-basins and even 0.8–1 in the upland sub-basins), and they therefore have a real potential for generating floods in these regions. This result is in agreement with the observation of Sui and Koehler (2001) that RoS events play a more important role in runoff generation than pure rainfall events for topographical elevations above about 400 m a.s.l. during the wintertime. In all lowland sub-basins, the quantile ranges are strongly influenced by the antecedent conditions of the stream in the wintertime, which is due to the fact that winter floods are rain-dominated in these areas, since snowfall occurs only infrequently. In the lowland sub-basins, winter floods strongly depend on antecedent moisture conditions (soil saturation and groundwater tables; Nied et al., 2013). Therefore RoS events do not necessarily cause floods in these regions, but they can exacerbate a flood. For example, the Elbe river basin generated discharges corresponding to return periods of up to 100 years during the January 2011 event not only because of the RoS event but also because of the very wet autumn 2010, which led to already very high water levels and discharges at the beginning of January (e.g. Kohn et al., 2014).

One challenge in the trend analysis of extreme RoS events is the censored data; i.e. events do not occur every year. Most methods for trend analysis try to statistically disclose outliers, as for example the Sen slope method. In our case, however, the outliers are often exactly the values that need to be considered. Therefore, linear regression was found to be the method better suited for the trend analysis. The zeros also explain why many trends were not statistically significant. This is a well-known problem in hydrology. Kundzewicz et al. (2012) observed for example that the strong natural variability of hydrologic events can alter trend detection, and IPCC (2012) pointed out that, due to the fact that extreme events are per definition rare, long record lengths are required to allow for detection of trends in extremes. However, the lack of statistical significance does not mean that the trends do not exist, but that the hypothesis of no existing trends could not be rejected. As discussed, e.g., by Stahl et al. (2010) in more detail, the application of trend tests has been criticised extensively in the literature because many assumptions are not met by hydrological time series data. As suggested in other large-scale studies, a systematic regional consistency of trend direction and magnitude is therefore a more relevant result than the number of statistical significances. In this study, the value of the trend analysis has its main value not in the absolute numbers estimated but in the comparison of the consistency in the trends of the individual components involved in rain-on-snow events.

The estimated magnitudes and frequencies revealed a different importance of RoS events in the different elevation zones represented by the sub-basins. Nearly half of the total winter precipitation (rain and snow) and even two-thirds of the late winter precipitation contribute to RoS events in the alpine sub-basins, while this contribution is one-third in the upland sub-basins, and only one-fifth in the lowland sub-basins. The largest changes in frequencies were observed in the upland basins, where late winter RoS events have become less frequent since the 1990s. This change in frequencies can be explained by the decreasing trends in snow depth observed in the late winter season in these areas (Figs. 6 and 7) and therefore by a decreasing probability of rain falling on a snow cover in the late winter season. In contrast, the trends in rainfall are positive in the early winter season, increasing the probability for RoS events, especially in January and February. These results agree well with the observations of many studies worldwide (e.g. Birsan et al., 2005; Knowles et al., 2006; Ye, 2008; Ye et al., 2008). Trends in magnitude vary from one basin to another. The upland Elbe and Rhine sub-basins show positive trends in the early and late winter seasons, while the other upland sub-basins have negative trends. This difference can be explained by the more frequent and stable snow cover in the upland basins of the Elbe and Rhine. The corresponding positive trends in the measured discharges of RoS events were positive from January to March in the alpine and upland sub-basins, and especially in March for the short-term trend analysis, suggesting an impact of this increasing magnitude.

Trends have to be interpreted carefully since they depend on the choice of the time period, which can be influenced by many climatic factors and also by extreme values. The analysis of long-term (1950–2011) vs. short-term (1990–2011) trends showed different, even opposite, results for all variables. Overall, long-term trends are smaller than short-term trends. The greatest difference between both periods is in the trend in the mean SWE. In the long-term analysis, snowpack has declined in all basins, while it has increased in the short-term period in all basins except the alpine sub-basins. Fricke (2006) observed that the average snowpack in Germany for the period 1961–1990 was substantially higher than for the period 1991–2000. The 21st century was characterised by extreme events. The winter 2005–2006 and December 2010, for example, were identified as extremely snow rich in Germany (Fricke, 2006; Pinto et al., 2007; Böhm et al., 2011). This explains the differences in the trends, since the extreme events of the 21st century will have more weight in a shorter time series, especially if they occurred in the beginning of the period such as for the lower observed snow depth.

The alpine sub-basins show different trends to the upland and lowland sub-basins. While the upland and lowland sub-basins have negative trends in the long-term analysis and positive trends in the short-term analysis, SWE in the alpine sub-basins decreased in both the long- and short-term trend analyses. This difference can be explained by the climatic conditions specific to the alpine regions, which are very different to those in the upland and lowland sub-basins. For example, while exceptionally great SWE was measured all over Germany in December 2010, the Swiss Alps experienced snowpacks below average (e.g. Trachte et al., 2012; Techel and Pielmeier, 2013). In another study in the Swiss Alps, Beniston (2012) found that the wintertime precipitation declined between 15 and 25 % over the 1931–2010 period and that the number of snow-sparse winters has increased in the last 40 years, while the number of snow-abundant winters has declined. But in the meantime, some winters since the 1990s have experienced record-breaking snow amounts and durations (Beniston, 2012). Rainfall also shows opposite trends, increasing for the long time series and mostly decreasing for the short time series in the wintertime. As snowpack and rainfall both influence the occurrence of RoS events, trends in RoS days are therefore difficult to identify for the long-term analysis, since rainfall is increasing but snow depth is decreasing.

For upland and lowland sub-basins, the results of the long-term analysis are opposite and show dominating positive trends in RoS days for the upland sub-basins and negative trends for the lowland sub-basins. In the short-term analysis in contrast, clear positive trends are detected for all RoS days in upland and lowland regions. The trends for the selected potentially flood-generating RoS days are even more positive, leading to the conclusion that they have become an important factor for the winter discharge and that the occurrence of maximum peak flow from RoS events between January and March has become more frequent since the 1990s.

5 Conclusions

In a context of climate change, snowpack and precipitation in the wintertime are very likely to change and therefore may influence the frequency and magnitude of the flood hazard from rain-on-snow events in central Europe. The analysis of causes and trends for past RoS events is challenging since these events depend on many influences. Defining threshold values to characterise RoS events allowed identifying the events with potentially high impact on river discharge on a large scale and analysing them for trends in frequency and magnitude. The results showed an elevation dependence of RoS events and suggest they have the strongest impacts in upland regions, where an increasing magnitude of these events was observed. However, the frequency of RoS events decreased in the second half of the time period 1950–2011 in the late winter season in upland and lowland basins and can be related to decreasing trends in snowpack in the late winter season. Increasing trends in rainfall in the early winter season as well as increasing trends in equivalent precipitation depth during RoS events in some upland sub-basins suggest that these events have become more important at this elevation class.

The results show the importance of the choice of the analysed period for the detection of trends, with opposite trends found for snow water equivalent in the long-term and short-term periods. The 21st century has been affected by several extreme events, which makes the analysis even more difficult. As the example of January 2011 in Germany and central Europe showed, rain can release a large amount of water stored in the snowpack and RoS events can cause very widespread flood events, delivering a large amount of water to the streams within a very short time. If such events are likely to become more frequent in the future in certain basins and elevation ranges, the winter flood hazard will increase. Therefore, there is a real need for an improved understanding of the relation between RoS events and flooding, and more analysis is needed on their occurrence at different scales.

Acknowledgements. In part, the study was funded by the Bundesanstalt für Gewässerkunde as part of a reanalysis of the hydrological extremes during the year 2011. The article processing charge was funded by the German Research Foundation (DFG) and the Albert Ludwig University Freiburg in the funding programme Open Access Publishing. The authors thank J. U. Belz for his support with data and advice. The authors would also like to acknowledge the E-OBS data set from the EU-FP6 project ENSEMBLES (http://ensembles-eu.metoffice.com) and the data providers ECA&D project (http://www.ecad.eu) for providing precipitation and temperature data. The authors also thank the environmental agencies of the German federal states for providing discharge data. Finally, the authors acknowledge two anonymous reviewers for their thoughtful comments that helped improve the manuscript.

Edited by: G. Di Baldassarre

References

Bastian, D., Göbel, K., Klump, W., Kremer, M., Lipski, P., and Löns-Hanna, C.: Das Januar-Hochwasser 2011 in Hessen, Hydrologie in Hessen, Heft 6, Hessische Landesamt für Umwelt und Geologie, Wiesbach, Germany, 2011 (in German).

Bavay, M., Grünewald, T., and Lehning, M.: Response of snow cover and runoff to climate change in high Alpine catchments of Eastern Switzerland, Adv. Water Resour., 55, 4–16, 2013.

Beniston, M.: Is snow in the Alps receding or disappearing?, WIREs Clim. Change, 3, 349–358, doi:10.1003/wcc.179, 2012.

Besler, C.: Dokumentation Elbehochwasser Januar 2011, Teil 1: Meteorologische Situation und hydrologischer Verlauf des Hochwassers, Report, Staatliches Amt für Landwirtschaft und Umwelt (StALU) Westmecklenburg, Schwerin, Germany, 2011 (in German).

Birsan, M.-V., Molnar, P., Burlando, P., and Pfaundler, M.: Streamflow trends in Switzerland, J. Hydrol., 314, 312–329, 2005.

Blöschl, G., Kirnbauer, R., and Gutknecht, D.: Modelling snowmelt in a mountainous river basin on an event basis, J. Hydrol., 113, 207–229, 1990.

Böhm, U., Fiedler, A., Machui-Schwanitz, G., Reich, T., and Schneider, G.: Hydrometeorolgische Analyse der Schnee- und Tauwettersituation im Dezember 2010/Januar 2011 in Deutschland, Report, Deutscher Wetterdienst (DWD), Offenbach, Germany, 2011 (in German).

Collins, E. H.: Relationship of degree-days above freezing to runoff, EOS T. Am. Geophys. Un., 15, 624–629, 1934.

Corps of Engineers: Summary report of the snow investigations, snow hydrology, US Army Engineer Division (North Pacific, 210 Custom House, Portland, Oregon), 437 pp., 1956.

Fell, E.: Kurzbericht: Hochwasser im Rheingebiet – Januar 2011, Report, Landesamt für Umwelt Wasserwirtschaft und Gewerbeaufsicht (LUWG) Rheinland-Pfalz, Mainz, Germany, 2011 (in German).

Finsterwalder, S. and Schunk, H.: Der Suldenferner, Zeitschrift des Deutschen und Oesterreichischen Alpenvereins, 18, 72–89, 1887 (in German).

Floyd, W. and Weiler, M.: Measuring snow accumulation and ablation dynamics during rain-on-snow events: innovative measurement techniques, Hydrol. Process., 22, 4805–4812, 2008.

Fricke, W.: Wieder mehr Schnee im Winter?, Global Atmosphere Watch, Brief 33, Deutscher Wetterdienst (DWD), Germany, March 2006 (in German).

Garvelmann, J., Pohl, S., and Weiler, M.: From observation to the quantification of snow processes with a time-lapse camera network, Hydrol. Earth Syst. Sci., 17, 1415–1429, doi:10.5194/hess-17-1415-2013, 2013.

Haylock, M. R., Hofstra, N., Klein Tank, A. M. G., Klok, E. J., Jones, P. D., and New, M.: A European daily high-resolution gridded dataset of surface temperature and precipitation for 1950–2006, J. Geophys. Res., 113, D20119, doi:10.1029/2008JD10201, 2008.

Hock, R.: Temperature index melt modeling in mountain areas, J. Hydrol., 282, 104–115, 2003.

IPCC: Summary for policymakers, in: Managing the risks of extreme events and disasters to advance climate change adaptation. A special report of working groups I and II of the Intergovernmental Panel on Climate Change IPCC, Cambridge University Press, Cambridge, UK and New York, NY, USA, 3–21, 2012.

Karuse, P.: Gewässerkundlicher Monatsbericht Januar 2011, Report, Bayerisches Landesamt für Umwelt und Geologie (HLUG), Wiesbaden, Germany, 2011 (in German).

Kattelmann, R. C.: Macropores in snowpacks of Sierra Nevada, Ann Glaciol, 6, 272–273, 1985.

Kattelmann, R. C.: Flooding from rain-on-snow events in the Sierra Nevada, in: Destructive water: Water-caused natural disasters, their abatement and control, edited by: Leavesley, G. H., Proceedings of the conference, Anaheim, California, June 1996, IAHS Publication, 239, 59–65, 1997.

Knowles, N., Dettinger, M. D., and Cayan, D. R.: Trends in snowfall versus rainfall in the Western United States, J. Climate, 19, 4545–4559, 2006.

Kohn, I., Freudiger, D., Rosin, K., Stahl, K., Weiler, M., and Belz, J.: Das hydrologische Extremjahr 2011: Dokumentation, Einordnung, Ursachen und Zusammenhänge, Report, Mitteilung Nr. 29, Bundesanstalt für Gewässerkunde (Hrsg.), Koblenz, Germany, 2014 (in German).

Köplin, N., Schädler, B., Viviroli, D., and Weingartner, R.: Seasonality and magnitude of floods in Switzerland under future climate change, Hydrol. Process., 28, 2567–2578, doi:10.1002/hyp.9757, 2014.

Kundzewicz, Z. W., Plate, E. J., Rodda, H. J. E., Rodda, J. C., Schnellnhuber, H. J., and Strupczewski, W. G.: Changes in flood risk – setting the stage, in: Changes in Flood Risk in Europe, edited by: Kundzewicz, Z. W., IAHS Press, Wallingford, 11–26, 2012.

Lefebvre, C. and Becker, A.: Das Klima des Jahres 2011 im globalen Massstab, Klimastatusbericht, Report, Deutscher Wetterdienst (DWD), Offenbach, Germany, 2011 (in German).

LHW: Bericht über das Hochwasser Januar 2011, Report, Landesbetrieb für Hochwasserschutz und Wasserwirtschaft (LHW) Sachsen-Anhalt, Magdeburg, Germany, 2011 (in German).

Mann, H. B.: Nonparametric test against trend, Econometrica, 13, 245–259, 1945.

Marks, D., Kimball, J., Tingey, D., and Link, T.: The sensitivity of snowmelt processes to climate conditions and forest cover during rain-on-snow: a case study of the 1996 Pacific Northwest flood, Hydrol. Process., 12, 1559–1587, 1998.

Martinec, J. and Rango, A.: Parameter values for snowmelt runoff modelling, J. Hydrol., 84, 197–219, 1986.

McCabe, G. J., Clark, M. P., and Hay, L. E.: Rain-on-snow events in the western United States, B. Am. Meteorol. Soc., 88, 319–328, doi:10.1175/BAMS-88-3-319, 2007.

Merz, R. and Blöschl, G.: A process typology of regional floods, Water Resour. Res., 39, 1340, doi:10.1029/2002WR001952, 2003.

Nied, M., Hundecha, Y., and Merz, B.: Flood-initiating catchment conditions: a spatio-temporal analysis of large-scale soil moisture patterns in the Elbe River basin, Hydrol. Earth Syst. Sci., 17, 1401–1414, doi:10.5194/hess-17-1401-2013, 2013.

Ohmura, A.: Physical basis for the temperature-based melt-index method, J. Appl. Meteorol., 40, 753–761, 2001.

Pinto, J. G., Brücher, T., Fink, A. H., and Krüger, A.: Extraordinary snow accumulations over parts of central Europe during winter of 2005/06 and weather-related hazards, Weather, 62, 16–21, 2007.

Pradhanang, S. M., Frei, A., Zion, M., Schneiderman, E. M., Steenhuis, T. S., and Pierson, D.: Rain-on-snow events in New York, Hydrol. Process., 27, 3035–3049, 2013.

Renard, B., Lang, M., Bois, P., Dupeyrat, A., Mestre, O., Niel, H., Sauquet, E., Prudhomme, C., Parey, S., Paquet, E., Neppel, L., and Gaillhard, J.: Regional methods for trend detection: Assessing field significance and regional consistency, Water Resour. Res., 44, W08419, doi:10.1029/2007WR006268, 2008.

Rössler, O., Froidevaux, P., Börst, U., Rickli, R., Martius, O., and Weingartner, R.: Retrospective analysis of a nonforecasted rain-on-snow flood in the Alps – a matter of model limitations or unpredictable nature?, Hydrol. Earth Syst. Sci., 18, 2265–2285, doi:10.5194/hess-18-2265-2014, 2014.

Singh, P., Spitzbart, G., Hübl, H., and Weinmeinster, H. W.: Hydrological response of snowpack under rain-on-snow events: a field study, J. Hydrol., 202, 1–20, 1997.

Stahl, K., Hisdal, H., Hannaford, J., Tallaksen, L. M., van Lanen, H. A. J., Sauquet, E., Demuth, S., Fendekova, M., and Jódar, J.: Streamflow trends in Europe: evidence from a dataset of near-natural catchments, Hydrol. Earth Syst. Sci., 14, 2367–2382, doi:10.5194/hess-14-2367-2010, 2010.

Sui, J. and Koehler, G.: Rain-on-snow induced flood events in Southern Germany, J. Hydrol., 252, 205–220, 2001.

Surfleet, C. G. and Tullos, D.: Variability in effect of climate change on rain-on-snow peak flow events in temperate climate, J. Hydrol., 479, 24–34, 2013.

Techel, F. and Pielmeier, C.: Schnee und Lawinen in den Schweizer Alpen, Hydrologisches Jahr 2010/11, Report, WSL-Institut für Schnee- und Lawinenforschung SLF, Davos, Switzerland, 2013 (in German).

Trachte, K., Obregón, A., Bissolli, P., Kennedy, J. J., Parker, D. E., Trigo, R. M., Barriopedro, D., Kendon, M., Prior, J., Achberger, C., Gouvela, C., Sensoy, S., Hovsepyan, A., and Grigoryan, V.: State of the Climate in 2011, Regional Climates – Europe, B. Am. Meteorol. Soc., 93, 186–199, 2012.

Venables, W. N. and Ripley, B. D.: Modern Applied Statistics Using S, 4th Edn., Springer, New York, United States, 2002.

Wetterchronik: http://www.wetterzentrale.de/cgi-bin/wetterchronik/home.pl, access: 14 June 2013, 2001.

Ye, H.: Changes in precipitation types associated with air temperature over the northern Eurasia, J. Climate, 14, 3140–3255, 2008.

Ye, H., Yang, D., and Robinson, D.: Winter rain on snow and its association with air temperature in northern Eurasia, Hydrol. Process., 22, 2728–2736, 2008.

12

Drainage area characterization for evaluating green infrastructure using the Storm Water Management Model

Joong Gwang Lee[1], Christopher T. Nietch[2], and Srinivas Panguluri[3]

[1]Center for Urban Green Infrastructure Engineering (CUGIE Inc), Cincinnati, OH 45255, USA
[2]Office of Research and Development, US Environmental Protection Agency, Cincinnati, OH 45268, USA
[3]Independent Consultant, Olney, MD 20832, USA

Correspondence: Joong Gwang Lee (jglee@ugiengineering.com)

Abstract. Urban stormwater runoff quantity and quality are strongly dependent upon catchment properties. Models are used to simulate the runoff characteristics, but the output from a stormwater management model is dependent on how the catchment area is subdivided and represented as spatial elements. For green infrastructure modeling, we suggest a discretization method that distinguishes directly connected impervious area (DCIA) from the total impervious area (TIA). Pervious buffers, which receive runoff from upgradient impervious areas should also be identified as a separate subset of the entire pervious area (PA). This separation provides an improved model representation of the runoff process. With these criteria in mind, an approach to spatial discretization for projects using the US Environmental Protection Agency's Storm Water Management Model (SWMM) is demonstrated for the Shayler Crossing watershed (SHC), a well-monitored, residential suburban area occupying 100 ha, east of Cincinnati, Ohio. The model relies on a highly resolved spatial database of urban land cover, stormwater drainage features, and topography. To verify the spatial discretization approach, a hypothetical analysis was conducted. Six different representations of a common urbanscape that discharges runoff to a single storm inlet were evaluated with eight 24 h synthetic storms. This analysis allowed us to select a discretization scheme that balances complexity in model setup with presumed accuracy of the output with respect to the most complex discretization option considered. The balanced approach delineates directly and indirectly connected impervious areas (ICIA), buffering pervious area (BPA) receiving impervious runoff, and the other pervious area within a SWMM subcatchment. It performed well at the watershed scale with minimal calibration effort (Nash–Sutcliffe coefficient $= 0.852$; $R^2 = 0.871$). The approach accommodates the distribution of runoff contributions from different spatial components and flow pathways that would impact green infrastructure performance. A developed SWMM model using the discretization approach is calibrated by adjusting parameters per land cover component, instead of per subcatchment and, therefore, can be applied to relatively large watersheds if the land cover components are relatively homogeneous and/or categorized appropriately in the GIS that supports the model parameterization. Finally, with a few model adjustments, we show how the simulated stream hydrograph can be separated into the relative contributions from different land cover types and subsurface sources, adding insight to the potential effectiveness of planned green infrastructure scenarios at the watershed scale.

1 Introduction

Conventional stormwater modeling has focused on the design of urban drainage systems and flood control practices that achieve fast drainage and reduce risk of flooding (NRC, 2009; WEF-ASCE, 2012). These objectives focus attention on larger storms, such as 2- to 10-year return period storms for designing drainage systems and 25- to 100-year storms for designing flood control practices (WEF-ASCE, 2012). Conversely, nearly 95 % of pollutant runoff from urban areas is produced from events smaller than a 2-year storm (Guo and Urbonas, 1996; Pitt, 1999; NRC, 2009). It is well recognized that the best way to resolve this pollution problem is to implement controls as close to the source of runoff generation as possible (Debo and Reese, 2002; WEF-ASCE, 2012).

Green infrastructure (GI) practices were developed to correct this water pollution problem and restore the natural hydrologic cycle (WEF-ASCE, 2012; USEPA, 2014). GI includes structures like green roofs, rain barrels, bioretention areas, buffer strips, vegetated swales, permeable pavements, and infiltration trenches, or practices, such as disconnecting downspouts. The specific design objectives for GI include minimizing the impervious areas directly connected to the storm sewer, increasing surface flow path lengths or time of concentration, and maximizing on-site depression storage and infiltration at the lot level (WEF-ASCE, 2012). This translates operationally to individual stormwater management practices that are relatively small but densely distributed in space (USEPA, 2009). Although GI is distributed at higher spatial densities, each unit is relatively inexpensive if the unit can be considered as part of landscaping and, in total, may provide a cost-effective alternative to more traditional larger centralized practices, like detention ponds especially in cases where land is not available or very expensive.

There is a great deal of interest in modeling GI effects at watershed scales to help inform regional stormwater management planning and design decisions. However, from a stormwater modeling perspective, the approach taken for model representations of GI requires different methodological considerations compared to the traditional large-size, low spatial density of the more centralized and regional control features (Fletcher et al., 2013; USEPA, 2012; Guo, 2008). While conventional stormwater management practices have focused on end-of-pipe controls at the downstream end of the drainage area (i.e., centralized systems), GI practices focus on on-site controls at the upstream side (i.e., distributed systems). The surface hydrologic properties (e.g., land cover, slope, overland flow path) remain the same before or after applying centralized systems, but they are altered with on-site GI systems. GI practices aim to amend the landscape hydrologic properties to reduce the negative impacts of stormwater (USEPA, 2007). Hence, modeling approaches for evaluating GI should be able to account for the changing surface hydrologic properties that come with GI implementation.

The Storm Water Management Model (SWMM) of the United States Environmental Protection Agency (USEPA) is one tool that has a large user base and a broad application history for informing stormwater management projects around the world (Niazi et al., 2017). In the current version of the model, GI effects are simulated using low-impact development (LID) algorithms. LID is largely synonymous with GI in SWMM vernacular. The LID modeling options were added in 2010 (Rossman, 2015; Rossman and Huber, 2016). Since then, while numerous LID/GI modeling studies have been introduced, best modeling practices for simulating GI in SWMM have received comparatively little attention in the literature (Niazi et al., 2017).

This study was intent on evaluating approaches to modeling GI effects at a watershed scale using SWMM. In the setup of a SWMM model, the urban area of interest is divided into smaller spatial units, referred to as subcatchments. To implement a traditional stormwater control feature like a retention pond, it is usually acceptable to provide minimal detail of the drainage area (Rossman and Huber, 2016). This leads to spatial aggregation which tends to produce larger subcatchment areas that aggregate land cover types and simplify the existing storm sewer system to realize a more cost-effective model setup and output data management. If the simulated hydrographs are matched with the observed data during model calibration, the model is considered a sound representation of the drainage processes important for designing the retention pond. The trade-off is that coarser schematization requires more decisions on how to aggregate catchment properties (Rossman and Huber, 2016). In contrast, for simulation of GI, the construction reality is that GI is built as part of building and landscape arrangements all upgradient of the drainage network (Dietz, 2007; Montalto et al., 2007; USEPA, 2009; Zhou, 2014). Therefore, to accurately examine GI alternatives, a drainage area for modeling should be defined as an area that drains runoff to a storm sewer inlet with no or minimal spatial aggregation of landscape features affecting hydrologic properties. In SWMM, this drainage area is modeled as single or multiple subcatchments. Since SWMM version 4.4H, overland flow routing is allowed from one subcatchment to another (Huber, 2001; Huber and Cannon, 2002). To minimize confusion between real and modeled drainage areas, we coin the term hydrologic response element (HRE) in this study. An HRE is the drainage area (i.e., a real spatial element of the landscape being modeled) where GI practices may be implemented to control the element's surface runoff prior to discharge to the stormwater collection system. There are many alternatives for configuring the HRE for GI modeling in SWMM. We evaluate six different options in this study.

In SWMM, the subcatchment representation of the HRE is comprised of one or more homogeneous subareas, such as impervious or pervious area, impervious area (IA) with or without depression storage, directly connected impervious area (DCIA) or indirectly connected impervious area (ICIA), or LID area (Rossman, 2015; Rossman and Huber, 2016). In urban landscapes, DCIA discharges runoff to the existing storm sewer system without any control, while ICIA discharges to the adjacent pervious area (PA). The PA that receives runoff from ICIA works like a buffer strip or swale, therefore acting like an existing GI practice, albeit not intentionally designed as such. This is a real characteristic of urban areas that is termed buffering pervious area (BPA) in this study. The other pervious area is called standalone pervious area (SPA) that does not receive or control any runoff from impervious area. An HRE may consist of all or part of these subareas – DCIA, ICIA, BPA, and SPA – and implementing GI practices can change subarea ratios and properties, surface runoff processes, and flow pathways within the HRE. Each subarea may also consist of different land cover components. For example, DCIA may include paved streets, building rooftops, driveways, or sidewalks. In this study, we questioned how these spatial and hydrologic realities should be modeled using SWMM.

After the HRE delineation is performed in SWMM, each HRE undergoes a model parameterization procedure that defines the relative proportions of impervious and pervious subareas, how they interact in terms of surface flow pathways, and their hydrologic properties (Rossman, 2015). The subcatchment/subarea configuration of each HRE ultimately specifies the physical conditions used by the model's mathematical algorithms to simulate the dynamics of hydrologic loading to the drainage network. The more subcatchments there are, the more input and output values there are to be managed by the modeler. When setting up a SWMM model using the conventional objectives, such as deriving hydrographs for designing storm collection systems and/or detention–retention systems, the subcatchment parameterization remains the same before and after simulation of the management practice; however, for GI simulation, the internal properties of a subcatchment change, as mentioned earlier. Pending the type of GI, changes may need to be made to the hydrologic properties of subareas or individual land cover components, the proportions of impervious and pervious area, the specification for the routing of runoff between them, the flow path length, and infiltration or the depression storage properties. Adequately rationalizing and tracking these changes can become a problem for the modeler when the total area being modeled is relatively large and aggregated, the GI scenarios are not the same among HREs, or the internal properties among HREs are heterogeneous. A systematic approach to characterizing HREs would help make SWMM GI simulation projects more efficient.

The question of how to best parameterize SWMM is not new, especially when it comes to spatial resolution and scaling, but as mentioned, GI modeling, in particular, requires special considerations. This paper describes a suggested method for modeling GI in SWMM and provides critical evaluation. Primary objectives of this study are to (1) examine how to configure HREs for GI modeling and (2) develop a methodology for parameterizing a SWMM model that reflects this configuration with the goal of demonstrating an urban watershed spatial discretization approach that optimizes model performance in terms of tracking model input values and presumed accuracy of the results. We hypothesized that conventional modeling approaches to subcatchment delineation are likely aggregating at too-coarse resolution in space and hydrologic response to be appropriate for highly spatially distributed modern GI. We also questioned how the SWMM setup could not only allow for modeling the effects of various GI scenarios, but also facilitate the scaling of GI scenarios from a small HRE, representing the parcel or lot level, to a watershed level. To answer these ques-

tions, we examine several acceptable approaches to representing spatial reality in SWMM when the modeling objective is to inform decisions about GI implementation. We use a hypothetical HRE-based analysis of spatial discretization alternatives to test our hypothesis related to the appropriateness of spatial and hydrologic response resolution. The hypothetical HRE represents a typical residential area that drains to a storm sewer inlet. The hypothetical HRE is modeled in SWMM by combining six conceivable options in spatial representation and eight design storms that represent the full spectrum of runoff events. From this analysis, an appropriate option is selected for GI modeling in SWMM, and a baseline SWMM model is developed using it for representing the existing condition of a well-characterized 100 ha urban watershed in a headwater area east of Cincinnati, Ohio. To examine the proposed approach, another SWMM model is developed for simulating a GI implementation scenario at the study watershed. Also, an approach for hydrograph separation is presented using the developed SWMM models, which can provide insight for arranging GI implementation scenarios. Currently, there is no provision for accomplishing this in SWMM.

2   Materials and methods

2.1   Study area

An experimental urban watershed drained by a natural headwater stream that does not have any surface stormwater inflows from outside its topographic boundaries was used for this study (Fig. 1). The Shayler Crossing watershed (SHC) is located east of Cincinnati, Ohio, and occupies approximately 100 ha that is characterized as 62.6 % urban or developed, 25.6 % agriculture, and 11.8 % forested based on the 2011 National Land Cover Database (Homer et al., 2015). The native soils of the watershed are characterized with high silty clay loam content and therefore are naturally poorly infiltrating.

2.2   The baseline spatial database

2.2.1   Data from the county GIS

Spatial data for the study area was provided by the Clermont County Office of Environmental Quality, which included a detailed GIS of the existing stormwater drainage system and surface topography. The drainage system consists of storm sewer inlets (or catch basins), manholes, pipes, wet–dry detention ponds, and channel network. The county GIS contains the location of the drainage system, invert elevations for inlets and manholes, and pipe sizes. Two types of surface topography data were also available; 0.76 m (2.5 feet) lidar (light detection and ranging) data and 0.3 m contours. High-resolution aerial orthophotographs were also provided by the county. Existing databases that include the details for

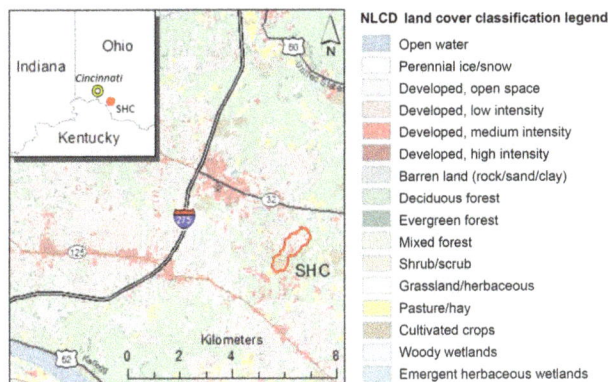

Figure 1. Location of the Shayler Crossing watershed. I-275 is an interstate highway around the Cincinnati metropolitan area.

the stormwater infrastructure like in this watershed are not always available to the modeler. In these cases, to adopt the subsequently described approach to GI scenario modeling in SWMM could require considerable ground-truthing and site surveying. In lieu of on-site visits, and as will become apparent from the descriptions below, what would be most important is determining the spatial location of storm sewer inlets. These are often visible from readily available aerial photographs; note that the visibility depends on the underlying image quality and the presence of obstacles such as trees or cars. When elevation data for the storm sewer network is unavailable, much can be inferred using surface elevation data and assuming local construction codes for stormwater infrastructures were applied, such as catch basin depths and conveyance pipe diameters and slopes. Such approximations would suffice for GI scenario analysis considerations and where storm sewer design is not the primary focus.

2.2.2 Detailed land cover and subarea categorization

To obtain a high-resolution digital characterization of spatial reality in the study watershed, 16 unique land cover types were identified and digitized using ArcGIS 10.2 (ESRI, 2013) spatial analysis tools on the aerial orthophotographs of the study area. These 16 types are later aggregated to 10 for setting up the watershed SWMM model (see Appendix). The resulting baseline spatial database included individual records of the watershed surface that could be used to access the location, pattern, and extent of the following sixteen land cover types: streets, parking areas, sidewalks, driveways, main buildings, miscellaneous buildings, paved walking paths, patios, other miscellaneous impervious areas, landscaped or lawn areas, agriculture, forest, dry ponds, stormwater detention area (in SHC this is created by the addition of a control structure to the stream channel itself), swimming pools, and wet ponds. Each spatial record has its own attributes (i.e., fields in the database) representing the current conditions (e.g., area, land cover) and was characterized

based on its future potential for GI implementation (e.g., to evaluate the potential of downspout disconnection for a main building). The initial parameterization and GI modeling approaches described below for the SWMM model are based on content extracted from this land cover database created using ArcGIS tools. This database is often reused to perform model adjustments during calibration and GI scenario analysis. The developed land cover database for SHC contains a total of 3682 records and the median area of each record is 23.5 m^2.

Each surface record in the database is further classified into four types based on its hydrologic characteristics including (1) DCIA, (2) ICIA, (3) pervious area (PA), or (4) water. The PA is subsequently split into two subcategories called BPA and SPA after the HRE delineation procedure for SWMM modeling is completed (see below). All main buildings are DCIA because the rooftop downspouts in the existing condition are plumbed to directly discharge to the stormwater collection system through buried pipes or street gutters. All the miscellaneous buildings (e.g., storage sheds) are considered ICIA. Streets with curb-and-gutter drainage systems are identified as DCIA. Any directly connected up-gradient impervious areas to these streets are initially considered as DCIA. These areas include directly connected driveways, parking areas, and sidewalks. However, if both sides of a sidewalk are surrounded by a pervious area, the sidewalk is categorized as ICIA. Streets without curb-and-gutter drainage are ICIA. The remaining miscellaneous impervious areas are ICIA.

Figure 2 contains a sample GIS representation of the 16 land cover types along with a corresponding attribute table, which indicates hydrologic characteristics representing the baseline classification and a GI scenario-related classification. In the attribute table shown in Fig. 2, the first column contains the record identifier, the second column defines the land cover type, the third column defines how it was classified for modeling the baseline condition, the fourth column defines how it was classified or reclassified for modeling a specific GI scenario, and the fifth column specifies the contributing area. For example, the record ID 36 contained in the table is initially classified as DCIA, but, after the rooftop drains were disconnected in the modeled GI scenario, the unit was reclassified as ICIA (in the fourth column). This methodology allows for GI-related hydrology evaluation to be performed without impacting the overall SWMM model structure and setup. A companion USEPA report (Lee et al., 2017) has been prepared to provide the relevant details on the applied spatial analysis techniques such as clip, intersect, union, buffer, and manipulating attribute data.

2.2.3 Configuring the BPA and SPA

BPA is not considered explicitly in a traditional urban stormwater modeling analysis using SWMM. Instead the modeler usually sets up PA within a subcatchment to re-

Figure 2. Sample GIS classified representation of the land cover and hydrologic characteristics.

Figure 3. Depiction of the different distances applied for the estimation of BPA in the baseline condition using ArcGIS.

ceive a certain percentage of runoff from impervious areas; this is how ICIA is distinguished from DCIA. However, in reality, not all of the PA receives runoff from ICIA, rather just the part of the PA that is immediately adjacent to the ICIA. When evaluating GI scenarios, one strategy might be to enlarge the size of the buffering area adjacent to ICIA, or engineer GI structures (e.g., cascading filtering or bioretention systems) around this buffering area (a.k.a. BPA) to reduce the direct runoff from impervious surfaces by routing them over grassy areas to slow down runoff and promote soil infiltration. Draining paved areas onto porous areas can reduce runoff volumes, rates, pollutants, and cost for drainage infrastructure (NRC, 2009; WEF-ASCE, 2012). Therefore, because of the nuanced, yet important differences, in the geospatial relationship of PA in different GI scenarios, we rationalized the need for retaining the ability to model this aspect while evaluating GI scenarios by splitting the PA into BPA and SPA for GI modeling in SWMM.

Characterizing the precise "physical" extent of BPA is a complicated process that would have to be defined from highly resolved surface topography around ICIA and an understanding of the unsaturated zone processes such as how infiltration and depression storage interact across the pervious surface types to influence flow path length. The physical extent of BPA is also affected by storm intensity, with higher intensity storms creating a larger spread of water across the surface and thereby increasing the extent of available adjacent buffering areas. Lacking the ability to infer flow path length without extensive physical measurements, we instead treat the width of the BPA from ICIA as a calibration parameter. In preparation for this, BPA based on different buffer widths was established during the development of the spatial database. This was done in ArcGIS using the geoprocessing tools "Buffer" and "Intersect". The Buffer tool established a

separate BPA area around all existing ICIA based on arbitrarily chosen distances that serve as equivalent buffer widths of 0.30, 0.61, and 1.52 m (Fig. 3). The Intersect tool establishes the area for the BPA and adjusts the area of the original pervious area from which it was subtracted, which is now SPA (Lee et al., 2017). Using this spatial information, we arranged three SWMM models that represent three different sizes of BPA. We determined which one among the three cases of BPA sizes provided the more accurate simulation compared to the observed flow data and as part of model calibration. In this way the BPA width was treated as a calibration parameter in this study.

2.3 HRE delineation

Urban HREs were delineated manually within the GIS using the surface topography (0.76 m lidar) and the layout of the storm sewer system (Rossman and Huber, 2016). Because GI is designed to capture and control stormwater runoff before it discharges to the storm sewer system, the HRE for GI analysis should be delineated as the area that drains runoff to an actual storm sewer inlet. With GI implementation, some inlets can be removed or combined for economic benefits because the peak and volume of stormwater discharge will be decreased after implementing GI practices (Sample et al., 2003; Braden and Johnston, 2004; USEPA, 2012). Based on this, two HREs are combined into one HRE if the two HREs were located side-by-side at one street location and one of the two HREs was smaller than 2023.4 m^2 (0.5 acre). For undeveloped or agricultural areas in the study watershed, the HRE boundaries were generally selected with an intent to keep all

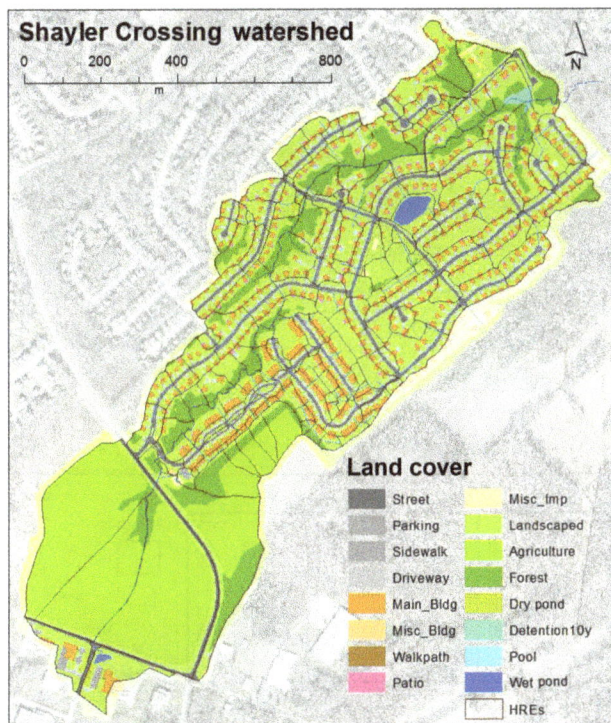

Figure 4. Detailed spatial representation of the Shayler Crossing watershed.

HREs a similar size to help maintain hydrologic continuity among them. The result of the HRE delineation for the entire SHC watershed is shown in Fig. 4.

2.4 SWMM parameterization

SWMM, developed by the USEPA, is a comprehensive mathematical model for analyzing hydraulics, hydrology, and water quality process dynamics in the urban environment (Huber and Dickinson, 1988; Gironás et al., 2009; Rossman, 2015; Rossman and Huber, 2016, Niazi et al., 2017). Here version 5.1.007 of SWMM was used. SWMM generates runoff when rainfall depth exceeds surface depression storage and infiltration capacity at the subcatchment scale. SWMM has extensive routing capability that can simulate the runoff through a conveyance system of pipes, channels, storage and treatment devices, pumps, and regulators. SWMM can also estimate the quality of runoff discharging from subcatchments and route it through the conveyance system. The model can be used within a continuous or event-based framework.

Unique to our application of the SWMM model is the setup of the BPA. This process is described in detail in the Appendix. Because the natural stream draining the study area receives lateral inflow through subsurface soil media (a.k.a. subsurface flow), SWMM's groundwater modeling options were implemented. The groundwater component of the SHC SWMM setup is also described in the Appendix.

A subcatchment is a fundamental hydrologic component of a SWMM application and can be defined as an area that drains runoff to a storm sewer inlet, open channel, or another subcatchment. The SWMM subcatchments in this study will represent the HREs that were delineated during the development of the spatial database described above. Each SWMM subcatchment is configured with a specific drainage area, % imperviousness, width, and slope. Subareas divide each subcatchment into impervious, pervious, and/or LID areas that are used to account for internal heterogeneity. These areas are modeled in the abstract based on the relative percentage of the subcatchment each occupies; i.e., subareas have no real spatial reference. Therefore, all pervious areas within one subcatchment, for example, are lumped and modeled as one contributing hydrologic entity no matter how disconnected or patchy the actual physical reality may be. This establishes a relationship between the subcatchment size and the spatial resolution of the model. The larger the subcatchment area, especially in the urban environment, the more spatial lumping that results, and the more abstracted from reality the model becomes. The size of the subcatchment and the heterogeneity among land covers and their organization within each subcatchment or subareas interact to effect model complexity as well as accuracy. In most cases, modelers try to strike a balance between these when configuring a SWMM project. Subareas are parameterized by setting values characteristic of each, such as n and DS for both IA and PA. The Green–Ampt option for infiltration modeling was used in this study, and this requires three parameters per subcatchment's PA, including the saturated hydraulic conductivity (K_{sat}), capillary suction head (Suct), and initial soil moisture deficit (IMD). Internal flow between the subareas can be routed from pervious to impervious, impervious to pervious, or directly to the outlet. LID areas have their own set of parameters.

The spatial database that included land cover digitization and HRE delineations was used to parameterize the SWMM model. With the land cover data spatially overlaid with the HRE delineation in ArcGIS (Fig. 4), the characteristics of each SWMM subcatchment could be defined using the detailed land cover status per subcatchment and unique hydrologic parameters per land cover component presented in Table 1. Each land cover type is either all impervious or all pervious. "Length" represents a typical distance for overland flow before it turns into a concentrated flow path, which is controlled by the hydrologic design features of the land cover type. For example, overland flow at a rooftop is maintained only from the roof crest to the gutter because flow through a gutter is considered concentrated. The same regime change in flow (i.e., from overland flow to concentrated flow) may happen at any place where more than one instance of impervious land cover converge hydrologically, e.g., at a street gutter where overland flows from streets and driveways intersect. Using ArcGIS, the initial values for length were determined by averaging multiple field measurements of perceived overland flow lengths for each land cover type. More-detailed procedures for the SWMM modeling methods used in this study are presented in the Appendix.

2.5 Model setup options for a hypothetical HRE in SWMM

As mentioned earlier, an HRE can be modeled as a single subcatchment or multiple subcatchments in SWMM. In the SWMM model setup just described we used a single subcatchment setup that was based on the results of an analysis done with the goal of determining which among a series of plausible HRE configuration options strikes a balance among the degree of spatial and hydrologic aggregation, output uncertainty, and computational effort. The most spatially refined approach to a SWMM setup (Option 1 in this study as presented below) would be to discretize every piece of impervious and pervious surface as an independent subcatchment. This promises a decrease in model output uncertainty (Krebs et al., 2014; Sun et al., 2014), but requires specifying all the modeling parameters and unique flow directions among all subcatchments, results in longer computational times, and produces data management burdens that are typically not practical. The opposite extreme would be a highly generalized subcatchment characterization where the entire area is modeled as one subcatchment with just two subareas, lumping all the spatial heterogeneity into a fictional space that has no basis in physical reality. Within this continuum, we chose to consider six plausible options for representing urban spatial constructs that are constrained by the SWMM subcatchment/subarea paradigm were examined (Fig. 5). As shown in the legend of Fig. 5, each rectangle represents a subcatchment in SWMM, and the dotted line divides subareas within the subcatchment. A rectangle without a dotted line means the subcatchment consists of a single (homogeneous) subarea, either 100 % impervious or pervious. The arrows represent flow routing directions. Conducting this assessment at the watershed scale would not only be tedious and time consuming to configure, but could be inappropriate because of potentially confounding effects introduced when the drainage network and groundwater algorithm are included in the simulation. We felt it more rational to base our assessment of HRE setup options at the scale of an HRE, i.e., the area that drains to a storm sewer inlet, and judge the results in comparison to the most spatially explicit option. Note, we do not have supporting observational data at this scale to prove this assumption. This would require flow data at the point of entry to a storm sewer inlet, which is very difficult to obtain in practice.

Instead, a hypothetical representation of a typical urbanscape was defined as the HRE and used to model eight synthetic single storm events for each of the six setup options (Fig. 5). The hypothetical HRE is meant to represent a typical 4041 m^2 (1 acre) residential area consisting of 809.4 m^2 (0.2 acre) DCIA, 1214.1 m^2 (0.3 acre) ICIA, and 2023.4 m^2 (0.5 acre) PA. The DCIA consists of 607.0 m^2 (0.15 acre)

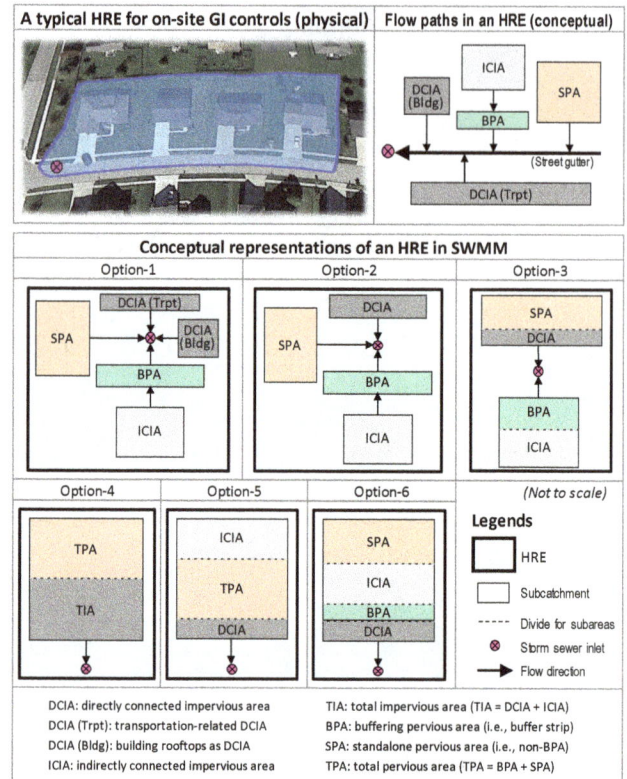

Figure 5. A conceptual representation of the hypothetical HRE (20 % DCIA, 30 % ICIA, 10 % BPA, and 40 % SPA) and the six options considered for representing this area in the setup of a SWMM model.

transportation-related surfaces (e.g., streets, driveways) and 202.3 m^2 (0.05 acre) building rooftops. The runoff from ICIA discharges through 404.7 m^2 (0.1 acre) BPA, thus the SPA of the area is 1618.7 m^2 (0.4 acre).

Referring to Fig. 5, in Option 1, five subcatchments are arranged for modeling the hypothetical HRE, separately modeling transportation DCIA (Trpt) and building DCIA (Bldg) along with ICIA, BPA, and SPA. DCIA is modeled with two subgroups because buildings have slanted rooftops while paved areas for transportation are basically flat in a typical residential area. This is the lowest level of spatial aggregation among the six options. This option would result in the highest number of subcatchments and, therefore, number of data requirements. Option 2 combines the two DCIA subcatchments in Option 1, resulting in four subcatchments set up for one HRE. In Option 3, the four subcatchments in Option 2 are aggregated into two subcatchments, and each subcatchment is configured with two subareas. The imperviousness and the flow direction between subareas per subcatchment need to be specified in SWMM with "% imperv", "subarea routing", and "percent routed". The impervious option for subarea routing means runoff from pervious area flows to impervious area, whereas pervious does the oppo-

Table 1. Initial and calibrated modeling parameters for the Shayler Crossing watershed. "n/a" indicates the parameter is not applicable to the land cover type.

Land cover	Length (m)		Slope (%)		n		DS (mm)		K_{sat} (mm h^{-1})	
	Initial	Calibrated	Initial	Calibrated	Initial	Calibrated	Initial	Calibrated	Initial	Calibrated
Main building	9.1	7.6	10	15	0.014	0.01	2.0	1.3	n/a	n/a
Misc. building	4.6	4.6	10	15	0.014	0.01	2.0	1.3	n/a	n/a
Street	3.0	3.0	2	2.5	0.011	0.01	2.5	1.3	n/a	n/a
Driveway	4.6	3.7	2	1.5	0.012	0.01	2.5	1.3	n/a	n/a
Parking	3.0	3.0	1	1.5	0.012	0.01	3.0	1.3	n/a	n/a
Sidewalk	0.9	0.9	1	1.5	0.012	0.01	3.0	1.3	n/a	n/a
Other impervious	3.0	2.4	1	1.5	0.012	0.01	3.0	1.3	n/a	n/a
Lawn	24.4	24.4	2	2	0.2	0.3	5.1	5.1	1.6	0.89
Forest	24.4	24.4	3	2	0.6	0.6	10.2	7.6	1.6	1.52
Agriculture	30.5	30.5	2	2	0.3	0.3	7.6	5.1	1.6	1.02

site (Rossman, 2015). Percent routed should be specified as 100 for both subcatchments. In options 4 through 6, the areas are further aggregated to a single subcatchment representation for an HRE in SWMM. Option 4 configures the single subcatchment with only two subareas, impervious and pervious areas. The runoff from pervious area discharges through impervious area (i.e., TIA = DCIA). In Option 5, DCIA and ICIA are independently modeled by specifying the subarea routing option as pervious and the percent routed as the ratio of ICIA/TIA. This option may be considered an unrealistic "green" development condition where runoff from ICIA is evenly distributed throughout the entire pervious area, which means the entire pervious area works like a buffer (i.e., TPA = BPA). Finally, in Option 6, LID controls in SWMM are used for modeling BPA and ICIA. BPA is modeled as a vegetated swale with a very small berm height, 2.54 mm (0.1 inch). In the "LID Usage Editor", the "area of each unit" specifies the size of BPA and the "% of impervious area treated" is the fraction ICIA/TIA. With this configuration, the four hydrologically homogenous subareas – DCIA, ICIA, BPA, and SPA – are accounted for.

Lengths for overland flow (or sheet flow) were assumed to be 4.57 m (15 feet), 9.14 m (30 feet), 12.19 m (40 feet), and 15.24 m (50 feet) for transportation-related DCIA, building rooftops as DCIA, ICIA, and pervious area, respectively. The surface slopes of these were assumed to be 3, 11, 5, and 2 %, respectively. Surface dimensions and slopes of typical urban land cover components are based on construction codes or were inferred based on the GIS. The values selected are meant to represent typical residential areas in the United States. For example, the assumed values were derived using overland flow from the center of the street to the curb in a crowned 9.14 m (30 feet) wide neighborhood street with 3 % cross-sectional slope for the crown, 18.29 m (60 feet) wide gable houses with 11 % cross-sectional slope for the rooftops, and pervious surfaces with 2 % slope on average. Every IA is modeled with 0.01 for Manning's roughness co-

Table 2. Profile of the selected eight 24 h single storm statistics.

Rain (mm)	Frequency	Percentile	Cumulative
12.7	<1 month	64.8 %	32.7 %
25.4	1–2 months	87.4 %	63.1 %
36.8	3 months	95.0 %	80.7 %
48.3	6 months	97.7 %	89.2 %
61	1 year	99.2 %	95.3 %
73.7	2 years	99.6 %	97.3 %
108	10 years	100 %	99.8 %
149.8	50 years	100 %	100 %

The percentile and cumulative percentage-based statistics qualify the exceedance probability of each event and the relative contribution of events of similar size or lower to the annual rainfall, respectively.

efficient (n) and 2.54 mm (0.1 inch) for depression storage (DS). Pervious area is modeled with 0.1 for n and 5.08 mm (0.2 inch) for DS. Identical infiltration parameters were applied to all the options. In this hypothetical HRE model setup analysis, all of the six options were arranged using the same spatial and hydrologic characteristics. However, the ways DCIA, ICIA, BPA, and SPA were parameterized in SWMM were different among the options (Fig. 5).

Rainfall–runoff response is also affected by storm size, so we applied eight different 24 h single storms (Table 2) selected from a regional rainfall frequency report produced by the National Oceanic and Atmospheric Administration (NOAA) and the Illinois State Water Survey (Huff and Angel, 1992). Another data set was used to estimate the percentile and cumulative rainfall depths per year, i.e., annual statistics per 24 h storm. This data set covered about 35 years of hourly precipitation records from a local weather station in Milford, Ohio. A certain percentile rainfall event represents a precipitation amount that the same percent of all rainfall events for the period of record does not exceed (USEPA, 2009). The percentile values in Table 2 were estimated using the method presented in the same report (USEPA, 2009).

For example, the 90th percentile rainfall event is defined as the measured precipitation depth accumulated over a 24 h period for the period of record that ranks as the 90th percentile rainfall depth based on the range of all daily event occurrences during this period. Values in the cumulative column of Table 2 represent the percentage of annual cumulative precipitation depth, which are less than or equal to the specific rainfall depth during a 24 h period. In SWMM, the selected storms were distributed with 5 min intervals by applying the Natural Resources Conservation Service (NRCS) Type-II distribution (USDA, 1986).

2.6   Calibration of the SHC watershed SWMM model

Stream flows were measured at the outlet from a rating curve using water depth recorded at 10 min intervals. A tipping bucket rain gauge measured rainfall depths at 10 min intervals, with a minimum detectable rainfall depth of 0.254 mm (0.01 inch). The SWMM model for SHC (Fig. 6) was run for a 6-month period (1 April 2009 to 31 August 2009) where the first 4 months of this period were used to stabilize the continuous simulation, in particular for the groundwater simulation. This is defined as the model warm-up period, which is the time period required to achieve a stable condition wherein the groundwater level ceases to increase or decrease by a specified initial parameter threshold value. After the warm-up period, the last 2 months, from July to August 2009, were used for model calibration. Model calibration was done manually by adjusting the initial values for the 10 land cover types and using the different sets of BPA (see Fig. 3). Changes were integrated one at a time into every subcatchment using the area-weighting approach in an Excel spreadsheet. The calibrated modeling parameters for individual land cover types are given in Table 1 alongside their initial values. An Excel worksheet was created with embedded lookup and averaging functions so that changes made to the original values in Table 1 or switches between BPA sets configured using the different buffer distances could be easily propagated to changes in the related parameter values used in the SWMM model using the SWMM Excel Editor function. With this approach, the calibration effort is evenly applied to the urban land cover types, which in turn are propagated to the parameterization of all subcatchments, instead of calibrating parameters individually for each subcatchment. This methodology assumes that urban land cover components are generalizable and independent from scale even though the subcatchments themselves are not generalizable or easily scalable. Also, notable about this approach, the parameter calibration domain remains the same even if the total number of subcatchments is increased and/or the size of watershed area is increased. If a land cover type does not maintain a sufficient level of homogeneity across the watershed under study, we would need to divide the land cover into subcategories and use more than one set of parameters for the land cover type in each category. For example, the main building

can be divided into two subcategories that represent slanted rooftops and flat rooftops independently, and the lawn area can be divided into multiple subcategories based on different surface slopes and/or soil infiltration properties. This can be handled by spatial analysis in GIS by overlaying land use, topography, or soil property data with the land cover layer.

Sensitivity analysis was conducted for the modeling parameters of width, slope, n and DS for IA, n and DS for PA, K_{sat}, and the size of BPA. Each parameter was decreased and increased 5, 10, and 20 %, respectively, one at a time, and in separate model runs. The sensitivity of each parameter was estimated as

$$\text{Sensitivity} = (\Delta MR/MR)/(\Delta p/p), \qquad (1)$$

where MR is the modeling result in units of flow volume from the SWMM run, ΔMR is the change in SWMM modeling result based on change in parameter value, p is the parameter value, and Δp is the change in parameter value.

2.7   Modeling GI scenarios

GI scenarios are added to the model using the land cover database, soils, storm sewer systems, and GIS techniques to derive relevant BPA and may require some field investigation to ground-truth the options. The general workflow for GI modeling is presented in the bottom half of Fig. A1. Implementing GI can be achieved by adjusting the hydrologic properties of individual land cover components, such as converting lawn area to shrub or forest (Lee et al., 2005). This sort of GI implementation can reduce the volume, peak, and speed of surface runoff and be modeled by adjusting DS, slope, n, or overland flow length for the converted land cover component. The one scenario we examined was decreasing DCIA by disconnecting the directly connected rooftop downspouts that directly route flow from the main buildings to the sewer system. This effectively reclassifies main buildings as ICIA. After the downspouts are disconnected, the PA that receives stormwater runoff from the disconnected rooftop now works as additional BPA. To model this additional buffering capacity, the size of BPA is re-estimated and the percent of IA routed to BPA is changed in SWMM. The increase in size of the BPA under this GI scenario was estimated again using the spatial analysis tools in ArcGIS by changing the buffering distance value from the calibrated baseline value of 0.61 m (2 feet) to 3.1 m (10 feet) around ICIA, including the disconnected main buildings. As a result, the modeled GI scenario includes two types model changes: one that reflects the downspout disconnection and another the buffering area extension.

The characteristic width per subcatchment is a computed value that is usually treated as a calibration parameter in SWMM (see Appendix). Under conventional stormwater management modeling approaches, once the width value is set, it is not adjusted during management scenario analysis. However, GI, by design, changes the flow path lengths and

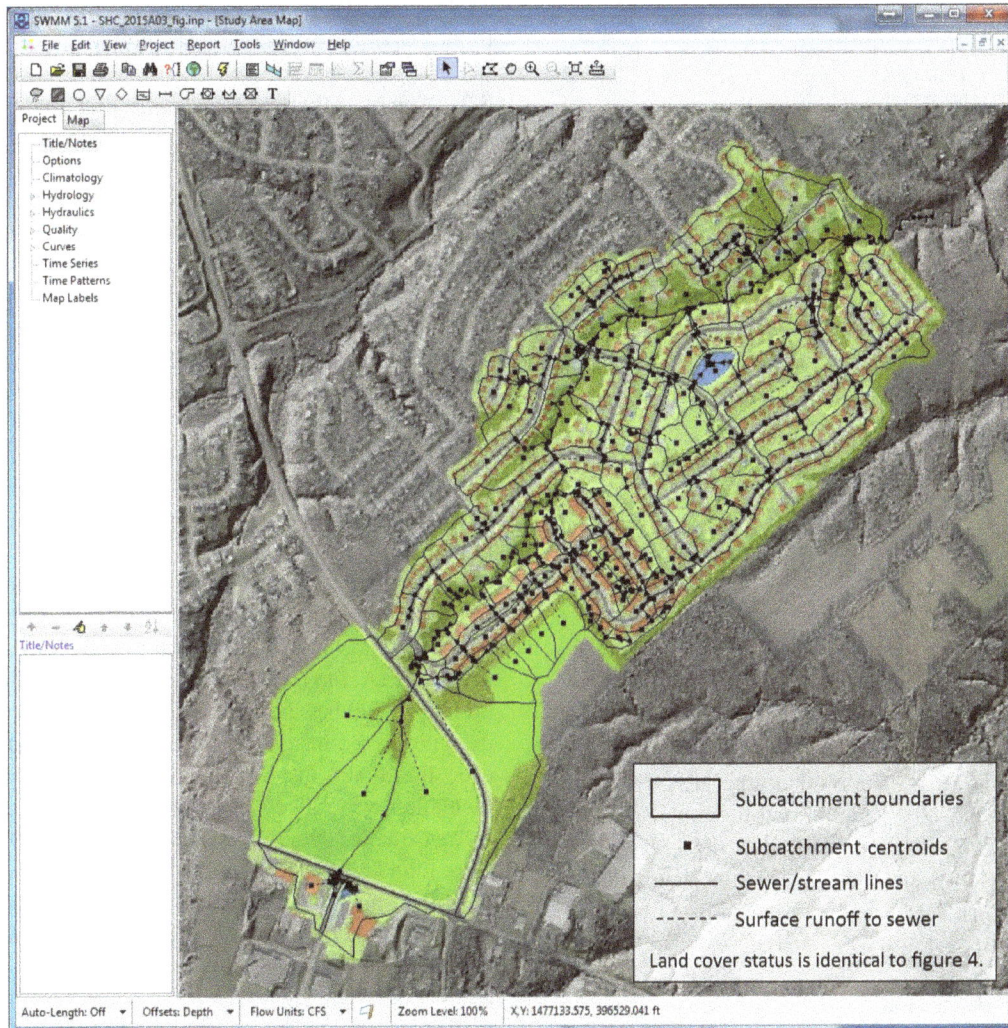

Figure 6. Diagram of the developed SWMM model for the Shayler Crossing watershed.

therefore the computed value of the width parameter as represented in SWMM should also change. The methodology we present here provides a systematic way of changing the width parameter in a rational and objective manner to account for the modeled GI scenario. Unfortunately, the suitability of this modeling approach cannot be determined until a high density of GI has been implemented at a watershed scale with before and after field observations.

2.8 Hydrograph separation

With the approach taken for the SWMM setup for both the baseline and GI scenario analysis adjustments can be made to apportion the simulated storm hydrologic loading from the watershed among the dominant sources: DCIA, ICIA+BPA, SPA, and subsurface flow. This can provide further insight into the effects of GI on watershed hydrology. For this purpose, the output from four SHC-SWMM runs were generated:

- Run 1: every subcatchment is specified as described under Option 6 (as conceptually represented in Fig. 7a) with groundwater options parameterized to represent the base SWMM model.

- Run 2: groundwater options were excluded from the base model setup to remove any subsurface flow contributions to the stream flow hydrographs. The difference between (1) and (2) represents the stormwater contributions to the stream as subsurface flow from the watershed.

- Run 3: to estimate surface runoff from all impervious areas (i.e., runoff from DCIA, plus ICIA through BPA) in the models without the groundwater, the SPA was also omitted from every subcatchment (Fig. 7b).

- Run 4: to estimate surface runoff from DCIA, only DCIA was modeled in this run (Fig. 7c).

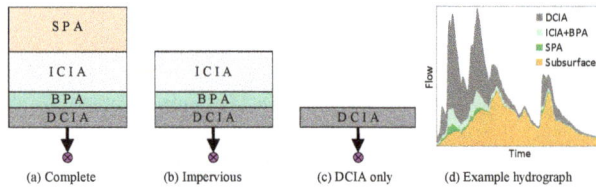

Figure 7. Conceptual representations of discrete SWMM models for hydrograph separation.

An example result of the hydrograph flow pathway separation is presented in Fig. 7d, and the process is summarized mathematically as follows:

$$Q_{\text{total}} = Q_{\text{DCIA}} + Q_{\text{ICIA+BPA}} + Q_{\text{SPA}} + Q_{\text{subsurface}}, \tag{2}$$

$$Q_{\text{surface}} = Q_{\text{DCIA}} + Q_{\text{ICIA+BPA}} + Q_{\text{SPA}}, \tag{3}$$

$$Q_{\text{subsurface}} = Q_{\text{total}} - Q_{\text{surface}}, \tag{4}$$

$$Q_{\text{SPA}} = Q_{\text{surface}} - Q_{\text{imperv}}, \tag{5}$$

$$Q_{\text{ICIA+BPA}} = Q_{\text{imperv}} - Q_{\text{DCIA}}, \tag{6}$$

where Q_{total} is the total runoff with groundwater flow in SWMM, Q_{surface} is the surface runoff without groundwater flow, Q_{imperv} is the runoff from impervious area (DCIA and ICIA through BPA), Q_{DCIA} is the runoff from DCIA only, $Q_{\text{ICIA+BPA}}$ is the runoff from ICIA and BPA, Q_{SPA} is the runoff from SPA only, and $Q_{\text{subsurface}}$ is the runoff through groundwater flow (i.e., subsurface or lateral flow).

3 Results and discussion

The SHC watershed model parameter values pre- and post-calibration are presented in Table 1.

3.1 Spatial analysis

Table 3 reveals the results of the detailed spatial analysis conducted using the described GIS techniques. The fractional DCIA for buildings, streets, driveways, parking areas, and sidewalks are 96.1, 79.5, 94.2, 42.8, and 14.2 %, respectively. Overall, the study watershed is covered by 18.8 % DCIA, and three sets of BPA were derived for 0.30, 0.61, and 1.52 m buffer lengths. After calibration, the 0.61 m buffer around ICIA was selected for SHC. This means that the runoff from ICIA is discharged to the adjacent pervious area with 0.61 m buffer width, based on runoff volume and timing in hydrographs (see Figs. 10 and 11). This existing buffer covers 22 683.5 m^2 of the pervious area, which is 2.3 % of the entire watershed and 3.0 % of the pervious area. As the baseline, the SHC watershed consists of 18.8 % DCIA, 5.2 % ICIA, 2.3 % BPA, 73.1 % SPA, and 0.6 % water. Under the modeled GI scenario of disconnecting rooftop drains and extended BPA, the DCIA is reduced to 9.6 %, the ICIA increases to

14.4 %, the BPA increases to 17.2 %, and the SPA is reduced to 58.2 % of the total area, respectively.

3.2 The hypothetical HRE modeling analysis

The eight single storm hypothetical HRE modeling analysis with the six discretization options resulted in 48 SWMM runs. As explained earlier, this was done to determine which HRE configuration option best balances model complexity and presumed accuracy. Each simulated storm was assumed to last from midnight to midnight. Results are presented as hydrographs between 11:00 and 13:00 where most concentrated rainfall occurs in the NRCS Type-II distribution (USDA, 1986) (Fig. 8). In large storms, larger than a 5-year storm in particular, all six types of spatial discretization produce very similar hydrographs as shown in Fig. 8g, h. The modeled flow rates and total runoff volumes are almost identical.

In the large storm situation, all of the PAs are saturated in the early stage of the storm. Once saturated, the PAs are not able to provide any additional on-site hydrologic control and behave as IA. In view of this, any of the spatial discretization options would be suitable for analyzing flood controls and in designing a drainage system based on a 10-year storm. However, this is not the relevant case for evaluating GI implementation, which focuses on controlling smaller storms. For storms smaller than a 2-year event, considerable differences were found among the simulated hydrographs (Fig. 8a–e).

In the smallest storm situation (Fig. 8a) the options for spatial discretization result in almost identical hydrographs except Option 4, where only DCIA discharges runoff, as the total impervious area (TIA) is modeled as DCIA. Rainfall onto PA is completely captured by DS and/or infiltrated to the soils. Because Option 4 ignores the difference between DCIA and ICIA, the entire impervious area (subarea IA) is modeled the same as DCIA, which means all of the runoff is discharged to the storm drainage system directly with no abatement. Under a small storm (like <1-month storm), runoff occurs only from IA – more specifically, only from DCIA. For small storms, runoff from ICIA is completely controlled by BPA (if ICIA exists), but no ICIA is modeled under Option 4. Because of this, modeled runoff from this option is higher than any of the other options under the small storms. DCIA is modeled explicitly in the other five options. The relative difference in runoff estimates caused by modeling TIA as DCIA contribution diminished as larger storms are modeled, Fig. 8a–c. Option 4 is not suitable to modeling GI alternatives because it ignores the significance of characterizing DCIA and ICIA within an HRE. Option 5 shows the most significant variation among the simulated hydrographs. This option estimates lower flow rates than the others for smaller storms, but higher peaks in medium-size storms (such as 6-month to 2-year return period storms; Fig. 8d–f). Option 5 is configured to simulate the "ideal" green implementation scenario of surface grading for stormwater discharge, in which

Table 3. Land cover status of Shayler Crossing watershed.

Surface components		DCIA (m^2)	ICIA (m^2)	Sum (m^2)	Fraction
Impervious areas	Building	91 770.0	3756.2	95 526.2	9.6 %
	Street	57 610.5	14 897.2	72 507.7	7.3 %
	Driveway	33 554.7	2083.7	35 638.4	3.6 %
	Parking	2362.7	3154.1	5516.8	0.6 %
	Sidewalk	1646.9	9990.3	11 637.2	1.2 %
	Miscellaneous	–	17 766.8	17 766.8	1.8 %
	Sum of IA	186 944.7	51 648.4	238 593.1	24.0 %
Pervious areas	Lawn			400 667.4	40.3 %
	Agriculture			219 430.4	22.1 %
	Forest			128 558.1	12.9 %
	Sum of PA			748 655.9	75.4 %
Water	Wet pond			5014.2	0.5 %
	Swimming pool			998.9	0.1 %
	Sum of water			6013.0	0.6 %
Sum				993 262.0	100 %

the entire pervious area works like BPA. The expanded on-site pervious buffer can thoroughly control runoff from ICIA until the DS and infiltration capacity of BPA are fully saturated. Once the hydrologic capacities for on-site controls are fully saturated, the entire PA hydrologically responds more or less like IA. Once a subcatchment DS fills and exceeds infiltration capacity, this unrealistic green development condition may result in higher peak discharges than the other options.

From the hypothetical modeling analysis, it can be surmised that an extensive on-site green infrastructure implementation could result in more frequent local flooding, e.g., water intrusion into basements. This may be especially the case when evaluating scenarios for locations where medium-size storms have a long duration, like during the wet season of the Pacific Northwest of the United States. The comparatively high runoff estimated for Option 5 (Fig. 8d–f) would be maintained until all PA is saturated by increased rainfall intensity. If a smaller portion of PA is modeled as BPA, while all the other conditions are kept the same, the BPA reaches the saturated condition under a smaller storm. Once the BPA is saturated the area hydrologically responds like IA. However, SPA (i.e., non-buffering pervious area) can still control rainfall within the area. This analysis suggests that it is important to properly define the area of BPA especially when analyzing GI alternatives for on-site stormwater controls, as we surmised originally. Therefore, Option 5 is not suitable for modeling a GI scenario because it ignores the actual significance of variance in BPA. It is a common modeling practice in SWMM to treat all pervious area the same (as in options 4 or 5 in Fig. 5), even though only the BPA can receive water from ICIA. As shown in Fig. 8, simulated runoff by

options 4 or 5 would presumably be inaccurate, especially for the <1-year small storms.

Figure 9 contains graphs comparing the results among the six options (Fig. 5), showing the relative difference for peak flow, average flow, and total runoff volume for each of the five other options compared to Option 1, presumably the most accurate one, in terms of output uncertainty because the level of spatial lumping is the lowest.

The relative differences reported in Fig. 9 ("variation from the result of 5-subs" in the y axis) were estimated as $(\mathrm{MR}^k_j - \mathrm{MR}^k_{5\mathrm{subs}})/\mathrm{MR}^k_{5\mathrm{subs}}$, where MR^k_j represents a modeling result from the kth synthetic storm with the jth discretization option and 5subs means the discretization with five subcatchments (i.e., Option 6). Options 1, 2, and 3 types of multi-subcatchment discretization present similar hydrologic responses for all storm sizes. In comparison, both options 4 and 5 result in significantly different hydrologic outcomes, particularly for smaller storms. Again, this is due to the unresolved spatial delineation of DCIA from TIA and BPA from the total pervious area (TPA), respectively, whereas Option 6 is based on a single-subcatchment approach, but produces similar results to the multi-subcatchment discretization approaches under options 1, 2 and 3, for all storm classes tested. The differences between Option 6 and Option 1, though worth noting, are marginal for the three important hydrologic characteristics (Fig. 9). This outcome of the hypothetical analysis supports our original rationale for the relevance of characterizing the BPA. Under Option 6, the four critical hydrologic components (i.e., DCIA, ICIA, BPA, and SPA) are distinctly modeled in SWMM within a single subcatchment that is delineated based on the actual drainage area to a storm sewer inlet (termed an HRE in this study). Based on the results, Option 6 balances the combination of discretiza-

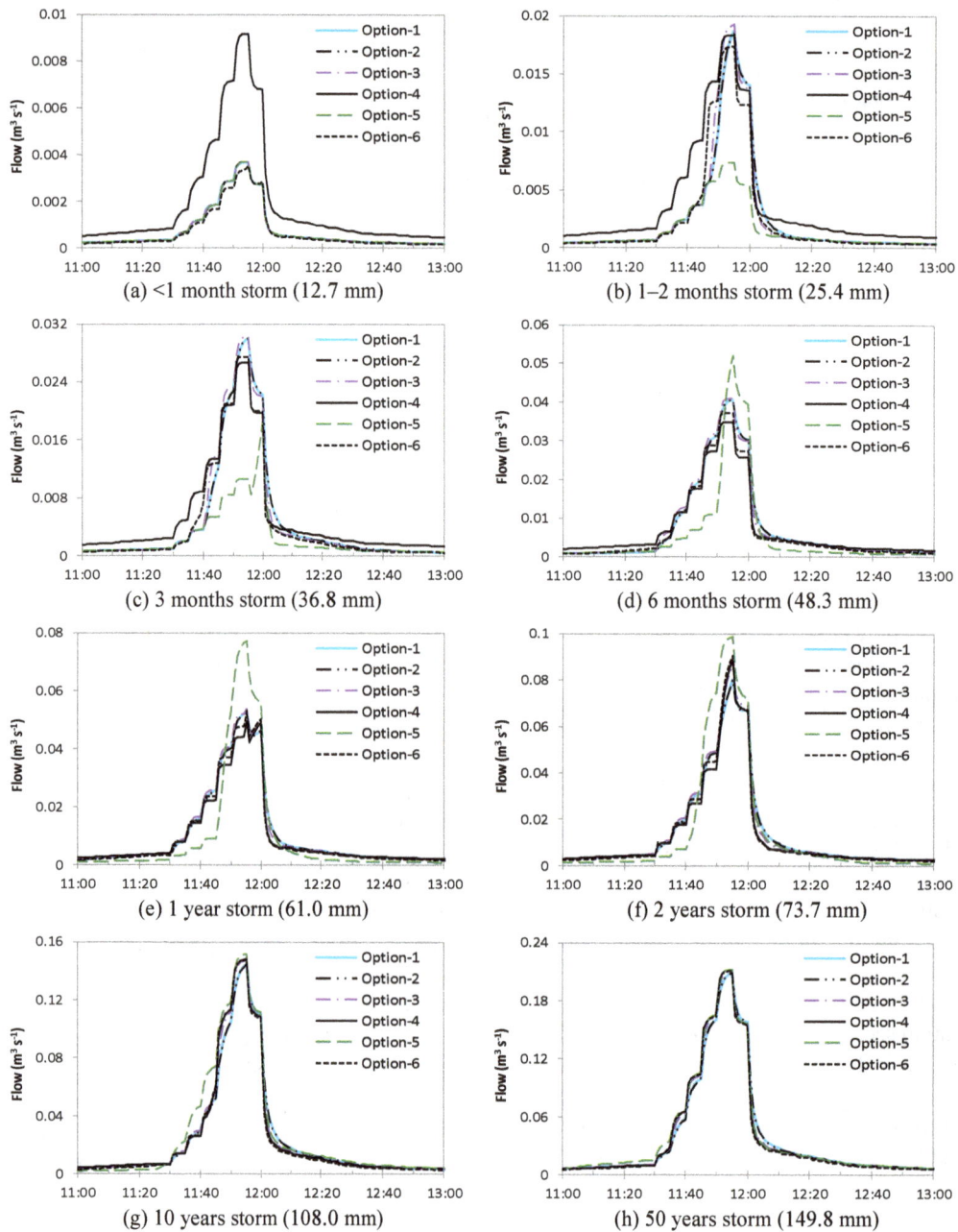

Figure 8. Hypothetical HRE SWMM modeling results.

tion criteria, especially in terms of the level of effort required in model setup, configuring parameter values and output uncertainty.

3.3   SHC watershed-scale modeling results

Option 6 (Fig. 5) was used to set up the SWMM model for the SHC watershed as described above. The SHC model consists of 191 subcatchments and 269 junctions and con-

duits (Fig. 6). The model also includes two wet ponds, two dry ponds, and a 10-year detention area modeled as storage structures with orifice-type hydrologic controls. The results of the model sensitivity analysis were summarized for the period 22 to 24 July 2009, using the total runoff volume as the endpoint being assessed with Eq. (2) (Fig. 10). There was a total of 164.6 mm rainfall during the 3 days of this period; this storm is smaller than the 1-year return period design storm (61.0 mm d^{-1}) but larger than the 6-month storm

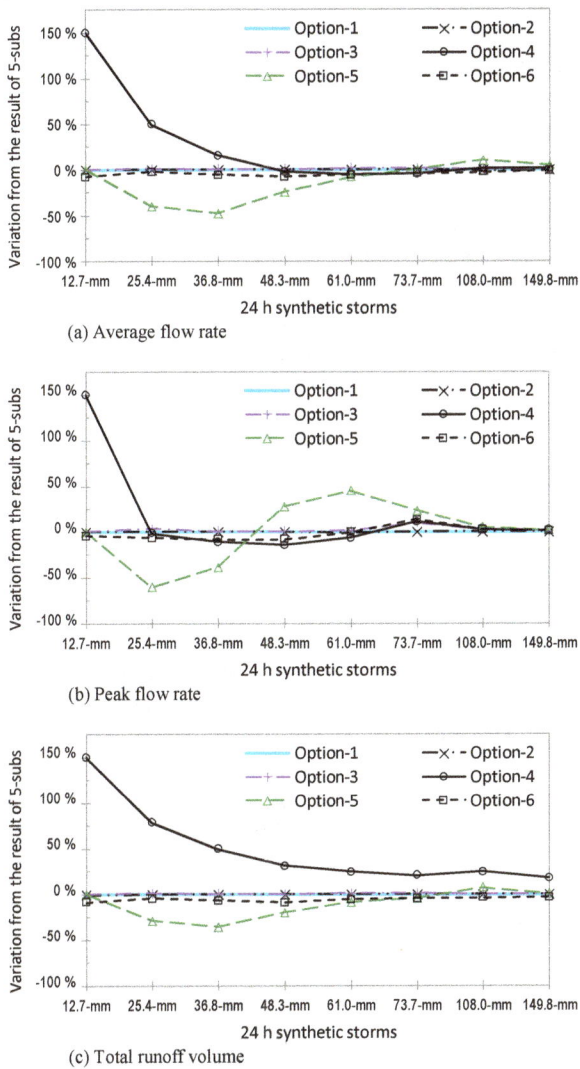

(a) Average flow rate

(b) Peak flow rate

(c) Total runoff volume

Figure 9. Comparison of the hypothetical HRE modeling results.

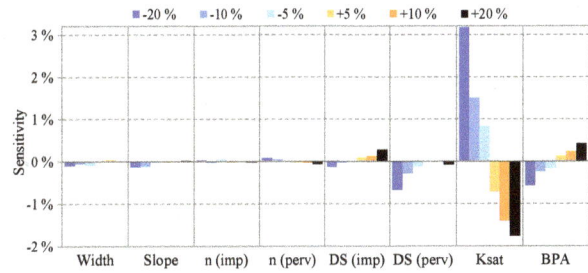

Figure 10. Sensitivity analysis of the SWMM parameters at SHC.

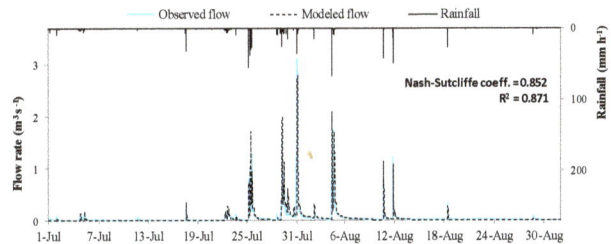

Figure 11. Watershed-scale SWMM modeling results from 1 July 2009 to 31 August 2009.

(48.3 mm d^{-1}) based on the storm statistics for the study area (see Table 2). While the 3 % change in total runoff is not significant in sensitivity, the most sensitive parameter was K_{sat}, followed by BPA and DS. Whereas the changes in K_{sat} affect the entire PA (75.4 % of SHC), the changes in BPA affect a much smaller area (2.3 % of SHC for the baseline condition) than PA. The other parameters (i.e., width, slope, and n) were found not to be as sensitive, with negligible changes in results $\leq \pm 0.15$ % even for ± 20 % change in the individual parameter value. When land cover status is represented accurately in a SWMM model, certain parameters will be less sensitive because of the underlying hydraulic and spatial realities are well represented. For example, the parameters representing the impervious land cover types in this modeling analysis were found to be less sensitive than pervious area parameters.

Model calibration was conducted by adjusting the land-cover-based modeling parameters and BPA to the entire study watershed. As shown in Table 1, parameters for the impervious land cover types changed little and were made equivalent for n and DS. As expected, parameters for the pervious land cover types needed more adjustment than those for the impervious. The initial value of K_{sat} was defined using the site-specific soil types (mainly silty loam clay), but the values for the individual pervious land cover types were varied by the model calibration effort. Whereas K_{sat} for forest area was adjusted only slightly (i.e., 1.6 initially to 1.52 for the final calibration), the values for lawn (or landscaped area) and agriculture required more adjustment (from 1.6 initial to 1.02 for agriculture, and from 1.6 initial to 0.89 for lawns). The relatively large changes for K_{sat} are indicative of more soil compaction for urban and agricultural soils compared to the expected native soil condition.

The measured rainfall intensities and stream flow rates, along with the calibrated model results are presented in Fig. 11. The modeled hydrographs are well matched with the measured data at the watershed scale with a Nash–Sutcliffe coefficient = 0.852 and $R^2 = 0.871$.

After making the model adjustments for the GI scenario, the relative percentages of the four classified subareas changed (Fig. 12a). Using the hydrograph separation approach, the relative contributions of the primary hydrologic components with and without GI implementation were estimated for the period 1 July 2009 to 31 August 2009 (Fig. 12b). A more-detailed representation for the hydrograph separation is presented in Fig. 13, which covers 72 h from 22 to 24 July 2009. It is interesting to note from Fig. 13 that the peak flow for the event depicted in the figure is

(a) Land cover components   (b) Hydrologic components

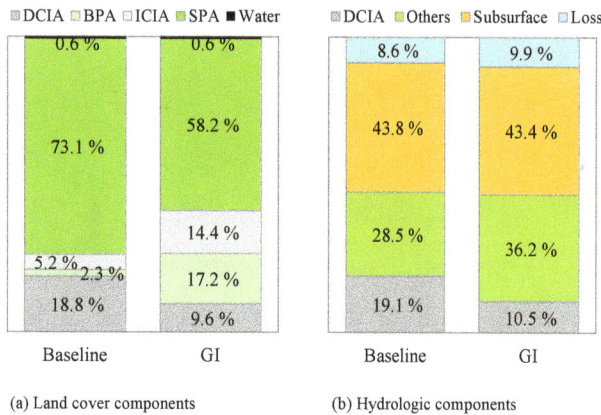

Figure 12. Relative percentages of (a) land cover and (b) hydrologic components computed for the period 1 July 2009 to 31 August 2009. In (b) "others" represents surface runoff from areas other than DCIA, "subsurface flow" is the subsurface contribution, and "loss" is rainfall loss by evaporation or deep percolation.

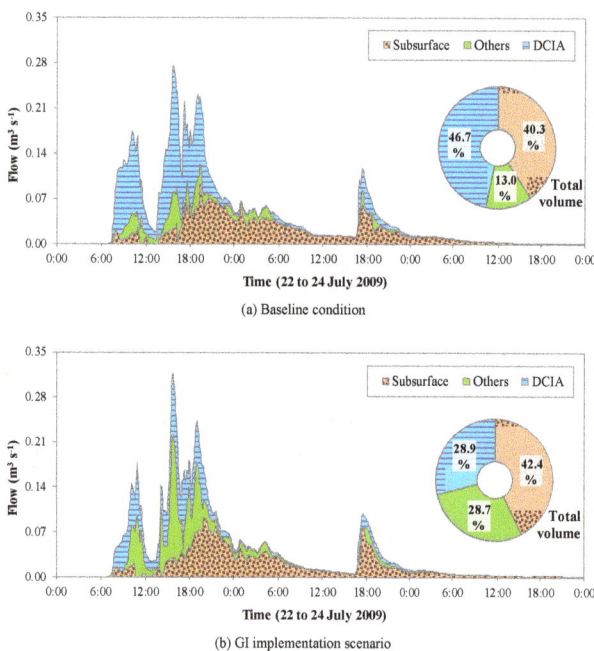

(a) Baseline condition

(b) GI implementation scenario

Figure 13. Hydrograph separation and volumetric percentages contributing to stream flow for the period 22 to 24 July 2009.

slightly higher in the GI scenario, but that the duration of flows slightly smaller than this peak is longer in the baseline scenario.

While the results from applying the hydrograph separation cannot be validated without extensive field measurements, the exercise provides insight to the potential effectiveness and rationale for developing strategies for GI in the watershed. For instance, about 48 % of the volumetric stream flow was contributed through subsurface flow over the simulation

period, even though the study watershed is characterized with poorly infiltrating soils. After applying the GI scenario, although the subsurface flow contributed a similar fraction to the stream flow, the fractional contributions of surface runoff from DCIA and the other areas are significantly changed (Figs. 12b and 13). This situation arises not from a change in land cover but the internal flow paths taken by the runoff. The result is reduced runoff from DCIA but increased runoff from the other areas (i.e., ICIA, BPA, and SPA).

From a water quality management perspective, it is necessary to consider hydrologic and contaminant discharge processes with respect to their sources and transport pathways. For example, if the watershed has water quality issues related to nutrients, the management effort might pay more attention to the stormwater discharge from pervious areas that include fertilizer applications. If GI were designed to intercept runoff from DCIA in the watershed, an unintended consequence could result from increased runoff volume traveling through a pervious area with elevated standing stocks of soluble or erodible nutrients. In this case, it would be important to consider turf management practices.

Another example of how the hydrograph separation approach (Fig. 13) provides additional opportunities for interpreting hydrodynamics before and after applying the GI scenario is revealed by considering that disconnecting downspouts reduced the total runoff volume, but also resulted in a higher peak flow (note the 16:00 time point on 22 July 2009 in Fig. 13). This result is like the single storm analysis using Option 5 (Fig. 5). Overall the flow volume is reduced from the GI scenario. However, when the peak occurred around 15:30 (shown in Fig. 13), the capacity of the GI for controlling stormwater was already exceeded because of controlling runoff during the previous rainfall that occurred between 7:00 and 14:00. Under this saturated condition, even the direct rainfall to the GI area will be discharged with minimum abatement. If there is no GI (as in the baseline condition), the same area receives only direct rainfall, there is no additional runoff from the impervious area, and that rainfall is controlled by still-available surface depression storage and not-saturated infiltration capacity. In the 22 July 2009 situation, the stormwater control capacity (mainly DS and infiltration) of the extended BPA is saturated by earlier rainfall. Once saturated, the BPA discharges higher runoff. The modeled GI contributes much higher runoff volume from PA, which might be nutrient enriched. With the hydrograph separation analysis, we gain insight to the consideration of stormwater management objectives and extend the utility of SWMM.

4   Summary and conclusions

We demonstrate how high-resolution spatial data can be applied to spatially discretize a watershed and develop a methodology that should decrease model output uncertainty with reduced calibration effort. The suitability of the spatial discretization approach for GI modeling was initially verified with a hypothetical urbanscape analysis using eight synthetic storms of various sizes. We evaluated our approach to SWMM subcatchment parameterization using the hypothetical analysis that allowed for the qualification of five different options relative to one that would be considered the most spatially explicit and, therefore, result in the least amount of output uncertainty (see Fig. 9). From the hypothetical analysis, the best option was selected to develop a watershed-scale SWMM model at the study area. The simulated hydrographs by the developed watershed-scale SWMM model were well matched with observed data over a 2-month continuous simulation (Nash–Sutcliffe coefficient $= 0.852$; $R^2 = 0.871$) after minimal calibration effort. A GI scenario that modeled downspout disconnection from all the main buildings that are DCIA was described. We demonstrate how simple model adjustments can be made to separate the total and surface runoff among primary pathways that runoff takes before discharging to the natural stream network. This hydrograph separation procedure can shed light on GI design requirements and water quality management.

The optimal spatial discretization scheme distinguishes DCIA from ICIA, and BPA from SPA, and explicitly models these as subareas within each subcatchment parameterization in SWMM. This approach is particularly useful when modeling the impact of small storms, i.e., when BPA can control all or most of ICIA runoff. The land-cover-based spatial discretization approach is scale-independent, can be applied directly to a larger watershed as long as any heterogeneity in landscape properties is accounted for in the GIS setup (e.g., by dividing land cover components into multiple subgroups such as flat and slanted rooftops, high and low sloped urban hillslopes, or B and C type hydrologic soil types), and affords the opportunity to evaluate urban stormwater management strategies with presumably decreased output uncertainty for small storms and expanded applicability to GI planning, design, and implementation. Parameters are adjusted per SWMM subcatchment in a typical calibration approach, which is scale-dependent and requires more effort in larger watersheds. In our approach, a SWMM model is calibrated by adjusting parameters per land cover component, which are categorized by urban development codes or general construction specifications for land uses. Overall this study demonstrates the relative effectiveness of different approaches in drainage area characterization using highly resolved spatial data to the setup and analysis of a SWMM model that should improve its utility for simulation of GI.

Competing interests. The authors declare that they have no conflict of interest.

Disclaimer. The US Environmental Protection Agency, through its Office of Research and Development, funded, managed, and collaborated in the research described herein. It has been subjected to the Agency's administrative review and has been approved for external publication. Any opinions expressed in this paper are those of the author(s) and do not necessarily reflect the views of the Agency; therefore, no official endorsement should be inferred. Any mention of trade names or commercial products does not constitute endorsement or recommendation for use.

Acknowledgements. The authors would like to thank Bill Mellman of Clermont County; Paul Weaver of APTIM; and Michael Tryby, Michael Elovitz, and William Shuster of the USEPA. They provided critical data, suggestions, and critical reviews.

Edited by: Louise Slater

References

Braden, J. B. and Johnston, D. M.: Downstream economic benefits from storm-water management, J. Water Res. Pl., 130, 498–505, 2004.

Debo, T. N. and Reese, A. J.: Municipal Stormwater Management. 2nd Edition. Lewis Publishers, CRC Press, ISBN 9781566705844, 2002.

Dietz, M. E.: Low impact development practices: A review of current research and recommendations for future directions, Water Air Soil Poll., 186, 351–363, https://doi.org/10.1007/s11270-007-9484-z, 2007.

ESRI: ArcGIS Desktop: Release 10.2., Environmental Systems Research Institute (ESRI), Redlands, CA, USA, 2013.

Fletcher, T. D., Andrieu, H., and Hamel, P.: Understanding, management and modelling of urban hydrology and its consequences for receiving waters – A state of the art, Adv. Water Res., 51, 261–279, 2013.

Gironás, J., Roesner, L. A., and Davis, J.: Storm Water Management Model Applications Manual, EPA/600/R-09/077, National Risk Management Research Laboratory, Office of Research and Development, U.S. Environmental Protection Agency (USEPA), Cincinnati, OH, USA, 2009.

Gregory, J. H., Dukes, M. D., Jones, P. H., and Miller, G. L.: Effect of urban soil compaction on infiltration rate, J. Soil Water Conserv., 61, 117–124, 2006.

Guo, J. C.: Volume-based imperviousness for storm water designs, J. Irrig. Drain. E., 134, 193–196, 2008.

Guo, J. C. and Urbonas, B.: Maximized detention volume determined by runoff capture ratio, J. Water Res. Pl.-ASCE, 122, 33–39, 1996.

Homer, C. G., Dewitz, J. A., Yang, L., Jin, S., Danielson, P., Xian, G., Coulston, J., Herold, N. D., Wickham, J. D., and Megown, K.: Completion of the 2011 National Land Cover Database for the conterminous United States-Representing a decade of land cover change information, Photogramm. Eng. Rem. S., 81, 345–354, 2015.

Horton, R., Ankeny, M. D., and Allmaras, R. R.: Effects of compaction on soil hydraulic properties, in: Soil Compaction in Crop Production, Elsevier Science B.V., Amsterdam, Netherlands, 479–500, 1994.

Huber, W. C.: New options for overland flow routing in SWMM, in: Urban Drainage Modeling, edited by: Brashear, R. W. and Maksimovic, C., Proc. of the Specialty Symposium of the World Water and Environmental Resources Conference, ASCE, Environmental and Water Resources Institute, Orlando, FL, 22–29, 2001.

Huber, W. C. and Dickinson, R.: Storm Water Management Model User's Manual, Version 4. EPA/600/3-88/001a (NTIS PB88-236641/AS), Environmental Research Laboratory, Office of Research and Development, U.S. Environmental Protection Agency (USEPA), Athens, GA, USA, 1988.

Huber, W. C. and Cannon, L.: Modeling non-directly connected impervious areas in dense neighborhoods, in: Global Solutions for Urban Drainage, Proc. Ninth International Conference on Urban Drainage, edited by: Strecker, E. W. and Huber, W.C., Portland, OR. American Society of Civil Engineers, Reston, VA, CD-ROM, 2002.

Huff, F. A. and Angel, J. R.: Rainfall Frequency Atlas of the Midwest. National Weather Service, National Oceanic and Atmospheric Administration and Illinois State Water Survey, A Division of the Illinois Department of Energy and Natural Resources, Bulletin 71 (MCC Research Report 92-03) available at: http://www.sws.uiuc.edu/pubdoc/B/ISWSB-71.pdf (last access: 14 February 2015), 1992.

Krebs, G., Kokkonen, T., Valtanen, M., Setälä, H., and Koivusalo, H.: Spatial resolution considerations for urban hydrological modelling, J. Hydrol., 512, 482–497, 2014.

Lee, J. G., Heaney, J. P., and Lai, F. H.: Optimization of integrated urban wet-weather control strategies, J. Water Res. Plan. Man., 131, 307–315, 2005.

Lee, J. G., Nietch, C. T., and Panguluri, P.: SWMM Modeling Methods for Simulating Green Infrastructure at a Suburban Headwatershed – User's Guide, EPA/600/R-17/414, Water Systems Division, Office of Research and Development, U.S. Environmental Protection Agency. Cincinnati, OH 45268, available at: https://nepis.epa.gov/Exe/ZyPDF.cgi/P100TJ39.PDF?Dockey=P100TJ39.PDF, (last access: 15 February 2018), 2017.

Montalto, F., Behr, C., Alfredo, K., Wolf, M., Arye, M., and Walsh, M.: Rapid assessment of the cost-effectiveness of low impact development for CSO control, Landscape Urban Plan., 82, 117–131, https://doi.org/10.1016/j.landurbplan.2007.02.004, 2007.

Niazi, M., Nietch, C., Maghrebi, M., Jackson, N., Bennett, B. R., Tryby, M., and Massoudieh, A.: Storm Water Management Model (SWMM) – Performance review and gap analysis, J. Sustain. Water B. Environ., 3, https://doi.org/10.1061/JSWBAY.0000817, 2017.

NOAA: Mean Monthly, Seasonal, and Annual Pan Evaporation for the United States. NOAA Technical Report NWS 34. U.S. Dept. of Commerce, National Oceanic and Atmospheric Administration, National Weather Service. Washington, D.C. December 1982, available at: http://www.nws.noaa.gov/oh/hdsc/PMP_related_studies/TR34.pdf (last access: 15 February 2015), 1982.

NRC: Urban Stormwater Management in the United States. Committee on Reducing Stormwater Discharge Contributions to Water Pollution. National Research Council (NRC). National Academies Press., Washington, DC, ISBN: 978-0-309-12539-0, 2009.

Pitt, R.: Small storm hydrology and why it is important for the design of stormwater control practices, Adv. Mod. Manag. Stormw., 7, 61–91, 1999.

Rossman, L. A.: Storm Water Management Model User's Manual, Version 5.1, EPA/600/R-14/413b, Revised September 2015, U.S. Environmental Protection Agency, Office of Research and Development, Water Supply and Water Resources Division, Cincinnati, OH, USA, 2015.

Rossman, L. A. and Huber, W. C.: Storm Water Management Model Reference Manual, Volume I – Hydrology (Revised). EPA/600/R-15/162A, Revised January 2016, U.S. Environmental Protection Agency, Office of Research and Development, Water Supply and Water Resources Division, Cincinnati, OH, USA, 2016.

Sample, D. J., Heaney, J. P., Wright, L. T., Fan, C. Y., Lai, F. H., and Field, R.: Costs of best management practices and associated land for urban stormwater control, J. Water Res. Plan. Man., 129, 59–68, 2003.

Sun, N., Hall, M., Hong, B., and Zhang, L.: Impact of SWMM catchment discretization: case study in Syracuse, New York, J. Hydrol. Eng., 19, 223–234, https://doi.org/10.1061/(ASCE)HE.1943-5584.0000777, 2014.

USDA: Urban Hydrology for Small Watersheds. Technical Release 55 (TR-55). Natural Resources Conservation Service, Conservation Engineering Division, U.S. Department of Agriculture (USDA), 1986.

USEPA: Reducing Stormwater Costs through Low Impact Development (LID) Strategies and Practices, EPA 841-F-07-006, Nonpoint Source Control Branch (4503T), U.S. Environmental Protection Agency, Washington, DC, 20460, 2007.

USEPA: Technical Guidance on Implementing the Stormwater Runoff Requirements for Federal Projects under Section 438 of the Energy Independence and Security Act. United States Environmental Protection Agency, Office of Water (4503T), Washington, DC 20460, EPA 841-B-09-001, December 2009, available at: http://www.epa.gov/oaintrnt/documents/epa_swm_guidance.pdf (last access: 8 June 2015), 2009.

USEPA: Terminology of Low Impact Development – Distinguishing LID from other Techniques that Address Community Growth Issues, EPA-841-N-12-003, Office of Water, U.S. Environmental Protection Agency, Washington DC, 20460, 2012.

USEPA: What is Green Infrastructure? (Last updated on 13 June 2014) Office of Water, U.S. Environmental Protection Agency. Washington, DC 20460, available at: http://water.epa.gov/infrastructure/greeninfrastructure/gi_what.cfm (last access: 8 May 2015), 2014.

WEF-ASCE: Design of Urban Stormwater Controls. WEF Manual of Practice (MOP) No. 23. ASCE Manuals and Reports on Engineering Practice No. 87, Water Environment Federation (WEF), Environmental & Water Resources Institute, American Society of Civil Engineers (ASCE), ISBN-13: 978-0071704441, ISBN-10: 0071704442, 2012.

Zhou, Q.: A review of sustainable urban drainage systems considering the climate change and urbanization impacts, Water, 6, 976–992, https://doi.org/10.3390/w6040976, 2014.

13

On the non-stationarity of hydrological response in anthropogenically unaffected catchments: an Australian perspective

Hoori Ajami[1,2], Ashish Sharma[1], Lawrence E. Band[3], Jason P. Evans[4], Narendra K. Tuteja[5], Gnanathikkam E. Amirthanathan[6], and Mohammed A. Bari[7]

[1]School of Civil and Environmental Engineering, University of New South Wales, Sydney, Australia
[2]Department of Environmental Sciences, University of California Riverside, Riverside, CA, USA
[3]Department of Geography and Institute for the Environment, University of North Carolina, Chapel Hill, NC, USA
[4]Climate Change Research Centre, University of New South Wales, Sydney, Australia
[5]Environment and Research Division, Bureau of Meteorology, Canberra, Australian Capital Territory, Australia
[6]Environment and Research Division, Bureau of Meteorology, Melbourne, Victoria, Australia
[7]Environment and Research Division, Bureau of Meteorology, Perth, Western Australia, Australia

Correspondence to: Hoori Ajami (hoori.ajami@ucr.edu)

Abstract. Increases in greenhouse gas concentrations are expected to impact the terrestrial hydrologic cycle through changes in radiative forcings and plant physiological and structural responses. Here, we investigate the nature and frequency of non-stationary hydrological response as evidenced through water balance studies over 166 anthropogenically unaffected catchments in Australia. Non-stationarity of hydrologic response is investigated through analysis of long-term trend in annual runoff ratio (1984–2005). Results indicate that a significant trend ($p < 0.01$) in runoff ratio is evident in 20 catchments located in three main ecoregions of the continent. Runoff ratio decreased across the catchments with non-stationary hydrologic response with the exception of one catchment in northern Australia. Annual runoff ratio sensitivity to annual fractional vegetation cover was similar to or greater than sensitivity to annual precipitation in most of the catchments with non-stationary hydrologic response indicating vegetation impacts on streamflow. We use precipitation–productivity relationships as the first-order control for eco-hydrologic catchment classification. A total of 12 out of 20 catchments present a positive precipitation–productivity relationship possibly enhanced by CO_2 fertilization effect. In the remaining catchments, biogeochemical and edaphic factors may be impacting productivity. Results suggest vegeta-tion dynamics should be considered in exploring causes of non-stationary hydrologic response.

1 Introduction

Increases in atmospheric CO_2 concentration are impacting the terrestrial water cycle through changes in radiative forcings (affecting precipitation and temperature) as well as plant physiological and structural responses (Betts et al., 2007; Wigley and Jones, 1985). As a result, projections of future changes in water resources become complicated due to the tight coupling between the terrestrial biosphere and hydrologic cycle (Band et al., 1996; Baron et al., 2000; Gedney et al., 2006; Ivanov et al., 2008). There is a growing body of evidence showing that increases in CO_2 often lead to decreases in leaf stomatal conductance (Field et al., 1995; Medlyn et al., 2001) and lower leaf-scale transpiration rates. However, the impact of reducing stomatal conductance on canopy-scale evapotranspiration (ET) and vegetation productivity (biomass and leaf area index (LAI) increases) is uncertain. In some ecosystems, decline in leaf-scale ET rates increases soil available water (Leuzinger and Körner, 2010). At the canopy scale, leaf-scale decline of ET might be compensated by increases in plant productivity and changes in ecosystem structure in terms of increases in LAI and changes in species composition (Kergoat et al., 2002). Due to this "compensatory response", the impact of elevated CO_2 on catchment-scale water balance is uncertain and expected to vary from region to region (Field et al., 1995; Kergoat et al., 2002). Moreover, terrestrial vegetation productivity is often limited by availability of nutrients, mostly nitrogen and phosphorous (Vitousek and Howarth, 1991) and light (Huxman et al., 2004; Schurr, 2003), which further increases uncertainty of projecting terrestrial ecosystem response to climate change (Wieder, 2014).

Understanding spatial and temporal variability of catchment-scale water yield in relation to precipitation variability and ecosystem productivity is challenging, as it requires long-term observational records from unimpaired catchments. A plethora of modeling studies have been performed to predict the climate change impacts on vegetation productivity (Kergoat et al., 2002; Leuzinger and Körner, 2010) and global runoff (Betts et al., 2007; Piao et al., 2007). However, projections depend on the underlying model assumptions and structure, process representation and scale of application (Medlyn et al., 2011). Similarly, assessing climate elasticity of streamflow has shown that the degree of sensitivity of streamflow to various factors depends on the model structure and calibration approach (Sankarasubramanian et al., 2001).

Here, we investigate the nature and frequency of non-stationary hydrologic response as evidenced through water balance studies over 166 anthropogenically unaffected catchments in Australia. Our assessment assumes the non-stationarity to manifest itself through the annual water balance, and more specifically, through the annual runoff ratio (Q / P). Our primary objective is to investigate whether there is evidence for such non-stationarity in the runoff ratio, and if there is, what could explain its existence.

Non-stationarity of runoff ratio is caused by complex interactions between precipitation, climate variability, plant physiological and structural responses to elevated CO_2 (Leuzinger and Körner, 2010; Chiew et al., 2014) and landscape characteristics (soil and topography) of a catchment. The question is whether patterns of similarities and differences across space and time exist as a result of these interactions to provide a framework for hydrologic prediction (Sivapalan et al., 2011). Instead of assuming vegetation as a static component of the hydrologic system (Ivanov et al., 2008), catchment-scale vegetation dynamics will be an integral component of this classification framework. Therefore, our secondary objective is to formulate a catchment classification framework based on catchment-scale ecohydrologic response. This first-order grouping of catchments helps to generalize catchment behavior in terms of changes in runoff ratio and vegetation productivity due to changes in precipitation. Previous catchment classification efforts have mostly considered hydrologic signatures related to precipitation,

temperature and streamflow (Sawicz et al., 2011; Wagener et al., 2007). Sawicz et al. (2014) illustrated that changes in climate characteristics of catchments can mostly explain hydrologic change which was characterized by changes in groupings of 314 catchments in the USA. Due to the lack of information, temporal changes in land use were not considered in characterizing hydrologic change in this approach. We argue that, in the context of climate change and to improve hydrologic prediction under change (Sivapalan et al., 2011), developing a catchment classification framework that incorporates the role of vegetation dynamics on catchment-scale water partitioning is required. This framework can inform future modeling experiments for determining the relative importance of contributing factors to non-stationary catchment response.

An assessment of the non-stationarity of the runoff ratio across 166 anthropogenically unaffected catchments in Australia is presented next using long-term ground- and satellite-based observational records.

2 Data and methods

2.1 Data

Daily stream discharge data are obtained from the Australian network of hydrologic reference stations (HRSs) that consists of 221 gauging stations (http://www.bom.gov.au/water/hrs/). Out of 221 catchments, 166 catchments have complete daily discharge time series covering the 1979–2010 period and these are the catchments used in the study (Fig. 1). The anthropogenically unaffected catchments cover a range of spatial scales with their areas ranging from 6.6 to 232 846 km^2. Catchment-averaged daily precipitation, actual and potential evapotranspiration, and temperature are obtained from the Australian Water Availability Project (AWAP) gridded time series products at 0.05° resolution (Raupach et al., 2009, 2012). AWAP potential evapotranspiration is calculated based on the Priestley–Taylor equation (Raupach et al., 2009). The monthly fraction of photosynthetically active radiation (fPAR) absorbed by vegetation is obtained from Donohue et al. (2008) at 0.08° resolution for the 1984–2010 period. Total fPAR (F_{tot}) values are approximately related to fractional vegetation cover and range between 0.0 (no vegetation cover) and 0.95 (maximum vegetation cover). The monthly F_{tot} dataset version 5 is derived from the Advanced Very High Resolution Radiometer (AVHRR) sensor, and F_{tot} values were used as a measure of vegetation productivity in this study, assuming energy-use efficiency is constant (https://data.csiro.au). This dataset has been previously used to assess trends in vegetation cover across Australia (Donohue et al., 2009).

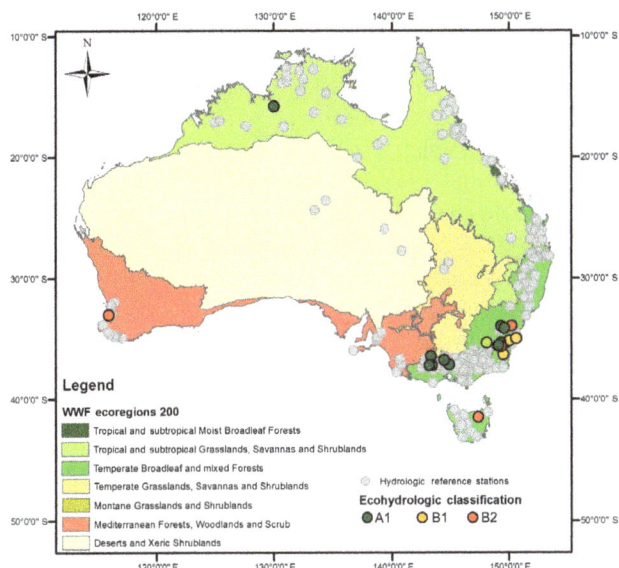

Figure 1. Distribution of hydrologic reference stations across Australia. Colored circles represent catchments with a significant trend in annual runoff ratio. Colors represent catchment grouping based on the classification framework of Fig. 6. The catchments with non-stationary hydrologic response span over three significant ecoregions of the continent. Ecoregion boundaries are from the World Wildlife Fund (http://maps.tnc.org/gis_data.html).

2.2 Methods

We used ground- and satellite-based observations to detect and investigate causes of non-stationarity of runoff ratio across HRS catchments. Our methodology consists of (1) detecting trends in annual runoff ratio, fractional vegetation cover, annual precipitation and precipitation seasonality indices at a catchment scale; (2) assessing long-term (27 years) water balance patterns across all catchments with non-stationary hydrologic response using hydrologic indices such as the Horton index (Troch et al., 2009); (3) exploring annual runoff ratio's sensitivities to water balance components and fractional vegetation cover at an individual catchment scale; and (4) formulating an ecohydrologic catchment classification framework.

2.2.1 Non-parametric trend analysis to detect non-stationarity

The modified Mann–Kendall non-parametric test (Hamed and Rao, 1998; Kendall, 1970; Mann, 1945) that accounts for serial autocorrelation in the time series is performed to detect significant trends in annual runoff ratio at a 0.01 significance level across the 166 HRS catchments. The first and last 5 years of data are removed from the record to reduce the impact of edge effect for trend analysis (1984–2005). Similar trend analysis is performed for annual precipitation and average fractional vegetation cover of each catchment.

Changes in precipitation seasonality across catchments with non-stationary hydrologic response are explored by assessing the trends in two measures of precipitation seasonality: the seasonality index (SI) (Walsh and Lawler, 1981) and days of a year at which the 10th, 25th, 50th, 75th and 90th percentiles of annual precipitation are reached (Pryor and Schoof, 2008). The SI is calculated based on monthly precipitation values (Walsh and Lawler, 1981):

$$SI = \frac{1}{P} \sum_{n=1}^{12} \left| X_n - \frac{P}{12} \right|, \tag{1}$$

where P is annual precipitation and X_n is monthly total precipitation in month n.

2.2.2 Hydrologic similarity across catchments with non-stationary hydrologic response

Similarity of ecohydrologic response across all catchments with non-stationary response is explored by examining the overall relationships of long-term mean (1984–2010) annual fractional vegetation cover with annual runoff ratio, precipitation and the Horton index (Troch et al., 2009). The Horton index, the ratio of evapotranspiration to catchment wetting, presents efficiency of catchments in using plant available water and is reflective of water and energy availability in the catchment. The Horton index ranges between 0 and 1, and incorporates the role of soil and topography in the catchment wetting (Brooks et al., 2011; Troch et al., 2009; Voepel et al., 2011). To estimate catchment-averaged ET and wetting, the water balance equation ($dS / dt = P - Q - ET$) is used assuming that changes in annual storage (dS / dt) are zero:

$$ET = P - Q \tag{2}$$
$$W = P - S, \tag{3}$$

where P is annual precipitation, Q is the total stream discharge, ET is annual actual evapotranspiration, W is catchment wetting and S is the quick flow component of stream discharge. A one-parameter recursive filter of Lyne and Hollick (1979) for baseflow separation is used to estimate quick flow and baseflow components of daily discharge:

$$b_k = ab_{k-1} + \frac{1-a}{2}(Q_k + Q_{k-1}) \tag{4}$$
$$S = Q - b_k, \tag{5}$$

where b_k is baseflow and a is the filter parameter and typically is set to 0.925 (Xu et al., 2012). In arid and semi-arid catchments, as quick flow constitutes most of the total streamflow, HI is approaching 1. In humid catchments, quick flow runoff is smaller than the total streamflow and HI is less than 1. In catchments with limited storage, HI is undefined (0/0) (Troch et al., 2009). Next, inter-annual variability of catchment-scale ecohydrologic response is explored.

2.2.3 Normalized sensitivities of annual runoff ratio to changes in water balance and vegetation

Sensitivities of annual runoff ratio to inter-annual variability of precipitation, ET and fractional vegetation cover in the 1984–2010 period are computed to identify factors that exert the largest sensitivity on annual runoff ratio. Normalized sensitivity of annual runoff ratio to precipitation is computed by estimating the slope of a linear regression between runoff ratio and precipitation, and multiplying it by the ratio of mean precipitation to runoff ratio (Fatichi and Ivanov, 2014; Hsu et al., 2012). Similarly, normalized sensitivity of runoff ratio to water balance ET and annual fractional vegetation cover is also computed. Normalized sensitivity of annual runoff ratio is equivalent to the streamflow elasticity approach of Zheng et al. (2009) that defined streamflow elasticity as the linear regression coefficient between the proportional changes in streamflow and a climatic variable (precipitation or potential evapotranspiration). Results of these analyses are used as the basis for formulating an ecohydrologic catchment classification.

3 Results

3.1 Non-stationary hydrologic response

Results of the modified Mann–Kendall trend test across 166 catchments indicate that 20 catchments (with areas that range between 18.7 and 5158.3 km^2; Table 1) have significant decreasing trends in annual runoff ratio ($p < 0.01$) except for the East Baines River in northern Australia (Fig. 1). An increasing trend for runoff ratio in the East Baines River (0.009 yr^{-1}) is consistent with annual precipitation increases (13.2 mm yr^{-1}). Moreover, this catchment has the smallest fractional vegetation cover (0.26) amongst the catchments with non-stationary hydrologic response. The North Esk catchment in Tasmania is the only catchment amongst the catchments with non-stationary response in which runoff ratio declined despite increases in annual precipitation (6.4 mm yr^{-1}) (Table 1). In the Tasmanian catchment, the increasing trend in fractional vegetation cover (0.009 yr^{-1}) is significant and results in ET increase and subsequently lower runoff ratio during the 1984–2005 period. In the rest of the catchments with non-stationary hydrologic response, total annual precipitation decreased between −1.9 and −24.7 mm yr^{-1} in the 1984–2005 period, which is consistent with the decreasing trend in annual runoff ratio (−0.0008 to −0.016 yr^{-1}). However, most catchments present an increasing trend in annual fractional vegetation cover, with the exception of three catchments in the eastern Australia temperate forests (410705, 410761 and 412066). Trees are the dominant vegetation cover in all the catchments with non-stationary hydrologic response, except for the Avoca River at Coonooer (408200) and Mollison Creek

(405238) catchments in Victoria in which grasslands are dominant (Supplement Table S1). The question is the following: what causes the increasing trends in annual fractional vegetation cover despite decreasing trends in annual precipitation of these catchments?

Based on the mean seasonality index using data from 1984 to 2010, only two catchments exhibit a seasonal climate ($0.6 < SI < 0.8$) (Table S2). However, all catchments have some degree of rainfall seasonality ($SI > 0.39$) (Walsh and Lawler, 1981). Using the modified Mann–Kendall trend tests, no significant trends in the 1984–2005 SI values are observed in the catchments with non-stationary hydrologic response ($\alpha = 0.01$). Few significant trends in precipitation seasonality indices using the percentiles are observed in the catchments with non-stationary hydrologic response including catchments 410061, 410731, 405238 and 212260. In these catchments, a significant trend in the timing of the 25th percentile are detected, except for catchment 212260 where the trend was significant for the 50th percentile. The seasonal shifts in precipitation can impact vegetation dynamics particularly when they occur between the growing and non-growing seasons, such as in catchment 410731. This result suggests that other factors besides precipitation are contributing to observed non-stationarity. Next, we explore hydrologic similarity across catchments with non-stationary hydrologic response.

3.2 Long-term patterns of hydrologic similarity across catchments with non-stationary hydrologic response

Long-term annual average dryness index (PET / P) (1984–2010) of the 166 study catchments illustrates presence of energy- and water-limited catchments in the region (Fig. S1 in the Supplement). The North Esk catchment in Tasmania is the only energy-limited catchment amongst the catchments with non-stationary response. Across the catchments with non-stationary hydrologic response, increases in mean annual precipitation (1984–2010) increase mean annual fractional vegetation cover particularly in catchments with mean annual precipitation of less than 800 mm (Fig. 2). After that, increases in F_{tot} reach an asymptote with mean annual precipitation greater than 800. Similarly, runoff ratio and its variability increases due to precipitation increase particularly across catchments with mean annual precipitation of greater than 800 mm and mean annual fractional vegetation cover of greater than 0.7. The exception is the East Baines River in northern Australia, which has the smallest fractional vegetation cover but has large variability in runoff ratio. In drier catchments, the Horton index is close to 1 and exhibits smaller variability compared to the wetter catchments.

To explore differences between catchments with non-stationary or stationary behavior, the cumulative absolute differences between consecutive annual values of precipitation, fractional vegetation cover and runoff ratio for each catch-

Table 1. Mean annual precipitation (P), discharge (Q), runoff ratio (Q / P) and fractional vegetation cover (F_{tot}) of catchments with non-stationary hydrologic response during the 1984–2005 period. Slopes of the trend lines are obtained from a linear regression model fitted to each time series.

Station	Area	Mean P	Mean Q	Mean Q / P	Mean F_{tot}	Slope of the trend (1984–2005)		
	(km^2)	(mm)	(mm)	(–)		Q / P	P	F_{tot}
1. 212260	713	886.8	189.4	0.19	0.75	−0.013*	−13.3*	0.0052*
2. 215002	1382.2	803.2	155.6	0.17	0.72	−0.012*	−15.5*	0.0041
3. 215004	165.6	935.8	292.5	0.29	0.69	−0.011*	−15.6	0.0049
4. 216004	95.7	1125.8	204.3	0.16	0.71	−0.010*	−24.7*	0.0048
5. 218001	90.6	815.1	266.9	0.3	0.72	−0.016*	−10	0.0008
6. 406214	237	580.3	45	0.07	0.54	−0.005*	−6.9	0.0018
7. 408200	2677.3	508.5	6.5	0.01	0.48	−0.0008*	−6.4	0.0014
8. 408202	82.6	605	48.1	0.07	0.7	−0.006*	−4.9	0.005
9. 410061	146.1	1004	245.1	0.24	0.77	−0.008*	−10.5	0.0029*
10. 410705	508.6	744.8	65.1	0.08	0.64	−0.006*	−11.5	−0.002
11. 410731	671.6	897.8	84.9	0.09	0.64	−0.006*	−11.9	0.0003
12. 410734	563.7	816.4	93.6	0.1	0.67	−0.008*	−13.2*	0.0018
13. 410761	5158.3	742.2	56.7	0.07	0.59	−0.005*	−7.7	−0.0002
14. 412028	2630.7	778.6	97.2	0.11	0.69	−0.006*	−12	0.0007
15. 412066	1629.7	785.9	100.1	0.12	0.68	−0.008*	−12	−0.0002
16. 415207	304.5	645.7	52.2	0.07	0.66	−0.006*	−6.4	0.0013
17. 613146	18.7	1019.4	209.6	0.2	0.69	−0.006*	−1.9	0.0086*
18. G8110004	2443.1	811.7	128.1	0.15	0.26	0.009*	13.2	0.0015
19. 318076	379.8	1156.9	383.3	0.33	0.7	−0.005*	6.4	0.0090*
20. 405238	164.1	734.5	112	0.14	0.64	−0.008*	−7.7	0.0035

* The trend is significant at 0.01 significance level using the modified Mann–Kendall trend test.

ment are calculated and normalized by the total absolute difference. In Fig. 3, the differences between catchments with non-stationary and stationary hydrologic response are illustrated by presenting the mean and standard deviations of normalized cumulative differences for each group. As can be seen in Fig. 3, normalized cumulative differences in annual precipitation and fractional vegetation cover between the catchments with non-stationary and stationary hydrologic response are very similar. However, large differences in the normalized cumulative differences of annual runoff ratio exist between these catchments. The catchment area ranges from 18.7 to 5158.3 km^2 in catchments with non-stationary hydrologic response (Table 1). While increases in runoff ratio, $P - Q$ and mean fractional vegetation cover with increases in mean catchment slope are observed in catchments with non-stationary hydrologic response, no distinct differences between catchments with stationary and non-stationary hydrologic response are observed.

Although consistent patterns are observed in catchments' ecohydrologic response due to differences in mean annual precipitation in catchments with non-stationary hydrologic response, characterizing catchment-scale terrestrial ecosystem response to inter-annual precipitation variability is important for hydrologic predictions.

3.3 Normalized sensitivities of annual runoff ratio at a catchment scale

Normalized sensitivity of annual runoff ratio to annual fractional vegetation cover, ET and precipitation indicates greater sensitivity of runoff ratio to fractional vegetation cover than precipitation in most of the catchments with non-stationary hydrologic response (Fig. 4a). While runoff ratio sensitivities to precipitation are positive across all catchments with non-stationary hydrologic response, these sensitivities become negative in some catchments with increases in fractional vegetation cover. These results indicate the importance of incorporating vegetation dynamics in examining non-stationary hydrologic response.

Normalized sensitivity of annual fractional vegetation cover to precipitation (Fatichi and Ivanov, 2014; Hsu et al., 2012) is plotted against mean aridity index (PET / P) (Fig. 4b). As can be seen in Fig. 4b, fractional vegetation cover presents both positive and negative sensitivities to precipitation inter-annual variability. Across catchments with positive precipitation–fractional vegetation cover relationships, fractional vegetation cover sensitivities approach zero in catchments with an aridity index of 1.5. Fractional vegetation cover sensitivity is highest in the semi-arid catchments with lower mean annual precipitation compared to the rest of the catchments with non-stationary hydrologic response.

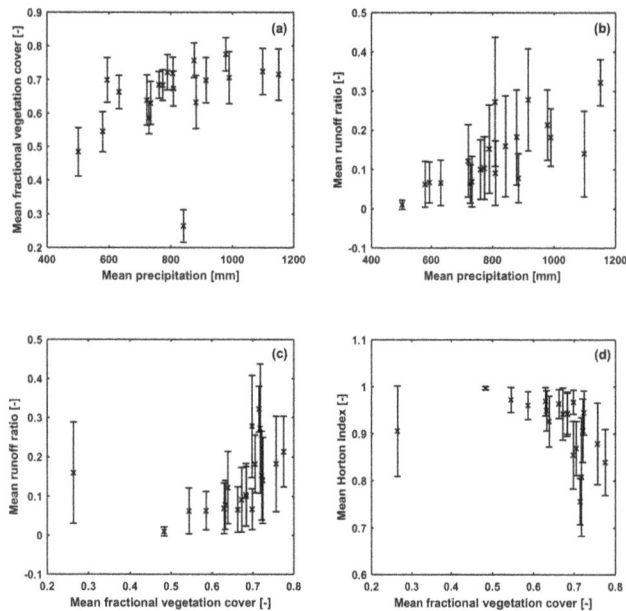

Figure 2. Mean and standard deviation of catchment-averaged **(a)** annual fractional vegetation cover and **(b)** annual runoff ratio, against mean annual precipitation in catchments with non-stationary hydrologic response. **(c)** Mean and standard deviation of annual runoff ratio and **(d)** Horton index versus catchment-averaged annual fractional vegetation cover.

In catchments with mean annual precipitation of 800 mm or higher (aridity index < 1.5), slopes of fractional vegetation cover to mean annual precipitation are zero or negative. A negative slope indicates increases in fractional vegetation cover despite precipitation decrease.

As vegetation productivity is controlled by plant available water (Brooks et al., 2011), fractional vegetation cover sensitivity to the Horton index is explored. Both positive and negative sensitivities between the fractional vegetation cover and the Horton index are observed in catchments with non-stationary hydrologic response (Fig. 4c). Positive sensitivities indicate increases in fractional vegetation cover as the Horton index increases. As a higher Horton index is indicative of a drier condition, removal of limiting factors like nutrient limitation is the likely cause of fractional vegetation cover increase in these catchments (Brooks et al., 2011). In a few of these catchments, light limitation may decrease fractional vegetation cover in wet years (positive correlations of sunshine hours with fractional vegetation cover; Table S3). In catchments with negative Horton index–fractional vegetation cover sensitivities, in which drier conditions decrease productivity, water availability is the primary factor in controlling vegetation growth. The annual runoff ratio's sensitivity to fractional vegetation cover was similar to the Horton index but with the opposite sign (Fig. 4d). This means that, in catchments with negative fractional vegetation cover–Horton index sensitivity, sensitivity of runoff ratio to fractional vege-

tation cover is positive, and vice versa. Across water-limited catchments (positive runoff ratio–fractional vegetation cover relationship), the runoff ratio sensitivities are smallest in catchments with the highest vegetation cover. As periods of higher productivity coincide with higher precipitation (positive precipitation–fractional vegetation cover relationship) in these catchments, runoff ratio increases in years with higher precipitation. It should be noted that the percentage of tree cover in these drier catchments are more than 60 %, with a few exceptions (Table S1). Negative runoff ratio–fractional vegetation cover sensitivities become more negative in catchments with higher fractional vegetation cover. Overall, mean annual runoff ratio and its variability (standard deviation) are smaller in drier catchments with smaller mean fractional vegetation cover (Fig. 2).

We used baseflow as a measure of catchment storage response to inter-annual precipitation variability. Baseflow sensitivities to mean annual aridity index are highest in drier catchments with non-stationary hydrologic response (Fig. 5a). Normalized fractional vegetation cover sensitivities to the baseflow decrease in catchments with higher annual baseflow index and even become negative at higher baseflow indices (Fig. 5b). This result suggests that, in catchments where groundwater constitutes significant component of streamflow, fractional vegetation cover exhibits smaller variability to changes in baseflow as vegetation roots have access to deeper water storage for transpiration and have less sensitivity to changes in baseflow.

Consistent patterns of fractional vegetation cover sensitivities to precipitation, baseflow and the Horton index across catchments with non-stationary hydrologic response present two distinct catchment response behaviors. We hypothesize plausible mechanisms to describe the likely causes of fractional vegetation cover sensitivity to inter-annual precipitation variability in order to distinguish between alternate catchment ecohydrologic responses.

3.4 Formulating catchment-scale ecohydrologic response

At the global scale, precipitation is the main driver of vegetation productivity particularly in arid and semi-arid environments (Huxman et al., 2004). However, mean annual vegetation productivity becomes less sensitive to mean annual precipitation in humid environments (Schuur, 2003) as biogeochemical factors (nutrients, light, soil oxygen availability) or biotic factors (Yang et al., 2008) limit productivity (Fig. 6a). This is consistent with observed precipitation–fractional vegetation cover patterns across all the catchments with non-stationary hydrologic response (Fig. 2a). At a catchment scale, catchments can be classified into two main groups based on the annual precipitation and vegetation productivity relationship. We hypothesize four plausible mechanisms to explain catchment-scale ecohydrologic response to inter-

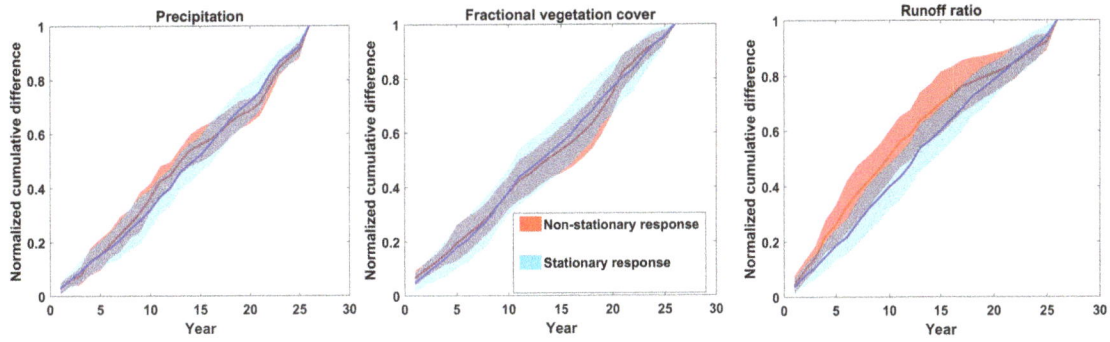

Figure 3. Mean normalized cumulative absolute differences in annual precipitation, fractional vegetation cover and runoff ratio between catchments with non-stationary (20 catchments) and stationary (146 catchments) hydrologic response. The shaded areas represent standard deviations.

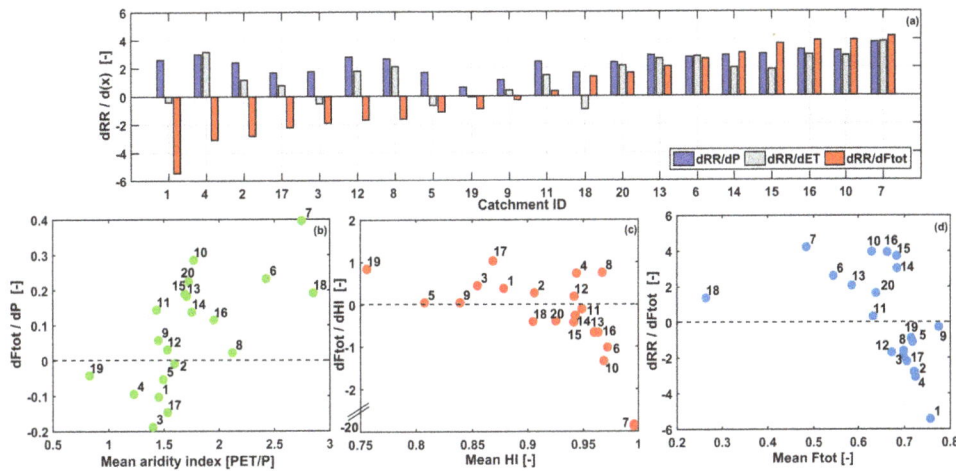

Figure 4. (a) Normalized sensitivities of runoff ratio (RR) to precipitation (P), water balance ET and fractional vegetation cover (F_{tot}); normalized sensitivities of annual **(b)** fractional vegetation cover to precipitation against catchments' mean aridity index, **(c)** fractional vegetation cover to the Horton index (HI) against catchments' mean Horton index and **(d)** runoff ratio to fractional vegetation cover against catchments' mean fractional vegetation cover in catchments with non-stationary hydrologic response. Data labels refer to the station identification number in Table 1.

annual climate variability in water- and energy-limited environments (Fig. 6b).

In group (A) catchments, a positive relationship between vegetation productivity and precipitation increases exists and can be either caused by (1) direct CO_2 fertilization effect in which increases in CO_2 enhance photosynthesis and increase LAI, and cause ET to increase due to precipitation and LAI increase (Fig. 6b – class A1), or by (2) indirect CO_2 fertilization effect in which increased CO_2 gradient between the atmosphere and leaf enhances photosynthesis but LAI does not increase. Therefore, reduction in the stomatal conductance reduces ET (Fig. 6b – class A2) (Ainsworth and Long, 2005). In group (A) catchments, changes in runoff ratio depend on the hydroclimatic condition. In years where precipitation increase is higher than ET, an increase in productivity is followed by increases in runoff ratio, while in drier-than-average years, increases in ET reduce the runoff ratio. It is

expected that under future warming, CO_2 increases will continue to increase productivity unless decreases in plant water availability limit plant growth, or changes in stomatal conductance, plant respiration rates (Wu et al., 2011) and nutrient availability impact productivity.

In group (B) catchments, vegetation productivity decreases in response to annual precipitation increases. This negative feedback is most likely due to biogeochemical constraints such as light, nutrients, temperature and soil characteristics (Bai et al., 2008) despite changes in the stomatal conductance due to CO_2 increases (Paruelo et al., 1999). In these catchments, productivity is likely constrained by (1) nutrients (class B1) or (2) light availability (class B2). In catchments where increases in precipitation are followed by ET increases, nutrient limitation (Norby et al., 2010; Schuur, 2003) is the likely cause of decline in productivity (Fig. 6b – class B1). In class B2 catchments, light and

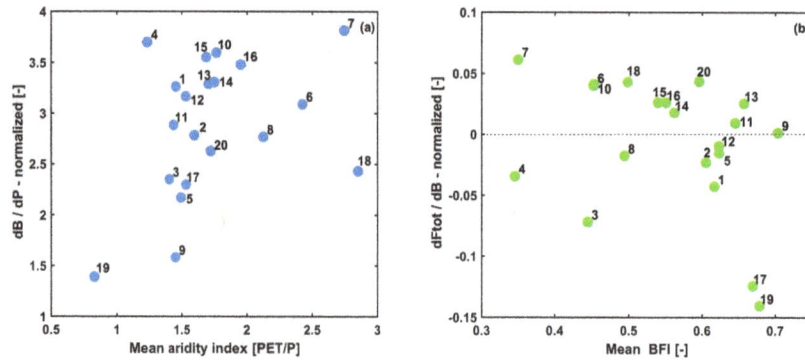

Figure 5. (a) Normalized baseflow (B) sensitivities to annual precipitation (P) in each catchment with non-stationary hydrologic response against its mean aridity index (1984–2010), (b) normalized annual fractional vegetation cover (F_{tot}) sensitivities to annual baseflow of each catchment with non-stationary hydrologic response against its mean baseflow index (BFI) (1984–2010). Data labels refer to the station identification number in Table 1. In general, annual baseflow sensitivities to mean annual precipitation decrease in wetter catchments (smaller aridity index). Positive sensitivities of fractional vegetation cover to baseflow decrease in catchments with higher baseflow index. Negative fractional vegetation cover sensitivities to baseflow become more negative in catchments with higher baseflow index, indicating larger contribution of groundwater to streamflow.

other factors (anoxic conditions, temperature) limit productivity and decline ET despite increases in precipitation. As these catchments are in the wetter regions, nutrient limitation might be caused by increased nutrient leaching in wet soils (Schlesinger, 1997) or increases in nutrient-use efficiency due to water availability, which subsequently leads to nutrient limitation (Paruelo et al., 1999). Similar to group (A), changes in runoff ratio depend on the catchment's hydroclimatic condition. In these catchments, future changes in vegetation productivity are likely dependent on the rate of nutrient mineralization (Brooks et al., 2011), nitrogen deposition and changes in disturbance regimes such as fire and drought. A flowchart illustrates how catchment classification is performed by computing Spearman rank correlations between two variables at each step (Fig. 6c).

The prevalence of the four classes identified above is presented using time series of annual precipitation, water-balance-derived ET and runoff ratio as well as catchment-averaged fractional vegetation cover for the 1984–2010 period. Three constitutive relationships are established for every catchment at an annual scale between (1) precipitation, (2) runoff ratio and (3) ET versus catchment-averaged fractional vegetation cover. Catchment-scale transpiration data are not available for this classification. According to these relationships and Spearman rank correlations, catchments with non-stationary hydrologic response are grouped in three classes (A1, B1 and B2; Fig. 1). None of the catchments with non-stationary hydrologic response presented a relationship proposed for class A2 catchments. Figure 7 shows Spearman rank correlation values for one example catchment in each class.

As presented in Fig. 1 and Table 2, 12 catchments are classified as class A1. The Spearman rank correlations between annual precipitation and fractional vegetation cover in class

A1 catchments are positive and typically larger than class B1 and B2 catchments, and in 8 out of 12 catchments the correlation is significant ($p < 0.05$). Only one catchment in class B1 (total of three) has a significant negative correlation between precipitation and fractional vegetation cover.

While data on catchment-scale nutrient availability are not available, general ET–fractional vegetation cover relationships in group (B) catchments can be further explained by annual precipitation–temperature relationships. In wetter years, despite lower vegetation cover, ET will likely increase due to higher water availability in warmer years in B1 catchments (positive precipitation–temperature correlations) (Table S3). In the B2 class with negative precipitation–temperature relationships, cooler temperatures and light limitation decline ET.

Groupings of all A1, B1 and B2 catchments illustrate significant correlations for all three constitutive relationships of Fig. 7 ($p < 0.05$; Table 2) in class A1 and group (B) catchments except between fractional vegetation cover and ET. Therefore, precipitation–fractional vegetation cover relationships present first-order groupings of the catchments. Further distinction within a group is speculative, as it depends on catchment-derived annual ET.

4   Discussions

According to our analysis, catchments with non-stationary hydrologic response present three distinct behaviors as a result of inter-annual variability in catchment water balance and vegetation fractional cover. In the following, we discuss whether the proposed catchment classification is consistent once other measures or data are used.

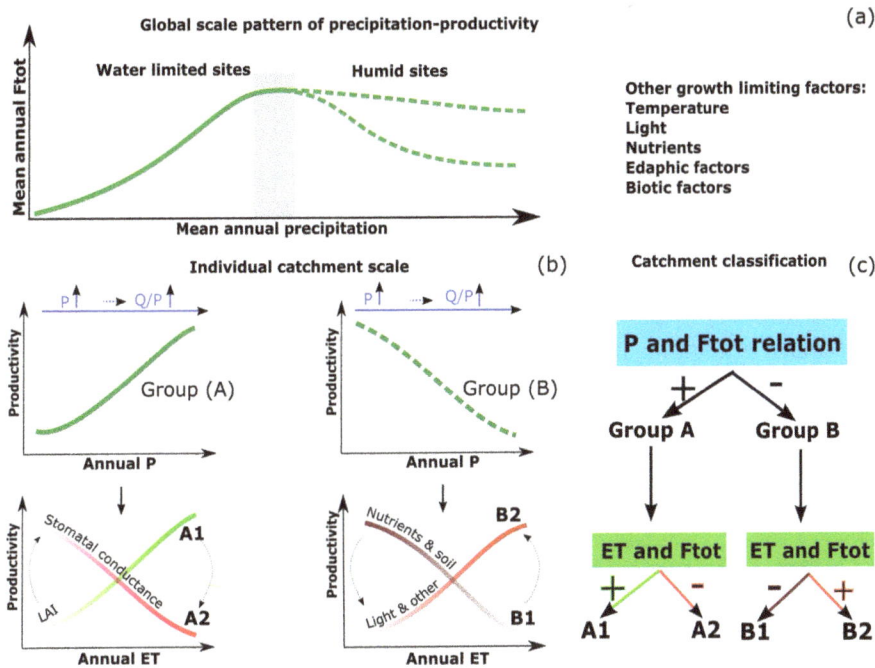

Figure 6. (a) Global pattern of annual productivity (F_{tot}) and mean annual precipitation relationship. While precipitation is the primary factor for vegetation growth in water-limited sites, productivity reaches an asymptote in humid areas or decreases (e.g., tropical forests) with increases in precipitation due to biogeochemical or edaphic constraints. The grey region corresponds to catchments in which productivity is insensitive to inter-annual precipitation variability. **(b)** A conceptual framework for characterizing changes in runoff ratio to changes in annual precipitation and vegetation productivity (F_{tot}) in relation to the catchment's hydroclimatic condition. In group (A) catchments, a positive relationship between annual precipitation and productivity exists and annual ET changes in relation to productivity depend on the dominance of structural control (increases in LAI, class A1) versus physiological control (decreases in stomatal conductance, class A2) in controlling productivity. In group (B) catchments, an inverse relationship between precipitation and productivity exists and productivity is likely constrained by biogeochemical factors. In B1 catchments, negative ET and productivity relation indicate productivity is likely controlled by nutrient availability as drier conditions induce nutrient mineralization. In B2 catchments, light availability and lower temperature reduce ET. In group (A) catchments, runoff ratio would increase as productivity increases, while in group (B), runoff ratio will likely decrease with increasing productivity (decreases in precipitation). Depending on the dominance of limiting resources, precipitation–productivity may shift between the two regimes. **(c)** The flowchart illustrates the classification procedure. The classification starts by assessing the correlations between annual precipitation and F_{tot} and then annual ET and F_{tot} in a catchment.

4.1 Did catchments with non-stationary hydrologic response experience similar changes in vegetation and water balance variables?

To explore whether HRS catchments have undergone similar changes during the period of analysis, regime curves based on daily runoff, precipitation and monthly fractional vegetation cover data for each catchment are developed using data from pre-drought (1984–1996) and drought (1997–2009) periods (Coopersmith et al., 2014). Regime curves are obtained by averaging daily values of precipitation or runoff for a given day over the length of the data. As daily fractional vegetation cover data are not available, monthly values are used to develop the regime curves. To summarize the differences between the regime curves for the pre-drought and drought periods, the Nash–Sutcliffe efficiency (NSE) criterion is calculated. As can be seen in Fig. 8, differences in daily precipitation and runoff and monthly fractional vegetation cover

regime curves are much higher (indicated by negative NSE) in catchments with non-stationary hydrologic response than the catchments that do not exhibit non-stationary behavior.

While the results of trend analysis are impacted by defining the significance level, the above analysis indicates that catchments with non-stationary behavior have undergone larger changes. To further assess the impact of significance level on the results of the trend analysis, the approach of Douglas et al. (2000) for computing the field significance of regional trend tests is implemented. In this approach, time series of runoff ratio for every catchment are resampled 10 000 times using the bootstrap approach. In the next step, Kendall's S is calculated for each bootstrap sample and the regional test statistics are calculated by averaging Kendall's S for each iteration and computing non-exceedance probability using the Weibull plotting position formula. Finally, the cumulative distribution function (CDF) of regional test statistics is compared with the historical Kendall's S calcu-

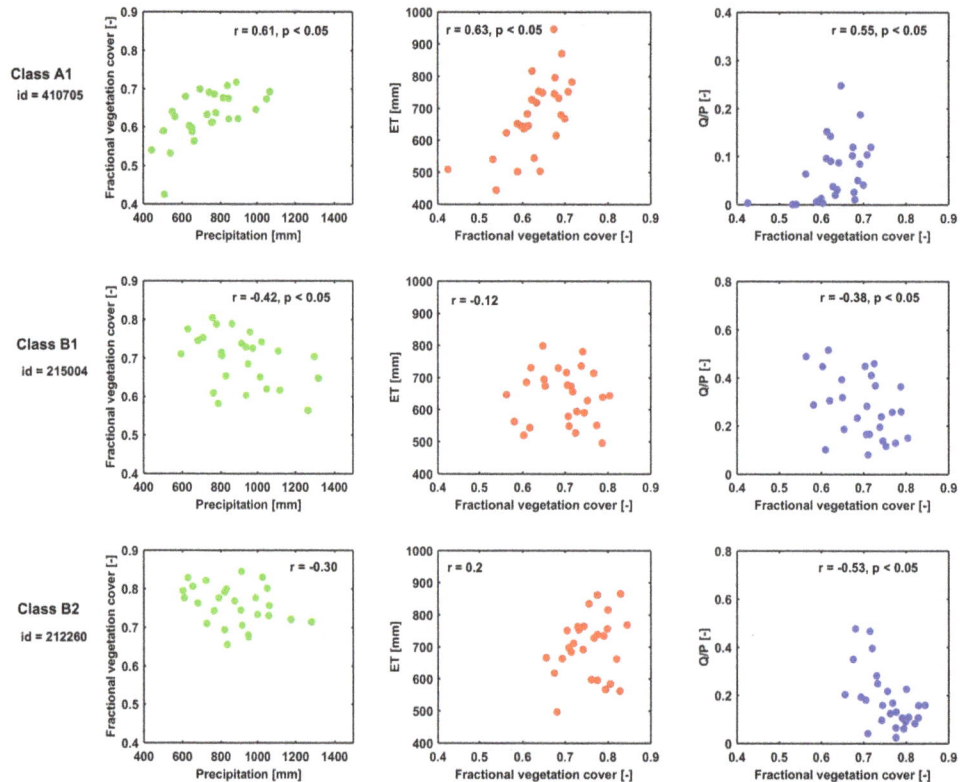

Figure 7. Relationships between catchment-averaged annual fractional vegetation cover and annual precipitation (left), water-balance-derived annual ET (middle) and runoff ratio (right) against mean annual fractional vegetation cover (1984–2010) across three catchments representative of each class in Fig. 6. The Spearman rank correlation (r) and p values are shown when correlation is significant.

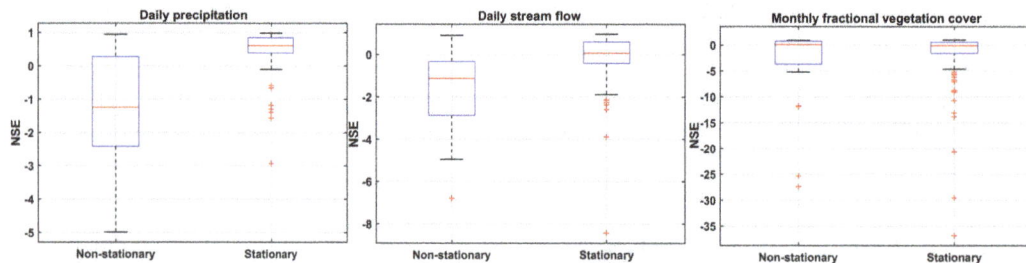

Figure 8. Box plots of Nash–Sutcliffe efficiency (NSE) values calculated between the regime curves of pre-drought and drought periods in catchments with non-stationary (20 catchments) and stationary (146 catchments) hydrologic response, respectively. Changes in daily precipitation and runoff and monthly fractional vegetation cover were larger in catchments with non-stationary hydrologic response.

lated for each station using 0.01 significance level. Indeed, the field significance level obtained from the bootstrap samples is 0.0239, which is more relaxed than the p value of 0.01 originally used. Using the new field significance level, 34 catchments are classified as non-stationary.

4.2 Is the ecohydrologic catchment classification consistent across other measures?

Positive precipitation–fractional vegetation cover relationships in class A1 catchments is consistent with positive normalized fractional vegetation cover sensitivities of individual catchments to annual precipitation (Figs. 4b and S2) and indicate that water availability primarily controls fractional vegetation cover increase in A1 catchments. A positive Spearman rank correlation between the coefficient of variation (CV) of annual fractional vegetation cover and CV of annual precipitation ($r = 0.34$, $p = 0.3$) across all A1 catchments further confirms this conclusion (Yang et al., 2008).

In group (B) catchments, negative normalized sensitivities of fractional vegetation cover to precipitation exist (Fig. 4b). This pattern is followed by a negative correlation between the CVs of these two factors across all group (B) catchments

Table 2. Catchment properties and Spearman rank correlations (r) for catchments with non-stationary hydrologic response in Australia. Data span the 1984–2010 period. Class categories refer to the catchment classification framework of Fig. 6.

Station	Mean P (mm)	Mean Q (mm)	r_1	r_2	$r_{2\text{-AWAP}}$	r_3	Class
406214	579.7	42.8	0.41*	0.47*	0.43*	0.21	A1
408200	501.2	5.9	0.65*	0.65*	0.47*	0.47*	A1
408202	594.5	43.4	0.11	0.17	0.19	0.01	A1
410705	734.9	58.1	0.61*	0.63*	0.70*	0.55*	A1
410731	882.7	75.2	0.17	0.24	0.29	0.06	A1
410761	729.6	51.0	0.45*	0.52*	0.47*	0.36	A1
412028	762.1	84.8	0.47*	0.53*	0.40*	0.29	A1
412066	774.4	88.9	0.55*	0.56*	0.41*	0.34	A1
415207	632.5	46.8	0.25	0.17	0.22	0.29	A1
G8110004	841.1	144.4	0.45*	0.23	0.41*	0.56*	A1
410061	979.1	221.2	0.06	0.18	0.38	−0.09	A1
405238	723.0	100.7	0.4*	0.4*	0.46*	0.29	A1
215004	915.3	273.3	−0.42*	−0.12	−0.42*	−0.38*	B1
216004	1099.5	179.0	−0.27	−0.01	−0.24	−0.39*	B1
218001	807.5	246.6	−0.23	−0.07	−0.15	−0.16	B1
212260	876.8	175.3	−0.30	0.20	−0.27	−0.53*	B2
215002	789.3	136.7	−0.01	0.29	0.06	−0.13	B2
410734	808.8	83.9	−0.03	0.17	0.2	−0.09	B2
613146	990.1	188.3	−0.23	0.22	−0.39*	−0.65*	B2
318076	1151.3	375.6	−0.01	0.29	−0.03	−0.46*	B2
Class A1			0.34*	0.32*	0.41*	0.33*	
Class B1			−0.23*	0.008	−0.13	−0.28*	
Class B2			−0.09	0.11	0.14	−0.16	
Group (B)			−0.14*	0.07	0.04	−0.21*	

* Correlation is significant ($p < 0.05$); r_1: correlation between mean annual fractional vegetation cover and annual precipitation; r_2: correlation between annual evapotranspiration (water balance approach) and mean annual fractional vegetation cover; $r_{2\text{-AWAP}}$: correlation between mean annual fractional vegetation cover and AWAP annual evapotranspiration; r_3: correlation between annual runoff ratio (Q / P) and mean annual fractional vegetation cover.

($r = -0.71$, $p = 0.06$), which highlights the role of biogeochemical factors in controlling productivity. Small or even negative sensitivities of vegetation cover to precipitation in group (B) might be due to the presence of perennial vegetation (shrubs and trees) as ecosystems with more perennial cover are less responsive to inter-annual precipitation variability (Jin and Goulden, 2014).

Our classification framework suggests that group (A) catchments are more sensitive to increases in CO_2 concentrations than the group (B) catchments that are in the humid zone ($P / \text{PET} > 0.65$). This result is consistent with Ukkola et al. (2016), as they showed greater sensitivities of annual ET and normalized difference vegetation index (NDVI) to increases in CO_2 concentrations in sub-humid and semi-arid catchments of Australia.

4.3 Are the inferred classification patterns artefacts of remote sensing data and catchment-scale ET?

To assess whether observed precipitation–productivity relationships are the artefacts of remote sensing data, two independent remote sensing vegetation products are used: vegetation optical depth (VOD) and enhanced vegetation index (EVI) (Huete et al., 2002, 2006). A global long-term (1988–2010) annual VOD dataset from passive microwave satellites with 0.25° resolution (Liu et al., 2011) is related to water content of leaf and woody components of aboveground biomass (Liu et al., 2015) and is able to detect structural differences in areas with near-closed canopy. Spearman rank correlations between VOD and annual precipitation across group (B) catchments were negative and consistent with the results of AVHRR fractional vegetation cover data (Sect. S2, Table S4). Moreover, the Australia coverage of the despiked EVI dataset (2001–2010) from the Moderate Resolution Imaging Spectroradiometer (MODIS) presented high correlations with F_{tot} data (2001–2010). Previous investigations have shown that EVI is more sensitive to net primary productivity compared to the normalized vegetation index (Huete et al., 2002, 2006). Analyses from these two independent datasets reduce uncertainty of identifying negative precipitation–productivity correlations at the catchment scale. However, further research is required to determine exact causes of the observed behavior.

Here, we assumed that changes in catchment storage at the annual scale are zero to compute annual water balance ET. However, this assumption is likely not correct in all years. Using AWAP actual annual ET similar relationships between fractional vegetation cover and annual ET are obtained, except in three catchments in class B2 (Table 2). AWAP ET is based on daily transpiration and soil evaporation values obtained from the WaterDyn model that simulates terrestrial water balance across Australia at 5 km resolution (Raupach et al., 2009). In addition to inter-annual water storage carry-over, inter-annual non-structural carbon storage across years (a wet year can result in greater biomass/leaf area in the following year) can impact precipitation–vegetation relationships.

4.4 Does the precipitation–fractional vegetation cover relationship depend on the period of analysis?

The period of analysis is limited to 1984–2010 in this study due to availability of AVHRR fractional vegetation cover data for Australia. To assess sensitivity of precipitation–fractional vegetation cover relationships to data length and catchment condition, these relationships are developed for two time periods: 1984–1996 and 1997–2009. It should be noted that 1997–2009 corresponds to the millennium drought in Australia (Chiew et al., 2014). Results indicate similar precipitation–fractional vegetation cover relationships in 1984–2010 in class A1 as well as in group (B) with a few exceptions (Fig. S2). Despite these exceptions, the drier conditions of 1997–2009 resulted in higher mean fractional vegetation covers in group (B) compared to the 1984–1996 period consistent with the classification framework. Results suggest that the record length is important in catchments where productivity is limited by resources besides water availability.

5 Summary and conclusions

We used precipitation–fractional vegetation cover relationships for first-order groupings of catchment-scale ecohydrologic response in 20 catchments with non-stationary hydrologic response, located in different hydroclimatic regions of Australia. Our results illustrate that fractional vegetation cover is more sensitive to increases in precipitation (stronger Spearman rank correlations) in class A1 catchments (12 catchments). This inference is consistent with the result of meta-analysis of productivity response to precipitation across the globe (Wu et al., 2011). The drawback of using precipitation as the main driver of vegetation productivity is that the impact of confounding variables that covary with precipitation is ignored (Wu et al., 2011). Fractional vegetation cover sensitivity to precipitation and Horton index provided consistent results with our catchment classification framework, except in two catchments. These catchments (408202, 410061) have smaller rank correlation between pre-

cipitation and fractional vegetation cover compared to the rest of class A1 catchments. A total of 8 out of 20 catchments with non-stationary hydrologic response present negative precipitation–fractional vegetation cover relationships impacted by nutrient or light availability.

While determining the exact causes of non-stationarity requires detailed modeling experiments, non-stationarity of runoff ratios could be attributed to changes in precipitation amount, intensity and seasonality, increases in air temperature and CO_2 concentrations (Chiew et al., 2014). The proposed framework provides a general guideline for projecting the likely changes in catchment water balance in response to climate change and designing simulation experiments. However, uncertainty still remains about the terrestrial ecosystem response as factors such as nutrients and light availability, vegetation developmental stage, space constraint and prevalence of pests may impact productivity (Körner, 2006). In addition, it is expected that frequency and duration of extreme events, such as fire, drought and floods, will increase, which can further alter ecosystem response and plant water availability (Medlyn et al., 2011).

6 Data availability

The datasets used in this research are publicly available from the following websites. The daily streamflow values are obtained from the hydrologic reference stations (HRS) website (http://www.bom.gov.au/water/hrs/). Daily precipitation, actual and potential evapotranspiration, and temperature are obtained from the Australian Water Availability Project (AWAP).

Acknowledgements. This research was funded by the Australian Research Council linkage grant (LP130100072), the Australian Bureau of Meteorology and WaterNSW. We acknowledge the Australian Bureau of Meteorology for providing the hydrologic reference station data supported by the Australian Government through the Water Information Program. We would like to acknowledge Yi Liu for providing the VOD data.

Edited by: A. Wei

References

Ainsworth, E. A. and Long, S. P.: What have we learned from 15 years of Free-Air CO_2 Enrichment (FACE)? A meta-analytic review of the responses of photosynthesis, canopy properties and plant production to rising CO_2, New Phytol., 165, 351–371, doi:10.1111/j.1469-8137.2004.01224.x, 2005.

Bai, Y., Wu, J., Xing, Q., Pan, Q., Huang, J., Yang, D., and Han, X.: Primary production and rain use efficiency across a precipitation gradient on the Mongolia Plateau, Ecology, 89, 2140–2153, doi:10.1890/07-0992.1, 2008.

Band, L. E., Mackay, D. S., Creed, I. F., Semkin, R., and Jeffries, D.: Ecosystem processes at the watershed scale: Sensitivity to potential climate change, Limnol. Oceanogr., 41, 928–938, doi:10.4319/lo.1996.41.5.0928, 1996.

Baron, J. S., Hartman, M. D., Band, L. E., and Lammers, R. B.: Sensitivity of a high-elevation rocky mountain watershed to altered climate and CO_2, Water Resour. Res., 36, 89–99, doi:10.1029/1999WR900263, 2000.

Betts, R. A., Boucher, O., Collins, M., Cox, P. M., Falloon, P. D., Gedney, N., Hemming, D. L., Huntingford, C., Jones, C. D., Sexton, D. M. H., and Webb, M. J.: Projected increase in continental runoff due to plant responses to increasing carbon dioxide, Nature, 448, 1037–1041, doi:10.1038/nature06045, 2007.

Brooks, P. D., Troch, P. A., Durcik, M., Gallo, E., and Schlegel, M.: Quantifying regional scale ecosystem response to changes in precipitation: Not all rain is created equal, Water Resour. Res., 47, W00J08, doi:10.1029/2010WR009762, 2011.

Chiew, F. H. S., Potter, N. J., Vaze, J., Petheram, C., Zhang, L., Teng, J., and Post, D. A.: Observed hydrologic non-stationarity in far south-eastern Australia: Implications for modelling and prediction, Stoch. Env. Res. Risk A., 28, 3–15, doi:10.1007/s00477-013-0755-5, 2014.

Coopersmith, E. J., Minsker, B. S., and Sivapalan, M.: Patterns of regional hydroclimatic shifts: An analysis of changing hydrologic regimes, Water Resour. Res., 50, 1960–1983, doi:10.1002/2012WR013320, 2014.

Donohue, R. J., Roderick, M. L., and McVicar, T. R.: Deriving consistent long-term vegetation information from AVHRR reflectance data using a cover-triangle-based framework, Remote Sens. Environ., 112, 2938–2949, doi:10.1016/j.rse.2008.02.008, 2008.

Donohue, R. J., McVicar, T. R., and Roderick, M. L.: Climate-related trends in Australian vegetation cover as inferred from satellite observations, 1981–2006, Glob. Change Biol., 15, 1025–1039, doi:10.1111/j.1365-2486.2008.01746.x, 2009.

Douglas, E. M., Vogel, R. M., and Kroll, C. N.: Trends in floods and low flows in the United States: Impact of spatial correlation, J. Hydrol., 240, 90–105, doi:10.1016/S0022-1694(00)00336-X, 2000.

Fatichi, S. and Ivanov, V. Y.: Interannual variability of evapotranspiration and vegetation productivity, Water Resour. Res., 50, 3275–3294, doi:10.1002/2013WR015044, 2014.

Field, C. B., Jackson, R. B., and Mooney, H. A.: Stomatal responses to increased CO_2: Implications from the plant to the global scale, Plant Cell Environ., 18, 1214–1225, doi:10.1111/j.1365-3040.1995.tb00630.x, 1995.

Gedney, N., Cox, P. M., Betts, R. A., Boucher, O., Huntingford, C., and Stott, P. A.: Detection of a direct carbon dioxide effect in continental river runoff records, Nature, 439, 835–838, doi:10.1038/nature04504, 2006.

Hamed, K. H. and Rao, A. R.: A modified Mann-Kendall trend test for autocorrelated data, J. Hydrol., 204, 182–196, doi:10.1016/S0022-1694(97)00125-X, 1998.

Hsu, J. S., Powell, J., and Adler, P. B.: Sensitivity of mean annual primary production to precipitation, Glob. Change Biol., 18, 2246–2255, doi:10.1111/j.1365-2486.2012.02687.x, 2012.

Huete, A., Didan K., Miura, T., Rodriguez, E. P., Gao, X., and Ferreira, L. G.: Overview of the radiometric and biophysical performance of the MODIS vegetation indices, Remote Sens. Environ., 83, 195–213, doi:10.1016/S0034-4257(02)00096-2, 2002.

Huete, A. R., Didan, K., Shimabukuro, Y. E., Ratana, P., Saleska, S. R., Hutyra, L. R., Yang, W., Nemani, R. R., and Myneni, R.: Amazon rainforests green-up with sunlight in dry season, Geophys. Res. Lett., 33, L06405, doi:10.1029/2005GL025583, 2006.

Huxman, T. E., Smith, M. D., Fay, P. A., Knapp, A. K., Shaw, M. R., Loik, M. E., Smith, S. D., Tissue, D. T., Zak, J. C., Weltzin, J. F., Pockman, W. T., Sala, O. E., Haddad, B. M., Harte, J., Koch, G. W., Schwinning, S., Small, E. E., and Williams, D. G.: Convergence across biomes to a common rain-use efficiency, Nature, 429, 651–654, doi:10.1038/nature02561, 2004.

Ivanov, V. Y., Bras, R. L., and Vivoni, E. R.: Vegetation-hydrology dynamics in complex terrain of semiarid areas: 1. A mechanistic approach to modeling dynamic feedbacks, Water Resour. Res., 44, W03429, doi:10.1029/2006WR005588, 2008.

Jin, Y. and Goulden, M. L.: Ecological consequences of variation in precipitation: Separating short-versus long-term effects using satellite data, Global Ecol. Biogeogr., 23, 358–370, doi:10.1111/geb.12135, 2014.

Kendall, M. G.: Rank correlation methods, Charles Griffin, London, 1970.

Kergoat, L., Lafont, S., Douville, H., Berthelot, B., Dedieu, G., Planton, S., and Royer, J. F.: Impact of doubled CO_2 on global-scale leaf area index and evapotranspiration: Conflicting stomatal conductance and LAI responses, J. Geophys. Res.-Atmos., 107, 4808, doi:10.1029/2001JD001245, 2002.

Körner, C.: Plant CO_2 responses: An issue of definition, time and resource supply, New Phytol., 172, 393–411, doi:10.1111/j.1469-8137.2006.01886.x, 2006.

Leuzinger, S. and Körner, C.: Rainfall distribution is the main driver of runoff under future CO_2-concentration in a temperate deciduous forest, Glob. Change Biol., 16, 246–254, doi:10.1111/j.1365-2486.2009.01937.x, 2010.

Liu, Y. Y., de Jeu, R. A. M., McCabe, M. F., Evans, J. P., and van Dijk, A. I. J. M.: Global long-term passive microwave satellite-based retrievals of vegetation optical depth, Geophys. Res. Lett., 38, L18402, doi:10.1029/2011GL048684, 2011.

Liu, Y. Y., van Dijk, A. I. J. M., de Jeu, R. A. M., Canadell, J. G., McCabe, M. F., Evans, J. P., and Wang, G.: Recent reversal in loss of global terrestrial biomass, Nature Climate Change, 5, 470–474, doi:10.1038/nclimate2581, 2015.

Lyne, V. and Hollick, M.: Stochastic time-variable rainfall-runoff modelling, in Hydrol. and Water Resour. Symp., publ. 79/10, Inst. Eng. Austr. Natl. Conf., Perth, Australia, 89–92, 1979.

Mann, H. B.: Non-parametric tests against trend, Econometrics, 13, 245–259, 1945.

Medlyn, B. E., Barton, C. V. M., Broadmeadow, M. S. J., Ceulemans, R., De Angelis, P., Forstreuter, M., Freeman, M., Jackson, S. B., Kellomäki, S., Laitat, E., Rey, A., Roberntz, P., Sigurdsson, B. D., Strassemeyer, J., Wang, K., Curtis, P. S., and Jarvis, P. G.: Stomatal conductance of forest species after long-term exposure to elevated CO_2 concentration: A synthesis, New Phytol., 149, 247–64, doi:10.1046/j.1469-8137.2001.00028.x, 2001.

Medlyn, B. E., Duursma, R. A., and Zeppel, M. J. B.: Forest productivity under climate change: A checklist for evaluating model studies, Wiley Interdisciplinary Reviews: Climate Change, 2, 332–355, doi:10.1002/wcc.108, 2011.

Norby, R. J., Warren, J. M., Iversen, C. M., Medlyn, B. E., and McMurtrie, R. E.: CO_2 enhancement of forest productivity constrained by limited nitrogen availability, P. Natl. Acad. Sci. USA, 107, 19368–19373, doi:10.1073/pnas.1006463107, 2010.

Paruelo, J. M., Lauenroth, W. K., Burke, I. C., and Sala, O. E.: Grassland precipitation-use efficiency varies across a resource gradient, Ecosystems, 2, 64–68, doi:10.1007/s100219900058, 1999.

Piao, S., Friedlingstein, P., Ciais, P., de Noblet-Ducoudré, N., Labat, D., and Zaehle, S.: Changes in climate and land use have a larger direct impact than rising CO_2 on global river runoff trends, P. Natl. Acad. Sci. USA, 104, 15242–15247, doi:10.1073/pnas.0707213104, 2007.

Pryor, S. C. and Schoof, J. T.: Changes in the seasonality of precipitation over the contiguous USA, J. Geophys. Res.-Atmos., 113, D21108, doi:10.1029/2008JD010251, 2008.

Raupach, M. R., Briggs, P. R., Haverd, V., King, E. A., Paget, M., and Trudinger, C. M.: Australian Water Availability Project (AWAP): CSIRO Marine and Atmospheric Research Component: Final Report for Phase 3, CAWCR Technical Report No. 013, 67 pp., 2009.

Raupach, M. R., Briggs, P. R., Haverd, V., King, E. A., Paget, M., and Trudinger, C. M.: Australian Water Availability Project. CSIRO Marine and Atmospheric Research, Canberra, Australia, http://www.csiro.au/awap (last access: January 2016), 2012.

Sankarasubramanian, A., Vogel, R. M., and Limbrunner, J. F.: Climate elasticity of streamflow in the United States, Water Resour. Res., 37, 1771–1781, doi:10.1029/2000WR900330, 2001.

Sawicz, K., Wagener, T., Sivapalan, M., Troch, P. A., and Carrillo, G.: Catchment classification: Empirical analysis of hydrologic similarity based on catchment function in the eastern USA, Hydrol. Earth Syst. Sci., 15, 2895–2911, doi:10.5194/hess-15-2895-2011, 2011.

Sawicz, K. A., Kelleher, C., Wagener, T., Troch, P., Sivapalan, M., and Carrillo, G.: Characterizing hydrologic change through catchment classification, Hydrol. Earth Syst. Sci., 18, 273–285, doi:10.5194/hess-18-273-2014, 2014.

Schlesinger, W. H.: Biogeochemistry: An Analysis of Global Change, 2nd Edn., Academic press, San Diego, California, 1997.

Schuur, E. A. G.: Productivity and global climate revisited: The sensitivity of tropical forest growth to precipitation, Ecology, 84, 1165–1170, doi:10.1890/0012-9658(2003)084[1165:PAGCRT]2.0.CO;2, 2003.

Sivapalan, M., Thompson, S. E., Harman, C. J., Basu, N. B., and Kumar, P.: Water cycle dynamics in a changing environment: Improving predictability through synthesis, Water Resour. Res., 47, W00J01, doi:10.1029/2011WR011377, 2011.

Troch, P. A., Martinez, G. F., Pauwels, V. R. N., Durcik, M., Sivapalan, M., Harman, C., Brooks, P. D., Gupta, H., and Huxman, T.: Climate and vegetation water use efficiency at catchment scales, Hydrol. Process., 23, 2409–2414, doi:10.1002/hyp.7358, 2009.

Ukkola, A. M., Prentice, I. C., Keenan, T. F., van Dijk, A. I. J. M., Viney, N. R., Myneni, R. B., and Bi, J.: Reduced streamflow in water-stressed climates consistent with CO_2 effects on vegetation, Nature Climate Change, 6, 75–78, doi:10.1038/nclimate2831, 2016.

Vitousek, P. and Howarth, R.: Nitrogen limitation on land and in the sea: How can it occur?, Biogeochemistry, 13, 87–115, doi:10.1007/BF00002772, 1991.

Voepel, H., Ruddell, B., Schumer, R., Troch, P. A., Brooks, P. D., Neal, A., Durcik, M., and Sivapalan, M.: Quantifying the role of climate and landscape characteristics on hydrologic partitioning and vegetation response, Water Resour. Res., 47, W00J09, doi:10.1029/2010WR009944, 2011.

Wagener, T., Sivapalan, M., Troch, P. A., and Woods, R.: Catchment classification and hydrologic similarity, Geography Compass, 1, 901–931, doi:10.1111/j.1749-8198.2007.00039.x, 2007.

Walsh, P. D. and Lawler, D. M.: Rainfall seasonality: Description, spatial patterns and changes through time, Weather, 36, 201–208, doi:10.1002/j.1477-8696.1981.tb05400.x, 1981.

Wieder, W.: Soil carbon: Microbes, roots and global carbon, Nature Climate Change, 4, 1052–1053, doi:10.1038/nclimate2454, 2014.

Wigley, T. M. L. and Jones, P. D.: Influences of precipitation changes and direct CO_2 effects on streamflow, Nature, 314, 149–152, doi:10.1038/314149a0, 1985.

Wu, Z., Dijkstra, P., Koch, G. W., Peñelas, J., and Hungate, B. A.: Responses of terrestrial ecosystems to temperature and precipitation change: A meta-analysis of experimental manipulation, Glob. Change Biol., 17, 927–942, doi:10.1111/j.1365-2486.2010.02302.x, 2011.

Xu, X., Yang, D., and Sivapalan, M.: Assessing the impact of climate variability on catchment water balance and vegetation cover, Hydrol. Earth Syst. Sci., 16, 43–58, doi:10.5194/hess-16-43-2012, 2012.

Yang, Y., Fang, J., Ma, W., and Wang, W.: Relationship between variability in aboveground net primary production and precipitation in global grasslands, Geophys. Res. Lett., 35, L23710, doi:10.1029/2008GL035408, 2008.

Zheng, H., Zhang, L., Zhu, R., Liu, C., Sato, Y., and Fukushima Y.: Responses of streamflow to climate and land surface change in the headwaters of the Yellow River Basin, Water Resour. Res., 45, W00A19, doi:10.1029/2007WR006665, 2009.

PERMISSIONS

The contributors of this book come from diverse backgrounds, making this book a truly international effort. This book will bring forth new frontiers with its revolutionizing research information and detailed analysis of the nascent developments around the world.

We would like to thank all the contributing authors for lending their expertise to make the book truly unique. They have played a crucial role in the development of this book. Without their invaluable contributions this book wouldn't have been possible. They have made vital efforts to compile up to date information on the varied aspects of this subject to make this book a valuable addition to the collection of many professionals and students.

This book was conceptualized with the vision of imparting up-to-date information and advanced data in this field. To ensure the same, a matchless editorial board was set up. Every individual on the board went through rigorous rounds of assessment to prove their worth. After which they invested a large part of their time researching and compiling the most relevant data for our readers.

The editorial board has been involved in producing this book since its inception. They have spent rigorous hours researching and exploring the diverse topics which have resulted in the successful publishing of this book. They have passed on their knowledge of decades through this book. To expedite this challenging task, the publisher supported the team at every step. A small team of assistant editors was also appointed to further simplify the editing procedure and attain best results for the readers.

Apart from the editorial board, the designing team has also invested a significant amount of their time in understanding the subject and creating the most relevant covers. They scrutinized every image to scout for the most suitable representation of the subject and create an appropriate cover for the book.

The publishing team has been an ardent support to the editorial, designing and production team. Their endless efforts to recruit the best for this project, has resulted in the accomplishment of this book. They are a veteran in the field of academics and their pool of knowledge is as vast as their experience in printing. Their expertise and guidance has proved useful at every step. Their uncompromising quality standards have made this book an exceptional effort. Their encouragement from time to time has been an inspiration for everyone.

The publisher and the editorial board hope that this book will prove to be a valuable piece of knowledge for researchers, students, practitioners and scholars across the globe.

LIST OF CONTRIBUTORS

D. Vrebos, J. Staes and P. Meire
Department of Biology, Universiteit Antwerpen, Universiteitsplein 1, 2610 Wilrijk, Belgium

T. Vansteenkiste
Department of Civil Engineering, KU Leuven, Kasteelpark Arenberg 40, 3001 Leuven, Belgium

P. Willems
Department of Civil Engineering, KU Leuven, Kasteelpark Arenberg 40, 3001 Leuven, Belgium
Department of Hydrology and Hydraulic Engineering, Vrije Universiteit Brussel, Pleinlaan 2, 1050 Brussel, Belgium

Daeha Kim and Jong Ahn Chun
APEC Climate Center, Busan, 48058, South Korea

Il Won Jung
Korea Infrastructure Safety & Technology Corporation, Jinju, Gyeongsangnam-do, 52852, South Korea

S. Gangopadhyay and T. Pruitt
Bureau of Reclamation, Technical Service Center, Denver, Colorado, USA

L. E. Condon
Bureau of Reclamation, Technical Service Center, Denver, Colorado, USA
Hydrologic Science and Engineering Program and Department of Geology and Geological Engineering, Colorado School of Mines, Golden, Colorado, USA

Yaning Chen, Weihong Li, Gonghuan Fang and Zhi Li
State Key Laboratory of Desert and Oasis Ecology, Xinjiang Institute of Ecology and Geography, Chinese Academy of Sciences, Urumqi 830011, China

Auguste Gires, Ioulia Tchiguirinskaia and Daniel Schertzer
HMCo, École des Ponts, UPE, Champs-sur-Marne, France

Susana Ochoa-Rodriguez and Rui Pina
Urban Water Research Group, Department of Civil and Environmental Engineering, Imperial College London, Skempton Building, London SW7 2AZ, UK

Patrick Willems, Li-Pen Wang and Damian Murla Tuyls
Hydraulics Laboratory, KU Leuven, 3001, Heverlee (Leuven), Belgium

Abdellah Ichiba
HMCo, École des Ponts, UPE, Champs-sur-Marne, France
Conseil Départemental du Val-de-Marne, Direction des Services de l'Environnement et de l'Assainissement (DSEA), Bonneuil-sur-Marne, 94381, France

Johan Van Assel
Aquafin NV, Dijkstraat 8, 2630 Aartselaar, Belgium

Guendalina Bruni and Marie-Claire ten Veldhuis
Water Management Department, Delft University of Technology, 2600 GA Delft, the Netherlands

P. J. Dillon
Chemical Sciences, Trent University, Peterborough, ON, Canada

J. Crossman
Chemical Sciences, Trent University, Peterborough, ON, Canada
Oxford University Centre for the Environment, Oxford University, Oxford, UK

P. G. Whitehead
Oxford University Centre for the Environment, Oxford University, Oxford, UK

M. N. Futter
Department of Aquatic Sciences and Assessment, Swedish University of Agricultural Sciences, Uppsala, Sweden

E. Stainsby
Ontario Ministry of Environment, Etobicoke, ON, Canada

H. M. Baulch
School of Environment and Sustainability and Global Institute for Water Security, University of Saskatchewan, Saskatoon, SK, Canada

L. Jin
Department of Geology, State University of New York College at Cortland, Cortland, NY, USA

S. K. Oni
Department of Forest Ecology and Management, Swedish University of Agricultural Science, Umeå, Sweden

R. L. Wilby
Department of Geography, Loughborough University, Leicestershire, UK

William Howcroft
School of Earth, Atmosphere and Environment, 9 Rainforest Walk, Monash University, Clayton, VIC 3800, Australia

Ian Cartwright
School of Earth, Atmosphere and Environment, 9 Rainforest Walk, Monash University, Clayton, VIC 3800, Australia
National Centre for Groundwater Research and Training, Flinders University, Adelaide, SA 5001, Australia

Uwe Morgenstern
GNS Science, 1 Fairway Drive, Avalon, Lower Hutt 5040, New Zealand

Remko Uijlenhoet
Hydrology and Quantitative Water Management Group, Department of Environmental Sciences, Wageningen University, 6708 PB Wageningen, the Netherlands

Hidde Leijnse
Research and Development Observations and Data Technology, Royal Netherlands Meteorological Institute, 3732 GK De Bilt, the Netherlands

Lotte de Vos and Aart Overeem
Hydrology and Quantitative Water Management Group, Department of Environmental Sciences, Wageningen University, 6708 PB Wageningen, the Netherlands
Research and Development Observations and Data Technology, Royal Netherlands Meteorological Institute, 3732 GK De Bilt, the Netherlands

Sara Castelar and Eugènia Martí
Integrative Freshwater Ecology Group, Center for Advanced Studies of Blanes (CEAB-CSIC), Blanes, 17300, Spain

Susana Bernal
Integrative Freshwater Ecology Group, Center for Advanced Studies of Blanes (CEAB-CSIC), Blanes, 17300, Spain
Departament de Biologia Evolutiva, Ecologia i Ciències Ambientals (BEECA), Universitat de Barcelona, Barcelona, 08028, Spain

Anna Lupon
Departament de Biologia Evolutiva, Ecologia i Ciències Ambientals (BEECA), Universitat de Barcelona, Barcelona, 08028, Spain

Department of Forest Ecology and Management, Swedish University of Agricultural Sciences, Umeå, 90183, Sweden

Núria Catalán
Department of Resources and Ecosystems, ICRA, Catalan Institute for Water Research, Girona, 17003, Spain

M. Adamovic
Irstea, UR HHLY, Hydrology-Hydraulics Research Unit, Lyon-Villeurbanne, France
CNRS, HydroSciences Laboratory, Place Eugene Bataillon, 34095 Montpellier, France

I. Braud and F. Branger
Irstea, UR HHLY, Hydrology-Hydraulics Research Unit, Lyon-Villeurbanne, France

J. W. Kirchner
Department of Environmental Systems Science, Swiss Federal Institute of Technology, ETH Zurich, Zurich, Switzerland
Swiss Federal Research Institute WSL, Birmensdorf, Switzerland

D. Freudiger, I. Kohn, K. Stahl and M. Weiler
Chair of Hydrology, University of Freiburg, Freiburg, Germany

Joong Gwang Lee
Center for Urban Green Infrastructure Engineering (CUGIE Inc), Cincinnati, OH 45255, USA

Christopher T. Nietch
Office of Research and Development, US Environmental Protection Agency, Cincinnati, OH 45268, USA

Srinivas Panguluri
Independent Consultant, Olney, MD 20832, USA

Hoori Ajami
School of Civil and Environmental Engineering, University of New South Wales, Sydney, Australia
Department of Environmental Sciences, University of California Riverside, Riverside, CA, USA

Ashish Sharma
School of Civil and Environmental Engineering, University of New South Wales, Sydney, Australia

Lawrence E. Band
Department of Geography and Institute for the Environment, University of North Carolina, Chapel Hill, NC, USA

Jason P. Evans
Climate Change Research Centre, University of New South Wales, Sydney, Australia

Narendra K. Tuteja
Environment and Research Division, Bureau of Meteorology, Canberra, Australian Capital Territory, Australia

Gnanathikkam E. Amirthanathan
Environment and Research Division, Bureau of Meteorology, Melbourne, Victoria, Australia

Mohammed A. Bari
Environment and Research Division, Bureau of Meteorology, Perth, Western Australia, Australia

Index